André B. de Haan, Johan T. Padding

Process Technology

Also of Interest

Industrial Separation Processes.
Fundamentals
De Haan, Real, Schuur, 2020
ISBN 978-3-11-065473-8, e-ISBN 978-3-11-065480-6

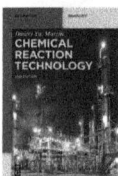

Chemical Reaction Technology
Murzin, 2022
ISBN 978-3-11-071252-0, e-ISBN 978-3-11-071255-1

Reactive Distillation.
Advanced Control using Neural Networks
Sakhre, Singh, 2022
ISBN 978-3-11-065614-5, e-ISBN 978-3-11-065626-8

Sustainable Process Engineering
Szekely, 2021
ISBN 978-3-11-071712-9, e-ISBN 978-3-11-071713-6

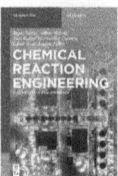

Chemical Reaction Engineering.
A Computer-Aided Approach
Salmi, Wärnå, Hernández Carucci, de Araújo Filho, 2020
ISBN 978-3-11-061145-8, e-ISBN 978-3-11-061160-1

André B. de Haan, Johan T. Padding

Process Technology

An Introduction

2nd Edition

DE GRUYTER

Authors

Prof. Dr. Ir. André B. de Haan
Department of Chemical Engineering
Delft University of Technology
Section Transport Phenomena
Van der Maasweg 9
2629 HZ Delft
Netherlands

Prof. Dr. Johan T. Padding
Faculty of Mechanical, Maritime and Materials Engineering
Delft University of Technology
Leeghwaterstraat 39
2628 CB Delft
Netherlands

ISBN 978-3-11-071243-8
e-ISBN (PDF) 978-3-11-071244-5
e-ISBN (EPUB) 978-3-11-071246-9

Library of Congress Control Number: 2021951791

Bibliographic information published by the Deutsche Nationalbibliothek
The Deutsche Nationalbibliothek lists this publication in the Deutsche Nationalbibliografie;
detailed bibliographic data are available on the Internet at http://dnb.dnb.de.

© 2022 Walter de Gruyter GmbH, Berlin/Boston
Cover image: Mimadeo/iStock/Getty Images Plus
Typesetting: Integra Software Services Pvt. Ltd.
Printing and binding: CPI books GmbH, Leck

www.degruyter.com

Preface

Multidisciplinary cooperation is one of the key contributors to successful innovation and project execution within the current and future process industry. The main challenge within multidisciplinary teams is communication, which will be much more effective when possessing basic understanding of each other's discipline. It is exactly this reason, enhancing the understanding of process technology by those without a background in (bio)chemical process engineering, that has motivated us to create the second edition of this book. The first edition served as the basis for the industrial course "Introduction into the World of Process Technology," which has been given for many years within multiple multinational companies. Participants included chemists (organic/catalytic/bio/physical/analytical), material technologists, economists, accountants, lawyers and many others working in a position where cooperation and communication with (bio)chemical process engineers was an essential part of their job. Since 2020 the course has been revived and is now provided as "Process Technology for Non-process Technologists" by the TU Delft Process & Product Technology Institute (www.tudelft.nl/Pro2Tech). Besides small revisions/corrections throughout all chapters and an update of references for further reading, this second edition now includes electrochemical conversion, a more complete overview of computational fluid dynamics approaches and more background on economic evaluation of projects.

The main objective of this book is to provide a general overview of chemical and biochemical process and product technology. It focuses on the structure and development of production processes, main technological operations and the most important aspects of product and process development, including economics. For the technological operations, the emphasis is on their operating principles, reasons for application and available industrial equipment. Design calculations and mathematics have been kept to a minimum required to understand why process technologists and engineers need certain information. All topics are extensively illustrated by representative examples.

The book is organized into seven distinct parts. An introduction to the structure of the chemical industry and (bio)chemical processes is treated in Chapters 1 and 2. Chapters 3–5 deal with (bio)chemical reaction engineering and reactor technology. The most frequently applied molecular separation techniques such as distillation, extraction, absorption, stripping, adsorption and ion exchange are discussed in Chapters 6–9. The part on mechanical separation technology presents an overview of the most important techniques for separating heterogeneous mixtures in Chapters 10–12. Technologies relevant for particles and final product manufacturing are treated in Chapters 13–15. Chapters 16–18 deal with the development, scale-up, design, engineering and safety of processes. The book concludes with three appendices in which major industrial processes for the production of base chemicals, polymers and fine chemicals are described. It should be noted by the reader that the appendices only represent a small selection of the numerous industrial processes in

https://doi.org/10.1515/9783110712445-202

operation and that the described processes have been selected with a bias on the companies where the course was lectured. For further reading, an extensive list of reference books is provided.

André B. de Haan
Johan T. Padding

Contents

1 The chemical industry

1.1 Introduction

The industry that applies the knowledge of chemical behavior is generally called the chemical process industry. Chemical reactions and separation of compounds are used to obtain products with desired properties. In reality, the chemical industry is a set of related industries with many diverse functions and products. Some of these different areas, divided into three general classes of products, are listed in Fig. 1.1:

1. Industrial chemicals and monomers such as acids, alkalis, salts, chlorine, ammonia, ethylene, propylene, caprolactam, acrylonitrile, industrial gases and other organic chemicals.
2. Polymers and end chemicals to be used in further manufacture such as synthetic resins, plastics, fibers, elastomers, dyes and pigments.
3. Finished chemical products for consumer applications as architectural paints, drugs, cosmetics and soaps or to be used as materials or supplies in other industries such as industrial paints, adhesives, fertilizers and explosives.

Figure 1.1 emphasizes that certain raw materials are used to prepare key chemicals, monomers and intermediates that may be sold independently or used directly in additional steps to give various polymers and end chemicals. These in turn can be formulated and fabricated into chemical products, which can sometimes be modified into finished products. Hence, the term chemicals and allied products accurately represents this diversity as well as the flow of materials and products from raw sources to finished formulations. Although the division is approximate, about 60% of the chemical industry manufactures industrial products that are further modified, whereas 40% of their products are sold directly to the consumer. Clearly, the chemical industry is part of the manufacturing industry and within this it plays a central part even though it is by no means the largest part of the manufacturing sector. Its key position arises from the fact that almost all the other parts of the manufacturing sector utilize its products.

The three major segments of the chemical industry depicted in Fig. 1.2 are related to commodities, fine chemicals and specialty chemicals. The substances representing these segments exhibit an increasing complexity in molecular architecture. In the business column starting from fossil fuels and culminating in the application products, fine chemicals take a position with their special characteristics between the commodities or base chemicals such as toluene, acetic acid, acetone, and methanol and specialty chemicals or desired products for various markets. Fine chemicals are sold on the basis of their chemical composition for use as intermediates in the production of other materials. They are often needed in relatively small quantities and

https://doi.org/10.1515/9783110712445-001

Chemical raw materials	→	Chemicals Monomers	→	Polymers End chemicals	→	Formulated chemical products	→	Finished products

Basic Petrochemicals
Sulfur
Salt
Phosphate rock
Minerals
Others

Organic industrial chemicals
Industrial gases
Inorganics
Ammonia

Plastics
Resins
Elastomers
Synthetic fibers
Dyes
Pigments
Surfactants

"Allied" industries

Textiles
Rubber processes
Photographic chemicals
Water and waste-treating
Perfume, flavor chemicals
Functional fluids
Carbon products
Explosives
Propellants
Fabricated plastics

Pharmaceuticals
Soap and detergents
Cosmetics
Paints, varnishes and lacquers
Inks
Adhesives
Pesticides
Fertilizers
Consumer and other industrial specialties

60% Industrial products

40% Consumer products

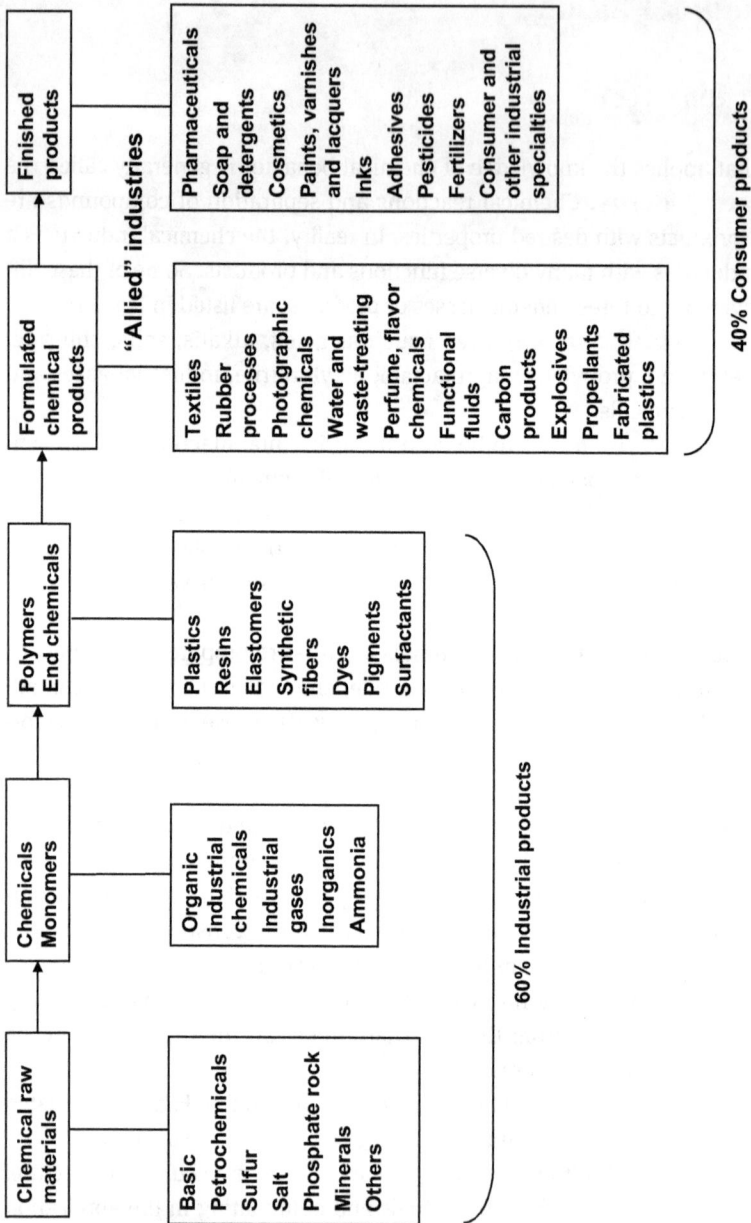

Fig. 1.1: Chemical process industry.

high purity for only one or a few end uses. Product applications can be found in pharmaceutical (50%), agrochemicals (20%) flavors and fragrances (5%), food additives (5%) and various other industries (20%). Specialty chemicals are purchased because of their effect rather than composition. There is some overlap in the definitions of fine and specialty chemicals. The worldwide market in fine chemicals is estimated to be approximately $100 billion in sales, compared to $1,000 billion for the commodity segment. With respect to the financial turnover (sales), the specialty chemicals industry is as large as the bulk chemical industry.

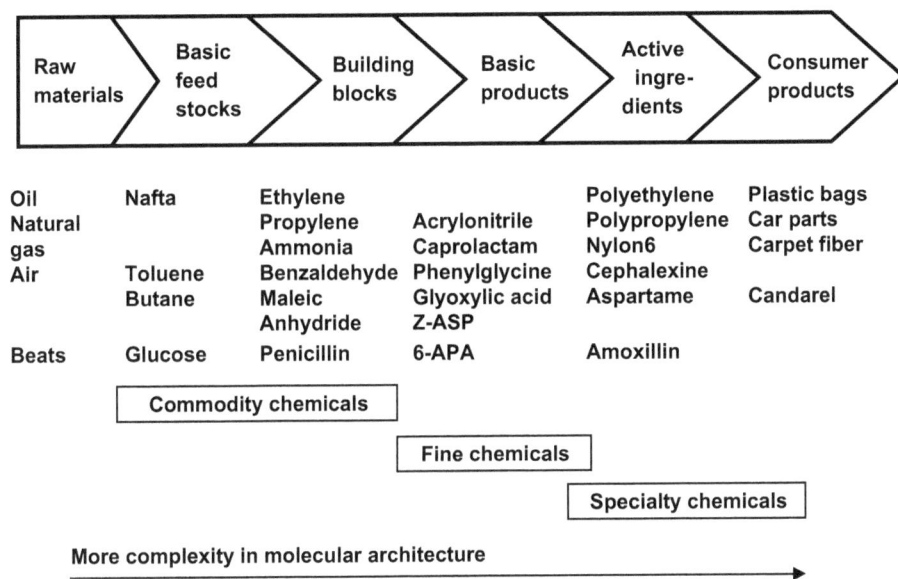

Raw materials	Basic feed stocks	Building blocks	Basic products	Active ingre-dients	Consumer products
Oil	Nafta	Ethylene		Polyethylene	Plastic bags
Natural		Propylene	Acrylonitrile	Polypropylene	Car parts
gas		Ammonia	Caprolactam	Nylon6	Carpet fiber
Air	Toluene	Benzaldehyde	Phenylglycine	Cephalexine	
	Butane	Maleic	Glyoxylic acid	Aspartame	Candarel
		Anhydride	Z-ASP		
Beats	Glucose	Penicillin	6-APA	Amoxillin	

| Commodity chemicals |

| Fine chemicals |

| Specialty chemicals |

More complexity in molecular architecture →

Fig. 1.2: Business column segmentation.

1.2 General characteristics of the chemical industry segments

The major sectors of the chemical industry are categorized on the basis of end product uses in Fig. 1.3. Going from commodities, through fine chemicals to specialty chemicals the products become higher priced, have a higher added value, a lower volume and generally a shorter life cycle than the products closer to the fossil fuels in the value chain. Within the chemical process industries, batch processing is focused on the fine and specialty chemicals sectors, while continuous processing is dominant in bulk chemicals production. The scale of operations ranges from quite small plants (a few tons per year) in the fine chemical area to the giants (100–1,000 thousand tons per year) of the petrochemical sector. Today, a typical base petrochemical plant is designed to produce enormous quantities (100–600 ktons/year) of a single product and operate 24 h a day all the year round. They are used to make key intermediates, which are turned into a very wide range of products by further processing. Clearly, such large and sophisticated plants require a very high capital investment and are characterized by high investment versus low labor components in the cost of manufacture. The investment per worker in a base petrochemicals olefins plant may well exceed a quarter of a million dollars. Although these plants take full advantage of the economy of scale effect, the losses due to running under design capacity can be extremely high if the balance between production capacity and market demand is disturbed. This is particularly evident when the economy is depressed, and the chemical industry's business tends to follow the cyclical pattern of the economy with periods of full activity followed by those of very low activity.

	Commodities:	Fine chemicals:	Specialties:
• Product life cycle	Long (>30 yrs)	Average(10-20 yrs)	Short(<10 yrs)
• Product spectrum	Narrow	Broad	Very Broad
• Product volume	>> 10.000 t/yr	< 10.000 t/yr	Variable
• Product price	< $ 5/kg	> $ 10/kg	Variable
• Product differentiation	None	Low	High
• Added value	Small	High	High
• R&D focus	Process improvement	Process development	Product development
• Technology	Continuous	Batch & Continuous	Batch & Continuous

Fig. 1.3: The three main segments of the chemical industry.

It is important to keep in mind that, although most of the discussion about the chemical industry tends to revolve around the multinational giants, the industry is very diverse and includes many small-sized companies as well. There is a similar diversity in the sizes of chemical plants. Batch type plants are used for the manufacture of relatively small amounts of fine and specialty chemicals, typically up to 100 tons per annum. They are therefore not dedicated to producing just a single product but are multipurpose and may be used to produce a number of different chemicals each year. In this part of the chemical industry the investment level may not exceed the order of 25,000 dollars per worker.

The research and development carried out in industry can be divided into product development, process development, process improvement and application development. The nature of research and development carried out varies significantly across the various sectors of the chemical industry. In the commodity chemicals sector, most of the R&D expenditure will be devoted to process improvement. Process improvement relates to processes, which are already operating. It may be due to problems arisen and hindering or stopping production. More commonly, however, process improvement will be directed at improving the profitability of the process. Improving the quality of the product, by process modification, may lead to new markets for the product. In recent years, the most important process improvement activity has been to reduce the environmental impact of the processes. At the other end of the scale lie the specialty chemicals. Here there are immense and continuous efforts undertaken to discover and develop new products, which exert the desired, specific effect. As such, the main focus in this sector is on new product development such as pharmaceuticals, agrochemicals and antioxidant additives. Process development absorbs considerable resources in the fine chemicals industry, in part because of the shorter life cycles of fine chemicals as compared to commodities. It covers developing new manufacturing processes for new as well as existing products. The push for the latter may originate from the availability of new technology or change in the availability and/or cost of raw materials. Process development for a new product depends on things such as the scale on which it is to be manufactured, the by-products formed and required purity.

1.3 Major raw materials

Inorganic chemicals are derived from many different sources. In contrast, the production of organic chemicals is almost entirely based on two raw material sources, crude oil and natural gas. At present, more than 90% (by tonnage) of all commercially important organic chemicals are produced from crude oil and natural gas via a multitude of petrochemical processes. Compared to crude oil, natural gas is less versatile because carbon–carbon bonds have to be built up. Apart from crude oil and natural gas, other raw materials are used on a smaller scale such as fatty oils, starch, sugar and molasses, wood and straw.

The inorganic chemical industry is based on a large variety of minerals, air and water. Minerals are converted into products like building materials and pigments. There is, however, a group of raw materials from which a limited number of rather important inorganic intermediates is made. Perhaps the most notable example is sulfur. Substantial quantities of sulfur are also removed and recovered from natural gas and crude oil. Over 80% of all sulfur is converted into sulfuric acid, and approximately half of this is then used in fertilizer manufacture. Sulfuric acid is the most important chemical of all in tonnage terms. Other important examples are air, the source of oxygen and nitrogen, and sodium chloride, the starting material for caustic soda and chlorine.

Refined chemicals and
consumer products
(≈30000)

Intermediates (≈300)

Basic products (≈20)

Raw materials (≈10)

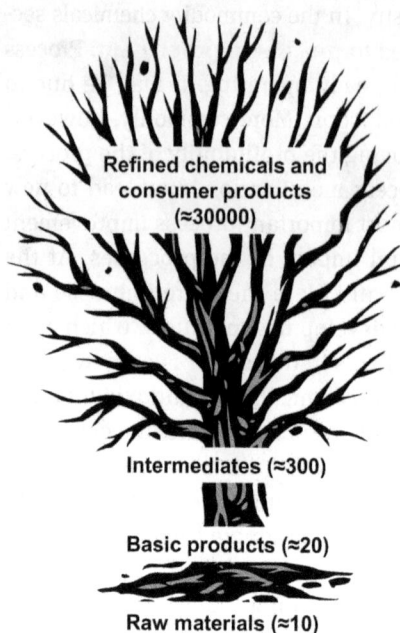

Fig. 1.4: Product family tree of the chemical industry.

1.4 Production structure of the chemical industry

As indicated, the chemical industry is concerned with converting raw materials, such as crude oil, firstly into chemical intermediates and then into a tremendous variety of other chemical products. In an earlier period of the chemical industry's development, chemical companies were generally production oriented, exploiting a process to produce a chemical and then selling it in rapidly expanding markets. As the industry has grown, there has been a strong tendency toward integration, both forward and backward. If today's production structure of the chemical industry is examined, it is seen that there are only a few hundred major basic products and intermediates that are produced on a scale of at least a few thousand to several million tons per annum worldwide. This relatively small group of key products, which are in turn produced from only about ten raw materials, forms a stable foundation on which the many branches of refining chemistry (dyes, pharmaceuticals, etc.), with their many thousands of often only short-lived end products are based. This has resulted in the well-known family tree, schematically depicted in Fig. 1.4, which can also be regarded as being synonymous with an integrated production system, with synergies that are often of critical importance for success.

A special characteristic of the major basic products and intermediates is their longevity. They are statistically so well protected by their large number of secondary products and their wide range of possible uses that they are hardly affected by

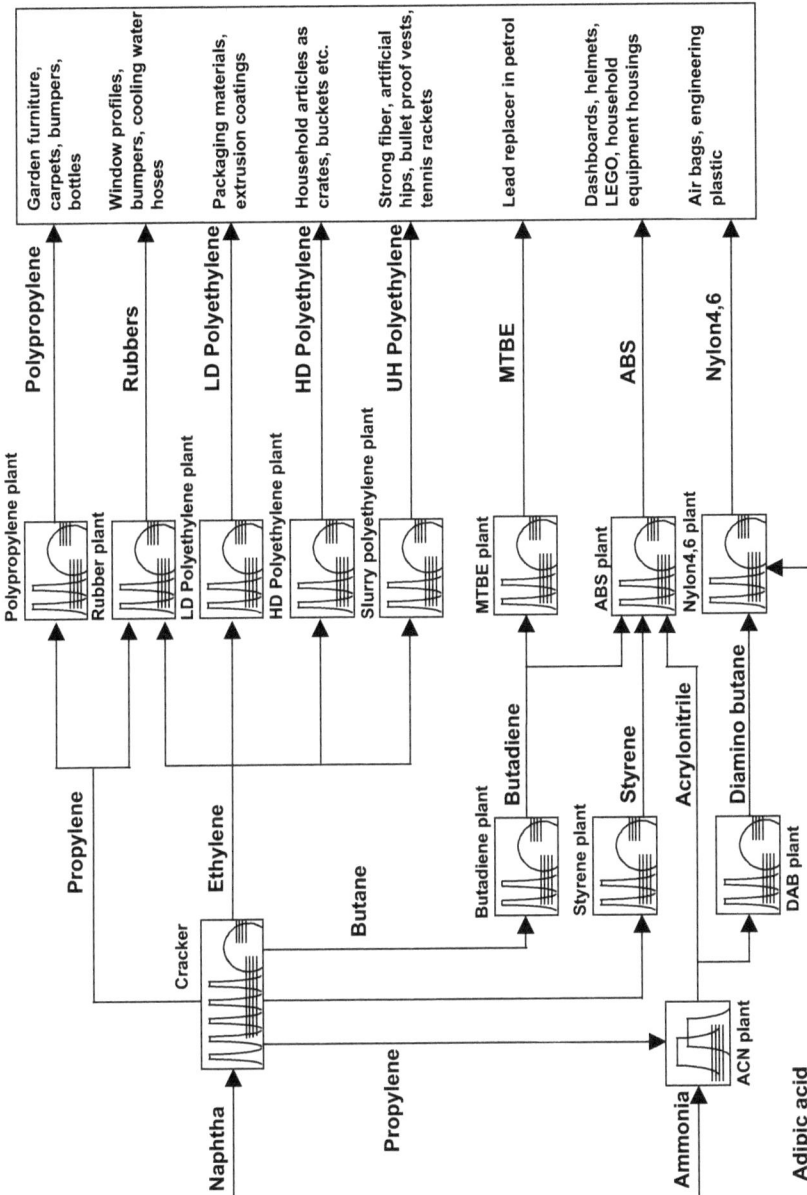

Fig. 1.5: Production chain for naphtha to polymers.

the continuous changes in the range of products on sale. Unlike many end products, which are replaced by better ones in the course of time, they do not themselves have a so-called life cycle. However, the processes for producing them are
subject to change. This is initiated by new technical possibilities and advances
opened up by research but also dictated by the current raw material situation. Here,
it is not the individual chemical product, but the production process or technology
which has a life cycle.

The relations between raw materials, intermediates, semimanufactured products and finished products are complex. For the chemical industry, cracking of
naphtha and gas oil is an important operation to produce raw materials, such as
ethylene, propylene and benzene. These are then used as raw materials for further
processing as shown in the product tree depicted in Fig. 1.5. The polymers sector
is the major user of petrochemical intermediates and consumes almost half of the
total output of produced organic intermediates. It covers plastics, synthetic fibers,
rubbers, elastomers and adhesives. Ammonia and fertilizers is a sector in which it
has been difficult to achieve a balance between capacity and demand. In tonnage
terms, it is one of the most important sectors based on the Haber process for ammonia. As shown in Fig. 1.6, a large variety of products are starting with ammonia as
the basic intermediate. Based on a combination of petrochemical intermediates and
natural raw materials such as glucose the fine chemicals sector (Fig. 1.7) produces
an amazingly wide range of products for an even wider range of applications.
Along with pharmaceuticals, agrochemicals is a very profitable area because the
demand for its products is unaffected by the world's economy and therefore remains high even during recessions. This contrasts with the situation for most other
sectors of the chemical industry. The product trees presented here are especially
useful in the development of new processes and products.

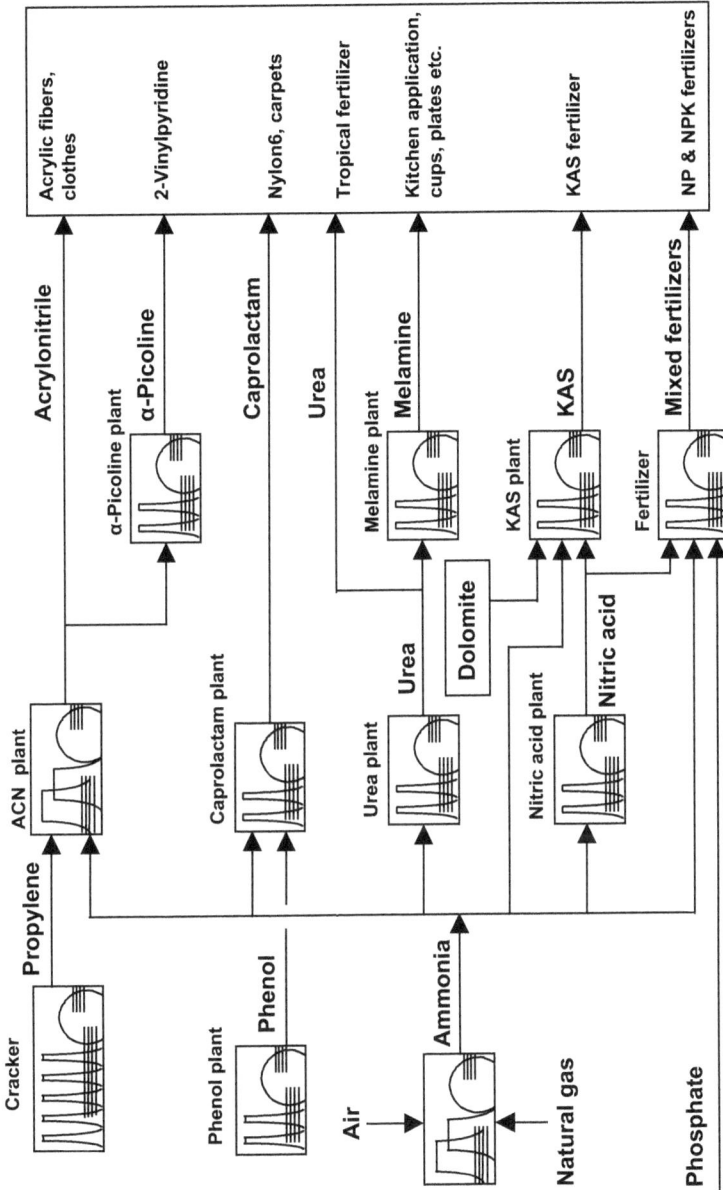

Fig. 1.6: Production chain from ammonia to chemical products and fertilizers.

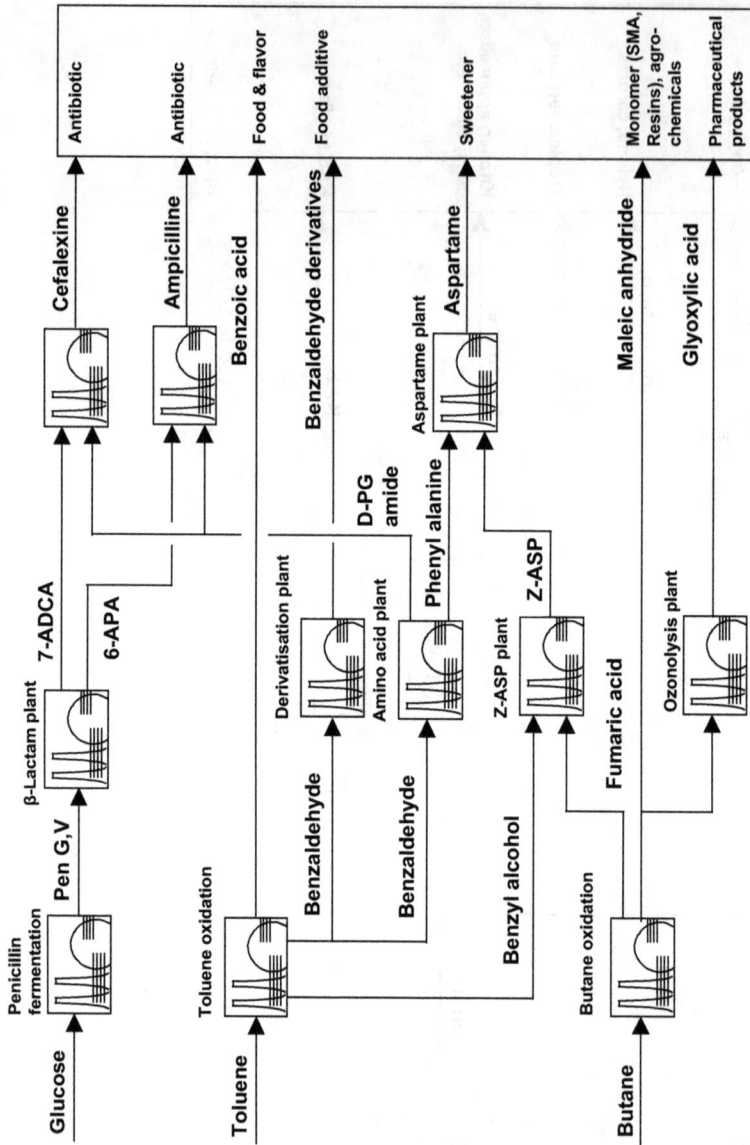

Fig. 1.7: Production chain for glucose, toluene and butane to fine chemicals.

2 The structure of chemical and biochemical process systems

2.1 Structure of chemical and biochemical processes

The route along which a raw material is converted to products is a logical coupling of interconnected operations: the process (Fig. 2.1). At least one of the process units is the chemical reactor in which chemical conversion takes place. In principle, every chemical reaction shows incomplete conversion and often the formation of by-products. Furthermore, auxiliary materials are often used, which must be separated in another process step.

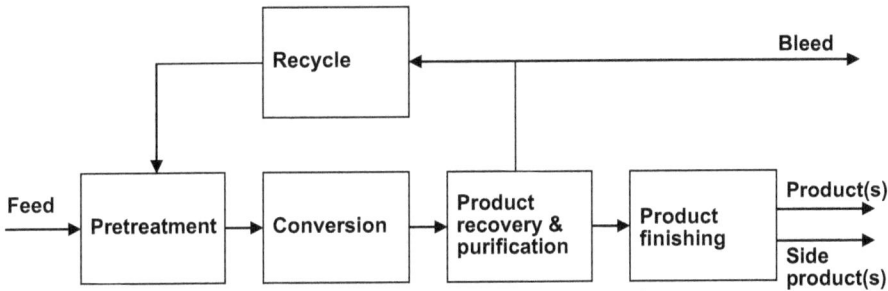

Fig. 2.1: General representation of a process.

Raw materials are generally impure, have the wrong physical form or even consist of mixtures of different compounds. Process units preceding the reactor prepare the feed with mechanical operations such as crushing or grinding, followed by physical treatments such as mixing, heating and evaporating. Operations succeeding the reactor treat the converted mass to recover and purify the product to the desired specification, generally utilizing a series of separation steps. Hence, in most cases, a plant contains a large number of separation steps, of which the investment typically accounts for 50–70% of the total plant investments. Unconverted feed components and auxiliary chemicals will generally be recycled after further purification in the recycle section. Often a part of the recycle is bled to avoid build-up of contaminants that are difficult to remove. Final product treatment consists usually of mechanical operations such as drying, granulation and packing.

Besides the desired product, usually several waste streams are produced, which must be brought into an acceptable condition before leaving the plant. Ideally, only air and water are emitted. After combining all process steps and streams, a complicated network results.

https://doi.org/10.1515/9783110712445-002

2.2 Characteristics of production processes

The practical way a product is prepared on a laboratory scale is in many ways different from the way it is done on an industrial scale. The main reasons for this are the huge differences in the amount of material that has to be processed. On a laboratory scale, one is usually satisfied with amounts that are sufficient for analytical purposes. Industrial production concerns the production of amounts that can vary from 1,000 kg/year up to over 1,000,000,000 kg/year for a single plant. It is clear that such large amounts should have significant effects on the way a production process is designed.

2.2.1 Batch production technology

Batch processing has been a part of man's activities throughout history and is still used most of the time on a laboratory scale. Batch processes are used to manufacture many of the products required for modern life. Within the chemical process industries, batch processing is focused on the fine and specialty chemicals sectors, while continuous processing is dominant in commodity chemicals production.

A batch process is one in which a series of operations are carried out over a period of time on a separate, identifiable item or parcel of material. It is different from a continuous process, during which all operations occur at the same time and the material being processed is not divided into identifiable portions. This definition of batch processing includes what has been called semibatch production, during which material is added continuously to a batch over some period. The sequence of events copies the sequence developed in the laboratory, but in larger-size vessels and batches. The raw materials are purified, perhaps by distillation or adsorption, and stored. Reactants are then pumped or poured into a reaction vessel. The agitation intensity, the rate of heating and/or cooling and the rates of flow of other reactants or catalysts are controlled in such a manner that the reaction proceeds as planned. When the reaction is completed, the reactant mass is removed to a separation system. The desired products are separated from unreacted feed materials and undesired byproducts. The reactants are usually recycled for use in the next batch. This is schematically represented in Fig. 2.2.

Batch processing is typically applied for small-volume products and in cases where the fundamental mechanisms of the reaction are not well known. This issue of robustness to incomplete knowledge is of extreme importance for the production of fine chemicals in a multipurpose environment. Batch processes often use complex chemistry of which substantially less is known than a typical continuous process. The cost of evaluating kinetics and physical parameters to the accuracy required to use standard chemical engineering design methods would be far too high. However, such detailed analysis is seldom necessary. It is usually sufficient to know how long an operation will take and even that need not be known to excessive accuracy. The use of rules of thumb for scale-up based on identification of the rate-controlling process

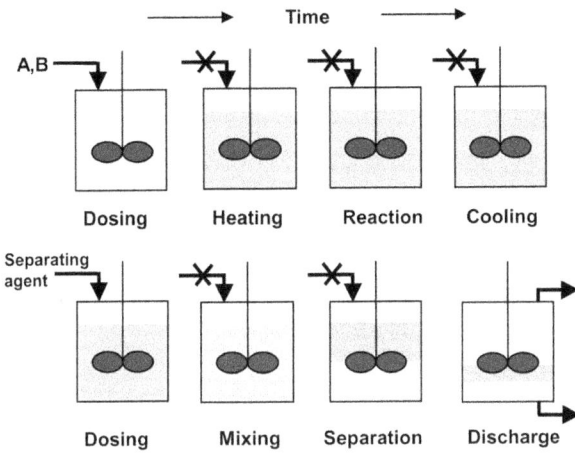

Fig. 2.2: Schematic representation of a batch process operation.

is a common way of assessing the time taken for an operation on the full scale. Discrepancies between the expected and actual times for each operation tend to average out over the many operations that constitute a single batch. Thus, the impact of inaccurate knowledge is much less serious than for a continuous process whose overall productivity is limited by the capacity of whichever unit has lowest throughput.

Besides robustness, batch production plants offer us the second advantage of extreme flexibility. In general, the plant used, typically employing stirred vessels of either stainless or glass-lined steel of 1–50 m³ in volume, is easily modified for use on new products. Such equipment is very versatile and can be used to blend reactants, heat them to reaction temperature, carry out the reaction, cool, distil off solvent and crystallize the product. In a continuous process, each operation would be carried out in a separate unit.

Small-scale processes and processes in which solids occur are likely to be batch. Numerous chemical process industries retain batch processing as their primary method of manufacture. Many of these continue to be made batch wise because quality is more important than price. The product has traditionally been made so, or the industry is not large or technically sophisticated enough to operate successfully in a continuous mode. Typical products manufactured by batch processes include pharmaceuticals, agrochemicals, dyestuffs, photosensitive materials, food additives, perfumes, vitamins and pigments.

2.2.2 Continuous processes

For large production capacities where the process reaction mechanism is better known and reaction rates are not too slow, continuous processing is often possible.

Here, the raw materials are prepared and fed to the reactor continuously. The reactor system is sized so that the materials reside in it long enough at the reaction conditions to achieve the desired extent of reaction. The reaction system may be a single vessel or a number of reactor vessels in series, each operating at different conditions. The product leaves the reaction zone continuously and passes to a sequence of separation steps where the desired products are obtained in continuous streams. Unreacted feed materials are obtained in other streams and continuously returned to the reactor. Any by-products are also removed.

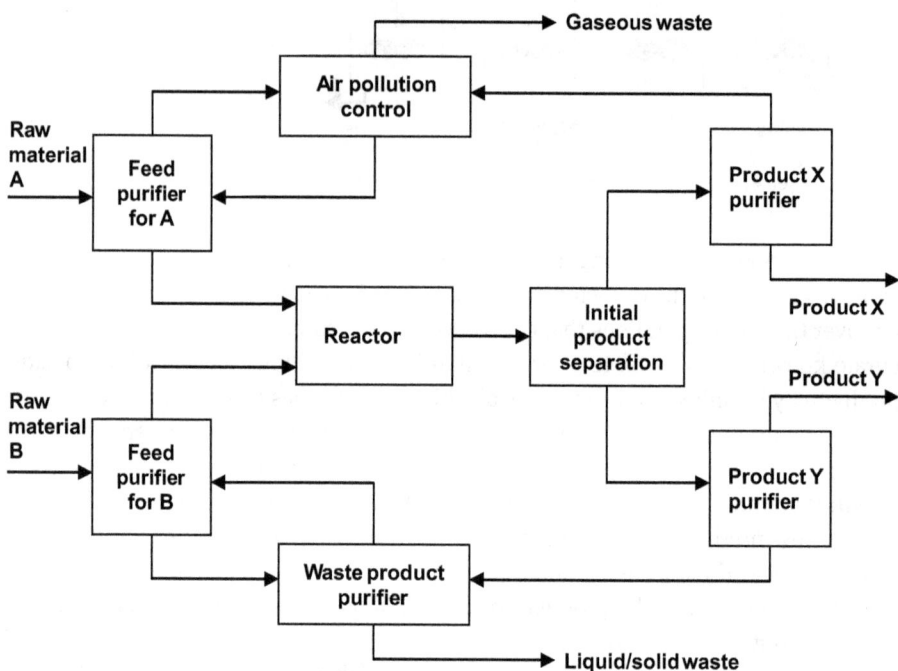

Fig. 2.3: Typical continuous process plant.

Figure 2.3 shows a possible configuration for a typical continuous process plant. Two raw materials are fed to a reactor after each is purified. The reactor effluent is separated in three separation steps, and all waste streams are treated before release to the environment. Product X might be the main product desired and product Y a salable by-product. Ideally, such a process operates under steady-state conditions, that is, a stable operating condition where none of the process parameters (temperature, pressure, process stream composition, flow rate, etc.) vary with time. In any real process, there will be a period of adjustment as the plant is started or stopped, and there will be disturbances as the process operates. The process control system attempts to minimize the effects of these process upsets.

In a plant operating under steady-state conditions, variables are different from point to point along the process path. But if you were to take a photograph of the control panel, where an array of instruments records the process conditions at all the crucial points in the process, you would see no differences between a picture taken at 8 a.m. and one taken at 8 p.m. Some basically continuous processes have components, for example, dryers, filters and ion exchange beds that operate cyclically. These units must be taken off-stream periodically for regeneration. The period of on-stream operation may be several hours to several days.

Most large-scale processes operate continuously, especially ones in which gases and low-viscosity liquids are handled. Petroleum refining, the manufacture of bulk chemicals and industrial gas manufacturing are typical examples.

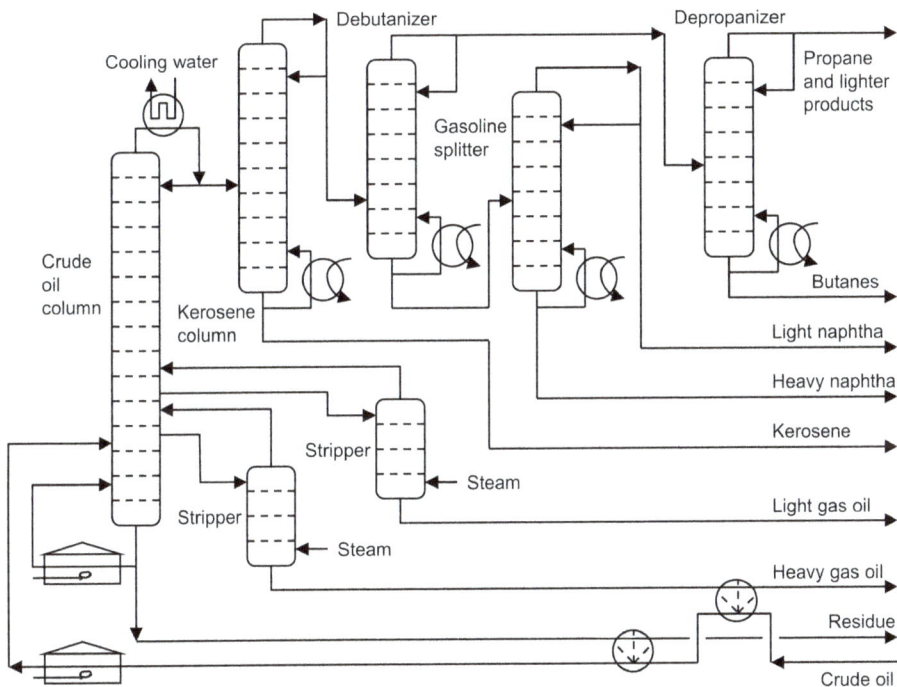

Fig. 2.4: Flowchart of an atmospheric crude oil distillation train.

2.3 Unit operations

Despite the very large number of chemical processes and products, there are only a modest number of different kinds of chemical process steps. This has allowed an economical method of organizing the field of chemical engineering into manageable segments. These segments are called unit operations and have been defined on two principles:

(1) Although the number of individual processes is great, each one can be broken down into a series of operations that appear in process after process.
(2) The individual operations are based on the same scientific principles. For example, in most processes, solids and fluids must be moved, heat or other forms of energy must be transferred from one substance to another, and tasks like drying, size reduction, distillation and evaporation must be performed.

Most of the unit operations are used to conduct the primarily physical steps of preparing the reactants, separating and purifying the products, recycling unconverted reactants and controlling the energy transfer into or out of the chemical reactor. Typical examples are operations such as distillation, evaporation, liquid–liquid extraction, filtration, drying, heat exchange, mixing, classification, crystallization and adsorption. When a process step involves a chemical change, it is sometimes called a unit process, or, more appropriately, a chemical reaction step.

The consequence of this way of thinking is that a process designer regards a plant first and foremost as a collection of operations connected by a network of pipes. It requires a certain amount of thinking in systems to design a process optimally. The resulting network is called a flowchart. Such a flowchart generally contains the mass and energy balances for all the operations. As illustrated in Fig. 2.4, a flowchart can be quite complicated, the more so since every unit process can be regarded as a subsystem.

Because the unit operations are a branch of engineering, they are based on both science and experience. Each process step can be carried out in a variety of equipment types. Theory and practice must combine to yield designs for equipment that can be fabricated, assembled, operated and maintained. Usually the apparatus selected is chosen because it has some particular advantage in light of the properties of the materials being processed or the goal of the process step.

2.3.1 Reactors

The reactor in which the chemical reaction takes place occupies a central position in the chemical process. In size and appearance, it may often seem to be one of the least impressive items of equipment, but its demands and performance are usually the most important factors in the design of the whole plant. The reactor provides the volume necessary for the reaction and holds the amount of catalyst required for the reaction. When a new chemical process is being developed, at least some indication of the performance of the reactor is needed before any economic assessment of the project as a whole can be made. An essential factor in this economic assessment is the formation of unwanted by-products, which directly affect the operating costs of the process. In most cases, the design of the reactor has a great effect on the amount of by-products formed and therefore the size of the separation equipment required.

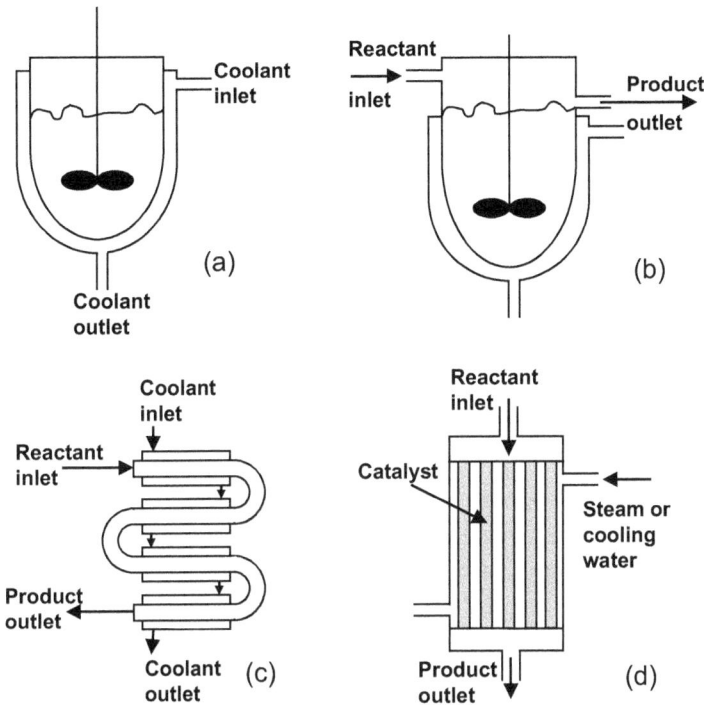

Fig. 2.5: Various common reactor types: (a) agitated batch reactor, (b) continuous stirred tank reactor, (c) tubular reactor and (d) multitubular packed bed reactor.

The design of a reactor and its mode of operation can thus have profound repercussions on the remainder of the plant.

Figure 2.5 shows several common types of reactors. The agitated batch reactor shown in (a) is an extremely common device. The jacket can be used to heat or cool the reactor, typically with steam or cooling water. The vessel may be built with thick walls so that the reactions can take place under pressure. There may be various ports for feed addition and product withdrawal. In many cases, baffles are used on the inside to prevent vortexing of the liquid and to improve mixing. Normal construction materials include glass-lined steel, stainless steel, carbon steel and various corrosion-resistant alloys. Internal cooling coils are sometimes used to provide additional heating or cooling capacity.

Sketch (b) shows a continuous stirred tank reactor. This is essentially identical to the agitated batch reactor but is operated continuously. Thus, feed and product are continuously added and removed. Obviously, it is not possible to operate this reactor in a way that every fluid molecule stays in for the same length of time. Fairly intense agitation is usually employed to keep the reactor content uniform. However, the degree of mixing decreases as the size of reactor increases.

The tubular reactor shown in (c) is the most common type of reactor for reactions where large amounts of heat have to be supplied or removed. This can be achieved by burners or a heating/cooling fluid on the outside of the tubes. The tubular reactor is also widely used for highly exothermic solid-catalyzed reactions as multitubular packed bed reactor (d).

2.3.2 Recovery, purification and fractionation technologies

Separation processes constitute more than half of the total equipment investment for the chemical and fuel industries. They are also widely used in pharmaceutical and food industries, mineral processing industry and a variety of other industries. Separation processes may have a number of purposes, which can be loosely categorized as follows:
– Recovery or concentration: increasing the desired substance concentration in a solution, usually by removal of a substantial fraction of solvent
– Purification: removal of impurities from the final product
– Fractionation: separation of desired substances from one another

Separating the desired product or products from the reaction mixture is often a complex and therefore expensive process. During the chemical and biochemical conversion, mixtures are generated that contain many different components including the product, undesired side products, unconverted raw materials, solvents, catalysts and so on. All these components can exist in different states: gas, liquid or solid. When all components are present in the same state, the mixture is called homogeneous. Typical examples are gas mixtures (air = oxygen + nitrogen + carbon dioxide) or liquid mixtures (sugar in water, ammonia in water, petrol, naphtha). If the components are present in different states the mixture is called heterogeneous. Solid–liquid mixtures (melting ice, water–sand slurries), solid–gas mixtures (dust in air) and liquid–gas mixtures (droplets in air, bubbles in water) are typical examples of heterogeneous mixtures.

The separation of homogeneous mixtures and the separation of heterogeneous mixtures require totally different techniques and separation principles. Separation technologies for homogeneous mixtures are used to separate on a molecular basis and therefore referred to as molecular separations. Technologies for heterogeneous mixtures are usually called mechanical separations because they are based on mechanical principles.

2.3.2.1 Molecular separations
Homogeneous mixtures with a molecularly dispersed distribution of individual components can only be separated with a molecular separation process. In these

processes, mass and often heat is exchanged between at least two phases of different composition. The phases are the mixture phase(s) and a selective auxiliary phase. This auxiliary phase is generated the introduction of an energy separating agent such as heat and refrigeration or by means of a mass separating agent such as a solvent, adsorbent or ion exchange resin. The separating agent serves to create the second phase and thereby form the required driving forces, concentration and temperature gradients. An overview of the main molecular separation technologies, their separation principle and used separation agent is given in Tab. 2.1.

Tab. 2.1: Overview of main molecular separation technologies.

Separation principle	Phase contact	Separation agent	Technology
Difference in volatility	Gas–liquid	Heat transfer	Evaporation Distillation Azeotropic distillation
		Solvent and heat transfer	Extractive distillation
		Liquid absorbent	Gas absorption
		Stripping gas	Stripping
	Liquid–solid	Heat transfer	Drying
Difference in solubility	Liquid–liquid	Solvent	Liquid–liquid extraction
	Solid–liquid		Solid–liquid extraction
	Solid–liquid		Leaching
Difference in solubility or melting point	Solid–liquid	Heat transfer	Crystallization
Difference in adsorbability	Gas–solid	Solid adsorbent	Gas adsorption
	Liquid–solid		Liquid adsorption
Difference in permeability and/or solubility	Liquid, vapor	Membrane	Membranes

2.3.2.2 Mechanical separations

The separation of heterogeneous mixtures is usually accomplished by exerting forces on the mixture that has to be separated. These forces move the components that comprise the mixture in such a way that a separation is obtained. Different forces are used in the different mechanical separation technologies:
- Gravity
- Centrifugal
- Impingement

- Electrostatic
- Magnetic

Many commercial separation operations also combine these separation principles, sometimes in the same apparatus or otherwise in different pieces of equipment in series. An overview of the different mechanical separation technologies and their basic separation principles is given in Tab. 2.2.

Tab. 2.2: Overview of the main mechanical separation technologies and their application areas.

Technology	Separating principle	Application area				
		Solid–solid	Liquid–solid	Gas–solid	Gas–liquid	Liquid–liquid
Sorting	Gravity	X				
Classifiers		X				
Sieves		X				
Settlers			X	X		
Filtration	Pressure		X			
Presses			X			
Centrifuges	Centrifugal		X			X
Cyclones			X	X	X	X
Fabric filters	Impingement			X	X	
Wet scrubbers				X	X	
Electrostatic precipitators				X		

2.3.3 Product finishing operations

In recent decades, the chemical industry has moved away from commodity chemicals toward products of higher added value such as specialty chemicals and consumer products. These materials are often complex multiphase materials such as pharmaceutical tablets and creams, cosmetic creams and lotions, ceramic and plastic products, ice cream and margarine, industrial paints and adhesives, fertilizer granules and so on. The quality and properties of these products is no longer determined solely by the concentrations achieved in the separation operations. End consumers generally judge products according to their end-use properties, such as taste, smell, feel and handling properties rather than their chemical composition. These end-use properties are typically linked to chemical and biological stability, degradability, chemical, biological and therapeutic activity, aptitude to dissolution, mechanical, rheological, electrical, thermal, optical, magnetic characteristics for solids and solid particles together with size, shape color, touch, handling, cohesion, friability, rugosity, tastes, succulence, esthetics, sensory properties and so on.

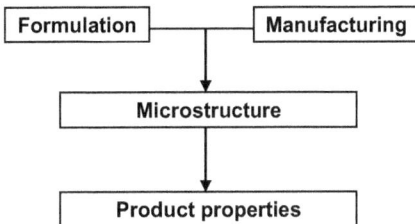

```
┌─────────────────┐           ┌─────────────────┐
│  Formulation    ├───────────┤  Manufacturing  │
└─────────────────┘     │     └─────────────────┘
                        ▼
            ┌───────────────────────┐
            │     Microstructure    │
            └───────────────────────┘
                        │
                        ▼
            ┌───────────────────────┐
            │   Product properties  │
            └───────────────────────┘
```

Fig. 2.6: Influence of formulation and processing on product properties.

An important characteristic of these industrial products is that they all possess a significant microstructure which is, as illustrated in Fig. 2.6, very dependent upon the product formulation (recipe) as well as the manufacturing conditions (technologies, process). Product finishing technologies are concerned with physical or physico-chemical principals, which add value to a product. Typical engineered products are:

- Structured solids such as catalyst carriers and coated pharmaceuticals. A catalyst carrier must be of a given shape and must have a defined porosity and inner surface with a given pore size distribution. Coated pharmaceutic granules may have an improved taste and often release their active ingredients in a controlled and retarded way.
- Particulate solid systems concern crystalline, polymeric or amorphous solids that represent over 60% of all products that chemical companies sell to their customers. Examples are fertilizers which have to be stored without caking and which have to be handled and applied without dust formation. Other solid systems are some pharmaceuticals and vitamins which have to be specially formulated to guarantee their bioavailability. These materials need to have a clearly defined physical shape in order to meet the designed and the desired quality standards.
- Emulsions, colloids, dispersions, suspensions, sprays, gels and foams concern complex media for which rheology and interfacial phenomena play a major role. Also involved are the so-called soft solids, systems which have a detectable yield stress, such as ceramic pastes, foods or drilling muds.

The main characteristics of product finishing technologies are that quite different unit operations are used, involving less reaction and separation tasks and more structuring and stabilization tasks. Structuring processes are the opposite of separation processes. Man-made structured products use assembly, structuring or texturizing processes, for example, crystallization and emulsification processes (e.g., margarines, mayonnaises, ice creams, paints and detergents), foaming (e.g., insulating materials, shaving cream and whipped creams), precipitation, granulation, agglomeration, drying, extrusion, compression, prilling, shaping, micronization, dough making and baking. The end product is often a complex microstructure of dispersed phases held together by binding forces and a continuous phase. The product microstructure leads to the desired product functionality in use. Stabilization processes are the opposite of transformation processes. Two major processes can be identified. The first, encapsulation, provides a barrier between two reacting species. For instance, to preserve the integrity of the ingredients or accomplish intelligent functions such as controlled reactivity or programmed release of active components that may be obtained by multiple layer coatings. The second, to combat spoilage, is rather typical for the food industry and pharmaceutical products. Naturally structured foods often use preservation or stabilization processes in which the main aim is to eliminate microbial, enzymatic or chemical spoilage of the raw materials, which are usually food tissues (fish, meat and vegetables).

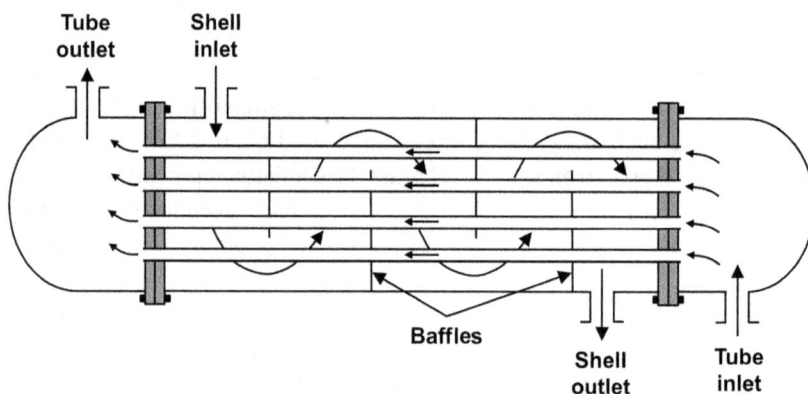

Fig. 2.7: Schematic of a single pass countercurrent shell-and-tube heat exchanger.

2.3.4 Other important process units

In addition to purification and reaction, chemical processes usually include steps to heat or cool process streams and devices to move fluids. Heating and cooling devices include furnaces, air coolers and heat exchangers. Furnaces and air coolers are familiar from your experiences with automobile radiators, household furnaces, refrigerators and air conditioning units. Fluid-to-fluid heat exchangers are not so familiar, but they are very common in chemical processes. The most common of these heat exchangers, the shell-and-tube heat exchanger, is shown in Fig. 2.7. In this unit, one fluid flows through the shell over the outside surfaces of the tubes. Baffles are usually used to increase the velocity of the fluid passing through the shell. A second fluid enters the tubes, flowing through them in parallel in one or more passes. As it flows through the exchanger, the cold fluid is heated by heat transferred through the tube walls. At the same time, the hot fluid is cooled. The figure shows inlet and exit nozzles for two fluid streams. These streams could be gas or liquid and could change phase within the heat exchanger.

Fluid motion in chemical processing is generally produced by pumps or, if gases are being moved, by blowers and compressors. The most common type of pump is the centrifugal pump shown in Fig. 2.8a. Liquid flows into this pump and is accelerated into high-speed circular motion by the vanes of the pump impeller. The liquid exits into the pump volute, where it slows down, converting at least part of the kinetic energy into pressure. Because of the mechanism of its operation, the performance of this pump is sensitive to the pressure that it develops. The higher the pressure rise across the pump, the lower the pump capacity is. Thus, the pressure that the pump produces will decrease as the throughput through the pump is increased. Pump efficiency also depends on flow rate. The other common type of liquid pump is a positive displacement pump illustrated in Fig. 2.8b. Capacity is directly related to the volume swept out by a piston operating in a cylinder or to the frequency with which a gear tooth enters the cavity between other like teeth. Pressure rise has much less of an effect on the capacity of positive displacement pumps.

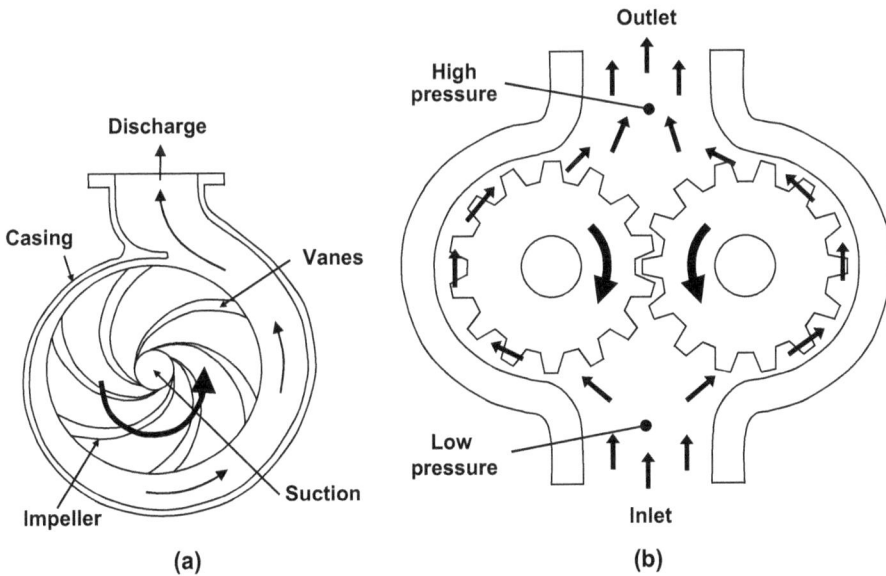

Fig. 2.8: Schematic of a (a) centrifugal and (b) positive displacement pump.

Gas compressors operate on the same principle as do liquid pumps but usually handle larger volumes and operate at higher speeds. Since large amounts of work are used to compress gases, cooling is usually needed during and after compression. Reciprocating, centrifugal and axial-flow compressors are commonly used in the chemical process industries. Figure 2.9 shows a schematic of a double piston gas compressor which are used in the high-pressure polyethylene process for the compression of ethylene in multiple stages up to pressures of 3,000 bar with cooling in between the stages.

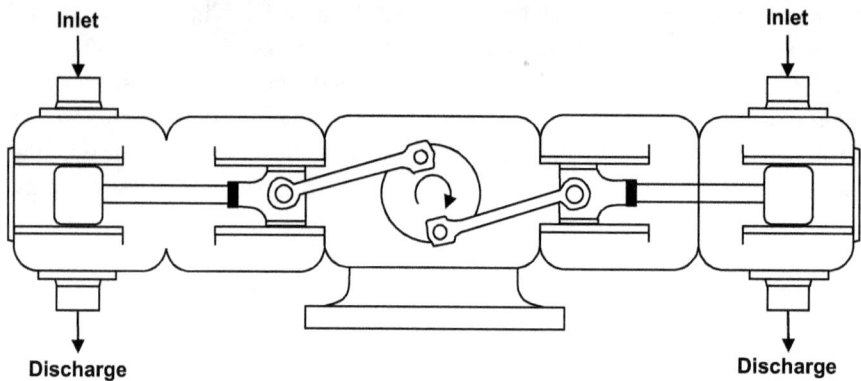

Fig. 2.9: Schematic of a double piston compressor.

2.4 Process synthesis

The development of a commercial product from a laboratory chemical involves the talents and efforts of many people. At the laboratory stage, the market potential of the product is estimated. The physical and chemical properties of the chemicals involved in the process are then determined, and the conditions under which the product can be produced in the laboratory are explored. But a long and complex effort still remains.

There are times when an engineer is called upon to devise a plausible process long before all of the development effort has taken place. Perhaps a decision whether to invest in the development effort is required. If a tentative process can be suggested, the cost and difficulty of the development of the process can be estimated. Also, a rough estimate of the cost of the product can be calculated, and the size of the market can be forecasted. Like most economic forecasts, these guesstimates are subject to an enormous amount of uncertainty. Accordingly, many technically and economically sound processes and products have been shipwrecked on the rocks of poor marketing forecasts.

The development of a process scheme involves coming up with that configuration of processing steps that efficiently and safely produces the desired product. An enormous amount of art, skill, intuition and innovation goes into developing the processing scheme. Literally millions of designs are possible. One of the tasks of a process engineer is to choose from these possibilities, taking into account the many conditions set by product markets, geographical location of the plant, the social situation, legal regulations and so on. This is not only important in choosing among existing processes but also in developing new processes. This process synthesis step is where vast amounts of money can be made or lost and where a good, innovative chemical engineer can be worth his or her weight in gold.

These activities are not merely a straightforward application of the scientific disciplines on which chemical technology is based (chemistry, physical transport processes, unit operations and reactor design). It is necessary to select relevant knowledge from these fields, combine different aspects and interpret these quantitatively. This means integration of knowledge from various fields of science. With such complicated systems as chemical processes, some of the questions will inevitably be answered in a semiquantitative or even qualitative way. Typical questions that must be answered as the engineer conceives and develops a process to make a marketable product:

1. What reaction steps are required to get the product?
2. What is the best type and size of reactor to use, and what are the optimum reactor operating conditions (temperature, pressure, agitation, catalyst concentration etc.)?
3. What is the optimum reactor conversion, and how does it affect the design and operation of the downstream separation steps? The optimization must incorporate the entire plant.
4. Will the catalyst degrade or be poisoned?
5. What side reactions occur? What will the likely by-products be? What effect will they have on yield and performance?
6. What are the best raw materials? Air, water, petroleum, natural gas, coal, minerals and agricultural products are the basic raw materials. However, a host of semiprocessed intermediates may also be used.
7. How can the raw materials be brought to conditions suitable for the reaction? Usually, several purification steps are needed, as well as heating and compressing to the appropriate conditions.
8. How can the products and by-products be separated and purified to meet market specifications?
9. Should cooling water be used, or should air-cooling be considered?
10. Are any of the materials used, produced or present in the process toxic or carcinogenic? Are there other health or safety hazards in the process? Is there a potential for hot spots or explosions in the reactor? Should the reactor be shielded and remotely operated?

All of these questions have several possible answers. Each choice the engineer makes has technical, economic, social and political repercussions. The engineer is expected to make prudent and wise choices, and the impact on the environment must be considered. In many cases, experience is a necessary asset to assist sound intuition and judgement.

3 Principles of chemical reaction engineering

3.1 Introduction

Design of a chemical reactor is no routine matter, and many alternatives can be proposed for a process. In searching for the optimum it is not just the cost of the reactor that must be minimized. One design may have low reactor cost, but the materials leaving the unit may be such that their treatment requires much higher cost than the alternative design. Hence, the economics of the overall process must be considered. Reactor design uses information, knowledge and experience from a variety of areas:
- thermodynamics
- chemical kinetics
- fluid mechanics
- heat transfer
- mass transfer
- mechanical engineering
- economics

Chemical reaction engineering is the synthesis of all these factors with the aim of properly designing a chemical reactor. The design of chemical reactors is probably the one activity which is unique to chemical engineering, and it is probably this function more than anything else which justifies the existence of chemical engineering as a distinct branch of engineering.

In reactor design, we want to know what size and type of reactor and method of operation are best for a given job. To establish the actual behavior of a reactor for a certain application, two major basic questions have to be answered:
- What changes can we expect to occur?
- How fast will these changes take place?

The first question concerns the thermodynamics, the second the various rate processes (chemical kinetics, heat transfer, etc.). Tying these all together and trying to determine how these processes are interrelated can be an extremely difficult problem. Since this is far beyond the scope of this course, the content of this chapter is limited to the most essential basics and some simple situations that provide a basis to discuss and appreciate the work of chemical reaction engineers.

https://doi.org/10.1515/9783110712445-003

3.2 Classification of reactions

There are many ways of classifying chemical reactions, see Table 3.1. In chemical reaction engineering, probably the most useful scheme is the breakdown according to the number and types of phases involved, the big division being between homogeneous and heterogeneous systems. A reaction is homogeneous if it takes place in one phase alone. A reaction is heterogeneous if it requires the presence of at least two phases to proceed at the rate that it does. It does not matter whether the reaction takes place in one, two or more phases, or at an interface, or whether the reactants and products are distributed among the phases or are all contained within a single phase. All that counts is that at least two phases are necessary for the reaction to proceed as it does.

Sometimes this classification is not clear-cut as with a large class of biological reactions, the enzyme substrate reactions. Since enzymes themselves are highly complicated large-molecular-weight proteins of colloidal size, enzyme-containing solutions represent a gray area between homogeneous and heterogeneous systems. Other examples for which the distinction is not sharp are the very rapid chemical reactions, such as the burning gas flame where large nonhomogeneity in composition and temperature may exist.

Cutting across this classification is the catalytic reaction whose rate is altered by materials that are neither reactants nor products. These foreign materials, called catalysts, need not be present in large amounts. Catalysts act somehow as go-betweens, either hindering or accelerating the reaction process while being modified relatively slowly if at all.

Tab. 3.1: Classification of chemical reactions useful in reactor design.

	Noncatalytic	Catalytic
Homogeneous	Most gas-phase reactions	Most liquid-phase reactions
_____ **Heterogeneous**	Fast reactions such as burning of a flame	Reactions in colloidal systems
		Enzyme and microbial reactions
	Burning of coal Roasting of ores	Ammonia synthesis Oxidation of ammonia to produce nitric acid
	Attack of solids by acids Incineration of waste Reduction of iron ore to iron and steel	Cracking of crude oil Electrochemical production of hydrogen and chlorine

3.3 Rate of chemical reactions

For the application of material balances to chemical reaction engineering problems an expression is required for the chemical reaction rate. The reaction rate of a component is defined as the number of moles produced or consumed per unit time and per unit volume. From this definition, it follows that the chemical reaction rate is negative for reactants and positive for products. In many cases, A and B are used as symbols for a reactant and P is the symbol for a product. In the reaction equation:

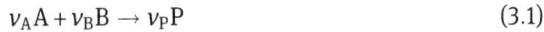

$$v_A A + v_B B \rightarrow v_P P \tag{3.1}$$

v_A, v_B and v_P are stoichiometric coefficients for the respective components A, B and P. These stoichiometric coefficients relate the relative reaction rates of the individual components:

$$(-R_A) = \frac{v_A}{v_B}(-R_B) = \frac{v_A}{v_P}(R_P) \tag{3.2}$$

It should be noted that the chemical reaction rate only reflects the chemical kinetics of the system. The reaction rate generally depends on the composition of the reaction mixture (most importantly the local concentrations of each of the reactants), its temperature and pressure and the properties of the catalyst.

3.3.1 Effect of concentration

For relatively simple reactions, it is usually postulated that the rate-controlling mechanism involves the collision or interaction of a single molecule of A with a single molecule of B. This means that the reaction rate is proportional to the concentration of each of the reactants, since the number of molecular collisions per unit time is proportional to the amount of molecules of each of the reactants. Such reactions are called elementary reactions. For example, in the case of a bimolecular elementary reaction:

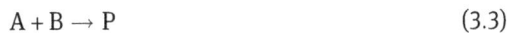

$$A + B \rightarrow P \tag{3.3}$$

the rate of disappearance of reactant A at a given temperature is proportional to the concentrations C_A and C_B of both reactants, with a proportionality constant called the reaction rate constant, k_A:

$$- R_A = k_A C_A C_B \tag{3.4}$$

The *order n* of a reaction is the power dependence of the reaction rate on the concentration of each reactant. The reaction in the example above is first order in A and first order in B. For nonelementary reactions, the reaction rate is a result of multiple elementary reactions, leading in general to a complicated dependence on the reactant concentrations. In such cases, for a limited range of concentrations, an *effective* reaction order can still be defined for each reactant. We will show later how the effective reaction order can be determined experimentally in a batch reactor.

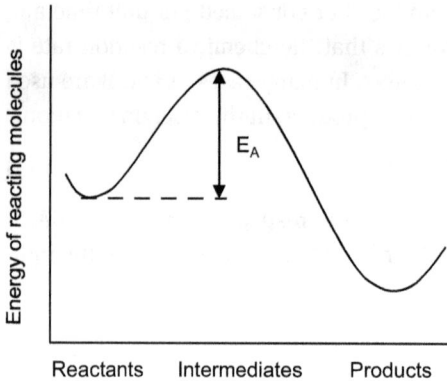

Fig. 3.1: The potential energy curve for a chemical reaction with activation energy E_A.

3.3.2 Effect of temperature

The effect of temperature on a chemical reaction is most easily explained by the transition state theory. Thermodynamics tell us that a chemical reaction will only proceed when an energetically more favorable situation is obtained. The reaction pathway toward this more stable situation usually proceeds through the formation of less stable intermediates by the breakage of chemical bonds, which subsequently decompose or form new bonds into the thermodynamically more stable products. This process is schematically represented in Fig. 3.1, where the potential energy of starting, intermediate and end situation is shown. From Fig. 3.1, it can be seen that activation energy, E_A, has to be provided to allow a chemical reaction to proceed. In reaction rate expressions, this activation energy is used to separate the influence of temperature and composition on reaction rate. The actual representation of the dependence of the reaction rate constant k_A on temperature T is done through an exponential probability function, better known as the Arrhenius relation:

$$k_A = k_{A,\infty} \exp\left(\frac{-E_A}{RT}\right) \tag{3.5}$$

In this equation, R represents the gas constant (8.3144 J/mol K) and the pre-exponential factor $k_{A,\infty}$ is called the frequency factor. Although the frequency factor has a weak dependence on temperature, this is usually ignored because of the overwhelmingly stronger temperature dependence of the exponential part. For most reactions, the activation energy lies in the range of 40–300 kJ/mol, resulting in an increase of the reaction rate constant by a factor of 2–50 for a temperature rise of 10 °C.

3.3.3 Chemical equilibria

So far, it has been assumed that the reactions are irreversible. In practice, many reactions are reversible, meaning that the products can be converted back into the reactants again:

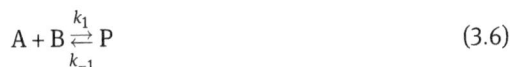

$$A + B \underset{k_{-1}}{\overset{k_1}{\rightleftarrows}} P \tag{3.6}$$

For an elementary equilibrium reaction, the rate of formation of P by the forward reaction and its rate of disappearance by the reverse reaction are now given by

$$R_{P,forward} = k_1 \, C_A \, C_B \tag{3.7}$$

$$-R_{P,reverse} = k_{-1} \, C_P \tag{3.8}$$

Because at equilibrium there is no net formation of P, the reaction equilibrium constant, K_R, can be obtained from the condition $R_{P,forward} + R_{P,reverse} = 0$:

$$K_R = \frac{C_P}{C_A \, C_B} = \frac{k_1}{k_{-1}} \tag{3.9}$$

What this equation shows is that the forward and reverse reaction rates are not independent, but related through the equilibrium constant. Thermodynamics allows the calculation of the equilibrium constant from the standard free energies of formation ΔG_f^o of the reacting components. For the considered reaction, this results in the following relation for the standard free energy of reaction ΔG_R^o and the reaction equilibrium constant K_R:

$$\Delta G_R^o = \Delta G_{f,P}^o - \Delta G_{f,A}^o - \Delta G_{f,B}^o = -R \, T \ln K_R \tag{3.10}$$

Note that the equilibrium constant K_R depends only on temperature and not on pressure. Using Eq. (3.10) and standard thermodynamic relations, it can be shown that the temperature dependence of the equilibrium constant is given by the Van 't Hoff equation:

$$\frac{d \ln K_R}{dT} = \frac{\Delta h^o_R}{RT^2} \tag{3.11}$$

Here, Δh^o_R is the standard enthalpy of reaction, which is the heat of reaction produced in an exothermic reaction ($\Delta h^o_R < 0$) or heat of reaction consumed in an endothermic reaction ($\Delta h^o_R > 0$). The Van 't Hoff equation shows that increasing the temperature decreases the equilibrium constant for an exothermic reaction, moving the equilibrium toward the reactant side, while it increases the equilibrium constant for an endothermic reaction, moving the equilibrium toward the product side.

With the equilibrium constant known, the expected maximum attainable yield of the products of the reaction can be estimated. An industrially important example where the maximum attainable conversion is limited by the thermodynamic equilibrium is the synthesis of ammonia from nitrogen and hydrogen:

$$N_2 + 3\,H_2 \rightleftharpoons 2\,NH_3 \ (\Delta H^o_{r,s} = -45.3\,kJ/mol\,NH_3) \tag{3.12}$$

As a result of the exothermic nature of the reaction and the volume contraction by conversion to ammonia, it is necessary to operate at relatively low temperatures and high pressures in order to move the equilibrium to achieve economically viable production rates. This is illustrated in Fig. 3.2, where the ammonia content of the equilibrium mixture is plotted as a function of pressure at various temperatures. Economically favorable conversion rates can only be obtained by catalyzing the reaction. Iron catalysts give good conversions at about 30 MPa and 500 °C in reactors of acceptable dimensions. Other typical examples of reactions in which the maximum attainable conversion is limited by the thermodynamic equilibrium are dehydrogenations (i.e., ethyl benzene to styrene), esterifications and polycondensations (i.e., polyesters, polyamides).

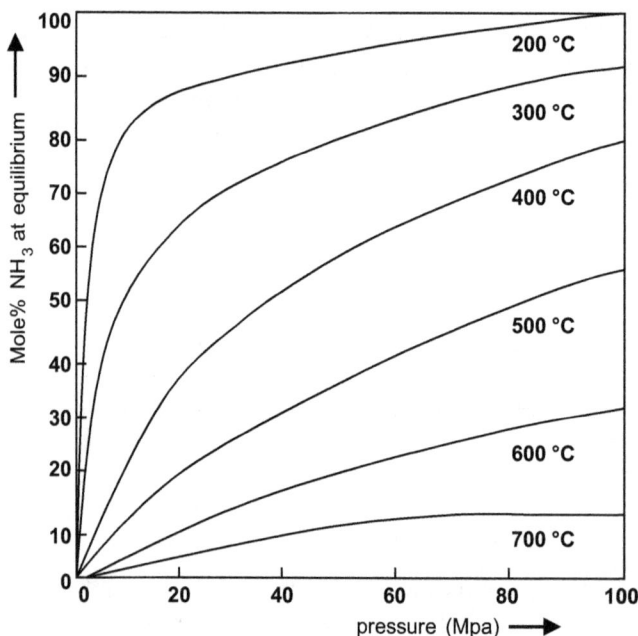

Fig. 3.2: Equilibrium yield of NH₃ from N₂ and H₂.

3.4 Catalysis

Catalysts are substances that influence the rate of a reaction without being consumed. In most cases, only trace amounts are needed to bring about or accelerate a reaction or reaction type. As illustrated in Fig. 3.3, the catalyst forms complexes with the intermediates that have significantly lower activation energy to provide a reaction pathway that proceeds in parallel with the existing thermal reaction. Hence, useful catalysis implies that the rate of the desired reaction considerably exceeds the rates of all other possible reactions. If a reaction is reversible, the reverse process is accelerated to the same extent as the forward. Furthermore, a catalyst in no way alters the thermodynamic properties of reactants and products. Hence, a catalyst cannot change the reaction equilibrium. It can only effect, or accelerate, a reaction that is thermodynamically feasible under practically attainable conditions.

Two broad classes of catalytic reactions are recognized, homogeneous and heterogeneous. When the catalytic material is in the same phase as the main reactants, for example, in a solution/liquid reaction mixture, the term homogeneous catalysis is applied. The term heterogeneous catalysis is used when the catalytic material exists as a distinct, usually solid phase with liquid or gaseous reactant. Some typical examples of homogeneous and heterogeneous catalysis systems are listed in Tab. 3.2.

3.4.1 Homogeneous catalysis

In homogeneous catalysis, the catalyst is usually dissolved in a liquid reaction mixture, though some or all of the reactants may be introduced as gases, or even as solids. In general, the nature of the catalytic material used in solution chemistry is well defined and usually highly reproducible kinetic behavior is obtained. In the case of transition metals, homogeneous catalysis has the advantage over heterogeneous catalysis that tailor made ligands can be attached to the catalyst metal to optimize the selectivity. Figure 3.4 illustrates that the cycle of a homogeneous catalytic reaction can be divided in four major steps:

1. Formation of a reactive species
2. Coordination of the reactants to the complex
3. Formation of the product on the catalyst
4. Elimination of the formed product under liberation of the reactive starting species

The major problem encountered in homogeneous catalysis, particularly when the catalytic material is expensive, lies in the separation of the reaction products from the catalyst such that the recovery process is efficient and does not impair catalytic activity. Thus, we can afford to lose small, catalytic quantities of mineral acid, alkalis or base metal compounds. But only minute losses can be tolerated if expensive noble

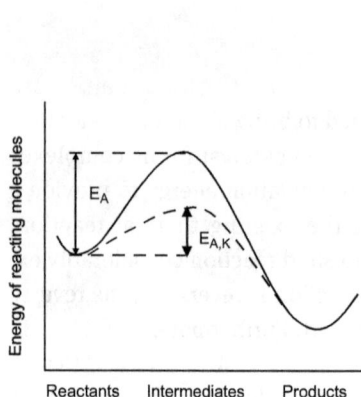

Fig. 3.3: Schematic representation of the reduction of the activation energy by a catalyst.

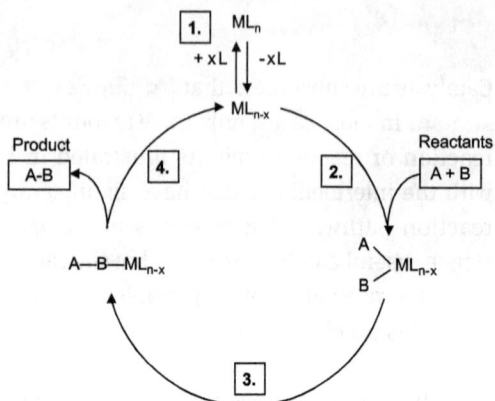

Fig. 3.4: Cycle of homogeneous catalysis.

metal catalysts are used. An illustrative example is the liquid phase oxidation process for the oxidation of toluene to benzoic acid. In this process, a homogeneous organic cobalt(II) complex is used to catalyze the oxidation reaction. After reaction, the oxidation products are recovered and purified by evaporation and distillation leaving the cobalt catalyst in the tar residue. Extraction with water is used to recover the cobalt from the tar and recycle the catalyst system to the reactor. Besides extraction, many other separation methods such as product and reactant evaporation, precipitation and membrane filtration have been developed to achieve a quantitative recovery of homogeneous catalysts. A different option is to use two immiscible liquids in the reactor where one phase contains the catalyst and the other the reactants and the product.

In seeking to apply any new catalytic material, the useful lifetime of the catalyst is important. Loss of catalytic activity may be an inevitable consequence of the chemical nature of the catalytic material, or a result of poisoning. Within the first category, loss by evaporation, thermal decomposition or internal rearrangement of the catalytic material may occur when very high reaction temperatures are used. Catalyst poisoning refers to materials that do not form part of the defined process chemistry, but which gain entry into the reaction mixture and lead to permanent or temporary catalyst deactivation. In general, poisons are impurities in the chemical feedstock or corrosion products from materials of construction.

Despite the high costs of noble metals, many mass chemicals are produced with homogeneous catalysis on a large scale. This is only possible when each molecule of catalyst produces a large amount of product (high turn over number) and the catalyst losses are kept to an absolute minimum. Important industrial applications of homogeneous catalysis are found in isomerization, hydrogenation, oligomerization, polymerization, metathesis, oxidation, hydroformulation and carbonylation. Typical examples are listed in Tab. 3.2.

Tab. 3.2: Some examples of industrial catalysis.

	Application	Catalyst system
Homogeneous catalysis	Toluene oxidation	Cobalt
	Benzoic acid oxidation	Copper
	Ziegler–Natta polymerization of polyethylene	Ti(IV) aluminum alkyls
	Ethylene oligomerisation	Ni complexes
	Carbonylation of butadiene and methanol	Pd complexes, RhI_3/HI
	Acid-catalyzed esterification	Sulfuric or nitric acid
Heterogeneous catalysis	Ammonia synthesis	Iron oxide
	Ammoxidation of propylene to acrylonitrile	
	Hydrogenation of benzaldehyde to benzyl alcohol	Palladium on coal
	Polypropylene slurry polymerization	
	Water–gas shift reaction	Copper complexes (low T)

3.4.2 Heterogeneous catalysis

Chemists have recognized the existence of heterogeneous catalysis for over 150 years, and practical applications increased dramatically from the beginning of the present century. A heterogeneous catalyst is a solid composition that can effect or accelerate reaction by contact between its surface and either a liquid-phase reaction mixture or gaseous reactants. In liquid-phase systems, one or more of the reactants may be introduced as a gas, but access of such reactants to the surface of the catalyst is almost invariably by dissolution in the reaction medium and subsequent diffusion.

Heterogeneous catalysis occurs on the surfaces of solid materials. To reach this surface, the reactants first have to be transported to the catalyst and – if porous catalyst particles are used – through the pores of the particle to reach the active material. The initiation of a chemical reaction involving an otherwise stable molecule requires a significant electronic disturbance in that molecule at the catalytic surface. The adsorption process that brings about such a chemical modification is usually referred to as chemisorption. Furthermore, when two or more molecules are involved in a reaction on a catalytic surface, as in hydrogenation or oxidation processes, we usually find evidence that the major reactions occur between chemisorbed or surface species derived from each of the reactant molecules. In many instances, it was recognized that only a relatively small proportion of the surface was catalytically active. Hence, the term active sites was introduced to describe those localities on the surface which would induce the desired chemical reaction. The combination of these complex processes makes heterogeneous catalysis one of the most complex branches of chemical kinetics. Rarely do we know the compositions, properties or concentrations of the reaction intermediates that exist on the surfaces covered with the catalytically effective material. Computational chemistry can provide some answers, and has therefore become one of the key components of catalysis research. However, it should be no surprise that the selection of a solid catalyst for a given reaction is to a large extent still empirical and based on prior experience and analogy.

Many heterogeneous catalysts in commercial use contain several components, often referred to as the active catalyst agent, promotor and the support. While all components contribute to the overall performance of the catalytic material, the active catalyst agent is essential for any activity in the type of reaction required. With metal catalysts, classification is usually reasonably straightforward. However, with oxide catalysts, none of the components may be individually active. Promotors are introduced to modify the crystal structure or electronic properties of the major component, to improve activity, selectivity or thermal stability. As illustrated in Tab. 3.3, solid catalytic materials can be divided into two major groups:

1. Bulk catalytic materials, in which the gross composition does not change significantly throughout the material, such as silver wire mesh or a compressed pellet of "bismuth molybdate" powder.
2. Supported catalysts, in which the active catalytic material is dispersed over the surface of a porous solid, such as activated carbon, alumina, silica or zeolites.

Tab. 3.3: Classification of heterogeneous catalysts.

Group	Catalyst classes	Examples
Bulk catalytic materials	Metals	Pt- or Ag-Nets, Raney nickel
	Metal alloys	Pt–Re, Ni–Cu, Pt–Au
	Acids	SiO_2/Al_2O_3, zeolites, montmorillonite
	Bases	CaO, K_2O, Na_2O
Supported catalysts	Oxides of transition metals	Cr_2O_3, Bi_2O_3/MoO_3, V_2O_5, NiO
	Oxides of other metals	Al_2O_3, SiO_2
	Metal sulfides	MoS_2, WS_2
	Metals on supports	Pt/Al_2O_3, Ru/SiO_2, Co/Kieselgur
	Metals and acids	Pt/zeolite, Pd/zeolite
	Other	Carbide, silicide

Bulk metals can be used in traditional engineering forms, more particularly as fine wire woven into gauzes. Such forms are generally used only in high-temperature processes, such as the partially oxidative dehydrogenation of methanol (over Ag) or ammonia oxidation (over Pt–Rh) at about 500–600 °C and 850–900 °C, respectively. Mechanical stability is of greater importance than high surface area.

A common feature of heterogeneous catalysis is the increase in rate with increasing subdivision of the catalytic material. This arises from increasing accessibility of the surface, and reduction in diffusional constraints between reactants and catalytic sites. The surface area, pore structure and chemical composition of the surface are extremely important parameters of any support material or solid catalyst. For high activity at modest temperatures, forms that present a high surface area to the reactants are highly desirable. Finely divided metal powders often show very high catalytic activity, but may present separation problems. Hence, we find the development of methods to produce coarser particles of metals in porous form, such as platinum sponge or, far more commonly, Raney nickel. Alternatively, larger shaped catalysts in the shape of cylinders, spheres, rings and so on can be produced from powder, by compression into molds or extrusion of a slurry and drying. An overview of the most common catalyst particles and production techniques is given in Tab. 3.4.

Tab. 3.4: Common catalyst particles.

Type	Method of manufacture	Shape	Size	Use
Pellets	High-pressure press	Cylindrical, very uniform, rings	2–10 mm	Packed tubular reactors
Extrudates	Squeezed through holes	Irregular lengths of different cross-section	>1 mm	Packed tubular reactors
Granules	Aging liquid drops	Spherical	1–20 mm	Packed tubular reactors, moving beds
Granules	Fusing and crushing or particle granulation	Irregular	>2 mm	Packed tubular reactors
Flakes	Powder encapsulated in wax	Irregular		Liquid-phase reactors
Powders	Spray-dried hydrogels		20–300 µm	Fluidized reactors
			75–200 µm	Slurry reactors

3.5 Conversion, selectivity and yield

3.5.1 Conversion

The degree of conversion is a dimensionless variable that represents the extent to which the reaction has proceeded. It is defined as the amount of reactant that has reacted divided by the initial amount of reactant that has been present. For a constant volume, the bimolecular elementary reaction:

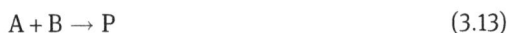

$$A + B \rightarrow P \tag{3.13}$$

the conversion for reactant A at a given time is now calculated from the initial amount of A present at time $= 0$, $N_{A,0}$, and the amount of A that is still present at time $= t$, $N_{A,t}$:

$$\xi_A = \frac{\text{amount of A converted (kmol)}}{\text{initial amount of A (kmol)}} = \frac{N_{A,0} - N_{A,t}}{N_{A,0}} \tag{3.14}$$

When the volume of reaction, V_R, can be considered constant, the conversion relates directly to the concentration by:

$$\xi_A = \frac{V_{R,0}\, C_{A,0} - V_{R,t}\, C_{A,t}}{V_{R,0}\, C_{A,0}} = \frac{C_{A,0} - C_{A,t}}{C_{A,0}} \tag{3.15}$$

3.5.2 Selectivity and yield

In most industrial chemical operations, we meet with the phenomenon that from a reactant A or reactants A and B not only desired products like P and Q, but also undesired products like X and Y, are formed. Therefore the concepts of selectivity and yield have been introduced in chemical reaction engineering. Both concepts are of utmost importance for the final economic results of an entire chemical operation.

Since the reactants in a reactor are meant to be converted into the desired product to the maximum extent, it is not sufficient to obtain a reasonably high degree of conversion, but this should be accompanied by a high selectivity of the reaction. The selectivity, σ_p, toward a product P with respect to a reactant A is defined as the ratio of the amount of A which reacted to P and the amount of A converted in total. By focusing on the amount of A that has reacted to P, instead of simply the amount of P formed, the stoichiometry of the reaction equation is taken into account and the selectivity ranges between 0 and 1. Note that when all possible products are taken into account, the sum of all selectivities must be 1, by definition. For the bimolecular elementary reaction, the selectivity of converting reactant A in product P is given by

$$\sigma_P = \frac{\text{amount of A reacted to P (kmol)}}{\text{amount of A converted (kmol)}} = \frac{N_{P,t}}{N_{A,0} - N_{A,t}} \tag{3.16}$$

Analogous to the degree of conversion, the selectivity can be related directly to the concentrations when the reaction volume, V_R, can be considered constant:

$$\sigma_P = \frac{V_{R,t}\, C_{P,t}}{V_{R,0}\, C_{A,0} - V_{R,t}\, C_{A,t}} = \frac{C_{P,t}}{C_{A,0} - C_{A,t}} \tag{3.17}$$

Ultimately, the merits of a complex reaction operation are closely related to the amount of desired product obtained with respect to the amount of key reactant A fed to the reactor. This ratio is called the product yield, η_P that is calculated by multiplying the selectivity and the degree of conversion:

$$\eta_P = \sigma_P \xi_A = \frac{C_{P,t}}{C_{A,0}} = \frac{\text{amount of A reacted to P (kmol)}}{\text{initial amount of A (kmol)}} \tag{3.18}$$

The yield can likewise vary between 0 and 1. It is high when both the selectivity and the relative degree of conversion are high, but it is low when either one of them is small. They apply equally well to simpler reactor systems as to systems with feed and/or discharge distribution, provided more general definitions of conversion are used in the latter case.

3.5.3 Multistep reactions

In most industrial chemical operations several undesired by-products are formed simultaneously next to the desired product. For these multistep reactions, the chosen reactor type and operating conditions are extremely important because they may have a tremendous effect on the ratio of the products and by-products formed. Although high conversion rates result in smaller reactors and thus in lower investment costs, the yield of the desired products is often more important for overall plant economics. Higher yields result in lower raw material costs per unit of desired product and thus in lower operating expenses for raw materials and disposal of waste products. Higher yields and selectivities indirectly also influence the investment costs for the separation of the reactor products and the disposal units.

Since multistep reactions are so varied in type and seem to have so little in common, we may despair of finding general guiding principles for design. Fortunately, this is not so because all multistep reactions can be considered to be combinations of two primary types: parallel reactions and series reactions.

3.5.3.1 Parallel reactions

An example of a parallel reaction system is the decomposition of reactant A by either one of two paths as encountered in the steam cracking of naphtha:

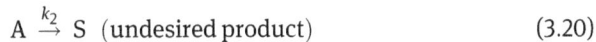

$$A \xrightarrow{k_1} P \text{ (desired product)} \tag{3.19}$$

$$A \xrightarrow{k_2} S \text{ (undesired product)} \tag{3.20}$$

with the corresponding rate equations in case of a first-order reaction in a batch reactor:

$$R_P = \frac{dC_P}{dt} = k_1 C_A \tag{3.21}$$

$$R_S = \frac{dC_S}{dt} = k_2 C_A \tag{3.22}$$

The selectivity for the formation of the desired product P from reactant A at any time and place in a reactor can now be calculated from

$$\sigma_P = \frac{\text{amount of A reacted to P}}{\text{amount of A converted}} = \frac{k_1 C_A}{k_1 C_A + k_2 C_A} = \frac{k_1}{k_1 + k_2} \tag{3.23}$$

Interestingly enough, the selectivity for two parallel first-order reactions does not depend on whether the reaction is carried out in a batch, tubular or continuous tank reactor. The ratio of P and S obtained is completely determined by the ratio of the reaction rate constants k_1 and k_2. It can only be influenced by temperature when both reaction paths have a difference in activation energy:

$$\frac{K_1}{K_2} = \frac{K_{1,\infty}}{K_{2,\infty}} \exp\left(-\frac{E_1 - E_2}{RT}\right) \tag{3.24}$$

3.5.3.2 Reactions in series

In practice, it is often encountered that the formed product reacts further to undesired by-products. Typical examples are oxidation, partial hydrogenation and chlorination reactions. Drawing conclusions about the optimal reactor choice and operating conditions for reactions in series is in general more complex than for parallel reactions. For a series of two first-order reactions, where the formed product P reacts on to the unwanted product S:

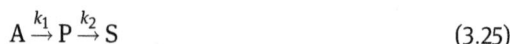

$$A \xrightarrow{k_1} P \xrightarrow{k_2} S \tag{3.25}$$

the corresponding rate equations in a batch reactor are:

$$R_A = \frac{dC_A}{dt} = -k_1 C_A \tag{3.26}$$

$$R_P = \frac{dC_P}{dt} = k_1 C_A - k_2 C_P \tag{3.27}$$

$$R_S = \frac{dC_S}{dt} = k_2 C_P \tag{3.28}$$

Introducing these reaction rates provides us with an equation for the (differential) selectivity toward product P formation in a batch reactor, which has become a function of concentration:

$$\sigma_P = \frac{\text{amount of A reacted to P}}{\text{amount of A converted}} = \frac{k_1 C_A - k_2 C_P}{k_1 C_A} \tag{3.29}$$

Calculation of the actual selectivity at a given time or place in a more general reactor is now only possible from the reactant's and products' concentration profiles as derived from the reactor mass balance equations, as will be discussed in the next subsection. The general form of these concentration profiles is illustrated in Fig. 3.5. From these concentrations profiles, it is directly seen that the selectivity approaches zero at long residence times and that an optimum exists between conversion and yield.

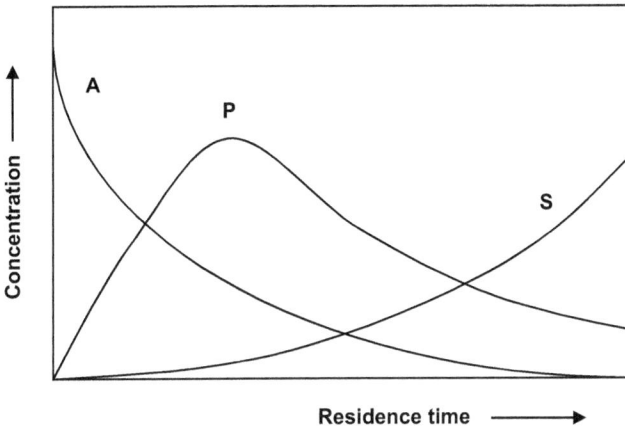

Fig. 3.5: Concentration profile for two first-order reactions in series as a function of the residence time in a reactor.

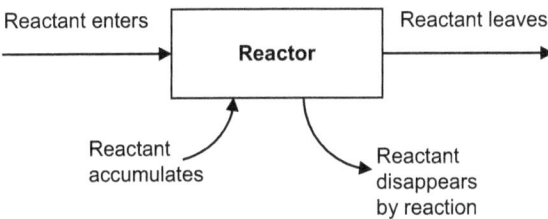

Fig. 3.6: Schematic representation of the molar balance for a reactor.

3.6 Basic design equations for model reactors

3.6.1 Molar balances

The starting point for any reactor calculation is the molar balance that uses the principle of conservation of matter to derive the basic design equations. Figure 3.6 illustrates that the basic form of the molar balance for a reactor has to take into account four contributions:

$$\text{Input} - \text{Output} + \text{Production} = \text{Accumulation} \tag{3.30}$$

or:

Rate of flow   Rate of flow   Rate of production of   Rate of accumulation
of component   −   of component   +   component by reaction   =   of component in
into reactor   out of reactor   in the reactor   the reactor

$$\tag{3.31}$$

Generally, for a volume V containing component A at concentration C_A, the molar balance is

$$\phi_{V,in} C_{A,in} - \phi_{V,out} C_{A,out} + R_A V = \frac{d(C_A V)}{dt} \tag{3.32}$$

Here, $\phi_{V,in}$ is the volumetric rate of the flow into V and $\phi_{V,out}$ the volumetric rate of the flow out of V. When the composition within the reactor is uniform (independent of position), a macrobalance can be made over the whole reactor ($V = V_R$). Where this is not the case, it must be made over a differential element of volume (a microbalance) and then integrated across the whole reactor. For various idealized reactor types, the general molar balance eq. (3.32) simplifies one way or another, and the resulting expression gives the basic performance equation for that type of unit. For example, the first two terms are zero in a batch reactor, while the accumulation term disappears in steady-state flow reactor such as a continuous stirred tank reactor or a plug flow reactor. We will now treat each of these idealized reactor types, which form the basis of many chemical reaction engineering calculations.

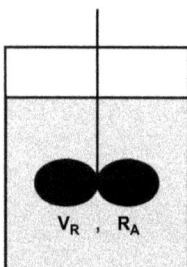

Fig. 3.7: Scheme of an ideal stirred batch reactor.

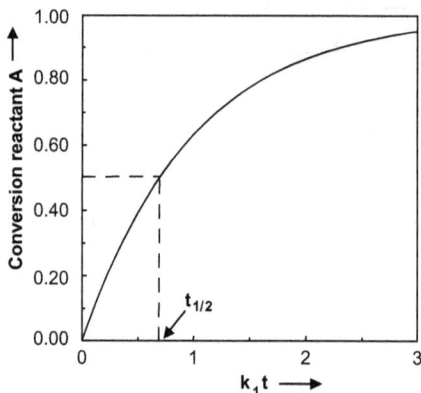

Fig. 3.8: Conversion of an irreversible first-order reaction versus time.

3.6.2 The ideal batch reactor

In a batch reactor, the feed material is treated as a whole for a fixed period of time without any inflow or outflow. This is schematically shown in Fig. 3.7. In most cases, the batch reactor is assumed to behave ideal, meaning that the contents are considered perfectly mixed. Under this condition, the component concentrations are uniform throughout the reactor at any instant of time and the molar balance can be made over the whole reactor. Noting that no fluid enters or leaves the reaction mixture during reaction, the general molar balance reduces to:

$$\text{Accumulation} \quad = \text{Production} \tag{3.33}$$

or

$$
\begin{array}{l}
\text{rate of accumulation} \qquad \text{rate of production of} \\
\text{of component in} \qquad = \quad \text{component by reaction} \qquad \qquad (3.34) \\
\text{the reactor} \qquad \qquad \text{in the reactor}
\end{array}
$$

For an irreversible first-order constant density reaction of reactant A into product P, the following molar balance equation is obtained:

$$V_R \frac{dC_A}{dt} = R_A V_R = -k_1 C_A V_R \tag{3.35}$$

Integration over the period of the reaction provides us with the relation between reaction time and reactant concentration:

$$\int_{C_{A,0}}^{C_A} \frac{1}{C_A} dC_A = -k_1 \int_0^t dt \quad \rightarrow \quad C_A = C_{A,0} e^{-k_1 t} \tag{3.36}$$

Here, $C_{A,0}$ is the initial concentration of A. Introducing this time dependent concentration in eq. (3.15) provides us with the equation for the conversion in an ideal batch reactor in case of a first-order reaction:

$$\xi_A = \frac{C_{A,0} - C_{A,t}}{C_{A,0}} = 1 - e^{-k_1 t} \tag{3.37}$$

The dependence of the conversion on the reaction time is illustrated graphically in Fig. 3.8 for a irreversible first-order reaction. The reaction time τ needed for a certain concentration decrease of reactant A can be obtained by rearranging eq. (3.36):

$$\tau = \frac{1}{k_1} \ln \left(\frac{C_{A,0}}{C_{A,\tau}} \right) \tag{3.38}$$

A characteristic value for a batch reaction is the time required to reduce the concentration of reactant A to half of its original value. This characteristic time is called the half value time and is obtained by introducing $C_{A,t} = \frac{1}{2} C_{A,0}$:

$$t_{1/2} = \ln \frac{2}{k_1} \tag{3.39}$$

Half value times are important to know, for example, for the characterization of the decomposition behavior of peroxides that are used as initiators in many radical polymerization processes. Because peroxide decomposition is a very exothermic reaction, it is absolutely essential to minimize spontaneous decomposition of the peroxide during storage. This is done by carefully controlling the temperature during storage, exploiting the strong temperature dependence of the rate of the decomposition reaction.

Half value times are also used to experimentally determine the effective order of a general, possibly nonelementary, reaction. For a reaction of effective order n in an ideal batch reactor, the molar balance is

$$V_R \frac{dC_A}{dt} = R_A V_R = - k_n C_A^n V_R \tag{3.40}$$

Solving this for an initial concentration $C_{A,0}$, the half value time is found to be equal to

$$t_{1/2} = \frac{2^{n-1} - 1}{k_n (n-1)} C_{A,0}^{1-n} \tag{3.41}$$

This shows that the half time for an effective nth order reaction depends on the initial concentration to the power $1 - n$. By measuring the half value time for a range of relevant initial concentrations in a batch reactor, and plotting the results on a double logarithmic scale, the slope $(1 - n)$ of a straight line through the data can be used to experimentally determine n for nonelementary reactions.

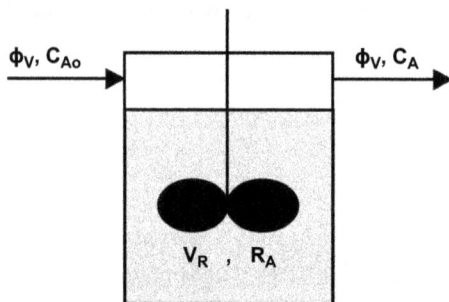

Fig. 3.9: Schematic representation of an continuous ideal stirred tank reactor (CISTR).

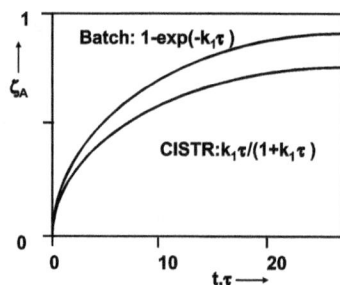

Fig. 3.10: Comparison of conversion for an irreversible first-order reaction in an ideal batch and ideal continuous stirred tank reactor.

3.6.3 The continuous ideal stirred tank reactor

In a continuous ideal stirred tank reactor (CISTR), reactants and products are continuously added and withdrawn from the reactor vessel. The term ideal reflects the assumption that the reactants are mixed with the reactor contents immediately upon entering and the composition is considered uniform throughout the reactor. As a result, the exit stream from the reactor has the same composition as the fluid within the reactor. Figure 3.9 illustrates that when the mass of withdrawn product mixture equals the mass of supplied reactants, the reaction volume remains constant and the reactor is said to operate in a steady state. In that case, the molar balance over the CISTR becomes

| Input | − | Output | + | Production | = 0 | (3.42) |

or

| Rate of flow of component into reactor | − | Rate of flow of component out of reactor | + | Rate of production of component by reaction = 0 in the reactor | (3.43) |

resulting in the following general form for the molar balance for reactant A:

$$\phi_{V,\,in} C_{A,0} - \phi_{V,\,out} C_A + R_A V_R = 0 \tag{3.44}$$

which becomes for an irreversible first-order constant density reaction of reactant A into product P:

$$\phi_V C_{A,0} - \phi_V C_A - k_1 C_A V_R = 0 \tag{3.45}$$

The average residence time τ is defined as the time that is needed to fill the reactor volume V_R with volume rate ϕ_V, in other words:

$$\tau = \frac{V_R}{\phi_V} \tag{3.46}$$

Rewriting eq. (3.45) provides us with the relations describing the concentration and conversion of reactant A as a function of the average residence time:

$$C_A = \frac{C_{A,0}}{1 + k_1 \tau} \tag{3.47}$$

$$\xi_A = \frac{C_{A,0} - C_A}{C_{A,0}} = 1 - \frac{1}{1 + k_1 \tau} = \frac{k_1 \tau}{1 + k_1 \tau} \tag{3.48}$$

This conversion is compared with the conversion of an ideal batch reactor in Fig. 3.10. It is seen that the volume of a CISTR must always be larger (requires a longer residence time) to achieve the same conversion for a first-order reaction. This is generally true for any reaction of order $n > 0$ due to the lower reactant concentrations caused by the continuous perfectly mixed operation.

3.6.4 Ideal continuous plug flow reactor

A plug flow reactor is a vessel through which flow is continuous, usually at steady state, and configured so that conversion and other dependent variables are functions

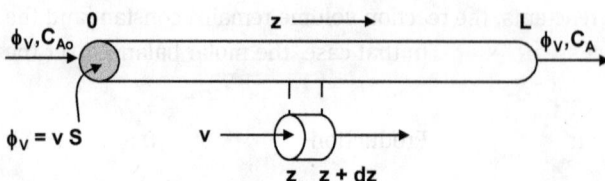

Fig. 3.11: Schematic representation of a differential element of a plug flow reactor.

of position within the reactor rather than time. In the ideal plug flow reactor, the reaction mixture passes through in a state of plug flow which, as the name suggests,
means that the fluid moves like a solid plug or a piston. Any backflow effects are neglected. Furthermore, it is assumed that the fluid properties, temperature, pressure and
composition are uniform across the section normal to the fluid motion (no radial gradients). Of course the compositions, and possibly also the temperature and pressure,
change between inlet and outlet of the reactor in the longitudinal direction. The basic
design equation for an ideal plug flow reactor is obtained by applying the general
molar balance over a differential element of volume dV_R, as illustrated in Fig. 3.11.
Under steady-state conditions, the accumulation term is zero and the molar balance for
a differential reactor element for reactant A yields:

$$\text{Input} - \text{Output} + \text{Production} = 0 \tag{3.49}$$

or

$$\phi_V C_A|_z - \phi_V C_A|_{z+dz} + R_A dV_R = 0 \tag{3.50}$$

Introduction of the reactor cross-sectional surface area S results in $\phi_V = v\,S$ and dV_R
$= S\,dz$, where v stands for the longitudinal velocity (m/s). For a first-order constant
density reaction of reactant A into product P:

$$v C_A|_z - v C_A|_{z+dr} - k_1 C_A dz = 0 \tag{3.51}$$

giving:

$$v\frac{dC_A}{dz} = -k_1 C_A \tag{3.52}$$

which can be integrated over the length of the reactor L to give the concentration
profile of reactant A over the length of the plug flow reactor:

$$\int_{C_{A,0}}^{C_A} \frac{1}{C_A} dC_A = -k_1 \int_0^L \frac{1}{v} dz \quad \rightarrow \quad C_A = C_{A,0} e^{-k_1 \tau} \tag{3.53}$$

where $\tau = L/v$ is the residence time in the plug flow reactor. Note that this result is
identical to the time-dependent conversion obtained for the ideal batch reactor.

Tube reactor = **moving small ideal batch reactors**

Total volume = τ

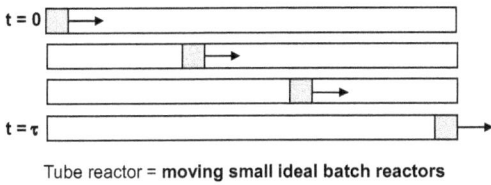

Fig. 3.12: Representation of a tubular reactor by a series of infinitely small ideal batch reactors.

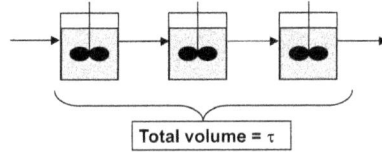

Fig. 3.13: Cascade of ideal stirred tank reactors with equal total volume V_R.

This leads us to the general conclusion that the ideal continuous plug flow reactor behaves identical to the ideal batch reactor. Schematically this is depicted in Fig. 3.12, which shows that a plug flow reactor can be considered to consist of a series of infinitely small ideal batch reactors that remain in the reactor for a residence time $\tau = L/v$.

3.6.5 Cascade of continuous ideal stirred tanks reactors

Higher conversions of reactants with stirred tanks can be obtained by using several of them with an equal total volume in series. From Fig. 3.13, it can be seen that the product stream from the first reactor is the feed for the second reactor. When each reactor has the same volume, the concentration and conversion of reactant A for a series of N reactors can be calculated from

$$C_A = \frac{C_{A,0}}{(1 + k_1 \tau)^N} \tag{3.54}$$

and

$$\xi_A = \frac{C_{A,0} - C_{A,t}}{C_{A,0}} = 1 - \frac{1}{(1 + k_1 \tau)^N} \tag{3.55}$$

When the number of reactors in series is increased while the total volume is kept constant, the characteristics of a plug flow reactor are approached. This is schematically depicted in Fig. 3.14.

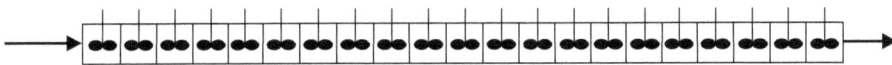

Fig. 3.14: Representation of plug flow reactor characteristics by a large number of continuous ideal stirred tanks (CISTR).

3.7 Heat effects in model reactors

Chemical reactions are always accompanied by heat effects. The amount of heat liberated or absorbed during reaction depends on the amount of converted material and the heat of reaction ΔH_R at the operating conditions. When this is not known, it can in most cases be calculated from known and tabulated thermochemical data on heats of formation or heats of combustion of the reacting chemicals. Heat-producing reactions are called exothermic and reactions where the addition of heat is required endothermic.

3.7.1 General heat balance

In nonisothermal operations, heat balances must be used in conjunction with molar balances. The general form of the heat balance for a reactor has a form very similar to the general molar balance:

$$
\begin{array}{ccccccc}
\text{rate of heat} & & \text{rate of heat} & & \text{rate of heat} & & \text{rate of heat} \\
\text{flow into} & - & \text{flow out of} & + & \text{production in} & = & \text{accumulation} \\
\text{reactor} & & \text{reactor} & & \text{the reactor} & & \text{in the reactor}
\end{array} \quad (3.56)
$$

In the inflow and outflow terms, the heat flow may be of two kinds. The first is the transfer of sensible heat or enthalpy Q (J/s) by the fluid entering and leaving. Assuming the density of the fluid ρ (kg/m³) and the specific heat of the fluid C_p (J/kg K) does not change significantly between inflow and outflow, the net amount of heat added to the system is:

$$ Q = \phi_V \, \rho \, C_P \, (T_{in} - T_{out}) \qquad (3.57) $$

The second is the heat transferred to or from the fluid across heat transfer surfaces such as cooling coils inside or jackets outside the reactor. In many cases, the temperature T_C on the cooling or heating side can be taken approximately constant and the rate of heat transfer Q across a surface of area A can be written as:

$$ Q = UA(T_C - T) \qquad (3.58) $$

where T is the temperature of the reaction mixture and U (W/m².K) is the heat transfer coefficient. The heat transfer coefficient is a function of reactor wall thickness and material, the fluid thermal properties and fluid flow velocity.

Combining the above terms results in the following general heat balance which can be applied as a macrobalance to a stirred tank reactor or as a microbalance to an element of a plug flow reactor:

$$\phi_V \rho C_P (T_{in} - T_{out}) + UA(T_C - T) + V_R R_A(-\Delta H_R) = V_R \rho C_P \frac{dT}{dt} \qquad (3.59)$$

difference between	heat transfer	heat produced by	rate of heat
enthalpy of inflowing	from cooling or	chemical reaction	accumulation
and outflowing	fluid heating elements		

The rate of heat production depends on the rate of reaction R_A, which in turn depends on the concentration levels in the reactor as determined by the general molar balance equation. Since the rate of reaction also depends sensitively on the temperature levels in the reactor, the molar and heat balances interact strongly with each other and have to be solved simultaneously.

3.7.2 Heat balance in a continuous ideal stirred tank reactor

For a stationary CISTR, the rate of heat accumulation is zero and the temperature T in the reactor is equal to that of the outflowing fluid. For a irreversible first-order reaction in a CISTR, the heat balance eq. (3.59) can therefore be written as

$$\rho C_P (T - T_{in}) + \frac{UA}{\phi_V}(T - T_C) = k_1 C_A \tau(-\Delta H_R) = \xi_A C_{A,0}(-\Delta H_R) \qquad (3.60)$$

where we used the conversion ξ_A of A in a CISTR, as given by eq. (3.48). Note that the heat produced by chemical reaction is directly proportional to the conversion in the reactor, and is a sensitive function of reactor temperature, as schematically shown by the solid line in Fig. 3.15. As the temperature in the reactor increases, the conversion increases rapidly according to the Arrhenius relation but then tends to an upper limit as the reaction approaches complete conversion. On the other hand, the rate of heat removal both by inflow–outflow and by heat transfer (left-hand side of eq. (3.60)) increases linearly with reactor temperature, as shown in Fig. 3.16. To satisfy the heat balance eq. (3.60), the point representing the actual operating temperature of the reactor is obtained from the intersection of the heat production curve (solid line) and the rate of heat removal line (dashed line). If the rate of heat removal is high, either due to low temperature T_{in} of the inflow or due to a high rate of heat transfer, there is only one point of intersection, corresponding to a low operating temperature. With a somewhat smaller rate of heat removal, there are three points of intersection corresponding to two stable operating conditions. The intermediate point of intersection is an unstable operating point where the slightest process disturbance will cause the system to pass into the lower or upper operating state. Such multiplicity of operating points should be avoided in practice, because it is difficult to operate or control such a reactor. The solution is to either increase the inflow temperature or to lower the heat transfer, which leads to only one operating temperature, corresponding to nearly complete conversion.

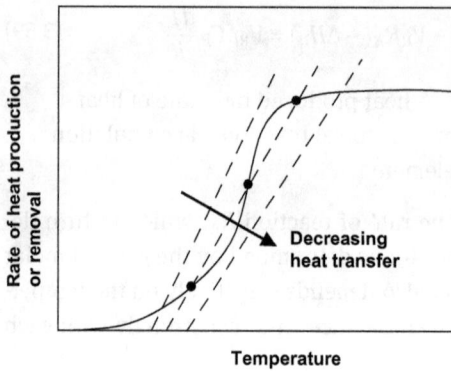

Fig. 3.15: Heat production (solid line) and removal (dashed line) for a continuous stirred tank reactor as a function of temperature in the reactor.

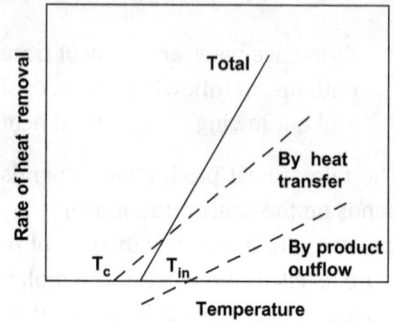

Fig. 3.16: Rate of heat removal from a continuous stirred tank reactor as a function of temperature in the reactor.

3.7.3 Heat balance in a batch reactor

The balance between the rate of heat production and rate of heat removal or addition determines the temperature profile of a batch reactor:

$$V_R \rho C_P \frac{dT}{dt} = UA(T_C - T) + V_R R_A(-\Delta H_R) \tag{3.61}$$

In general, the simultaneous solution of the molar and heat balances for a batch reactor is only possible by numerical simulation. For an exothermic reaction, the reactor temperature usually rises after starting the reaction. A typical requirement

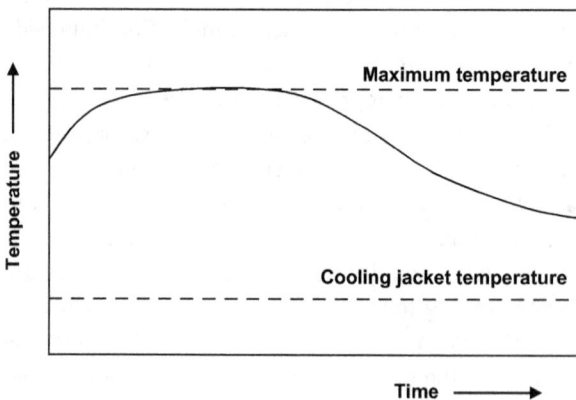

Fig. 3.17: Typical reactor temperature profile for an exothermic reaction in a nonisothermal batch reactor with just sufficient cooling.

is that the temperature shall not rise above the maximum allowable temperature to avoid by-products, catalyst deactivation or hazardous operation. This is achieved by using sufficient cooling as illustrated by Fig. 3.17.

Nomenclature

A	Surface area for cooling/heating	$[m^3]$
C	Concentration	$[mol/m^3]$
C_p	Specific heat	$[J/kg\ K]$
E_A	Activation energy	$[J/mol]$
ΔG°	Standard free energy	$[J/mol]$
ΔH°	Standard enthalpy	$[J/mol]$
k	Reaction rate constant (unit depends on reaction, first order =)	$[mol/s]$
k_∞	Frequency factor (unit depends on reaction, first order =)	$[mol/s]$
K_R	Reaction equilibrium constant	$[-]$
L	Reactor length	$[m]$
n	Order of a reaction	$[-]$
N	Number of moles	$[mol]$
N	Number of reactors	$[-]$
P_{tot}	Total pressure	$[N/m^2]$
Q	Enthalpy	$[J/s]$
R	Reaction rate	$[mol/s]$
R	Gas constant (8.3144)	$[J/mol\ K]$
S	Cross-sectional surface area	$[m^2]$
t	Time	$[s]$
T	Temperature	$[K]$
U	Heat transfer coefficient	$[W/m^2\ K]$
v	Longitudinal velocity	$[m/s]$
V_R	Volume of reaction, reactor volume	$[m^3]$
z	Position in plug flow reactor	$[m]$
ϕ_V	Volumetric flow rate	$[m^3/s]$
η	Yield	$[-]$
v	Stoichiometric coefficients	$[-]$
ρ	Density	$[kg/m^3]$
σ	Selectivity	$[-]$
τ	Reaction time, residence time	$[s]$
ξ	Degree of conversion	$[-]$

Indices

A	Reactant
B	Reactant
C	Cooling water
F	Formation
K	Intermediate complex
0	Initial (time = 0)
P	Product
R	Reaction

4 Chemical reactors and their industrial applications

4.1 Introduction

The reactor in which the chemical reaction takes place occupies a central position in the chemical process. This chapter provides an overview of commonly encountered reactor types, illustrated with various examples of their industrial applications. However, first of all, the different ways to classify chemical reactors are discussed.

4.1.1 Classification of reactors

Chemical reactors exist in such a wide range of forms and types that a complete systematic classification is impossible. Two main categories that can be distinguished are homogeneous and heterogeneous reactors. In homogeneous reactors, only one phase, usually a gas or a liquid, is present. If more than one reactant is involved, provision must be made for mixing them together to form a homogeneous mixture.

Another kind of classification, which cuts across the homogeneous–heterogeneous division, is the mode of operation, batchwise or continuous. Homogeneous batch reactions are carried out in vessels, tanks or autoclaves in which the reaction mixture is agitated and mixed in a suitable manner. This operation is familiar to anybody who has carried out small-scale preparative reactions in the laboratory. Continuous flow reactors for homogeneous reaction systems already show a much greater variety. Predominant forms are the tubular reactor and the mixed tank reactor, which have essentially different characteristics.

In heterogeneous reactors, two or more phases are present. The classification of reactors for heterogeneous systems shows a great number of possibilities. The dominant factor is the contact between the different phases. This leads to a classification of reactors as a contact apparatus. Common examples are gas–liquid, gas–solid, liquid–solid, liquid–liquid and gas–liquid–solid systems. In many cases, the solid phase is present as a catalyst. Gas–solid catalytic reactors comprise an important class of heterogeneous chemical reaction systems. Generally, heterogeneous reactors exhibit a greater variety of configurations and contacting patterns than homogeneous reactors.

4.1.2 Influence of heat of reaction on reactor type

Associated with every chemical change, there is a heat of reaction that is only in a few cases small enough to be neglected. The magnitude of the heat of reaction has often a major influence on the design of a reactor. With a highly exothermic reaction,

https://doi.org/10.1515/9783110712445-004

for example, a substantial rise in temperature of the reaction mixture will take place unless provision is made for heat to be removed as the reaction proceeds.

When feasible, adiabatic operation of a reactor without provision for heating or cooling (except for preheating or precooling the inflow stream) is preferred for simplicity of design. If the reactor cannot operate adiabatically, its design must include provision for heat transfer. Commonly used ways to heat or cool a reactor are external jackets, internal coils or an external heat exchanger through which the reactor contents is circulated. If one of the constituents of the reaction mixture is volatile, the external heat exchanger may be a reflux condenser, just as in the laboratory.

4.2 Stirred tank reactors

4.2.1 Description

Most stirred tanks used as chemical reactors are cylindrical, and are equipped with a centrally positioned stirrer. The volume of the tank is normally determined by the residence time required, but the aspect ratio of the tank (height/diameter, H/D) can be selected. Bulk mixing is favored by $H/D = 1$, while a large H/D ratio is preferred when heat must be transferred. So, a compromise is often necessary. Whenever possible, wall baffles are installed inside the vessel. These baffles prevent the rotation of the reactor contents with the stirrer. Wall baffles are to be avoided in systems that are prone to fouling or in systems in which gas or a floating material is to be entrained in the liquid.

Jackets are often fitted around the vessel to provide external heating or cooling of the reactor. In addition, an internal heating or cooling coil is often used to improve heat transfer. Stirrers are used to enhance the contacting, mixing, mass and heat transfer processes. A wide variety of stirrers is available. If the aspect ratio of the tank (H/D) is greater than 1.5, it is recommended to use additional stirrers on the shaft. Spacing of the stirrers should be approximately equal to the tank diameter. Selection of stirrers is initially based on the reaction mixture.

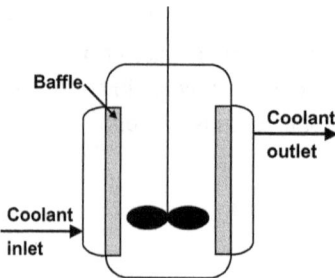

Fig. 4.1: Schematic of a jacketed batch reactor with baffles.

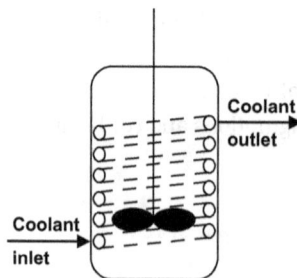

Fig. 4.2: Schematic of a batch tank with internal heating/cooling coil.

4.2.2 Batch stirred tank reactors

There is a tendency in chemical engineering to try to make all processes continuous. While continuous reactors are likely to be most economic for large-scale production, batch reactors may be preferred for small-scale production of multiple low-volume high-priced products in the same equipment, particularly if many sequential operations are employed to obtain high product yields. A batch reactor is one in which the feed material is treated as a whole for a fixed period of time without any inflow or outflow. A big advantage of batch reactors in the dyestuff, fine chemical and pharmaceutical industries is their versatility. A corrosion-resistant batch reactor with heating and cooling coils can be used for a wide variety of similar kinds of reactions. The relatively small batch reactors used for these kinds of applications generally require less auxiliary equipment, such as pumps, and their control systems are less elaborate and costly than those for continuous reactors. In addition, additional process steps upstream or downstream of the reaction can also be performed in the reactor.

Batch reactors may also be justified when long reaction times are required to achieve a desired conversion or when continuous flow is difficult, as with highly viscous or sticky solids-laden liquids. Because residence times can be more uniform in batch reactors, better yields and higher selectivity may be obtained than with continuous reactors. This advantage exists when undesired reaction products inhibit the reaction or the product is an unstable or reactive intermediate. In some processes, such as polymerization and fermentation, batch reactors are traditionally preferred because the interval between batches provides an opportunity to clean the system thoroughly. Some important industrial applications of batch reactors are listed in Tab. 4.1.

Tab. 4.1: Some examples of the industrial application of batch reactors.

Long reaction times	Biotechnological processes
	Polymer lattices
	Polycondensation resins
Multiple reaction steps	Pharmaceutical ingredients
	Agrochemicals
	Polypropylene slurry polymerization
	Acrylonitrile–butadiene–styrene (ABS)
	Emulsion polymerization

One of the main disadvantages of batch reactors is relatively high operating costs due to long downtimes between batches and high manpower requirements. A second important disadvantage is the quality difference between charges because reaction conditions are only partly reproducible. When an external jacket is used, ensuring a homogenous temperature distribution throughout the reactor is usually difficult, especially with highly endothermic or exothermic reactions.

Fine chemical products such as pharmaceutical ingredients are usually produced in multipurpose batch reactors with volumes between 1 and 10 m³. Many of the reactors are glass lined because of the corrosive conditions that are employed. As illustrated in Fig. 4.1, heating and cooling are accomplished through an external jacket. After dosing the liquid reactants and solvents from barrels or storage tanks, the solid reactants are in most of the cases added manually from bags. The reactor content is then heated to start the reaction that is allowed to proceed until the desired conversion is reached. In case of strongly exothermic or endothermic reactions, significant amounts of heat have to be supplied or removed. In those situations, the heating or cooling capacity of the external jacket often limits the maximum production rate.

Batch reactors with internal heating/cooling coils (Fig. 4.2) are employed in production of saturated and unsaturated polyester resins from dicarboxylic acids and diols. An example is the reaction of neopentyl glycol with terephthalic acid:

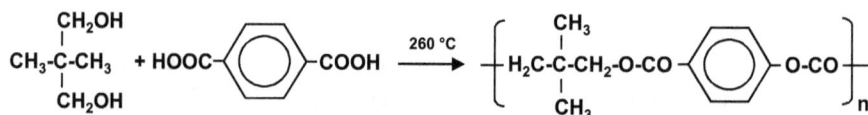

Neopentylglycol(l)   Terepthalic Acid(s)

The polyesterification reaction is normally carried out in stainless steel vessels ranging from 8 to 20 m³. The reactor is filled with liquid glycols from a metering tank and then heated to about 110 °C. At this temperature, the reaction can start and the solid dicarboxylic acids are added from big bags. Blade agitators are used to homogenize the reaction mixture during polymerization. The water formed in the reaction is first used to build up pressure to increase the reaction temperature gradually to about 260 °C. Additional water is evaporated and separated from the entrained glycol in a distillation column. Once the polymer is formed, the reaction mixture is cooled to below 180 °C before draining it on flake belts or into a cooled blend tank containing styrene monomer.

Batch reactors are also used for multiphase reactions. A good example is the hydrogenation of benzaldehyde to benzyl alcohol with the aid of an aqueous slurry of porous nickel catalyst particles:

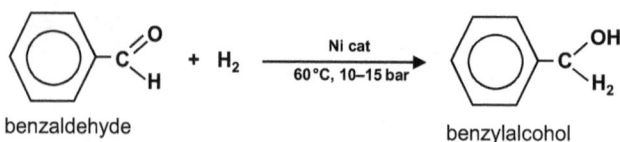

benzaldehyde   benzylalcohol

Vigorous stirring ensures that the catalyst is homogeneously dispersed in the liquid phase. As shown in Fig. 4.3, the hydrogen gas is fed to the reactor below the stirrer through a sparger. The stirrer is designed in such a way that gas is sucked into the stirrer shaft and recirculated to the liquid. At the surface of the catalyst particle, reaction takes place between the dissolved hydrogen and an adsorbed benzaldehyde molecule. The produced benzyl alcohol desorbs from catalyst surface and dissolves in the liquid. When benzaldehyde is fully converted, the hydrogen supply and the stirrer are stopped to allow the catalyst to settle to the bottom of the reactor.

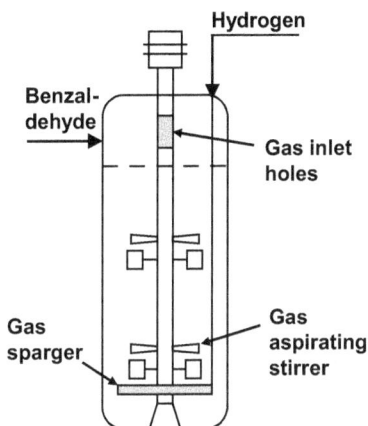

Fig. 4.3: Schematic drawing of batch reactor for benzaldehyde hydrogenation.

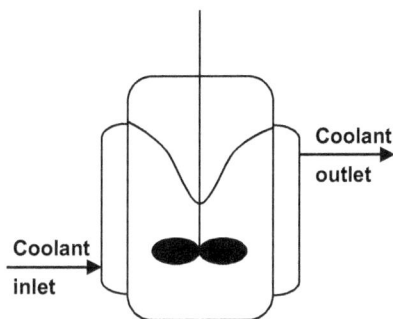

Fig. 4.4: Schematic of a polypropylene slurry polymerization reactor with a vortex.

Other applications of multiphase batch reactors involve the slurry polymerization of polypropylene, emulsion polymerization of polymer lattices (polystyrene, poly-butadiene, acrylonitrile–butadiene–styrene (ABS)) and suspension polymerization (styrene-acrylonitrile (SAN), polyvinyl chloride (PVC)). Interestingly, the reactors in the polypropylene slurry process are deliberately operated with a large vortex to enhance transfer of the gaseous propylene into the slurry where the polymerization takes place (Fig. 4.4). The reactor is jacketed to remove the heat of polymerization.

4.2.3 Continuous stirred tanks

The stirred tank reactor in the form of either a single tank or more often a series of tanks is particularly suited for liquid-phase reactions, and is widely used in the organic chemicals industry for medium- and large-scale production. Other often-encountered applications are gas–liquid reactions and gas–liquid reactions over suspended catalysts. It can form a unit in a continuous process, giving consistent product quality, ease of automatic control and low manpower requirements. Although the volume of a stirred tank must be larger than that of a plug flow tubular reactor for the same production rate, there is little disadvantage because large volume tanks are relatively cheap to construct. If the reactor has to be cleaned periodically, as happens sometimes in polymerization or in plants manufacturing different products over time, the open structure of a tank is an advantage. When high conversions of reactants are needed, several stirred tanks in series can be used. Equally good results can be obtained by dividing a single vessel into compartments while minimizing back-mixing and short-circuiting. The larger the number of stages, the closer performance approaches that of a tubular plug-flow reactor.

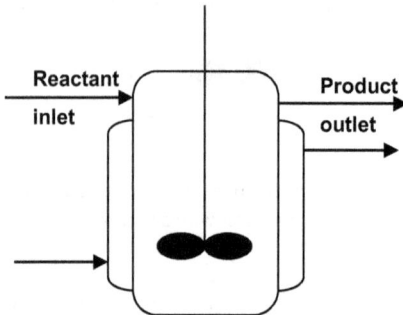

Fig. 4.5: Schematic of a continuous stirred tank reactor.

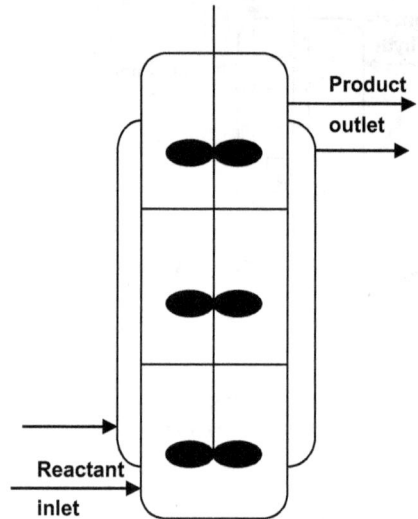

Fig. 4.6: Schematic of a continuous stirred tank reactor with compartments.

In a continuous-flow stirred tank reactor (CSTR), reactants and products are continuously added and withdrawn. In practice, mechanical or hydraulic agitation is required to achieve uniform composition and temperature. The reactants are mixed and therefore diluted immediately on entering the tank. In many cases, this favors the desired reaction and suppresses the formation of by-products. Because fresh reactants are rapidly mixed into a large volume, the temperature of the tank is readily controlled, and hot spots are much less likely to occur than in tubular reactors. Moreover, if a series of stirred tanks is used, it is relatively easy to hold each tank at a different temperature so that an optimum temperature sequence can be attained.

CSTRs in series are simpler and easier to design for isothermal operation than tubular reactors. Reactions with narrow operating temperature ranges or those requiring close control of reactant concentrations for optimum selectivity benefit from series arrangement. If severe heat transfer requirements are imposed, heating or cooling zones can be incorporated within or external to the CSTR. For example, impellers or centrally mounted draft tubes circulate liquid upward, then downward through vertical heat-exchanger tubes. In a similar fashion, reactor contents can be recycled through external heat exchangers. A different solution is used in the solution polymerization processes used for the production of polyethylene, ethylene propylene diene monomer (EPDM) rubber and styrene maleic anhydride (SMA) polymers. As illustrated in Fig. 4.7, the monomers are dissolved in the solvent and cooled to a temperature of −30 °C. The continuous stirred tank reactor is operated adiabatically during which the heat of polymerization is taken up by the reaction mixture and the temperature rises to 150–200 °C. The polymer is recovered from the solution by evaporating the solvent, which is recycled to the deep cooling.

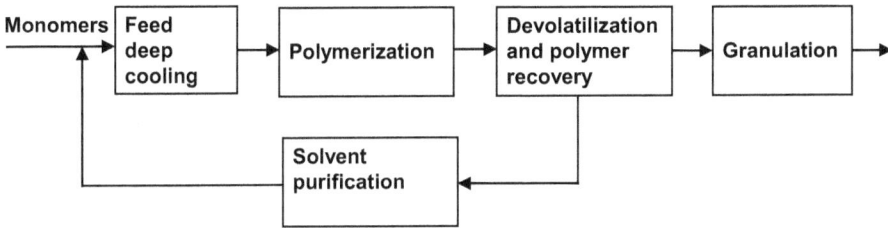

Fig. 4.7: Block scheme of a solution polymerization plant.

Continuous stirred tank reactors with several compartments in series are employed in many different processes such as the production of 5-oxohexane nitrile (5-OHN) in the α-picoline process, synthesis of urea, high-pressure polyethylene autoclave reactor, mass polymerization of styrene and suspension polymerization of ABS and SAN. Depending on the specific application, vertical reaction columns or horizontal reactors are used. Each compartment can contain a stirrer and cooling coil. In mass polymerization processes, each reactor compartment is equipped with a condenser to remove the heat of polymerization by evaporating solvent or monomer. This is schematically illustrated in Fig. 4.8.

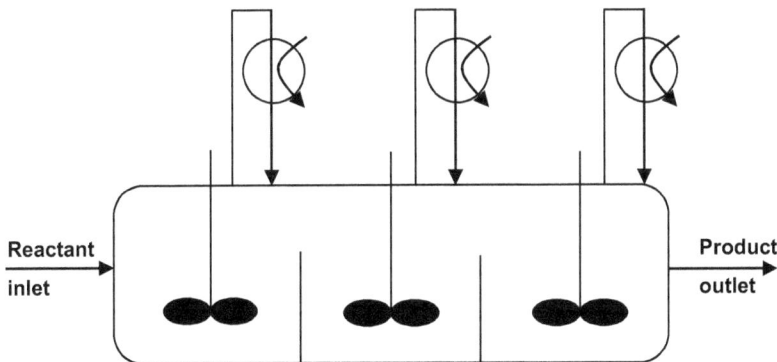

Fig. 4.8: Drawing of a stirred three-compartment tank reactor with condensers for heat removal.

4.2.4 Cascade of stirred tanks

A cascade of stirred tank reactors is employed in many different processes such as polymerizations and oxidations. One example is the uncatalyzed oxidation of cyclohexane to cyclohexyl hydroperoxide:

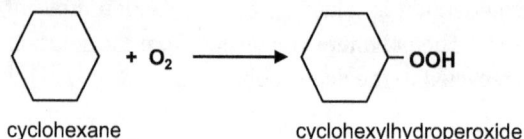

cyclohexane   cyclohexylhydroperoxide

As illustrated in Fig. 4.9, a cascade of stirred tanks is used for the oxidation to minimize back-mixing of the produced cyclohexyl hydroperoxide and allow better control of the air distribution into the liquid. All reactors contain an air sparger and are stirred to facilitate dissolution of oxygen. To keep the selectivity around 80%, the overall conversion is limited to 5–10%. The heat of reaction is removed by evaporation of the cyclohexane into the unconverted air. After oxidation, the cyclohexyl hydroperoxide is decomposed to yield the desired products, cyclohexanone and cyclohexanol:

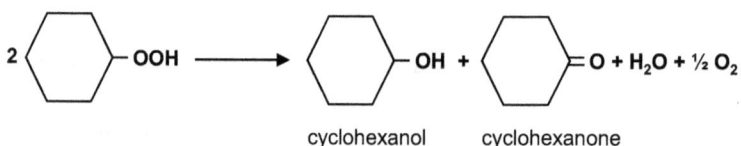

cyclohexanol   cyclohexanone

Besides the main products, significant amounts of side products such as acids, esters and acetals are formed.

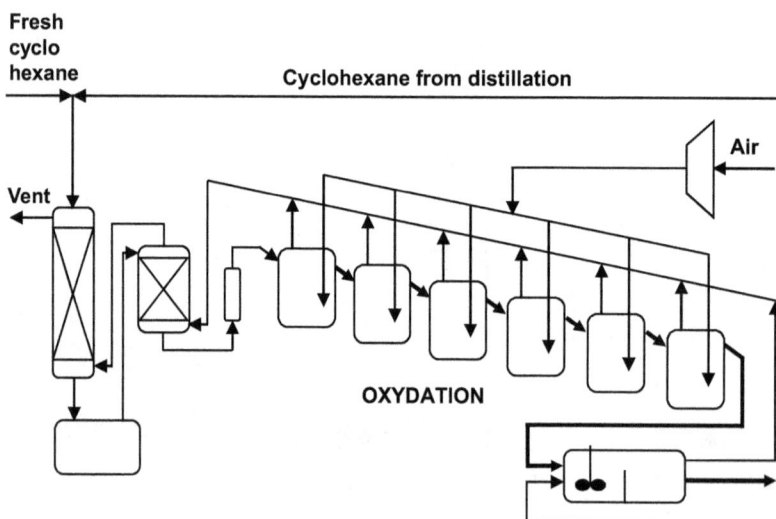

Fig. 4.9: Cyclohexane oxidation reactor section.

4.3 Tubular reactors

4.3.1 Introduction

Most industrial applications of tubular reactors involve homogeneous gas- and liquid-phase reactions. In these applications, a tubular flow reactor is chosen when it is desired to operate the reactor continuously but without back-mixing of reactants and products. In the ideal tubular reactor, the fluids flow as if they were solid plugs or pistons. Thus, in the idealized tubular reactor, all elements of fluid take the same time to pass through the reactor and experience the same sequence of temperature, pressure and composition changes. In practice, there is always some degree of departure from the ideal plug flow condition. The most serious are departures from a uniform temperature profile across the radius of the reactor. There will be local variations in reaction rate and therefore in the composition of the reaction mixture. Flow in tubular reactors can be laminar or turbulent. Generally, turbulent flow is preferred to laminar flow, because mixing and heat transfer are improved, less back-mixing is introduced in the direction of flow, and the flow profile is closer to the ideal plug flow.

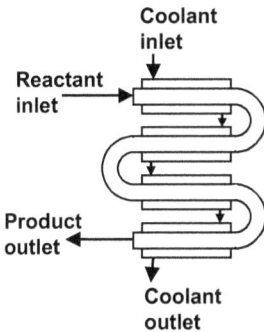

Fig. 4.10: Schematic of a tubular reactor.

When significant heat transfer to the reactor is required, a configuration with a high surface-to-volume ratio is employed (Fig. 4.10). In such reactors, the reaction volume is made up from a number of tubes that can be arranged in parallel or series. The parallel arrangements gives a lower velocity of the fluid in the tubes, which results in a lower pressure drop but also in a lower heat transfer coefficient. It is very suitable to use a second fluid outside the tubes for heat transfer. On the other hand, with tubes in series, a high fluid velocity is obtained inside the tubes and a higher heat transfer coefficient results. The series arrangement is therefore often more suitable if heat transfer is by radiation or when the main resistance to heat transfer is located inside the tube. From this, it is clear that a tubular reactor has the advantage of favorable conditions for temperature control by heat supply or removal. Another important advantage is the lack of moving mechanical parts, which makes tubular reactors especially suitable for high-pressure operation. Due to their length and relatively high velocities, usually high pressure drops are encountered. The flexibility of tubular reactors is limited because they are, in most cases, designed for a specific application with a high degree of specialization.

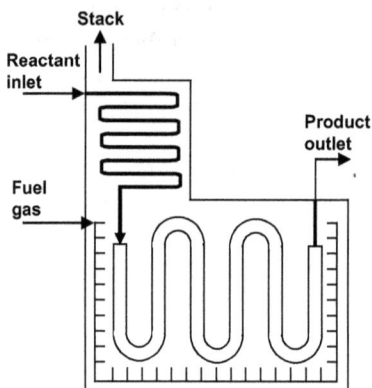

Fig. 4.11: Schematic representation of a steam cracker furnace.

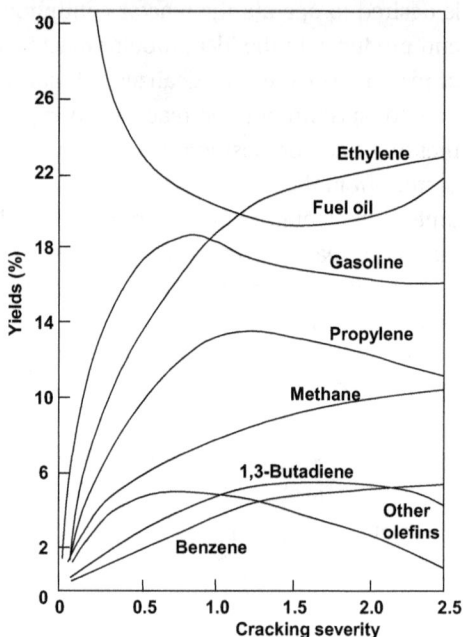

Fig. 4.12: Effect of cracking gas composition severity on product.

4.3.2 Gas-phase reactors

Production of olefins by steam cracking of hydrocarbons such as naphtha provides an excellent example of the use of tubular reactors for an important petrochemical process. The reaction is highly exothermic, and the highest selectivity toward the desired olefinic products is achieved at short residence times combined with high temperatures up to 900 °C. Despite the name steam cracking, steam is not involved in the reactions themselves. The role of steam is to lower the hydrocarbon partial pressure in the reactor to enhance the olefin yield. Other functions of steam are aiding heat transfer from the reactor wall and reaction with coke deposits on the reactor walls to reduce coke formation. The cracking of hydrocarbons occurs almost entirely by homogeneous gas-phase free-radical chain reactions. In the simplest case, ethane pyrolysis, the steps can be set out as follows:

$$C_2H_6 \quad\quad \rightarrow CH^*_3 + CH^*_3$$

$$C_2H_6 + CH^*_3 \rightarrow C_2H^*_5 + CH_4$$

$$C_2H^*_5 \quad\quad \rightarrow C_2H_4 + H^*$$

$$C_2H_6 + H^* \quad \rightarrow C_2H_4 + H_2$$

Besides these reactions, several hundreds of other reactions take place, yielding a large variety of other hydrocarbon products. Chemical simulation models that can simulate more than a hundred different species and several thousands of different chemical reactions are used to optimize the product distribution of large-scale steam crackers. The main objective is to maximize the yield to ethylene and propylene for each type of naphtha. As illustrated in Fig. 4.12, this can be accomplished by optimizing the cracking severity, which is a combination of residence time, temperature and pressure. Typically, the product gas from a naphtha steam cracker contains about 30% ethylene and 14% propylene.

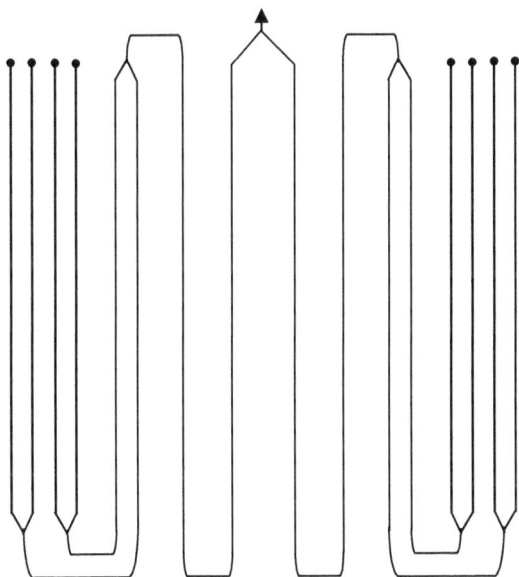

Fig. 4.13: Example of steam cracker reactor geometry giving improved heat transfer and lower pressure drop.

Since rapid heating to reaction temperature is required and the reaction is highly endothermic, it must be carried out in a tube reactor with a high surface-to-volume ratio. In the earliest tubular furnaces, the internal diameters of the reactor coils were usually uniform. The introduction of new steel alloys allowed the construction of

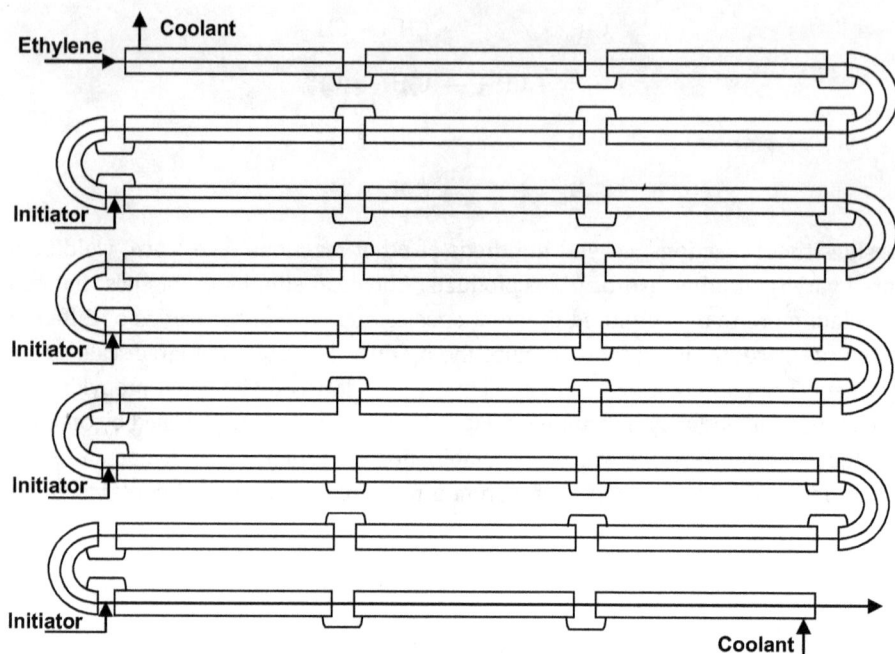

Fig. 4.14: Schematic drawing of high-pressure tubular reactor for polyethylene (LDPE) production.

more complex reactor coils that have the advantage of better heat transfer and lower pressure drop. In these reactor coils, the first one or two passes are made from smaller diameter tube. The resulting larger surface-to-volume ratio allows higher heat inputs and as a result, the full cracking temperature is reached more rapidly. In a typical arrangement, four parallel 50 mm tubes are followed by two 75 mm tubes and then a single 100 mm tube. Several of such coils are combined in a single firebox and are directly heated through natural gas burners. Virtually all heat transfer to the outside of the tubes occurs by radiation from the hot combustion gases.

4.3.3 Liquid-phase reactors

Low-density polyethylene is produced by polymerization of ethylene at very large pressures between 1,500 and 3,500 bar. These extreme operating conditions and the high heat of polymerization are strongly in favor of using a tubular reactor. The used reactors typically consist of 1,000–2,000 m of jacketed high-pressure tubing (Fig. 4.14). It is constructed from a large number of sections, each 10 m long, which are arranged in the shape of an elongated coil. Inner diameters range typically between 50 and 75 mm, depending on the capacity of the system. A ratio of outer to inner diameters of about 2.5 is used to provide the necessary strength for the high pressures involved.

In the first section of the reactor, the compressed (supercritical) ethylene is pre-heated to the temperature at which polymerization is initiated. This temperature de-pends on the initiator employed and ranges from 140 to 180 °C. After introduction of the peroxide initiator, the temperature of the reaction mixture rises to a peak of 300–350 °C and is then cooled back to about 250 °C (Fig. 4.15). The ethylene veloc-ity must be at least 10 m/s to provide sufficient heat transfer and reduce fouling of the reactor wall. In general, the reactor is long enough to provide more than one reaction zone and the polymerization is reinitiated. This is repeated three to four times until conversions between 30% and 40% are obtained.

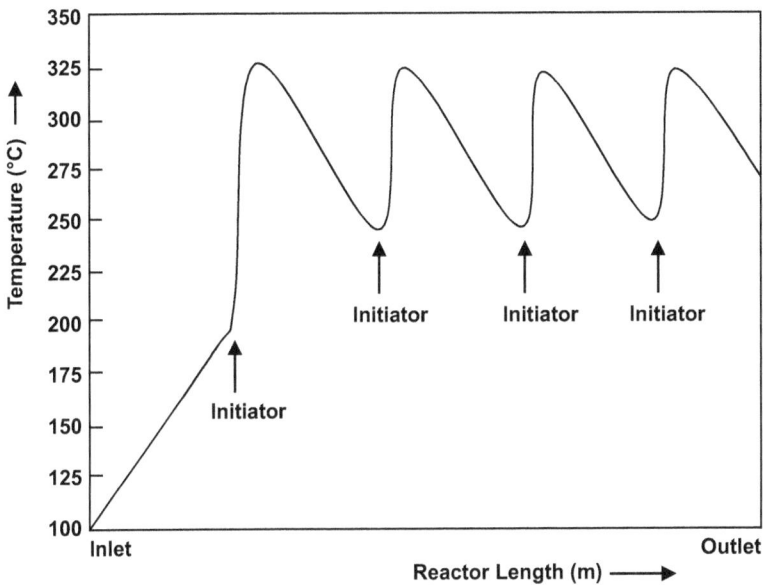

Fig. 4.15: Typical temperature profile in a high-pressure polyethylene tubular reactor.

4.4 Loop reactors

A loop reactor consists of a reaction vessel through which process fluids are recircu-lated. Often the reactor is combined with a pump for recirculation of the liquid, a heat exchanger, an injection device for reactants and a separation device. Loop re-actors can be operated in either batch or continuous configuration.

4.4.1 Continuous loop reactors

A continuous loop reactor has feed streams entering the loop and product streams leaving it. The rate of recirculation around the loop can generally be set independently of the throughput rate. This flexibility can be used to adapt the residence-time distribution of the overall reactor to the requirements of the reaction. A good example is the rearrangement of cyclohexanone oxime into caprolactam that includes the opening of the cyclohexyl ring and takes place according to the following equations:

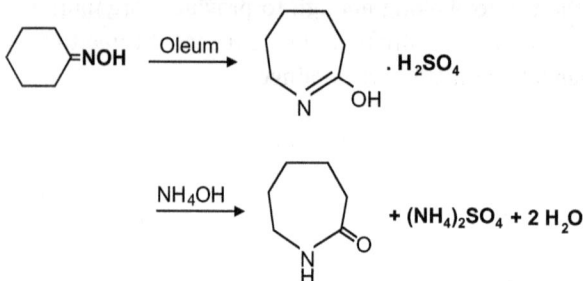

Since this so-called Beckmann rearrangement is very rapid and highly exothermic, the reaction is carried out in a loop reactor configuration with heat exchanger outside the reaction vessel (Fig. 4.16). The rate of reaction is so high that insufficient time is available to achieve complete mixing on a microscale. This problem is increased because a low-viscous reactant (cyclohexanone oxime) has to be mixed with a high-viscous product mixture of caprolactam in concentrated sulfuric acid. Because of the incomplete mixing, many side products are formed which must be removed in the subsequent purification train. To minimize side-product formation, the molten cyclohexanone oxime and concentrated oleum are introduced on carefully selected places into a relatively large amount of already rearranged product to provide an optimal distribution of the feed streams over the reaction mixture. For optimal heat removal and minimization of temperature differences in the reaction mixture, a high circulation rate is chosen. This way, the circulation time is small compared to the residence time in the reactor system and the overall behavior is close to that of an ideal stirred tank.

4.4.2 Buss loop reactor

A Buss loop reactor combines a reaction autoclave with a special design centrifugal pump, a shell and tube heat exchanger of variable size and a top mounted ejector in the autoclave (Fig. 4.17). In a batch loop reactor, the reaction vessel is initially charged with the reactants. For multiphase reactions, the choice of the proper recirculation rate may also be influenced by the need to disperse a second phase (gaseous or immiscible liquid reactant) in the reacting liquid. This is especially valuable in gas–liquid systems where the energy of a fast recirculating liquid stream can be used to entrain (draw in) a gas reactant by means of an ejector. The Buss loop reactor is based on this principle and is used for reactions where intensive contact between gas, liquid and/or catalyst is needed such as hydrogenation, hydroformylation, amination, ethoxylation, oxidation, epoxidation and carbonylation. The draft tube of the ejector dips into the liquid level of the reactor and the gas bubbles are thus distributed over the entire reactor contents, generating turbulence and good mixing. It has the advantage of rapid gas–liquid mass transfer in the initial reaction zone, combined with high heat-removal capability in the tubular heat exchanger. Because the heat exchanger is separated from the reactor, it has no influence on the mass transfer and mixing mechanisms inside the reactor. This results in easy scale up from laboratory size to industrial production because the dimensions of the reactor scale linearly with capacity.

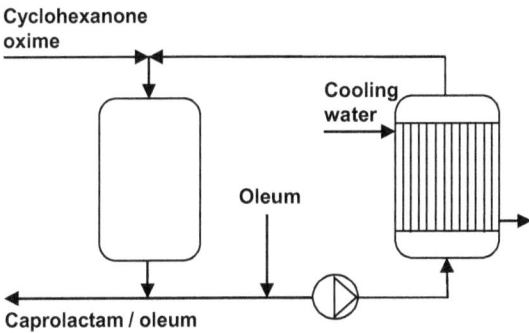

Fig. 4.16: Continuous loop reactor for the production of caprolactam from cyclohexanone oxime.

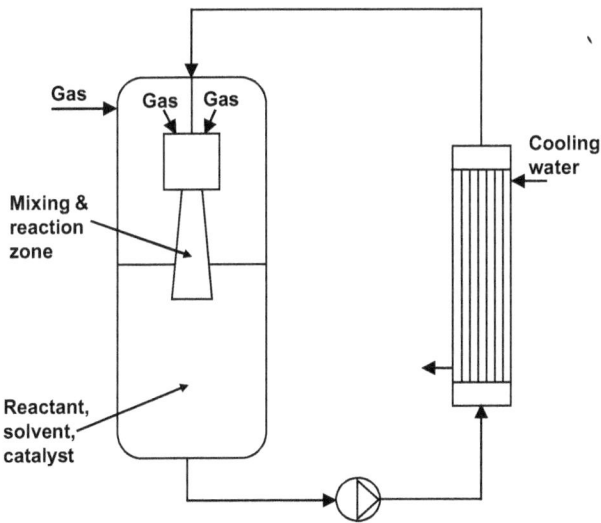

Fig. 4.17: Buss loop reactor.

4.5 Bubble columns

Bubble columns are devices in which gas bubbles are contacted with liquid. The most common purpose is to dissolve gaseous reactants in a liquid, often a catalyst containing reaction mixture, to form the desired products. Oxidation, hydrogenation, chlorination, phosgenation, alkylation and other processes have long been performed in bubble column reactors in the chemical industry. Industrial reactors for high-tonnage products have volumes of 100–300 m³. Larger bubble columns, with volumes up to 3,000 m³, are employed as fermentors. The largest units (20,000 m³) are found in wastewater treatment.

The simplest design is the bubble column where the gas is fed into the column at the bottom and rises through the liquid, escaping from it at the upper surface. Depending on the intensity of mixing and the rate of chemical reaction, the gas is consumed to a greater or lesser extent. When the off-gas contains high concentrations of valuable reactants, part of it is recycled to the reactor. This recycle design, however, lowers the concentration profile in the bubble column and must be optimized from an economic standpoint. In a simple bubble column, the liquid is led in either cocurrently or countercurrently to the upward gas stream and has a long residence time. The flow direction of the liquid phase has minimal effect on the gas-phase residence time, which is comparatively short.

Usually, the gas is dispersed to create small bubbles and distributed uniformly over the cross section of the equipment to maximize the intensity of mass transfer. In most cases, pores or holes are used to generate gas bubbles. Figure 4.18 shows typical forms of static gas spargers, in which bubble formation occurs without any additional energy supplied from outside. The simplest of these devices, the dip tube, only gives an acceptably uniform gas distribution over the cross section at a sufficiently large distance above the sparger. Perforated plates, perforated ring and perforated spider spargers are more effective. Both of them require a certain minimum gas flow rate to achieve uniform distribution and prevent the liquid from getting into the sparger. Very fine bubbles can be generated by the use of porous plates.

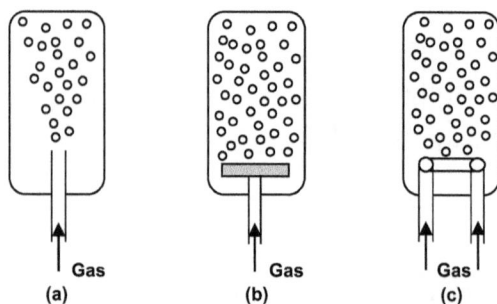

Fig. 4.18: Static gas spargers: (a) dip tube; (b) perforated or porous plate; and (c) perforated ring.

A large-scale industrial application of a bubble column for oxidation reactions is the conversion of toluene to benzoic acid:

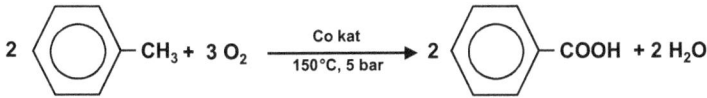

$$2 \; C_6H_5{-}CH_3 + 3\,O_2 \xrightarrow[150°C,\ 5\ bar]{Co\ kat} 2 \; C_6H_5{-}COOH + 2\,H_2O$$

In this oxidation reaction, an organic cobalt component is used as the catalyst. The reaction is performed under pressure (±5 bar) and a temperature of about 150 °C. A perforated ring sparger is used to introduce the air in the reactor (Fig. 4.19). The cobalt catalyst enhances the rate of reaction to such an extent that the oxygen reacts almost momentarily after dissolving in the liquid toluene and the off-gas contains only 1–2% residual oxygen. Evaporation of toluene is used to remove the heat of reaction and thereby control the reaction temperature. After condensing the evaporated toluene, it is recycled to the reactor. Due to the high gas velocity in the reactor, the contents are very well mixed by the bubbles and can be described as an ideal stirred tank reactor.

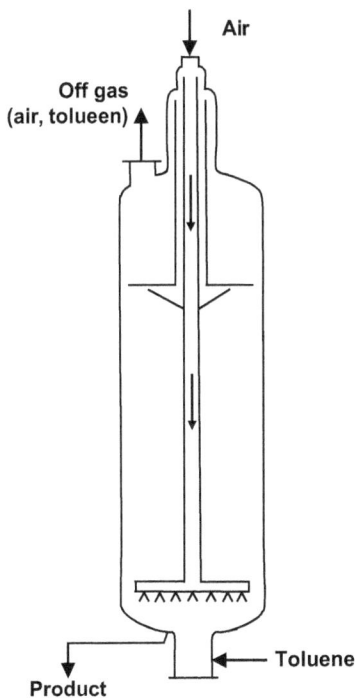

Fig. 4.19: Schematic of the toluene oxidation reactor.

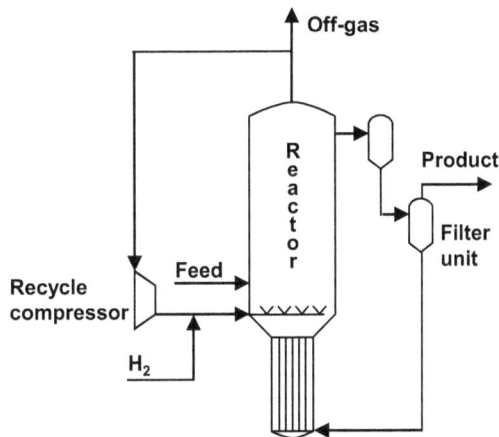

Fig. 4.20: Bubble column hydrogenation reactor for hydroxylamine production.

A bubble column with a heterogeneous catalyst (in the form of small solid particles) is used in the preparation of hydroxylamine for the caprolactam production. Hydrogen is used to reduce ammonium nitrate in an acidic environment over a Pd/C catalyst. Besides the desired main reaction, several parallel reactions are responsible for the formation of N_2 and N_2O:

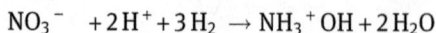

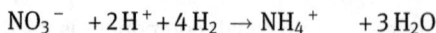

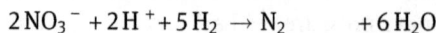

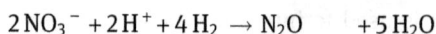

$$NO_3^- + 2H^+ + 3H_2 \rightarrow NH_3^+OH + 2H_2O$$
$$NO_3^- + 2H^+ + 4H_2 \rightarrow NH_4^+ \quad + 3H_2O$$
$$2NO_3^- + 2H^+ + 5H_2 \rightarrow N_2 \quad + 6H_2O$$
$$2NO_3^- + 2H^+ + 4H_2 \rightarrow N_2O \quad + 5H_2O$$

A large gas recycle is used to disperse the catalyst particles in the reactor. This way the reactor contents remain well mixed and a gas–liquid–solid or slurry reactor is obtained. The catalyst particles containing liquid are circulated over an external loop where candle filters are used to separate the catalyst particles from the product stream. These filters consist of porous metal tubes that are only permeable for the liquid. Almost no catalyst losses are observed. The heat of reaction is removed in the heat exchanger located at the bottom of the reactor.

4.6 Fixed and moving bed reactors

4.6.1 Fixed bed reactors

Catalytic fixed bed reactors are the most important types of reactor for the synthesis of large-scale base chemicals and intermediates. In these reactors, the reaction takes place in the form of a heterogeneously catalyzed gas reaction on the surface of, and often also inside the pores of, catalyst particles that are arranged as a so-called fixed bed in the reactor. In addition to synthesis of valuable chemicals, fixed bed reactors have been increasingly used to treat harmful and toxic substances. For example, the reaction chambers used to remove nitrogen oxides from power station flue gases constitute the largest type of fixed bed reactors regarding reactor volume and throughput. Automobile exhaust purification in the catalytic converter represents by far the most widely employed application of fixed bed reactors.

With regard to application and construction, it is convenient to differentiate between fixed bed reactors for adiabatic operation and nonadiabatic operation. Adiabatic reactors are used only where the heat of reaction is small, or where there is only one major reaction pathway. In these cases, no adverse effects on selectivity or yield due to adiabatic temperature development are expected. Reactions with a large heat of reaction as well as reactions that are extremely temperature sensitive are carried out in reactors with provisions for heat removal. The most common arrangement is the multitubular fixed bed reactor, in which the catalyst is arranged in many parallel tubes and the heat carrier is circulated externally outside the tubes.

4.6.2 Adiabatic fixed bed reactors

Adiabatic fixed bed reactors are the oldest fixed bed reactor configurations. In the simplest case, they consist of a cylindrical tube in which the catalyst is loosely packed with catalytic particles and the reactant gas traverses in the axial direction (Fig. 4.21). For the choice of particle size, a balance must be found between having a large specific surface area available for reaction, favoring smaller particles, and avoiding a large pressure drop caused by hydrodynamic drag between the gas and the particles, favoring larger particles. The pressure drop necessary to push a fluid through an adiabatic fixed bed reactor of length L, filled with approximately spherical particles of diameter d_p at a bed porosity ε (fraction of free space around the particles available for the fluid), can be estimated from the popular Ergun equation:

$$\frac{\Delta P}{L} = 150 \frac{\eta}{d_p^2} \frac{(1-\varepsilon)^2}{\varepsilon^3} v_s + 1.75 \frac{\rho}{d_p} \frac{(1-\varepsilon)}{\varepsilon^3} v_s^2 \tag{4.1}$$

The first laminar term, proportional to the fluid viscosity η and the superficial velocity v_s, is dominant for packed beds with viscous fluids or very small particles. The second turbulent term, proportional to the fluid density ρ and the square of the superficial velocity, is relevant for packed beds with large particles. The Ergun equation shows that when too small particles are used, the necessary pressure drop can become excessively large.

To avoid catalyst abrasion by partial fluidization, catalyst packings are always traversed from top to bottom. An alternative design is encountered in the oxidation of ammonia to nitrogen oxides, where extremely short residence times are required. This is achieved with a bed of large diameter and low height, followed by direct quenching of the reaction (Fig. 4.22). The fixed bed consists of several layers of platinum wire gauze.

Purely adiabatic fixed bed reactors are used mainly for reactions with a small heat of reaction. Such reactions are primarily involved in gas purification, in which small amounts of interfering components are converted to noninterfering compounds. The chambers used to remove NO_x from power station flue gases, with catalyst volumes of more than 1,000 m³, are the largest adiabatic reactors. Exhaust catalytic converters for internal combustion engines, with a catalyst volume of 1 L are the smallest. Typical chemical applications include the methanation of CO and CO_2 residues in ammonia synthesis gas and the hydrogenation of small amounts of unsaturated compounds in hydrocarbon streams.

Reactant
inlet

Product
outlet

Fig. 4.21: Adiabatic fixed bed reactor.

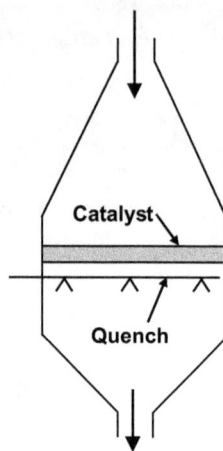

Catalyst

Quench

Fig. 4.22: Schematic of the ammonia
oxidation reactor.

4.6.3 Fixed bed reactors with supply or removal of heat

In the majority of fixed bed reactors for industrial synthesis reactions, direct or indirect supply or removal of heat in the catalyst bed is utilized to adapt the temperature profile. Here, a clear development trend can be observed. Starting with the adiabatic reactor, higher conversions were achieved at the same mean temperature level when several adiabatic stages were introduced with intermediate heating or cooling after each stage. The simplest form involves injecting hot or cold gas between the stages. The main disadvantages of this form of temperature control strategy are the. Increase in cross-sectional loading from stage to stage, and the mixing of hot and cold streams, which is energetically unfavorable. A further development was the replacement of injection cooling by interstage heat exchangers (Fig. 4.23), through which the required or released heat of reaction is supplied or removed. The development of reactors in which the heat-exchange surfaces are integrated in the fixed bed occurred parallel with the development of multistage adiabatic reactors with intermediate heating or cooling. The multitubular fixed bed reactor (Fig. 4.24) constitutes the oldest and still predominant representative of this class of fixed bed reactors, characterized by reaction tubes of 20–80 mm internal diameter. Here, the catalyst packing is located in the individual tubes of the tube bundle. Depending on the reactor capacity, the number of tubes varies between 30 and 30,000. The heat transfer medium is circulated around the tube bundle and through an external heat exchanger.

Adiabatic multistage fixed bed reactors with intermediate cooling or heating are nowadays used particularly where the reaction proceeds selectively to a single product but is limited by the thermodynamic equilibrium conditions. Intermediate

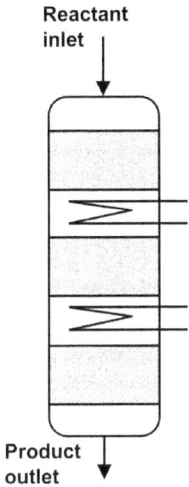

Fig. 4.23: Multistage fixed bed reactor.

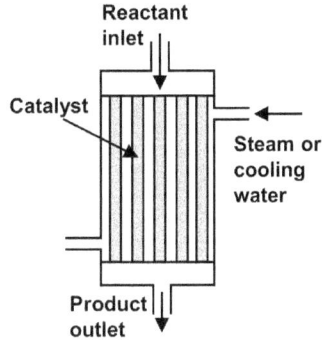

Fig. 4.24: Multitubular fixed bed reactor.

cooling or heating is used to shift the reaction equilibrium to higher conversion. Typical examples comprise the synthesis of ammonia, sulfur trioxide and methanol. In these exothermic reactions, the equilibrium conversion decreases with increasing temperature, but very low temperatures should also be avoided because of slow reaction kinetics. The kinetically optimum reaction pathway with the smallest required catalyst volumes is obtained by dividing the reactor in a large number of small stages. In practice, a balance must be found between equipment costs associated with a large number of stages and the savings in catalyst. Conventional multistage reactors for this class of reactions have three to five stages. Figure 4.25 shows the layout of an ammonia synthesis reactor designed on this basis. Reactor design for ammonia synthesis is subject to numerous constraints. Equilibrium favors high pressures (150–300 bar) and low temperatures (430–480 °C) while kinetics favor the reverse conditions. For structural reasons, the heat exchanger is incorporated between the inflow and outflow in the lowest part of the pressure casing. The reaction gas then flows upward in the annular gap between the pressure casing and the fixed beds, whereby it is further heated and at the same time protects the pressure-bearing structural components against excessively high fixed bed temperatures to prevent hydrogen embrittlement. The three adiabatic fixed beds are traversed from top to bottom. A part of the heat of reaction is utilized to generate steam in the two intermediate heat exchangers. To start up the cold reactor, hot gas must be added to the uppermost bed, for example, through an external start-up preheater.

Another way to circumvent limitations imposed by thermodynamic equilibrium conditions is to install membranes inside the fixed bed and selectively extract the product. Such fixed bed membrane reactors combine catalytic conversion with

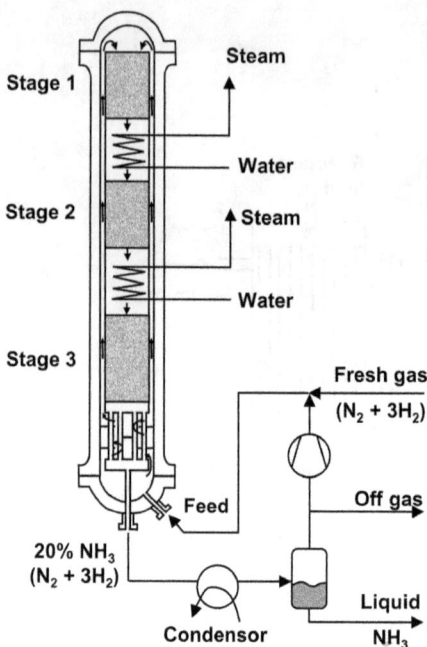

Fig. 4.25: Schematic of a multistage reactor for ammonia synthesis (adapted from [11]).

separation. Because the product is continuously removed, the system cannot reach thermodynamic equilibrium, leading to higher conversions.

Multitubular fixed bed reactors are used for many highly exothermic reactions (oxidation, alkylation and hydrogenation) or highly endothermic reactions (steam reforming, dehydrogenation and dehydration). Some typical examples of these reaction systems involve the hydrogenation of phenol to cyclohexanone, the dehydrogenation of cyclohexanol to cyclohexanone and the conversion of 5-OHN into α-picoline. The hydrogenation of phenol to cyclohexanone is normally carried out in the gas phase at 140–170 °C and atmospheric pressure with a noble metal catalyst.

4.6.4 Moving bed reactors

Moving bed reactors operate as continuous plug flow reactors for solid products. They use gravity to transport the solids continuously through the reactor. Because the gas or liquid has to flow through the bed of solids, mass and heat transport between the phases is relatively good. One of the reasons for using a moving bed reactor is the long residence times that can be attained for the solids. This is, for instance, used in the process for Stanyl (nylon 4,6) preparation from diaminobutane and adipic acid:

$$H_2N-(CH_2)_4-NH_2 + HOOC-(CH_2)_4-COOH \rightarrow$$
$$-\left[-HN-(CH_2)_4-NH-CO-(CH_2)_4-CO-\right]-_n + n\ H_2O$$

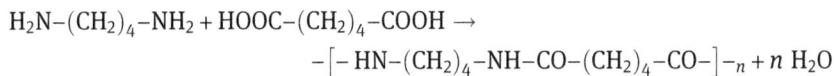

First, a relatively low viscous aqueous prepolymer solution is produced in a stirred tank reactor at a temperature between 180 and 210 °C and pressure of 10–19 bar. After removing the excess of water by flash evaporation, the prepolymer is obtained as small particles with a size of 30–1,000 µm. To avoid dust problems, the prepolymer particles are pelletized and continuously introduced in a vertical cylinder, which is the moving bed reactor. Hot nitrogen gas with a temperature of 240 °C is used to strip the residual water from the prepolymer pellets to force the equilibrium of the polycondensation reaction to the desired degree of polymerization. The produced Stanyl polymer is continuously withdrawn from the bottom of the reactor and cooled before packaging.

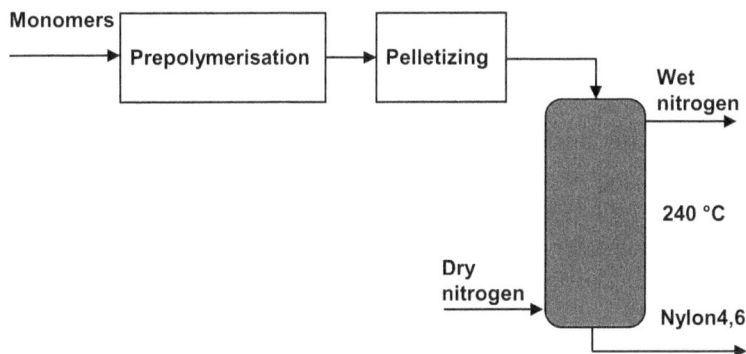

Fig. 4.26: Schematic of Stanyl production with a moving bed reactor.

4.7 Fluidized bed reactors

The fluidization principle was first used on an industrial scale in 1922 for the gasification of fine-grained coal. Since then, fluidized beds have been applied in many industrially important processes. The present spectrum of applications extends from a number of physical processes such as cooling, heating, drying, sublimation, adsorption, coating and granulation to many heterogeneous catalytic reactions as well as noncatalytic reactions.

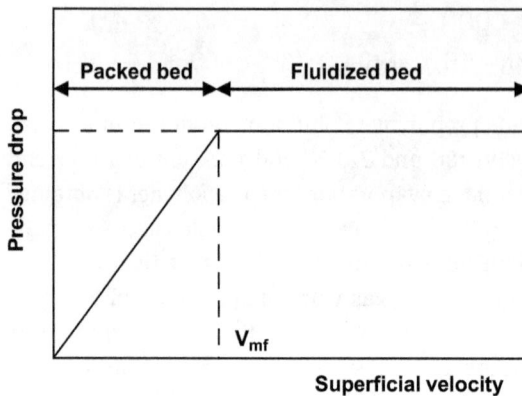

Fig. 4.27: Pressure drop as a function of the superficial velocity for a packed and fluidized bed.

4.7.1 The fluidization principle

A fluidized bed is created by the upward flow of a fluid (gas or liquid) through a bed of particles, where the fluid is distributed at the bottom of the bed through spargers or a perforated or porous plate. Hydrodynamic drag between the fluid and the particles causes a certain pressure drop that initially increases with superficial velocity of the fluid through the packed bed of particles, as predicted by the Ergun equation (4.1). However, at a certain superficial velocity, the so-called minimum fluidization velocity v_{mf}, the pressure drop is sufficiently large to overcome the full weight of the particle bed (Fig. 4.27) and the bed becomes unstable. As the superficial velocity of the fluid is increased beyond the minimum fluidization velocity, the pressure drop remains more or less constant and two things can happen, depending on the fluidizing medium.

If the fluidizing medium is a liquid, the bed begins to expand uniformly, with only very localized void formation and small-scale particle motion. As the superficial velocity is increased further, the particles move further apart to accommodate the flow and the bed expands further.

If the fluidizing medium is a gas, beyond the minimum fluidization velocity virtually solids-free gas bubbles begin to form (Fig. 4.28). Upon further increase of the superficial velocity, most of the excess gas passes through the bed in the form of bubbles and gives the bed the appearance of a violently bubbling liquid. The local mean bubble size increases rapidly with increasing height above the gas distributor because of coalescence of the bubbles. If the bed vessel is sufficiently narrow and high, the bubbles ultimately fill the entire cross section and pass through the bed as a series of gas slugs. As the gas velocity increases further, more and more solids are carried out of the bed, the original, sharply defined surface of the bed disappears, and the solids concentration comes to decrease continuously with increasing height. To achieve steady-state operation of such a turbulent fluidized bed, solids entrained in the fluidizing gas must be collected and returned to the bed. The simplest way to do this is with cyclone integration into the bed vessel and a standpipe dipping into the bed. A further increase in gas velocity finally leads to fast fluidization, which is made possible in a circulating fluidized bed and characterized by a much lower average solids concentration than the previous systems. The high solids entrainment requires an efficient external solids recycle system with a specially designed pressure seal.

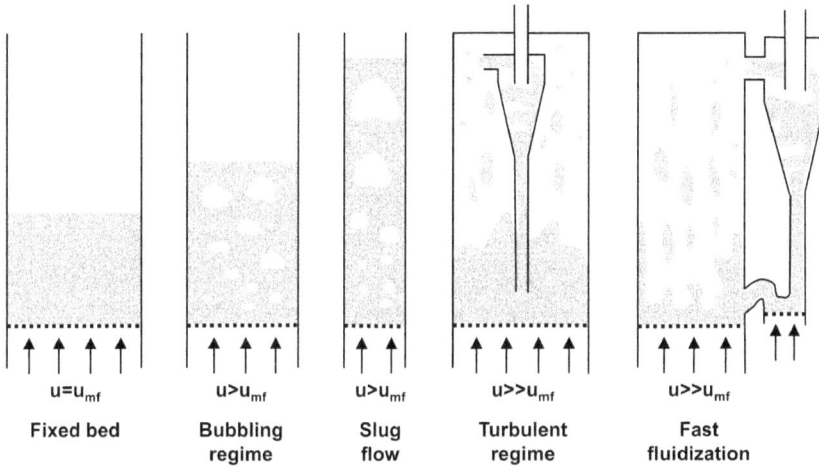

$u=u_{mf}$	$u>u_{mf}$	$u>u_{mf}$	$u>>u_{mf}$	$u>>u_{mf}$
Fixed bed	**Bubbling regime**	**Slug flow**	**Turbulent regime**	**Fast fluidization**

Fig. 4.28: Forms of gas–solid fluidized beds (adapted from [51]).

Figure 4.29 illustrates that the behavior of a fluidized bed in many respects resembles that of a liquid. The bed can be stirred like a liquid. Objects of greater specific gravity sink, whereas those of lower specific gravity float. If the vessel is tilted, the bed surface resumes a horizontal position. If two adjacent fluidized beds with different bed heights are connected to each other, the heights become equal and the fluidized bed flows out like a liquid through a lateral opening. Particularly advantageous features of the fluidized bed for use as a reactor are uniform temperature and composition due to intensive mixing of particles in the bed, excellent gas–solid contact, good gas–particle heat and mass transfer and high bed wall and bed internal heat transfer coefficients. These characteristics are a direct consequence of the presence of bubbles in the bed, because the bubbles cause the movement of particles. From this point of view, the bubbles are very desirable, but they also cause a reduction in gas contact efficiency with the particles. Therefore, the application of a fluidized bed in a particular process is always a compromise between the advantages of good solids mixing and the disadvantages of reduced contacting efficiency.

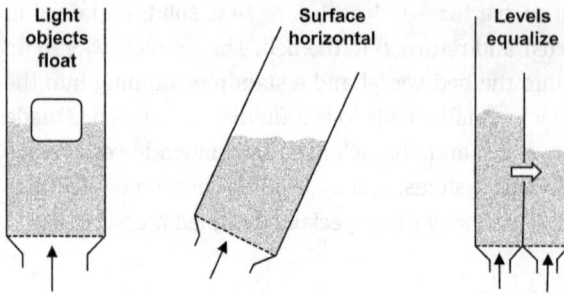

Fig. 4.29: Various characteristics of a fluidized bed (adapted from [51]).

4.7.2 Fluidization properties of typical bed solids

In fluidization with gases, solids display characteristic differences in behavior that can also affect the operating characteristics of fluidized bed reactors. As shown in Fig. 4.30, an empirical classification proposed by Geldart can be used to divide solids in four groups (A to D) with respect to fluidization behavior. Solids of group C are very fine-grained, cohesive powders (flour, fine dust) that cannot be fluidized without fluidization aids. The adhesion forces between the particles are so strong that the gas will form channels through the bed, bypassing most of the particles. Fluidization properties can be improved by the use of mechanical equipment (agitators, vibrators) or flow ability additives. Solids of group A have small particle diameters (±0.1 mm) or low bulk densities. This class includes catalysts used, for instance, in fluidized bed catalytic crackers. As the gas velocity increases beyond the minimum fluidization point, the bed of such a solid first expands uniformly (similar to a liquid fluidized bed) until bubble formation sets in. If the gas flow is cut off abruptly, the gas storage capacity of the fluidized suspension causes the bed to collapse rather slowly. Group B solids have moderate particles sizes and densities. Typical representatives are sands with mean particle diameters between 0.05 and 0.5 mm. Bubble formation begins immediately above the minimum fluidization point. Group D includes solids with large particle diameters or high bulk densities. The character of bubble flow is markedly different from that in group B solids. Group D solids are characterized by the formation of "slow" bubbles. On sudden stoppage of the gas flow, the bed also collapses suddenly.

4.7.3 Applications

High temperature homogeneity, even with strongly exothermic reactions, and easy solids handling are the main advantages of a fluidized bed reactor over fixed bed reactors. These advantages are achieved at the expense of high solids separation

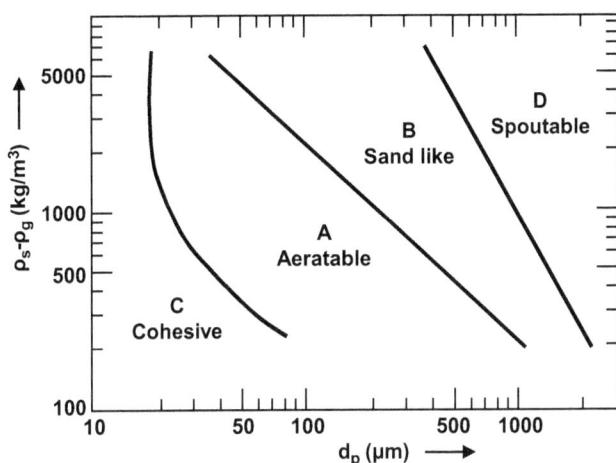

Fig. 4.30: Geldart diagram (adapted from [53]; for explanation, see text).

costs for gas purification, low conversion due to intensive mixing and significant erosion of the internals in a fluidized bed reactor.

The ease of solids handling was the basic reason for the success of long-chain hydrocarbons cracking with zeolite catalysts in the fluidized bed. Because the cracking reaction is endothermic and involves the deposition of carbon on the catalyst surface, the catalyst must be continuously discharged from the reactor and regenerated in an air fluidized regenerator bed. As illustrated by Fig. 4.31, the reaction is carried out in so-called riser cracking reactors where the oil is fed at the bottom of the riser, vaporized in contact with the hot catalyst and the mixture of oil vapors and cracking gas transports the catalyst up. In the reactor bed, solids are collected before passing through the stripper to the regenerator.

Another example of successful use of the fluidized bed reactor is for the synthesis of acrylonitrile by the ammonoxidation of propene, where fluidization is crucial for reliable control of this strongly exothermic reaction:

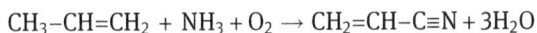

$$CH_3-CH=CH_2 + NH_3 + O_2 \rightarrow CH_2=CH-C\equiv N + 3H_2O$$

The reaction is carried out at a bed temperature of 400–500 °C and gas contact time of 1–15 s. As can be seen in Fig. 4.32, air is fed to the bottom of the reactor and enters the fluidized bed through an air distributor. This gas distribution device ensures uniform fluidization over the entire cross section of the bed and prevents solids from raining through the grid both during operation and after the bed has been shut off. Many different distributor designs are possible, as illustrated by Fig. 4.33. The reactants, ammonia and propene, are introduced through a separate distributor. This distributor or sparger is designed in such a way that the particles cannot enter the holes to prevent plugging. Catalyst regeneration by carbon burn off

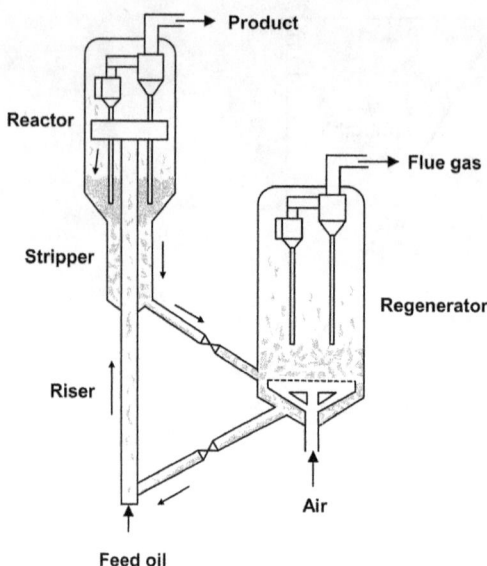

Fig. 4.31: Riser cracking reactor.

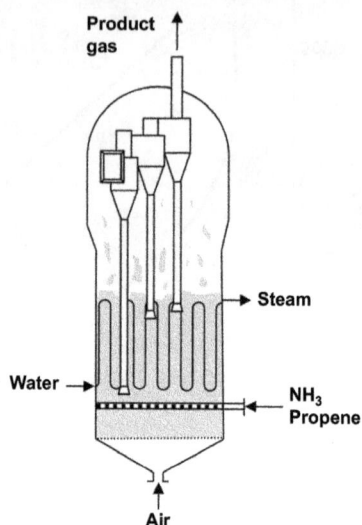

Fig. 4.32: Fluidized bed reactor for the synthesis of acrylonitrile (Sohio process).

occurs in the space between the air distributor and the feed–gas distributor. The heat of reaction is removed by bundles of vertical tubes inside the bed in which high-pressure steam is produced at a temperature of around 480 °C. Internal cyclones are used to prevent the entrainment of small catalyst particles from the reactor.

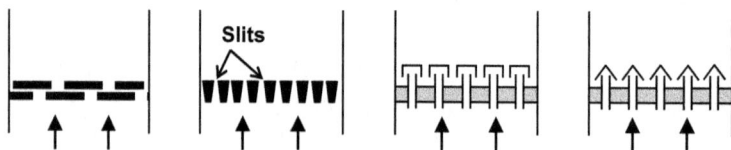

Fig. 4.33: Typical designs of various gas distributors.

Figure 4.34 shows the fluidized bed reactor used for the synthesis of maleic anhydride from n-butane by partial oxidation over a V_2O_5 catalyst:

$$n\text{-}C_4H_{10} + 3\tfrac{1}{2}O_2 \;\rightarrow\; C_4H_2O_3 + 4H_2O$$

Compressed air and butane are introduced separately into the bottom of the reactor. Heat from the exothermic reaction is removed from the fluidized bed through steam coils in direct contact with the bed of fluidized solids. Due to the high mixing in the bed, a uniform temperature profile is obtained which is an important precondition

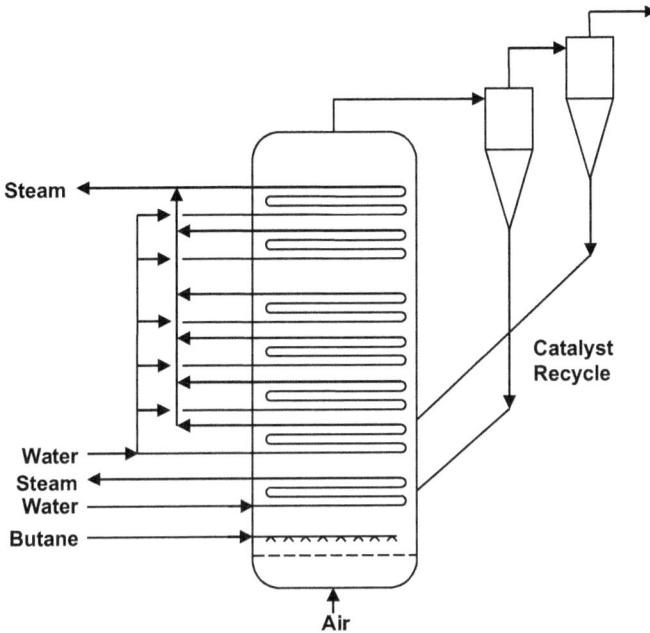

Fig. 4.34: Fluidized bed reactor for the oxidation of butane to maleic anhydride.

for high reaction selectivity. The solids are removed from the product gas using a combination of cyclones and filters. A particular problem of this process is mechanical stress on the catalyst, its abrasion and erosion at the heat-dissipating surfaces.

The gas-phase polymerization of ethylene in the fluidized bed shown in Fig. 4.35 was developed by Union Carbide. The reaction gas fluidizes the bed at a temperature of 75–100 °C and a pressure of 20 bar. Extremely fine-grained catalyst is metered into the bed. Polymerization occurs on the catalyst surface and yields a granular product with diameters ranging from 0.25 to 1 mm. Temperature control is achieved by keeping the conversion per pass low, around 2%, and recycling the gas over an external heat exchanger. The catalysts used have such a high activity that they are only present in trace amounts in the final product and removal from the final product is not necessary.

Other important fluidized bed reactor processes are the low-pressure synthesis of melamine from urea, oxidation of naphthalene to phthalic anhydride, the Exxon flexi-coking process, coal gasification, biomass torrefaction and gasification and many others.

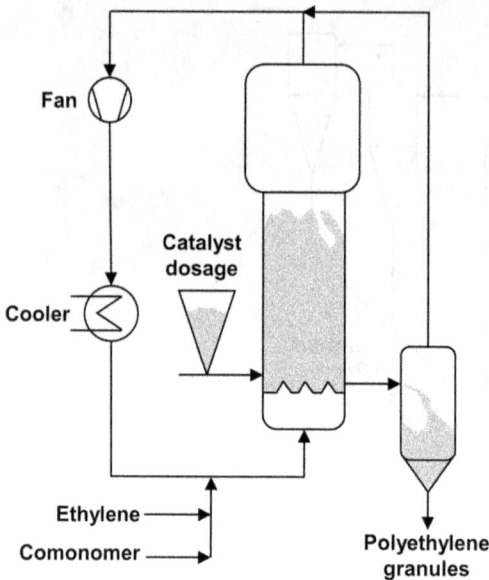

Fan

Catalyst dosage

Cooler

Ethylene ⟶

Comonomer ⟶

Polyethylene granules

Fig. 4.35: Gas-phase polymerization of ethylene.

4.8 Electrochemical reactors

In electrochemical reactors electric charge is used to enable and enhance the rate of redox reactions which would otherwise not happen spontaneously. Although historically, electrochemistry has not received much attention from the chemical engineering community, the role of electrochemical reactors is expected to increase significantly because of the global transition from fossil energy sources to renewable sources in the form of electrical energy.

4.8.1 Basic elements of an electrochemical reactor

In an electrochemical reactor, reactions take place on the surfaces of electrodes, where certain reactants oxidate (donate electrons) on a positively charged anode and other reactants reduce (accept electrons) on a negatively charged cathode. In between the electrodes is an electrolyte, which can be a solvent with dissolved ions or a molten salt. The ions are attracted to the electrode with opposite charge, enabling the electric current to flow through the electrolyte from one electrode to the other, closing the electric circuit necessary for the reactions to proceed in a continuous manner. In most applications, the anodic and cathodic halves of a reactor are separated by an ion-exchange membrane, allowing only certain ions to pass to the other half, while keeping other ions and neutral molecules within their respective halves. The membrane separator is often an expensive part of the electrochemical

reactor, and leads to additional energy losses. However, it allows the anodic and cathodic products to remain well separated, saving on potentially expensive separation steps further downstream in the process. It also opens up possibilities to use
different electrolytes for the anodic and cathodic halves, optimizing the anolyte and
catholyte compositions and pH for each of the respective half-reactions.

The voltage difference E applied to the two electrodes must be larger than the
equilibrium potential E_0 of the redox reaction to achieve a sufficiently large reaction rate and to overcome several types of potential losses. This so-called overpotential consists of three main contributions:

$$\eta = E - E_0 = \eta_{act} + \eta_{conc} + IR \tag{4.2}$$

The first contribution is the anodic and cathodic activation overpotential η_{act}, representing kinetic charge transfer limitations dominating at low current densities. They
are nonlinearly related to the current density through the Buttler–Volmer equation,
which may loosely be viewed as the electrochemical equivalent of the Arrhenius equation, and can be minimized by the right choice of electrocatalyst. The second contribution is the anodic and cathodic concentration overpotential η_{conc}, dominating at
high current densities. It represents the effects of mass transfer limitations of the elec-
troactive species reaching and leaving the electrode surfaces. The final type of loss is
related to the total electrical resistance R, which includes electrical contact resistance, electrolyte resistance and membrane resistance.

4.8.2 Design of an electrochemical reactor

The design of electrochemical reactors shares many similarities with that of the reactors discussed before. For industrial applications, there is a clear preference for
scalable continuous reactors such as parallel plate reactors which may be viewed
as a specific type of liquid–solid heterogeneous reactor, where the electrolyte fluid
containing the reactants is continuously flowing tangentially along the catalytically
active electrode surface. If the catalyst material is relatively cheap and sufficiently
conductive, the entire electrode plate can be made of catalyst material (such as the
case of carbon electrodes), while for expensive catalyst materials, active sites are
dispersed onto the surface of an electrode made of a cheaper conductive material.
The electrode plates can be flat and smooth, but to increase the conversion per unit
volume reactor also porous electrodes or even (possibly fluidized) suspension
electrodes are employed. If the solubility and diffusivity of the reactant in the liquid electrolyte is limiting, the reactant may be provided separately in the form of a
gas through a gas-diffusion electrode.

The most important challenges in the design of electrochemical reactors lie in
achieving high conversion rates (generally high current densities over 100 mA/
cm^2 electrode surface are preferred), while at the same time keeping overpotential

low and Faradaic efficiency high. Faradaic efficiency is the electrochemical equivalent of selectivity, and defined as the fraction of electrons ending up in the desired product rather than in side products. Mass transfer limitations may be partially alleviated by increasing the electrolyte flow velocity leading to thinner concentration boundary layers near the electrodes, but this is at the expense of a higher pressure drop. A large source of energy losses lies in the electrical resistance caused by the finite electric conductivity of the electrolyte and membrane. This naturally leads to a desire to minimize the thickness of flow channels and membranes. On the other hand, extremely narrow flow channels lead to higher pressure drops and the possibility of bubble-forming gaseous products getting stuck in the flow channels, while very thin membranes lead to issues with their mechanical stability.

Because of the necessary small dimensions perpendicular to the electrodes, upscaling electrochemical reactors is usually done by forming stacks of multiple plates. As illustrated by Fig. 4.36, there is a choice between a bipolar stack and a monopolar stack configuration. Advantage of the bipolar stack configuration is that electrical connections are easier (only the end electrodes are connected to the external power supply), but the two sides of each electrode must be able to deal with very different reactions and environments. The monopolar stack configuration has more complicated electrical connections, but each electrode has the same environment on both sides.

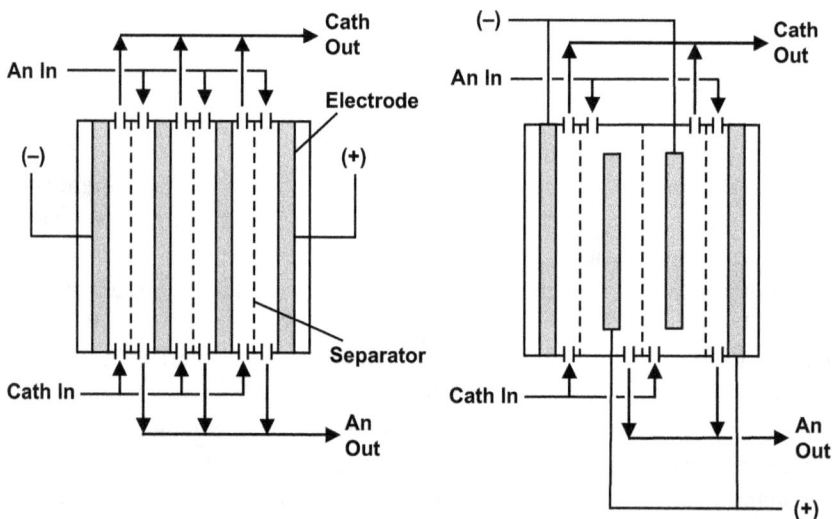

Fig. 4.36: The parallel plate reactor is the most common type of continuous flow electrochemical reactor. There are two ways of forming stacks: (a) bipolar stack and (b) monopolar stack. An, anolyte; Cath, catholyte (adapted from [55]).

4.8.3 Applications

Important examples of industrially relevant electrochemical reactions are given in Tab. 4.2.

Tab. 4.2: Some examples of the industrial application of electrochemical reactors.

Process	Anode (+) oxidation reaction	Cathode (−) reduction reaction	Total reaction
Chloralkali	$2Cl^- \rightarrow Cl_2 + 2e^-$	$2H_2O + 2e^- \rightarrow H_2 + 2OH^-$	$2NaCl + 2H_2O \rightarrow Cl_2 + 2NaOH + H_2$
Hall–Heroult	$2O^{2-} + C \rightarrow CO_2 + 4e^-$	$Al_2O_3 + 6e^- \rightarrow 2Al + 3O^{2-}$	$2Al_2O_3 + 3C \rightarrow 4Al + 3CO_2$
Water electrolysis	$2H_2O \rightarrow O_2 + 4H^+ + 4e^-$	$2H^+ + 2e^- \rightarrow H_2$	$2H_2O \rightarrow 2H_2 + O_2$

In the chloralkali process, electricity is used to convert an aqueous sodium chloride solution into chlorine gas at the anode and sodium hydroxide (caustic soda) at the cathode. The anode must be able to withstand the corrosive nature of chlorine production and is therefore typically made of platinum or titanium alloys or graphite. The ion-exchange membrane is often Nafion, a material which is able to tolerate high temperatures and a corrosive environment. Both chlorine gas and sodium hydroxide are important industrial chemicals. Chlorine gas is difficult to produce by other means, which explains the large-scale application of the chloralkali process despite its extremely high electricity consumption. Hydrogen gas is also produced, but this is usually not the primary product of interest.

Electrochemical reactors are also widely applied in the metals-processing industry. An important example is the Hall–Heroult process in which alumina (Al_2O_3) is converted to metallic aluminum (Al). To achieve this, alumina is dissolved in a molten salt of Na_3AlF_6 at a temperature of approximately 960 °C. Typically, carbon anodes and cathodes are used, where the anode electrode is slowly turned into CO_2 during the electrolysis.

Water electrolysis is a well-known process to convert water into hydrogen gas using electricity. Hydrogen is not only an important starting molecule for the chemical industry, but also projected to be one of the most important molecules for the energy transition because it has the potential to store large quantities of intermittent electric energy, generated by sustainable sources such as wind- and solar panel farms, for long periods of time. Large efforts are undertaken in both academia and industry to make the production of "green" hydrogen by water electrolysis economically competitive relative to the traditional production of "grey" hydrogen by steam reforming of natural gas. Reactor design plays a key part in achieving this goal. Currently, there are two main technologies, one based on alkaline water electrolysis, where electrodes are operating in alkaline potassium hydroxide or sodium hydroxide solutions using a standard membrane permeable to hydroxide ions, and the other based on a polymer electrolyte membrane, where the liquid electrolyte is replaced by a solid electrolyte membrane to conduct protons from the porous anode to the porous cathode, while water is supplied to the anode from the outside.

5 Biochemical reaction technology

5.1 Characteristics of biochemical processes

Biochemical reactions involve the transformation and production of biological and chemical substances. Typical of these reactions is their use of enzymes as biocatalysts. In addition to their natural substrates, many of these enzymes can utilize other structurally related compounds as substrates and therefore catalyze unnatural reactions upon addition of foreign substrates to the reaction medium. Thus, enzymatic conversion is a specific category of chemical synthesis. The enzymes can be present as cell constituents of living microorganisms, or they can be isolated in dissolved form or bound to inert supports. At present, fermenting microorganisms is still the dominant and least costly form of industrial biological catalyst production. Their great advantage is their versatility, as illustrated by the wide range of the processes in current use. However, fermentation requires complex and often expensive nutrients that are mainly used as an energy source. Continuous operation may also be difficult with fermentation because of strain mutation.

Table 5.1 presents a comparison between chemical processes and bioprocesses such as simple fermentation. The analysis is based on the characteristics of the catalysts. When compared with chemical catalysts, enzyme catalysts are both highly active and highly selective. Additional advantages include the variety of reactions catalyzed, the high degrees of conversion and the mild conditions employed which are especially important when labile reactants are used. Today, use of enzymes in the whole cell as the catalyst is preferred.

Tab. 5.1: Comparison of bioprocessing and chemical processing.

Criterion of comparison	Chemical process	Bioprocess
Catalyst	More or less active and selective Expensive regeneration	Enzymes highly active and selective Regeneration easy due to microbial growth
Reaction conditions	Mostly high temperatures, sometimes high pressures	Mainly 25 °C and 1 bar pressure
Raw materials	Pure in most cases	Impure, inactive and diluted
Process	Often multistage processing with recovery of intermediates Fast reaction rates at high concentration level and high yields	One stage possible without intermediate product recovery Mainly slow reaction rates Risk of infections and mutations

https://doi.org/10.1515/9783110712445-005

5.1.1 Fermentation

Fermentation is a long-established process, which has expanded over the years to become the basis of biotechnology and biochemical engineering. Many new applications have been discovered in the recent decades such as mass production of secondary metabolites (antibiotics), biotransformations of organic substances such as steroids, mass cultivation of microorganisms for enzyme production and the marked impact of genetic engineering. Biochemical conversions with the aid of a microorganism differ from the purely chemical process in several aspects:
- Complexity of the reactant mixture
- Increase in the mass of microorganisms simultaneously with the accomplishment of the biochemical transformation
- Ability of microorganisms to synthesize their own catalysts (enzymes)
- Mild conditions of temperature and pH involved, and its greater sensitivity to these conditions
- Restriction to the aqueous phase

5.1.1.1 Fermentation products

Microorganisms are exploited industrially to produce additional microorganisms and to achieve biochemical conversion of water-soluble organic compounds. The microbial cells are used as animal foodstuffs, for seeding batch processes and as a source of enzymes for both research and commercial applications. The biochemical products on the other hand may be beverages, medicinal compounds such as antibiotics and steroids and industrial chemicals, for instance, solvents or organic acids.

Microbial fermentations are important sources of biological products used in the pharmaceutical, food and chemical industries. During the last decade, there has been a large increase in the range of commercial products, especially secondary metabolites and recombinant proteins. Among these products cells, primary metabolites, secondary metabolites and enzymes can be distinguished as the main products. Yeast fermentation for brewing and baking represents the most traditional form of fermentation technology in which the cell is obtained as the main product. Primary metabolites such as citric acid, lactic acid, ethanol and glutamic acid are produced in very large-scale operations. Because of the relatively simple metabolics, these processes are relatively well characterized. Stirred tank fermenters of around 200 m^3 in scale are generally used. If it is desired to go to greater volumes, then it is simpler to switch to an airlift type design because of heat transfer limitations.

Since the widespread use of antibiotics began, this area has represented the most commercially significant fermentation activity for secondary metabolites. Products such as penicillin, tetracycline, erythromycin, cephalosporins, cephamycins and clavulanic acid are made at a very large scale in fermenters varying between 50 and 200 m^3 in volume. In most cases, these products serve a vital role in modern medicine. Secondary metabolite fermentations are very complex and have yet to yield many of their secrets.

Historically, industrial enzymes made by fermentation were mainly restricted to those produced extracellularly, such as amylases and proteases. With the development of mechanical techniques for protein release from microorganisms on a larger scale, intracellular enzymes have found wider application in the food, pharmaceutical and chemical industries. Some important examples are glucose isomerase for high-fructose syrup production and penicillin acylase for removal of the side chain of penicillin to allow subsequent manufacture of semisynthetic antibiotics. Fermenter sizes for enzyme production generally range from 30 to 220 m^3.

5.1.1.2 Microorganisms

The living cells used in bioreactors can be bacteria, yeasts, molds and plant or animal cells (Fig. 5.1). This broad classification is made on a basis of cell size and morphology when examined with an optical microscope. Bacteria are unicellular microorganisms with their smallest dimension in the range of 0.5 to 2 µm and they reproduce asexually by binary fission. Yeasts, on the other hand, while also unicellular, have a size range from 5 to 10 µm and reproduce either by asexual budding or by sexual processes. Molds may also reproduce by sexual or asexual means, but have a multicellular structure and may be 5 µm or considerably larger in size. Plant and animal cells are fragile and relatively large. They grow slowly, making great demands on nutrient supply. As a result of their large size (50–100 µm), complex structure and mechanical sensitivity, nutrient media have to satisfy high demands and the bioreactor must fulfil special requirements.

Bacteria have an optimum pH range for growth between 6.5 and 7.5 and display a wide variety of patterns of response to free oxygen. Some have absolute oxygen requirement while others can grow only in its complete absence. Intermediate species exist that develop with or without oxygen, though not necessarily at the same rates. Molds, however, grow most rapidly under aerobic conditions, but generally more slowly than bacteria. They can therefore be overgrown by bacterial contaminants. An abundant supply of oxygen is required for the growth of yeasts. The acid tolerance of yeast fermentations ranges from pH 2.2 to pH 8.

As illustrated in Fig. 5.2, microbial metabolism may be thought of as a series of interconnecting reaction loops or metabolic pathways arranged spatially throughout the cell. The basic unit in a metabolic pathway is a reaction catalyzed by an enzyme. The overall reaction path is controlled by the microorganism itself, largely by adjustment of the rate of enzyme synthesis, or alternatively by inhibition of the enzymes by the product itself. The industrial objective is to use a part of the overall metabolism for a particular biochemical conversion. An example is glutamic acid that can be obtained as a primary metabolite from the Krebs cycle shown in Fig. 5.2. This is achieved by the supply of a primary organic raw material, called

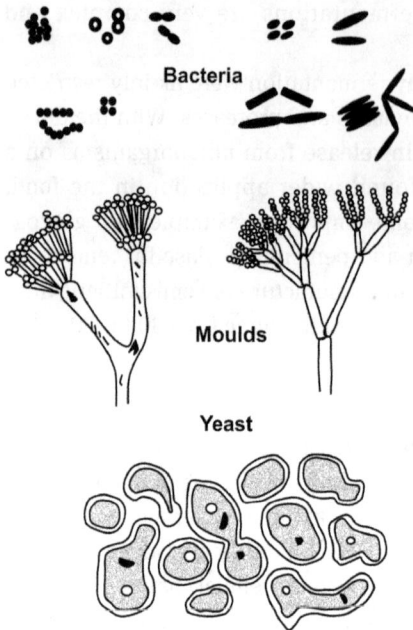

Fig. 5.1: Morphological characteristics of microorganisms.

substrate, in addition to those metabolites required by the microorganism for survival. If possible, an attempt is made to enhance the amounts and activity of the enzymes involved in the conversion, while those enzymes acting on the desired product are inhibited or repressed. These objectives are attained by the addition of appropriate chemical constituents to the reaction mixture or nutrient medium. In contrast to isolated enzymes or chemical catalysts, microorganisms adapt the structure and activity of their enzymes to the process conditions, whereby selectivity and productivity can change. Mutations of the microorganisms can occur under suboptimal biological conditions. Microorganisms are frequently sensitive to strong shear stress and to thermal and chemical influences.

The type and metabolic state of the microorganisms used are decisive for the initial choice of reactor. Most modern biotechnological methods are based on aerobic processes. In the cultivation of aerobic microorganisms, the culture medium must always contain sufficient dissolved oxygen to allow the biochemical reactions of the microorganisms to occur. Oxygen is only sparingly soluble in the fermentation medium, making the supply of the microorganisms with this nutrient technically very complex. In aerobic processes, an effective and adequate oxygen supply must be provided in the medium by dispersing and subsequent mixing in all zones of the reactor.

fat

NADH$_2$ NAD Pyruvate ← Carbohydrate

HS-CoA
coenzyme A → Alanine

CH_3-$\overset{O}{\overset{\|}{C}}$-S-CoA
Acetyl coenzyme A

Pyruvate + CO$_2$

O=$\overset{|}{C}$-COOH
$\overset{|}{C}H_2$
$\overset{|}{C}OOH$
Oxalo acetate

NADH$_2$
NAD

COOH
$\overset{|}{C}H_2$
HO-$\overset{|}{C}$-COOH
$\overset{|}{C}H_2$
$\overset{|}{C}OOH$
Citrate

H$_2$O

COOH
$\overset{|}{C}H_2$
$\overset{|}{C}$-COOH
$\overset{\|}{C}$-H
COOH
cis-Aconitate

H$_2$O

Phospho-pyruvate + CO$_2$

COOH
H-$\overset{|}{C}$OH
$\overset{|}{C}H_2$
$\overset{|}{C}OOH$
Malate

H$_2$O

COOH
$\overset{|}{C}H_2$
H-$\overset{|}{C}$-COOH
H-$\overset{|}{C}$-OH
COOH
Isocitrate

COOH
$\overset{|}{C}H$
$\overset{\|}{C}H$
COOH
Fumarate

NAD
NADH$_2$

2 H → Malonate block

Oxalosuccinate

COOH
$\overset{|}{C}H_2$
H-$\overset{|}{C}$-COOH
$\overset{|}{C}$=O
COOH

Propionate + CO$_2$

COOH
$\overset{|}{C}H_2$ Succinate
$\overset{|}{C}H_2$
$\overset{|}{C}OOH$

GTP + HS-CoA
GDP + P

COOH
$\overset{|}{C}H_2$
$\overset{|}{C}H_2$
CO-S-CoA
Succinyl coenzyme A

COOH
$\overset{|}{C}H_2$
$\overset{|}{C}H_2$
$\overset{|}{C}$=O
COOH
α-Ketoglutarate

CO$_2$

CO$_2$ + NADH$_2$ HS-CoA + NAD   Glutaric acid

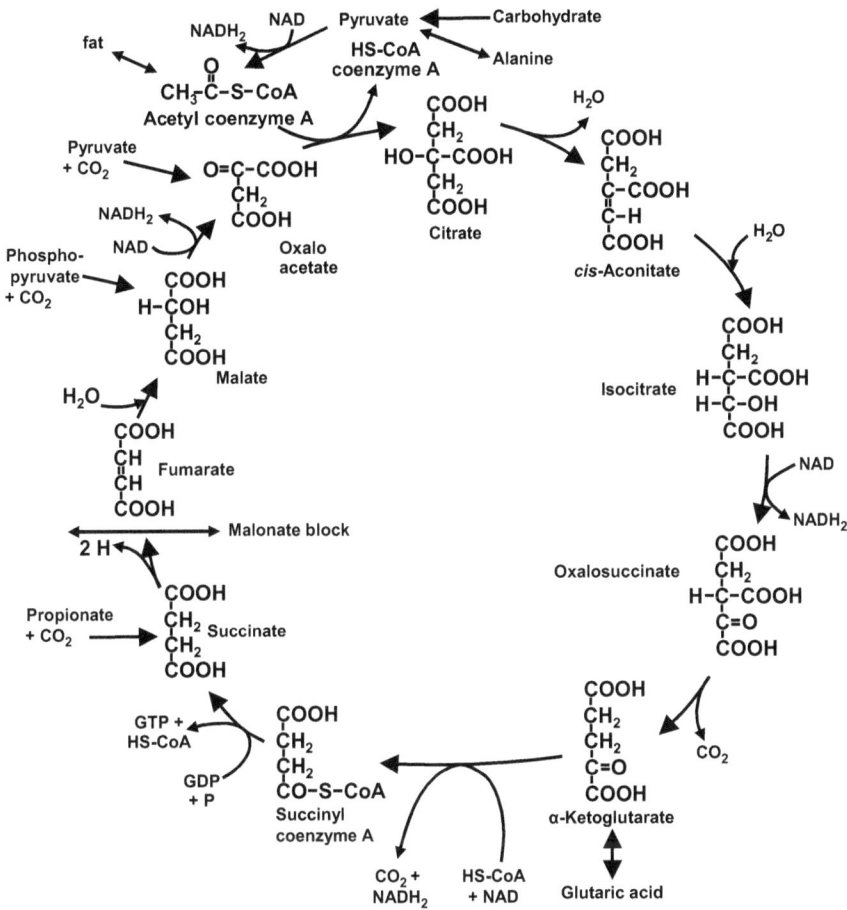

Fig. 5.2: Schematic of the Krebs cycle.

5.1.1.3 Requirements of fermenters

In most bioprocesses, product processing is the cost-determining step, with the result that the aim of the reactor design is to obtain the maximum product concentration. To achieve this goal, bioreactors with capacities up to thousands of cubic meters are used. Fermenters are classified into those employing anaerobic or aerobic conversions with free microorganisms. Many different reactor configurations are employed depending on the microbial structure. The basic reactor arrangement for the accomplishment of fermentations is the deep tank fermenter shown in Fig. 5.3. The basis of this reactor is the fact that microorganisms in their normal state contain a considerable amount of water and have a density that differs only slightly from water. Very little hydrodynamic drag is required to maintain them in suspension. The logical arrangement then becomes an essentially completely mixed reactor in which the motion is either by mechanical stirring or by air bubbling through the medium. This air bubbling provides the free oxygen demand by aerobic processes.

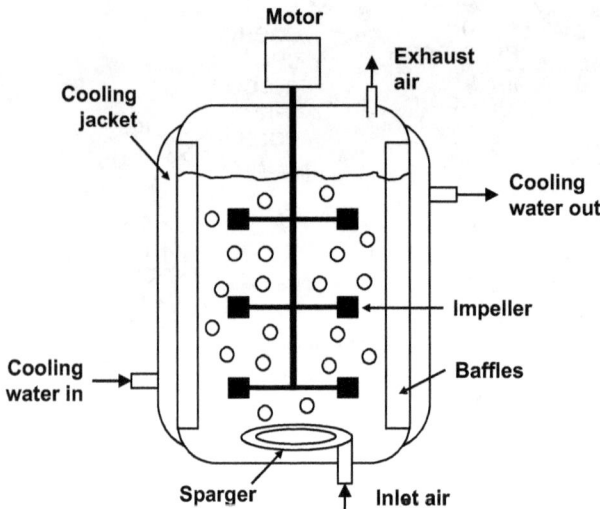

Fig. 5.3: Schematic of the deep tank fermenter.

The prerequisite for the use of living microorganisms in reactors is the provision of favorable living conditions to ensure that living microorganisms can exhibit their activity under defined conditions. This results in a series of special features in the reaction engineering of biocatalytic processes. The reaction rate, cell growth and process stability depend strongly on the environmental conditions in the bioreactor. Concentrations of substrates and products in the reaction mixture are frequently low and both of them may inhibit the process. Cell growth, the structure of intracellular enzymes and product formation depend on the nutritional needs of the cell and on the maintenance of optimum biological conditions within narrow limits. Certain substances (inhibitors, effectors, precursors, metabolic products) influence the rate and the mechanism of the reactions and intracellular regulation. Microorganisms can metabolize unconventional or even contaminated raw materials (cellulose, molasses, mineral oil, starch, ores, wastewater, exhaust air, biogenic waste). General fermenter requirements involve monoseptic conditions, containment, optimal mixing with low uniform shear, adequate oxygen transfer, feeding of substrate with prevention of under- or overdosage, suspension of solids, gentle heat transfer and compliance with design requirements.

5.1.2 Enzymatic conversions

5.1.2.1 Industrial applications

Enzymes accelerate hundreds of reactions taking place simultaneously in the cell and its immediate surroundings. Enzymatic processes enable natural raw materials to be upgraded and turned into finished products. They offer alternative ways of making products previously made only by conventional chemical processes. The detergent industry is the largest user of industrial enzymes. The starch industry, the first significant user of enzymes, developed special high fructose syrups that could not be made by means of conventional chemical hydrolysis. These were the first products made entirely by enzymatic processes. Foodstuffs and components of animal feed can be produced by enzymatic processes that require less energy, less equipment or fewer chemicals compared to traditional techniques.

During the last decade, the role of biocatalysis in organic synthesis, biochemical and biomedical processes has increased dramatically. An example of a very large-scale application is the isomerization of glucose to fructose with glucose isomerase. Other important processes where enzymes are used on scale over 1,000 t/year include the synthesis of aspartame and the enzymatic hydrolysis of penicillin G. In the aspartame process, the enzyme thermolysine is used for the enantioselective condensation of the two amino acid constituents (Z-L-Asp and L-Phe-Ome). Penicillin G acylase is used for the selective hydrolysis of penicillin G and penicillin V to obtain 6-aminopenicillanic acid, which is an important building block for semisynthetic antibiotics. The need for enantiomerically pure compounds in the pharmaceutical, food and crop protection industries owing to consumer and regulatory demands will continue to fuel the interest in biocatalysts and associated processes.

5.1.2.2 Distinctive features of enzymes

Enzymes are very powerful biological catalysts, catalyzing all reactions that constitute cellular metabolism. They consist of L-amino acids linked together by covalent bonds in a defined sequence that is coiled in a complex fashion. The active site consists of only a few amino acids that have a direct role in binding the substrate and catalyzing the reaction characteristic of each particular form of enzyme molecule.

Enzymes are very efficient, catalyzing reactions often 10^8–10^{11} times more rapidly than the corresponding chemical catalysts. Most of the reactions proceed in water at much less extreme conditions, such as temperature, pH and pressure. The range of reactions that can be catalyzed by enzymes is extremely broad. Most enzymes are very specific in terms of the type of reactions catalyzed and the structures of the substrate and product formed, such that often only a single chemical present in a mixture with very similar chemicals is transformed to a single product. This can result in higher product yields and fewer potentially polluting side products. This specificity is due to the ability of the enzyme to bind the substrate and organize reactive groups so that a specific reaction transition is particularly favored. Several main types of enzymatic catalysts can be recognized, including whole cells, organelles and enzymes used in both free and immobilized forms. Any combination of these systems may also be possible. Although enzymes are sufficiently large to be regarded as heterogeneous catalysts, they are usually classed as being homogeneous catalysts due to their solubility. Obviously, immobilized enzymes in which the particle size of the catalyst is at least an order of magnitude bigger than the enzyme are genuine heterogeneous catalysts.

5.1.2.3 Enzymatic catalysis

Enzymes are classified into six different groups depending on the type of reaction they catalyze:

- Oxidoreductases catalyze oxidation–reduction reactions involving oxygenation, such as C–H→C–OH, or overall removal or addition of hydrogen atom equivalents, for example CH(OH)→C=O
- Transferases mediate the transfer of various groups such as aldehyde, ketone, acyl, sugar, phosphoryl and so on from one molecule to another
- Hydrolases can act on a very broad array of hydrolysable groups. It includes esters, amides, peptides and other C–N containing functions, anhydrides, glycosides and several others
- Lyases catalyze the addition of groups to double bonds such as C=C, C=O and C=N or the reverse
- Isomerases transfer groups within molecules to yield isomeric form such as racemization
- Ligases are often termed synthesases. They mediate the formation of C–O, C–S and C–N bonds

Not all enzymes are capable of acting alone. Many require the presence of nonprotein cofactors to enable catalysis to be carried out. Such cofactors, which are in reality cosubstrates because they undergo chemical transformation during the reaction, include simple metal ions or organic molecules that may in some cases be covalently bound to the enzyme. They must be continuously reconverted back to their original form in order for catalysis to continue. This process, illustrated in Fig. 5.4, is usually referred to as regeneration. Regeneration can take place spontaneously in aerobic aqueous conditions by hydrolysis or oxidation reactions. However, in most cases, including ATP, coenzyme A, folic acid, NAD^+ and $NADP^+$, regeneration can only be achieved by directly coupling with the oxidation of high-energy substrate molecules, either by cytochromes or by enzymes. In the whole cells, regeneration is an aspect of the normal integrated metabolism of the cell, via substrate level and cytochrome linked exergonic reactions. However, when isolated enzymes or disrupted cells are used, regeneration may be a considerable problem.

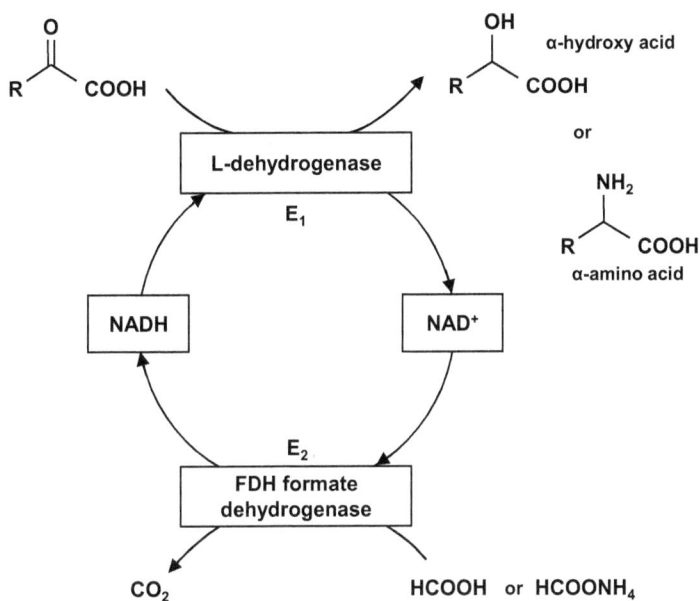

Fig. 5.4: Example of cofactor regeneration of NAD^+/NADH.

In industry, enzymes or cells are often used in an immobilized form to enable reuse or continuous use of the biocatalyst and easy product recovery. Continuous use is especially important to maintain a constant environment for the immobilized biocatalyst and thereby improve enzyme stability. Numerous immobilization methods are used. They can be divided into the following three groups:

- Entrapment or encapsulation in a porous polymer network
- Covalent, ionic or physical attachment to an appropriate water-insoluble solid support
- Aggregation of cells by physical or chemical cross-linking with glutaraldehyde or other agents

Although immobilization tends to increase the stability of the enzyme or cell, usually some enzyme activity is lost due to denaturation during immobilization. With immobilized cells, flow rates through the reactor and/or the concentration of the enzyme can be increased. As a result, a much smaller reactor can be used to achieve the same productivity, reducing the capital and operating costs of enzyme catalyzed processes. The choice between the use of immobilized enzymes or immobilized cells is similar to the choice between the use of purified or crude soluble enzyme as biocatalyst. Clearly, the immobilized cell or crude enzyme is cheaper, and larger quantities are available, but they are less specific catalysts than immobilized enzymes or the purified soluble enzymes.

5.2 Biochemical reaction engineering

5.2.1 Principles

As with chemical reactors the calculation of microbiological and biochemical processes is based on the balance equations of the bioreactor. Depending on their state of mixing, bioreactors are denoted as ideally mixed stirred tank reactors or ideal plug-flow reactors. To solve the balances, the stoichiometry, kinetics, and energetics of the biochemical process must be known and understood. In this modeling, the kinetics of microbial growth and product formation or material transformation by the enzymes or microorganisms are of central importance.

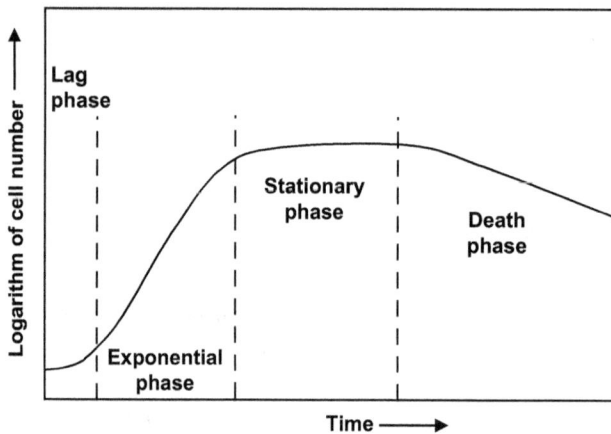

Fig. 5.5: Growth curve of a batch bacterial culture.

5.2.2 Kinetics of biochemical reactions

5.2.2.1 Microbiological processes

In many respects, the bacterial growth process is similar to a chemical reaction in which the components of the medium produce cells in addition to excreted products. Figure 5.5 shows that after inoculation of a batch reactor, the bacteria start to grow. Slowly in the beginning and exponentially at their maximum rate during the major part of the growth period. In this characteristic exponential phase, the growth rate is autocatalyzed by the bacterial population itself. Then, the rate of growth decreases and ceases, and finally the cells start to die.

Generally, the bacterial growth rate remains virtually constant until the medium is almost exhausted of the limiting nutrient. This seeming paradox is explained by the action of enzymes (permeases) which are capable of maintaining constant intracellular concentrations of substrates and nutrients over a wide range of external concentrations. Nevertheless, usually at extremely low concentrations of external nutrients, the permease enzymes are no longer able to maintain the intracellular concentrations and the growth rate falls. The curves relating growth rate to substrate concentration are typically hyperbolic in form (Fig. 5.6) and can be described by the Monod equation:

$$\mu = \mu_{MAX} \frac{S}{K_S + S} \tag{5.1}$$

where μ is the specific growth rate (s^{-1}), S the substrate concentration (mol/m^3), μ_{MAX} the growth rate at infinite concentration of the substrate and K_S (mol/m^3) is the Monod uptake constant for the substrate.

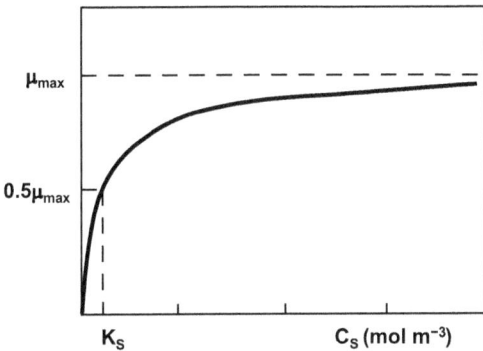

Fig. 5.6: Effect of substrate concentration on the specific growth rate in case of Monod kinetics.

5.2.2.2 Enzyme catalyzed reactions

Enzyme catalysts are subject to the same kinetic and thermodynamic constraints as are chemical catalysts. They alter the rate at which a reaction proceeds but not the final position of equilibrium between the substrate (S) and product (P). Catalysis is effected by enabling an alternative reaction mechanism and thus an alternative transition state with a lower free energy, by lowering the free energy of the conventional transition state intermediate, or by providing an environment that decreases the free energy of the product.

In the time course of an enzyme reaction, generally three phases are recognized. At the start most of the free enzyme combines with substrate in a dynamic equilibrium that persist as long as fresh substrate molecules are available. In the second phase, all the reactants including the enzyme molecules are in a dynamic equilibrium with the maximum activity being expressed. In the final stage of the reaction, the substrate concentration is strongly reduced, and thus the rate of the enzyme catalyzed reaction falls asymptotically. This phase of reaction is especially important in many industrial reactions where complete conversion is desired. It can often occupy the majority of the reaction time especially when product inhibition takes place, which is especially likely when high substrate concentrations are employed. The kinetics of enzyme reactions can be derived on basis of the following general reaction scheme for conversion of a single substrate (S) to a single product (P):

$$S + E \underset{k_{-1}}{\overset{k_1}{\rightleftharpoons}} ES \overset{k_2}{\longrightarrow} E + P \tag{5.2}$$

When the concentration of substrate is much greater than the concentration of enzyme (E), the rate of reaction depends only on the concentration of enzyme present. Thus, the velocity of the reaction remains essentially constant until nearly all the substrate has been consumed. This kinetic behavior is reasonably well described by the well-known Michaelis–Menten relationship (Fig. 5.7):

$$r_S = k_2 E_0 \frac{S}{K_M + S} = r_{S,\,max} \frac{S}{K_M + S} \tag{5.3}$$

where r is the reaction rate, E_0 the enzyme concentration and $K_M = (k_{-1} + k_2)/k_1$ the Michaelis constant that represents the substrate concentration which gives half the maximum rate of reaction.

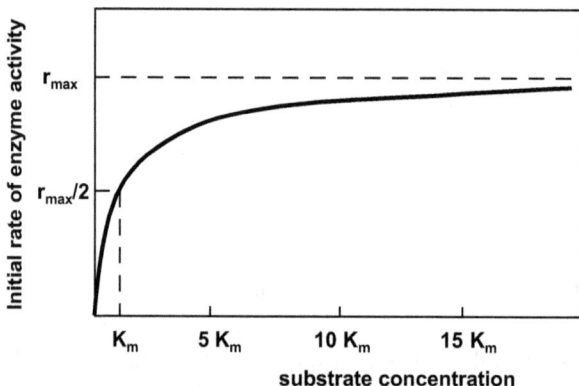

Fig. 5.7: Plot of the Michaelis–Menten enzymatic reaction rate.

5.2.2.3 Environmental effects

In general, the pH of the reaction medium has a pronounced effect on both enzyme and cell kinetics. The activity as a function of pH generally displays a bell-shaped profile as shown in Fig. 5.8. This definite optimum pH for enzyme activity is usually observed because enzymes contain many ionizable groups. As a result, pH changes alter the conformation of the enzyme and thereby the binding of the substrate and the catalytic activity. The overall effects may be observed by a change in the maximum reaction rate ($r_{s,max}$), a change in the affinity of the enzyme for the substrate (K_m) or an alteration in the stability of the enzyme. In a similar fashion, ionizable groups in the substrate can be affected by pH and alter the enzyme–substrate complex formation process.

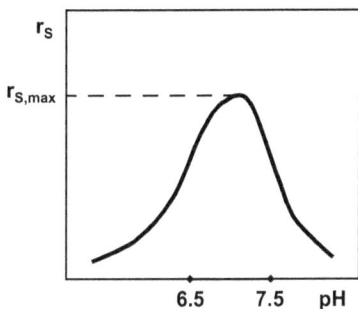

Fig. 5.8: Effect of pH on enzymatic activity.

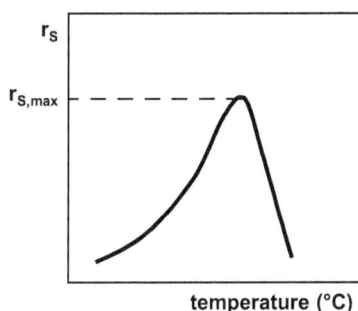

Fig. 5.9: Enzymatic activity as a function of temperature.

The effects of temperature on enzyme and cell kinetics generally exhibit a rather simple behavior. The overall rate passes through a maximum with a sharp decrease at higher temperatures (Fig. 5.9). This behavior is the result of a combined effect. Raising the temperature affects the enzyme-catalyzed reaction as well as the thermal inactivation of the enzymes. Although the underlying mechanisms are very complex, the rate of enzyme catalysis as well as the rate of enzyme denaturation can be described satisfactorily with the Arrhenius equation:

$$k_A = k_{A,\infty} \, \exp\left(\frac{-E_A}{RT}\right) \tag{5.4}$$

Activation energies for enzyme-catalyzed reactions are normally in the range of 4–20 kcal/mol such that the rate of reaction increases only slowly with an increase in temperature. Enzyme stability is much stronger when influenced by temperature because activation energies of enzyme denaturation range from 40 to 130 kcal/mol. The optimum temperature for reaction to take place is compromise between these dual effects of increased temperature.

5.2.2.4 Inhibition

Four common types of inhibition are recognized. Irreversible inhibition occurs when the inhibitor molecule combines irreversibly with the enzyme, chemically modifying its structure and destroying its catalytic activity. As illustrated in Fig. 5.10, three types of reversible inhibition can occur:

– Competitive, which can be reduced by increasing the substrate concentration

$$E + \begin{matrix} S & \Leftrightarrow & ES \rightarrow E + P \\ I & \Leftrightarrow & EI \end{matrix}$$

(5.5)

– Noncompetitive, which cannot be reduced

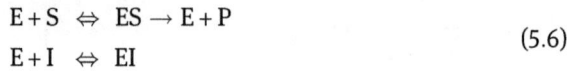

$$E + S \Leftrightarrow ES \rightarrow E + P$$
$$E + I \Leftrightarrow EI$$

(5.6)

– Excess substrate inhibition, caused by the formation of nonproductive complexes that can be decreased by using a lower substrate concentration

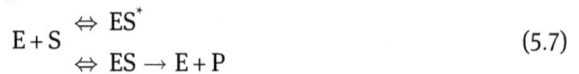

$$E + S \begin{matrix} \Leftrightarrow & ES^* \\ \Leftrightarrow & ES \rightarrow E + P \end{matrix}$$

(5.7)

Competitive inhibitors compete with the normal substrate molecules to occupy the active site of the enzyme forming a reversible enzyme–inhibitor complex which is characterized by the K_i value, the dissociation constant of this complex. With noncompetitive inhibition, the inhibitor also forms a reversible complex with the enzyme but at a site other than the active site that is not reduced by increasing the substrate concentration. Inhibition by substrate is less common but occurs at high substrate concentration even though inhibition is not readily apparent and Michaelis–Menten kinetics are obeyed at low substrate concentrations. Inhibitors are often present in low concentrations in many substrates and inhibition may only become serious when continuous high throughputs of substrate are used.

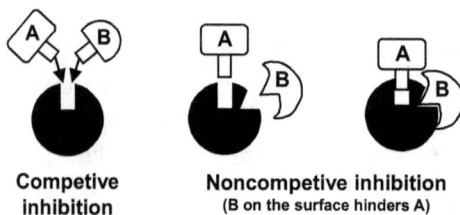

Competive inhibition

Noncompetive inhibition
(B on the surface hinders A)

Fig. 5.10: Schematic representation of inhibition.

5.2.3 Basic reactor operations

Like chemical reactors, bioreactors are operated in batch and continuous mode. In addition, several semicontinuous modes are frequently employed in biotechnological production:

- Fed-batch operation with varying feed rates and volume–time patterns in the reactor.
- Extended fed-batch operation with continuous feeding, the substrate concentration remaining constant
- Repeated (fed)-batch where after the batch reaction a small amount of the fermentation broth is left in the reactor as inoculum for the next batch or fed-batch process

Batch biotechnological processes are characterized by inoculation of the sterile culture medium with microorganisms, cultivated for a specific reaction period. During the reaction cell, substrate and product concentrations alter. Good mixing ensures that there are no significant local differences in composition and temperature of the reaction mixture. As the oxygen in the culture medium is slightly soluble, a continuous oxygen supply is needed for aerobic cultivation, while removing the CO_2 formed in the same way. This is generally done by aeration of the medium. Acid or alkali is added periodically to the system to control the pH value. Antifoaming agents are also added for chemical foam suppression. The main advantage of batch bioreactors is the low risk of infection and cell mutation due to relatively short cultivation periods. Other advantages of batch operation are low investment costs, greater flexibility, higher conversion levels. Disadvantages of batch operation are the nonproductive idle time, higher operating cost and the greater risk of contact with pathogenic organisms or toxic products. Hence, batch reactors are used in biotechnology when only small amounts of product are involved, there is a high risk of infection or microorganism mutation or when downstream processing is discontinuous.

In continuous operation, the culture medium is fed continuously into the bioreactor and the reaction mixture is also drawn continuously from the reactor. The medium added may be sterile or it may contain the microorganisms or enzyme used. All reaction variables and control parameters remain constant in time. As a result, a steady state is established in the reactor characterized by constant productivity and output. Main advantages of continuous bioreactors are lower reactor volume, constant product quality and less possible danger from pathogenic microorganisms or toxic materials. However, continuous operation suffers from high risk of infection or microorganism mutation due to long cultivation periods. Therefore, continuous operation is preferred for processes with high production rates, for gas, liquid or soluble solid substrates and when microorganisms with high mutation stability are involved.

Semicontinuous operation can be regarded as a combination of batch and continuous. Many variations of this type of process are practiced. The most popular involves starting the reactor as a batch process until the growth-limiting substrate is consumed, which is subsequently fed to the reactor in a specified manner (fed batch) or kept constant in concentration by extended culture. Moreover, programmed substrate addition is frequently practiced for secondary metabolite production in which cell growth and product formation occur in separate phases. Semicontinuous processes attempt to combine the advantages of continuous and batch operation. They have high flexibility and a high yield but still suffer from the nonproductive idle time and risk of contact with pathogenic microorganisms or toxic products. It is clear that semicontinuous processes are often used when continuous methods are not possible (due to slight mutation or infection of microorganisms) and batch production would result in low productivity figures. Like batch reactors, semicontinuous reactors are nonstationary.

5.3 Industrial bioreactors

5.3.1 Classification

Bioreactors are vessels in which raw materials are biologically converted into specific products, using microorganisms, plant, animal or human cells or individual enzymes. A bioreactor supports this process by providing suitable conditions such as optimum temperature, pH, sufficient substrate, nutritional salts, vitamins and oxygen. The design of bioreactors depends on biomass concentration, sterility requirements, mixing, suspension, aeration, heat transport and the generation of optimum shear conditions. For microbial as well as enzymatic conversions reactors, good heat removal systems have to be employed because the temperature of the fermentation process should be maintained at an optimum, constant value. Bioreactions with a sensitive pH optimum should be conducted in reactors with a high mixing intensity and a distributed feed system for agents that prevent local high concentrations of acids or bases. Labile or toxic intermediates or products can be withdrawn from the reaction mixture by an integrated separation step. Enzyme reactors differ from chemical reactors because they function at low temperatures and pressures, and comparatively little energy is consumed or generated during reaction. Enzyme reactors differ from fermentations in not behaving in an autocatalytic fashion.

Biochemical processes are either aerobic or anaerobic and each type demands a different set of conditions. Aerobic reactors can be subdivided into submerged and surface reactors. The difference is that in surface reactors, microorganisms are supplied with oxygen through the surface of a medium, which generally forms a thin film on the solid carrier. In submerged reactors, the necessary oxygen and substrate are supplied to microorganisms by air dispersion and subsequent intensive bulk mixing of the medium and dissolved oxygen. Submerged reactors are more suitable for mass production and have therefore become more popular over the last few decades.

A frequently used, practical classification of submerged bioreactors is based on the type of energy supply. There are a number of ways in which turbulence can be produced by energy dissipation. In some reactors, energy is generated by mechanically moving agitators. The stirred-tank reactor, the best known in this category, has been regarded as the classic example of a biotechnological reactor. However, it is also true to say that over the same period, cases have appeared for which the standard stirred tank reactor is not suitable. Thus, it might be uneconomic to operate or unsuitable for biological reasons. Large numbers of new types of reactors have been developed and modifications to existing models made in attempts to provide alternatives that offer technical or economic advantages for special production processes or aerobic effluent treatment. These new variations offer alternatives to mechanical energy input. Examples are reactors with an external liquid pump and reactors with no mechanical parts where gas expansion through a sparger provides the required energy for mixing.

5.3.2 Bioreactors with mechanical mixing

The stirred tank reactor is still the most important type of reactor in biochemical technology. Mechanical stirred bioreactors have the following functions: homogenization, suspension of solids, dispersion of liquid–liquid mixtures, gassing/aeration of the liquid, heat exchange and influencing microorganisms through shear. Homogenization is the most important aim of mechanical agitation in nonaerated submerged reactors. As the density of the microorganisms is almost the same as that of water, forming a suspension is not very difficult. On the other hand, gas-phase dispersion (usually air) in cultivation media is very significant aspect of aerobic submerged cultures. As the required reactor size increases, the more difficult it is to obtain uniform aeration and adequate homogenization and heat removal. In small aerobic reactors (<1 m^3), the main function of a mechanical agitator is gas dispersion. Above this level (>10 m^3), homogenization, aeration and heat removal are equal in importance. Heat is generated by biological and agitation processes, so the intensification of heat removal by mechanical agitators is a vital aspect.

A simple stirred-tank fermenter is presented in Fig. 5.11. As a rule, the tank is provided with four baffles. Typical stirrers for low-viscosity media are the propeller and pitched-blade turbine for axial motion and the disk and impeller stirrer for radial motion. In moderately to highly viscous media other stirrer types are used. Multiple stirring devices are advantageous for mixing highly viscous media such as mycelia fermentations used in producing antibiotics. Heat transfer in stirred-tank reactors is achieved by use of double jacket, helical or meander coils or by external heat exchangers.

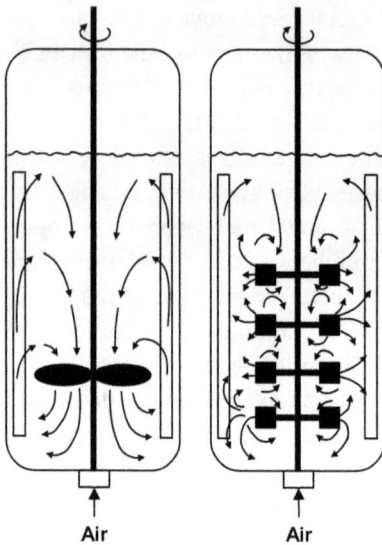

Fig. 5.11: Schematic of single and multiple stirrer reactors.

5.3.3 Bioreactors with pneumatic mixing

In this type of reactor, compressed air is expanded and dispersed through a gas distributor. As the density of the gas is considerably lower than that of the medium, gas bubbles rise up and liquid is entrained with them. The power supplied by the compressed gas disperses the gas and mixes the medium at the same time. Bioreactors with pneumatic mixing possess the advantage of simple design, good heat transfer, small base area, low-maintenance operation and very low shear forces. They can be classified into bubble column reactors and airlift reactors.

5.3.3.1 Bubble column reactors

These reactors (Fig. 5.12) comprise vertical columns fitted with a static gas distributor (air sparger) at the base of the column and different design variations to improve gas dispersion and residence time. To prevent too heterogeneous flow patterns in the lower compartment, the sparger nozzles have to be distributed over the cross section of the bottom. Although their construction is very simple, the flow patterns inside them are highly complex and detailed design specifications are required for optimum operation. This is largely due to the fact that the properties of the media change greatly during the reaction and cause process engineering problems due to foam formation, biomass flotation and bubble coalescence.

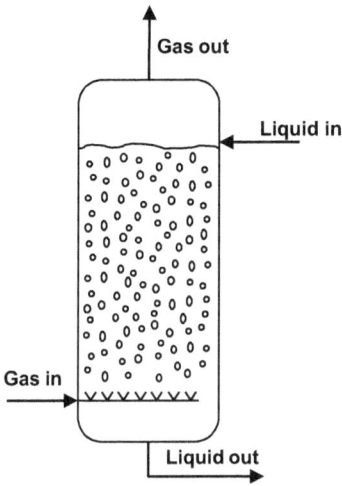

Fig. 5.12: Bubble column.

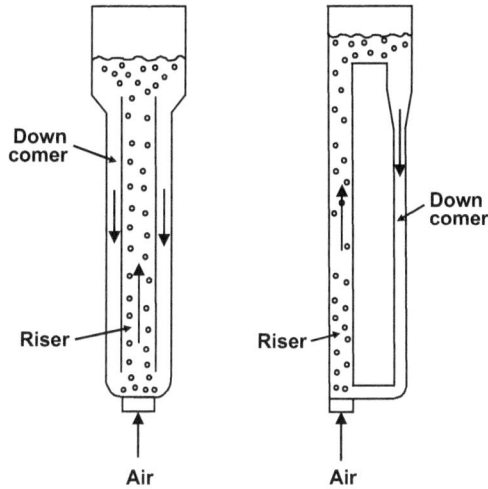

Fig. 5.13: Airlift reactor with internal and external loop.

5.3.3.2 Airlift reactors

In airlift reactors (Fig. 5.13), the circulation of the fermentation liquid is provided by the density difference between the aerated liquid in the tower and the nonaerated liquid in the loop. They may have internal or external circulation. Internal circulation can be produced by two coaxial draft tubes or flow inversion by gassing the outer cylinder space. Perforated plates, static mixers or packing may be installed for multiple dispersion of the gas bubbles. Airlift reactors with external circulation exhibit favorable flow behavior because heat exchangers or additional mixing elements can be installed in the downcomer. Airlift reactors are especially suitable for shear-sensitive, flocculating and foaming fermentation systems. Volumes can be up to thousands of cubic meters.

The airlift consists of two pipes, interconnected at top and bottom. In one of the pipes (riser), air is sparged at the bottom. The air rises and escapes at the top. Therefore, under most circumstances there is no air present in the other pipe (downcomer). The density difference between riser and downcomer causes an intensive liquid circulation. The aerated liquid with a lower density level reaches a higher surface level than the nonaerated liquid, which has a lower gas content and greater density. Hence, the surface of the aerated phase in the riser should be above that of the nonaerated liquid phase in the downcomer. Liquid passing through the closed loop moves from the higher to the lower lever, thus establishing a circulation pattern.

5.3.4 Bioreactors for immobilized enzymes and cells

Three techniques have been developed for increasing the biocatalyst concentration in the reactor and thus retention for continuous process operation. Separation of the biocatalyst at the reactor outlet, retention of the biocatalyst by micro- or ultrafiltration and immobilization. The reactors are considerably smaller than those for free enzymes and cells due to the concentration of the biocatalyst. Examples of bioreactors that employ immobilized enzymes and cells are shown in Fig. 5.14. They include fixed bed reactors (glucose isomerization, L-amino acid production), staged fixed bed reactors and stirred slurry reactors.

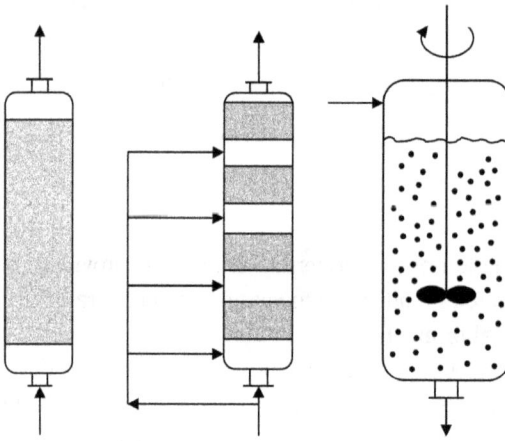

Fig. 5.14: Schematic of fixed bed, multiple fixed bed and slurry bioreactors employing immobilized enzymes and cells.

Nomenclature

E	Enzyme concentration	[mol/m^3]
E_A	Activation energy	[J/mol]
EI	Enzyme–inhibitor complex concentration	[mol/m^3]
ES	Enzyme–substrate complex concentration	[mol/m^3]
I	Inhibitor concentration	[mol/m^3]
k	Reaction rate constant	[s^{-1}]
K_M	Michaelis constant	[mol/m^3]
K_S	Monod substrate uptake constant	[mol/m^3]
P	Product concentration	[mol/m^3]
r	Reaction rate	[mol/m^3 s]
R	Gas constant (8.3144)	[J/mol K]
S	Substrate concentration	[mol/m^3]
T	Temperature	[K]
μ	Specific growth rate	[s^{-1}]

Indices

A	Species A
MAX	At infinite substrate concentration
0	Initial
S	Substrate
*	Nonproductive

6 Evaporative separations

6.1 Evaporative separation

6.1.1 Introduction

A large part of the separations of individual substances in a homogeneous liquid mixture or complete fractionation of such mixtures into their individual pure components is achieved through evaporative separations. Evaporative separations are based on the difference in composition between a liquid mixture and the vapor formed from it. This composition difference arises from differences in effective vapor pressures, or volatilities, of the components in the liquid mixture. The required vapor phase is created by partial evaporation of the liquid feed through adding heat, followed by total condensation of the vapor. Due to the difference in volatility of the components the feed mixture is separated into two or more products whose compositions differ from that of the feed. The resulting condensate is enriched in the more volatile components, in accordance with the vapor–liquid equilibrium (VLE) for the system at hand. When a difference in volatility does not exist, separation by simple evaporation is not possible.

The basis for planning evaporative separations is a knowledge of the VLE. Technically, evaporative separations are the most mature separation operations. Design and operating procedures are well established. Only when VLE or other data are uncertain a laboratory and/or pilot-plant study is necessary prior to the design of a commercial unit. The most elementary form is simple distillation in which the liquid mixture is brought to boiling, partially evaporated and the vapor formed separated and condensed to form a product. This technique is commonly used in the laboratory for the recovery and/or purification of products after synthesis in an experimental setup as illustrated in Fig. 6.1.

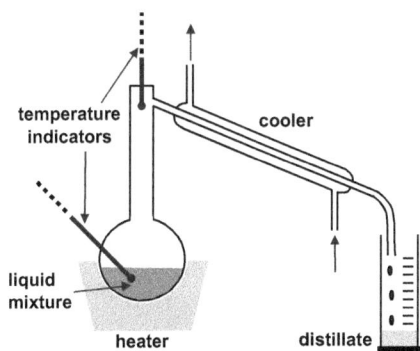

Fig. 6.1: Laboratory distillation setup.

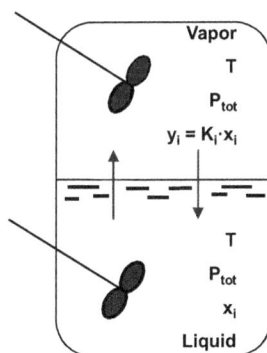

Fig. 6.2: Vapor–liquid equilibrium schematic.

https://doi.org/10.1515/9783110712445-006

6.1.2 Vapor–liquid equilibria

It is clear that equilibrium distributions of mixture components in the vapor and liquid phase must be different if separation is to be made by evaporation. At thermodynamic equilibrium the compositions are called VLE that may be depicted as in Fig. 6.2. For binary mixtures the effect of distribution of mixture components between the vapor and liquid phases on the thermodynamic properties is illustrated in Fig. 6.3. Figure 6.3a shows a representative boiling point diagram with equilibrium compositions as functions of temperature at a constant pressure. Commonly the more volatile (low boiling) components are used to plot the liquid and vapor compositions of the mixture. The lower line is the liquid bubble point line, the locus of points at which on heating a liquid forms the first bubble of vapor. The upper line is the vapor dew point line, representing points at which a vapor on cooling forms the first drop of condensed liquid. The region between the bubble and dew point lines is the two phase region where vapor and liquid coexist in equilibrium. At the equilibrium temperature T_e the liquid with composition x_e is in equilibrium with vapor composition y_e. Figure 6.3b displays a typical isobaric y–x diagram that is obtained by plotting the vapor composition that is in equilibrium with the liquid composition at a fixed pressure or a fixed temperature. At equilibrium, the concentration of any component present in the liquid mixture is related to its concentration in the vapor phase by the equilibrium constant, K_i:

$$K_i = \frac{y_i}{x_i} \tag{6.1}$$

where y_i is the mole fraction of component i in the vapor phase and x_i the mole fraction of component i in the liquid phase. The more volatile components in a mixture will have the higher K_i values, whereas less volatile components will have lower values of K_i. The key separation factor in distillation is the relative volatility, defined as

$$\alpha_{12} = \frac{K_1}{K_2} = \frac{y_1/x_1}{y_2/x_2} \tag{6.2}$$

The higher the value of the relative volatility, the more easily components may be separated by distillation.

In ideal systems the behavior of vapor and liquid mixtures obeys Dalton's and Raoult's laws. Dalton's law relates the concentration of a component present in an ideal gas or vapor mixture to its partial pressure:

$$p_i = P y_i \tag{6.3}$$

where p_i is the partial pressure of component i in the vapor mixture and P is the total pressure of the system given by the sum of the partial pressures of all components in the system:

$$P = \sum_{i=1}^{N} p_i \tag{6.4}$$

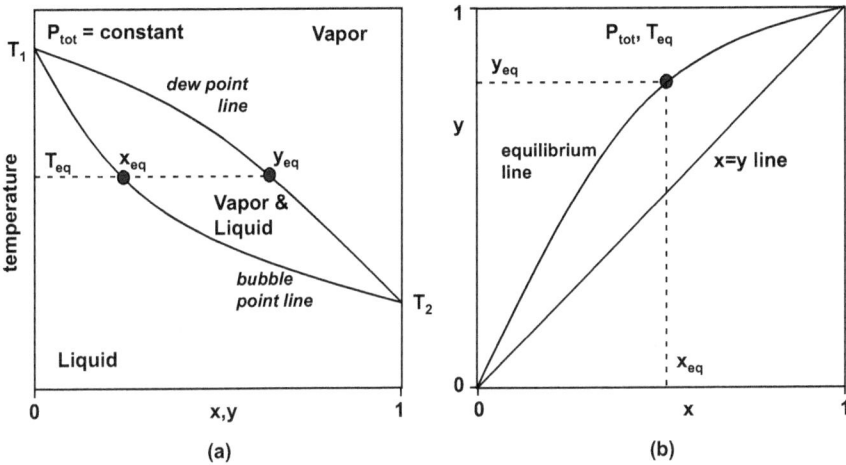

Fig. 6.3: Isobaric vapor–liquid equilibrium diagrams: (a) dew and bubble point and (b) y–x diagram.

Raoult's law relates the partial pressure of a component in the vapor phase to its concentration in the liquid phase:

$$p_i = P_i^o x_i \tag{6.5}$$

where P_i^o is the vapor pressure of pure component i at the system temperature. Combining eqs. (6.3) and (6.5) yields:

$$P y_i = P_i^o x_i \tag{6.6}$$

This results in the following relation between the pure component vapor pressure, its equilibrium coefficient and the relative volatility for an ideal mixture:

$$K_i = \frac{P_i^o}{P} \qquad \alpha_{12} = \frac{K_1}{K_2} = \frac{P_1^o}{P_2^o} \tag{6.7}$$

Equation (6.7) shows that for ideal systems the relative volatility is independent of pressure and composition. Vapor pressures for many components are published in the literature and often correlated as a function of temperature by the Antoine equation:

$$\ln P_i^o = A_i - \frac{B_i}{C_i + T} \tag{6.8}$$

where A_i, B_i and C_i are Antoine constants and T is the temperature in degrees Celsius or Kelvin. Since vapor pressures of components depend on temperature, equilibrium ratios are a function of temperature as well. However, the relative volatility is considerably less sensitive to temperature changes because it is proportional to the ratio of the vapor pressures. In general, the vapor pressure of the more volatile component tends to increase at a slower rate with increasing temperature than the less volatile component. Therefore, the relative volatility generally decreases with increasing temperature and increases with decreasing temperature. For a binary system, the relative volatility equation can be rearranged to give

$$y_1 = \frac{\alpha_{12} x_1}{1 + (\alpha_{12} - 1) x_1} \tag{6.9}$$

Equation (6.9) is used to express the concentration of a component in the vapor as a function of its concentration in the liquid and relative volatility. It is plotted in Fig. 6.4 for various values of relative volatility. When relative volatility increases, the concentration of the most volatile component in the vapor increases. When the relative volatility is equal to 1, the concentrations of the most volatile component in the liquid and vapor phases are equal and a vapor/liquid separation is not feasible.

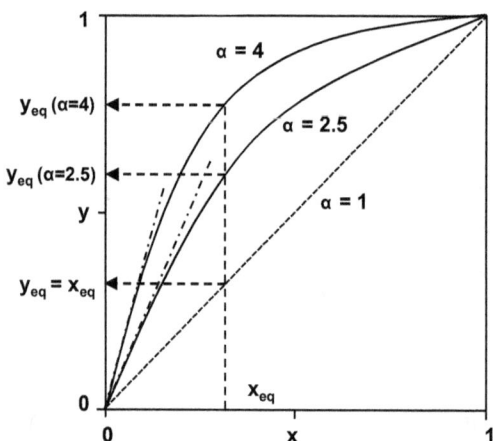

Fig. 6.4: Vapor–liquid equilibrium compositions as a function of relative volatility.

Most liquid mixtures are non-ideal and require Raoult's law to be modified by including a correction factor called the liquid phase activity coefficient:

$$p_i = \gamma_i P_i^o x_i \tag{6.10}$$

Although at high pressures, the vapor phase may also depart ideal vapor mixture, the common approach in distillation is to assume ideal vapor behavior and to correct non-ideal liquid behavior with the liquid phase activity coefficient. The standard state for reference for the liquid phase activity coefficient is commonly chosen as $\gamma_i \rightarrow 1$ for the pure component. The liquid phase activity coefficient is strongly dependent upon the composition of the mixture. Positive deviations from ideality ($\gamma_i > 1$) are more common when the molecules of different compounds are dissimilar and exhibit repulsive forces. Negative deviations ($\gamma_i < 1$) occur when there are attractive forces between different compounds that do not occur for either component alone. For nonideal systems, the relations for the distribution coefficient and relative volatility now become:

$$K_i = \frac{\gamma_i P_i^o}{P} \qquad \alpha_{12} = \frac{K_1}{K_2} = \frac{\gamma_1 P_1^o}{\gamma_2 P_2^o} \tag{6.11}$$

In nonideal systems, the coefficients, distribution coefficients and relative volatility are dependent on composition because of the composition dependence of the activity coefficients. When the activity coefficient of a specific component becomes high enough an azeotrope may be encountered, meaning that the vapor and liquid compositions are equal and the components cannot be separated by conventional distillation. Figure 6.5 shows binary vapor–liquid composition $(x–y)$, temperature–composition $(t–x)$ and pressure–composition $(P–x)$ diagrams for intermediate boiling, minimum azeotrope and maximum azeotrope systems. A minimum azeotrope boils at a lower temperature than each of the components in their pure states. When separating the components of this type of system by distillation, such as ethanol-water, the overhead product is the azeotrope. A maximum boiling azeotrope boils higher than either component in their pure states and is the bottom product of distillation. An example of this type of system is acetone–chloroform.

6.1.3 Separation by single-stage partial evaporation

Two main modes are utilized for single-stage separation by partial evaporation. The most elementary form is differential distillation, in which a liquid is charged to the still pot and heated to boiling. As illustrated in Fig. 6.6, the method has a strong resemblance with laboratory distillation, shown previously in Fig. 6.1. The vapor formed is continuously removed and condensed to produce a distillate. Usually the vapor leaving the still pot with composition y^D is assumed to be in equilibrium with perfectly mixed liquid in the still at any instant. The distillate is richer in the more volatile components and the residual bottoms are richer in the less volatile components. As the distillation proceeds, the relative amount of volatile components composition of the initial charge and distillate decrease with time. Because the produced vapor is totally condensed, $y_D = x_D$, there is only one single equilibrium stage, the still pot. Simple differential distillation is not widely used in industry, except for the processing of high value chemicals in small production quantities or for distillations requiring regular sanitation.

The continuous form of simple single-stage equilibrium distillations is called flash distillation. A flash is a single equilibrium stage distillation in which a continuous feed is partially vaporized to give a vapor richer in the more volatile components than the remaining liquid. In Fig. 6.7, a liquid feed is heated under pressure and flashed adiabatically across a valve to a lower pressure, resulting in the creation of a vapor phase that is separated from the remaining liquid in a flash drum. If the equipment is properly designed, the liquid and vapor leaving the flash drum are considered to be in equilibrium. Mechanically a demister is used to prevent droplets from leaving with the

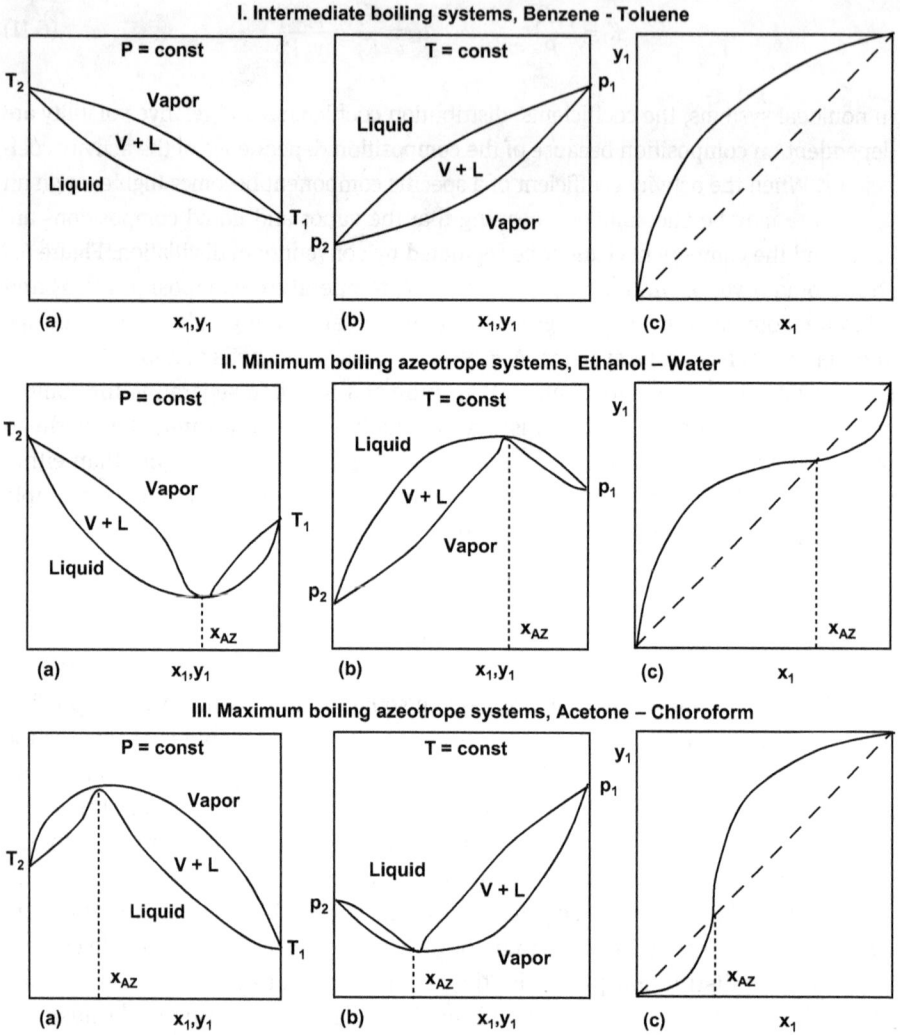

Fig. 6.5: Types of binary temperature–composition (a), pressure–composition (b) and x–y (c) phase diagrams for vapor–liquid equilibrium.

vapor. Modeling is done through simple mass balances combined with equilibrium data or equivalent expressions. The overall mass balance is

$$F = D + B \tag{6.12}$$

and the component i balance becomes

$$x_i^F F = y_i^D D + x_i^B B \tag{6.13}$$

Unless the relative volatility is very large, the degree of separation achievable between two components in a single equilibrium stage is poor. Therefore, flashing is

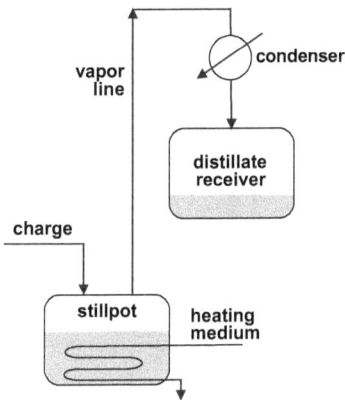

Fig. 6.6: Simple differential distillation.   Fig. 6.7: Flash distillation.

an auxiliary operation used to prepare streams for further processing and/or where a crude separation is adequate. Typically, the vapor phase is sent to a vapor separation system, while the liquid is sent to a liquid separation system.

6.2 Multistage distillation

6.2.1 Distillation cascades

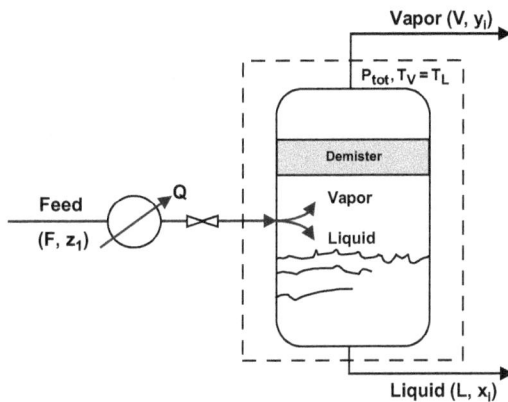

Flash distillation is a very simple unit operation that in most cases produces only a limited amount of separation. Increased separation is possible in a cascade of flash separators that produces one pure vapor and one pure liquid product. Within the cascade the intermediate product streams are used as additional feeds. The liquid streams are returned to the previous flash drum, while the produced vapor streams are forwarded to the next flash chamber. Figure 6.8 shows the resulting counter-current cascade, so called because vapor and liquid streams go in opposite directions. The advantages of this cascade are that there are no intermediate products and the two end products can both be pure and obtained in high yield.

Although a significant advance, this multiple flash drum system is seldom used industrially. Operation and design is easier if part of the top vapor stream is condensed and returned to the first stage, reflux, and if part of the bottom liquid stream is evaporated and returned to the bottom stage, boilup. This allows control of the internal liquid and vapor flow rates at any desired level by applying all of the heat required for the distillation to the bottom reboiler and do all the required cooling in the top condenser. Partial condensation of intermediate vapor streams and partial vaporization of liquid streams is achieved by heat exchange between all pairs of passing streams. This is most effectively achieved by building the entire system in a column

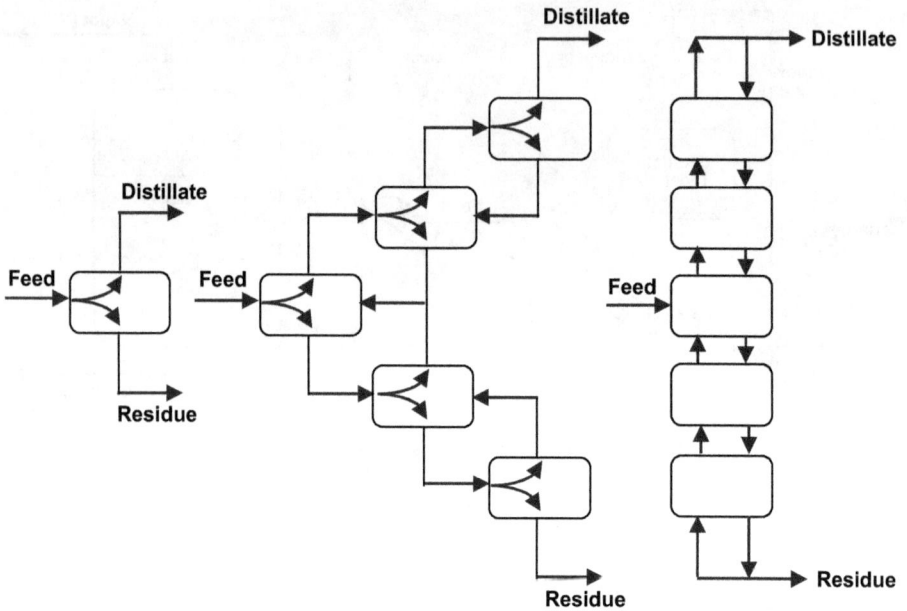

Fig. 6.8: Flash drum cascades.

instead of the series of individual stages as shown in Fig. 6.8. Intermediate heat exchange is the most efficient with the liquid and vapor in direct contact. The final result is a much simpler and cheaper device, the distillation column shown in Fig. 6.9.

6.2.2 Column distillation

Most commercial distillations involve some form of multiple staging in order to obtain better separation than is possible by single vaporization and condensation. It is the most widely used industrial method of separating liquid mixtures in the chemical process industry. As shown by Fig. 6.9, most multistage distillations are continuously operated column-type processes separating components of a liquid mixture according to their different boiling points in a more volatile distillate and a less volatile bottoms or residue. The feed enters the column at the equilibrium feed stage. Vapor and liquid phases flow counter currently within the mass transfer zone of the column where trays or packings are used to maximize interfacial contact between the phases. The section of column above the feed is called the rectification section and the section below the feed is referred to as the stripping section. Although Fig. 6.9 shows a distillation column equipped with only 12 sieve trays, industrial columns may contain more than 100. The liquid from a tray flows through a downcomer to the tray below and vapor flows upward through the holes in the sieve tray to the tray above. Intimate contact between the vapor and liquid phases

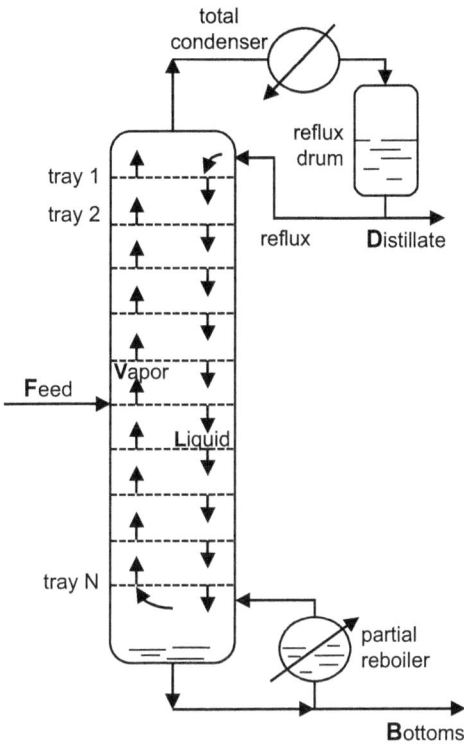

Fig. 6.9: Sieve plate distillation column.

is created as the vapor passes through the holes in the sieve tray and bubbles through the pool of liquid residing on the tray. The vapors moving up the column from equilibrium stage to equilibrium stage are increasingly enriched in the more volatile components. Similarly, the concentration of the least volatile components increases in the liquid from each tray going downward.

The overhead vapor from the column is condensed to obtain a distillate product. The liquid distillate from the condenser is divided into two streams. Part of is withdrawn as overhead product (D) while the remaining distillate is refluxed (L') to the top tray to enrich the vapors. The reflux ratio is defined as the ratio of the reflux rate to the rate of product removal. The required heat for evaporation is added at the base of the column in a reboiler, where the bottom tray liquid is heated and partially vaporized to provide the vapor for the stripping section. Plant-size distillation columns usually employ external steam powered kettle or vertical thermosyphon type heat exchangers (Fig. 6.10). Both can provide the amount of heat transfer surface required for large installations. Thermosyphon reboilers are favored when the bottom product contains thermally sensitive compounds, only a small temperature difference is available for heat transfer and heavy fouling occurs. The vapor from the reboiler is sent back to the bottom tray and the remaining liquid is removed as the bottom product.

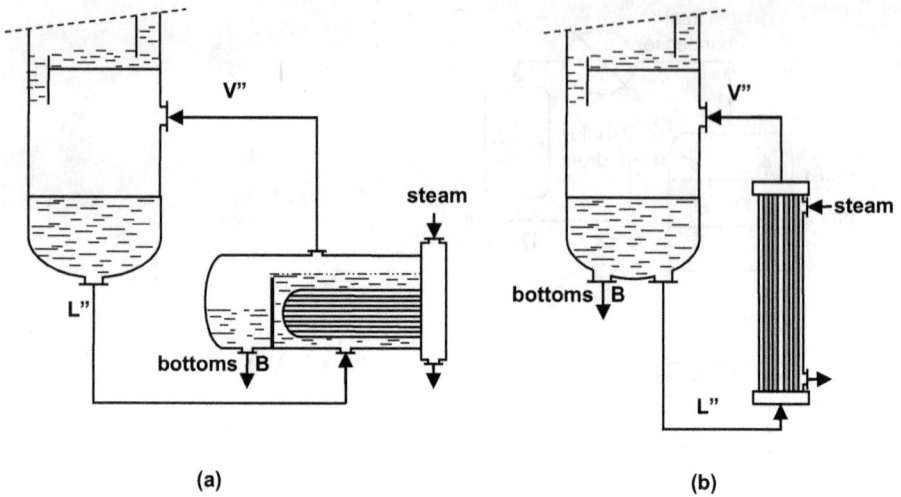

Fig. 6.10: Industrial reboilers: (a) kettle type and (b) vertical thermosyphon type.

6.2.3 Feasible distillation conditions

Industrial distillation processes are restricted by operability of the units, economic conditions and environmental constraints. An upper limit exists for feasible operating temperatures. One reason is the thermal stability of the species in the mixtures to be
separated. Many substances decompose at higher temperatures and some species are
not even stable at their normal boiling points. A second reason for a maximum temperature limitation is the means of heat supply. In most cases the required energy is supplied by condensing steam. The pressure of the available steam places an upper limit
on temperature levels that can be achieved. Only in special cases (crude oil, sulfuric
acid), higher temperatures may be realized by heating with hot oil or natural gas
burners. With high-pressure steam the maximum attainable temperatures are limited to 300 °C, which can be raised to temperatures as high as 400 °C when other
media are used. The temperature of the coolant for the overhead condenser dictates
the lower limit of feasible temperatures in distillation columns. In most cases water
is used, resulting to a minimum temperature in the column of 40–50 °C.

In most cases temperature constraints can be met by selecting a proper operating pressure. Operating temperatures can be decreased by the use of a vacuum. It is
technically feasible to operate distillation columns at pressures up to 2 mbar. Lower
pressures are seldom used because of the high operating and capital costs of the vac-
uum-producing equipment. A column can be operated at higher pressure to increase
the boiling point of low boiling mixtures. The upper limit for the operating pressure
lies in the range of the critical pressures of the constituents. In addition to temperature and pressure the feasible number of stages in industrial columns is also limited.

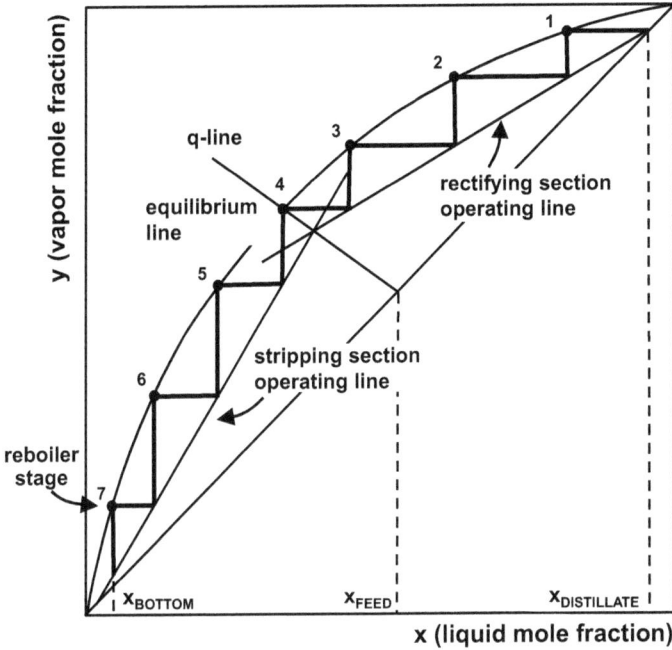

Fig. 6.11: Determination of number of equilibrium stages with the McCabe–Thiele graphical method.

Only in exceptional cases commercial distillation columns are constructed that contain more than 100 stages.

6.2.4 Basic design calculations

The separation of a feed F into a distillate product D and bottoms product B can be described by a mass balance over the entire column:

$$F = D + B \tag{6.14}$$

in combination with a component i balance over the entire column:

$$x_i^F F = y_i^D D + x_i^B B \tag{6.15}$$

A solution to the material balance and equilibrium relationships inside the column is provided by the graphical McCabe–Thiele design method. It employs the simplifying assumption that the molal overflows in the stripping and rectification sections are constant. As illustrated in Fig. 6.11, this assumption leads to straight operating lines in the rectifying and the stripping section:

$$y^{N-1} = \left(\frac{L'}{V'}\right)_R x^N + \left(\frac{D}{V'}\right)_R x^D \tag{6.16}$$

$$y^{N-1} = \left(\frac{L}{V}\right)_S x^N - \left(\frac{B}{V}\right)_S x^B \tag{6.17}$$

The constant liquid and vapor molal flows in each section are designated by L and V. Assuming constant molal overflow implies that the molal latent heats of the two components are identical, the sensible heat effects are negligible and the heat of mixing and heat losses equal zero. This simplified situation is closely approximated for many distillations. In Fig. 6.11, the operating lines relates the solute concentration in the vapor passing upward between two stages to the solute concentration in the liquid passing downward between the same two stages. The equilibrium curve relates the solute concentration in the vapor leaving an equilibrium stage to the solute concentration in the liquid leaving the same stage. This makes it possible to determine the required number of stages by constructing a staircase between the operating line and the equilibrium curve, as shown in Fig. 6.11.

For any distillation operation however, there are infinite combinations of reflux ratios and numbers of theoretical stages possible. The larger the reflux ratio, the fewer theoretical stages are required but the more energy is consumed. For a given combination of feed, distillate and bottom compositions, there are two constraints that set the boundary conditions within which the reflux ratio and number of theoretical stages must be. The minimum number of theoretical stages and the minimum reflux ratio. Both can be determined by simple analytical methods that are also applicable to multicomponent mixtures when the two key components are used. Key components are those between which the specified separation must be made. The minimum number of theoretical stages occurs when the operating lines coincide with the $y = x$ line, meaning that the system is at total reflux. The minimum number of stages may again be determined graphically, as illustrated in Fig. 6.12, but an approximate value of the minimum number of equilibrium stages at total reflux can also be obtained from the Fenske equation:

$$N_{MIN} = \frac{\ln\left[\left(\frac{x_{lk}}{x_{hk}}\right)_D \left(\frac{x_{hk}}{x_{lk}}\right)_B\right]}{\ln\left(\alpha_{lk/hk}\right)_{av}} \tag{6.18}$$

The Fenske equation is also applicable for multicomponent mixtures because the relative volatility is based on the light key relative to the heavy key. The average value of the relative volatility is generally calculated by taking the average of the relative volatility at the top of column and at the bottom of the column.

Column operation with minimum internal gas and liquid flow (i.e. minimum reflux) separates a mixture with the lowest energy input. On a McCabe–Thiele plot the minimum reflux ratio occurs when the upper and lower operating lines and the

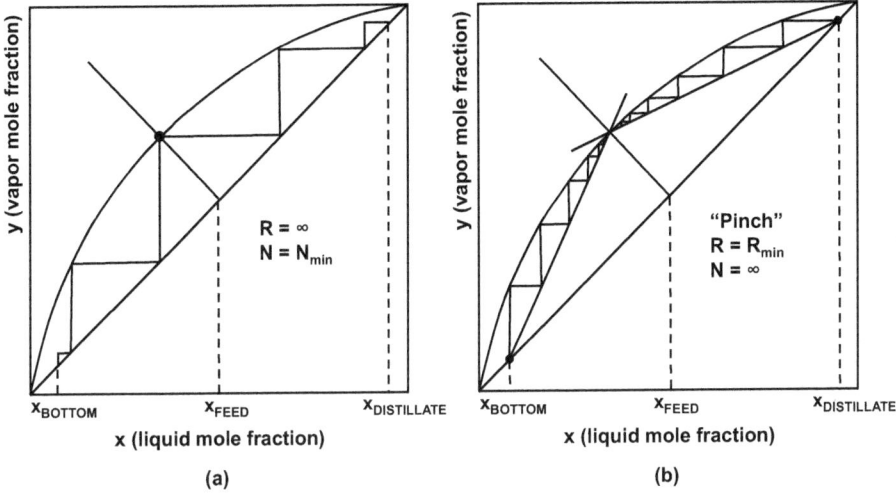

Fig. 6.12: Determination of (a) minimum reflux ratio and (b) minimum number of theoretical stages in a McCabe–Thiele diagram.

feed point (q-line) coincide at a single point on the equilibrium line shown in Fig. 6.12. Under this condition an infinite number of theoretical stages would be required to achieve the desired separation. The Underwood method provides an analytical expression for estimating the minimum reflux ratio:

$$R_{MIN} = \frac{1}{\alpha - 1}\left(\frac{x_D}{x_F} - \alpha\frac{1 - x_D}{1 - x_F}\right) \tag{6.19}$$

If the distillate product is required as a (nearly) pure substance ($x_D = 1$), this equation reduces to

$$R_{MIN} = \frac{1}{(\alpha - 1)\,x_F} \tag{6.20}$$

Both of these limits, the minimum number of stages and the minimum reflux ratio serve as valuable guidelines within which the practical distillation conditions must lie. The operating, fixed and total cost of a distillation system depend strongly on the ratio of operating reflux ratio to minimum reflux ratio. As shown in Fig. 6.13, first the fixed cost decrease by increasing the reflux ratio because fewer stages are required but then rises again as the diameter of the column increases at higher vapor and liquid loads. Similarly, the operating cost for energy increase almost linearly as the operating reflux ratio increases. For most commercial operations the optimal operating reflux ratios are in the range of 1.1–1.5 times the minimum reflux ratio.

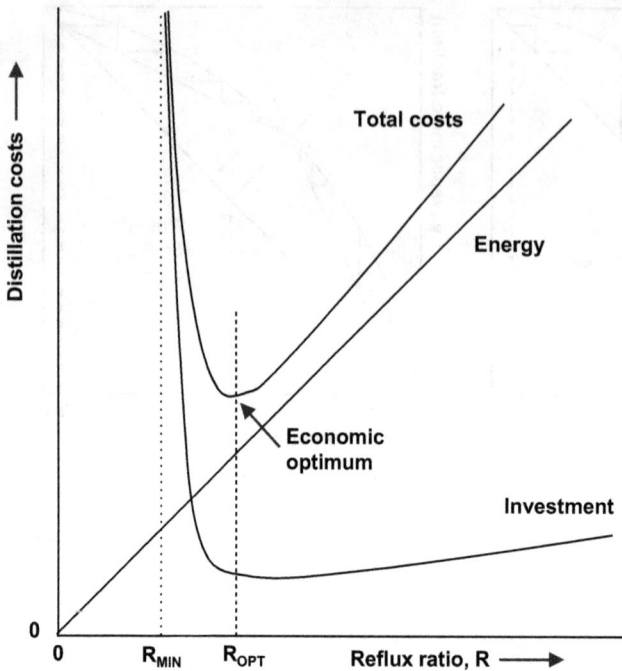

Fig. 6.13: Typical distillation fixed, operating and total costs as a function of reflux ratio.

6.2.5 Energy requirements

Following the determination of the feed condition, reflux ratio and number of theoretical stages estimates of the heat duties of the condenser and reboiler can be made. When energy losses to the environment are minimal, the feed is at the bubble point and a total condenser is used, the energy balance over the entire column gives

$$\text{In} = \text{Out} \tag{6.21}$$

$$Q_{\text{Reboiler}} = Q_{\text{Condensor}} \tag{6.22}$$

The energy balance can be approximated by applying the assumptions of the McCabe–Thiele method, yielding for the reboiler and condenser duty:

$$Q_{\text{Reboiler}} = Q_{\text{Condenser}} = D\,(R+1)\,\Delta H_{\text{vap}} \tag{6.23}$$

If saturated steam is the heating medium for the reboiler, the required steam rate becomes

$$m_{\text{Steam}} = \frac{Q_{\text{Reboiler}}}{\Delta H_{\text{Steam}}^{\text{vap}}} \qquad (6.24)$$

where ΔH^{vap} is the enthalpy of vaporization of steam (2100 kJ/kg). The cooling water rate for the condenser is:

$$m_{\text{CW}} = \frac{Q_{\text{Condenser}}}{C_{\text{P,water}} \left(T_{\text{out}} - T_{\text{in}}\right)} \qquad (6.25)$$

where $C_{\text{P,water}}$ is the specific heat capacity of water (kJ/kg K). In general the cost of cooling water can be neglected during a first evaluation because the cost of reboiler steam is an order of magnitude higher.

6.2.6 Batch distillation

In most large chemical plants distillations are run continuously. For small production units where most chemical processes are carried out in batches it is more convenient to distil each batch separately. In batch distillation, a liquid mixture is charged to a vessel (still) where it is heated to the boiling point. When boiling begins the vapor is passed through a fractionation column and condensed to obtain a distillate product, as indicated in Fig. 6.14. As with continuous distillation the purity of the top product depends on the still composition, the number of plates of the column and on the reflux ratio used. In contrast to continuous distillation, batch distillation is usually operated with a variable amount of reflux. Because the lower boiling components concentrate in the vapor and the remaining liquid gradually becomes richer in the heavier components, the purity of the top product will steadily drop. This is generally compensated by a gradual increase in reflux ratio during the distillation process to maintain a constant quality of the top product. To obtain the maximum recovery of a valuable component, the charge remaining in the still after the first distillation may be added to the next batch.

The main advantage of batch distillation is that multiple liquid mixtures can be processed in a single unit. Different product requirements are easily taken into account by changing the reflux ratio. Even multicomponent mixtures can be separated into the different components by a single column when the fractions are collected separately. An additional advantage is that batch distillation can also handle sludges and solids. The main disadvantages of batch distillation are that for a given product rate the equipment is larger and the mixture is exposed to high temperatures for a longer time. This increases the risk of thermal degradation or decomposition. Furthermore, it requires more operator attention, energy requirements are higher and its dynamic nature makes it more difficult to control and model.

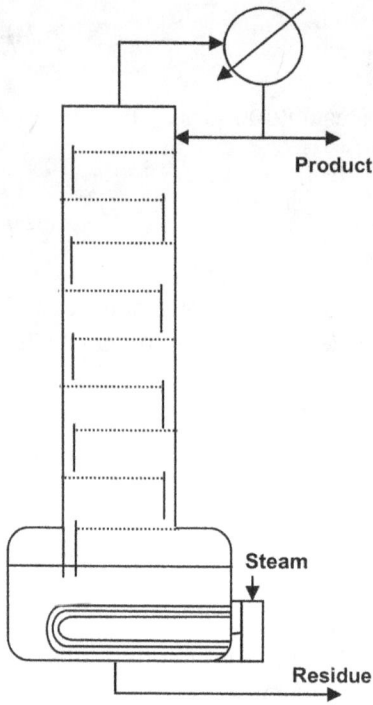

Product

Steam

Residue

Fig. 6.14: Schematic of a batch distillation unit.

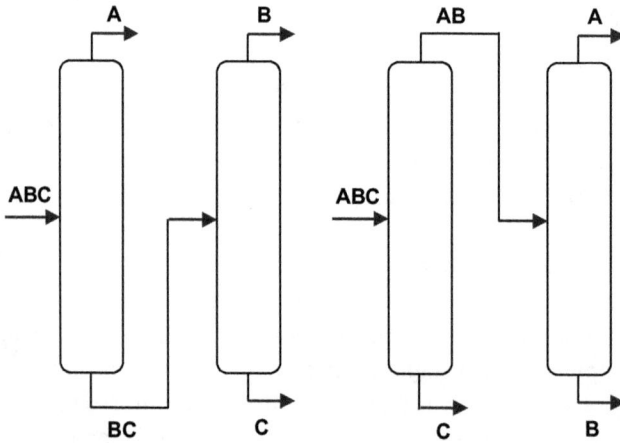

Fig. 6.15: Possible paths for complete separation of a ternary mixture into its pure products.

6.2.7 Continuous separation of multiple products

If each component of a multicomponent distillation is to be essentially pure when recovered, the number of required columns is equal to the number of components minus one. Thus, a three-component mixture requires two and a four-component mixture requires three separate columns. Those columns can be arranged in many different ways as illustrated in Fig. 6.15 for the possible separation paths of a ternary mixture. The more components in the mixture the more separation paths are possible. Fortunately, the separation sequence of multicomponent mixtures is often determined by other limiting conditions such as stability or corrosiveness of individual substances, danger of explosion and toxicity. In all these cases, the difficult substance should, if possible, be separated first.

6.2.8 Enhanced distillation techniques

When, due to minimal difference in boiling point and/or highly non-ideal liquid behavior the relative volatility becomes lower than 1.1, ordinary distillation may be uneconomic and in case an azeotrope forms even impossible. In that event, enhanced distillation techniques should be explored. For these circumstances the most often used technique is extractive distillation where a large amount of a relatively high-boiling solvent is added to increase the relative volatility of the key components in the feed mixture. In order to maintain a high concentration throughout the column, the solvent is generally introduced above the feed entry and a few trays below the top. It leaves the bottom of the column with the less volatile product and is recovered in a second distillation column as shown in Fig. 6.16. Because the high-boiling solvent is easily recovered by distillation when selected in such a way that no new azeotropes are formed, extractive distillation is less complex and more widely used than azeotropic distillation.

   In homogeneous azeotropic distillation an entrainer is added to the mixture that forms a homogeneous minimum or maximum boiling azeotrope with one or more feed components. The entrainer can be added everywhere in the column. If an entrainer is added to form an azeotrope, the azeotrope will exit the column as the overhead or bottom product leaving behind component(s) which may be recovered in the pure state. A classical example of azeotropic distillation is the recovery of anhydrous ethanol from aqueous solutions. Organic solvents such as benzene or cyclohexane are used to form desirable azeotropes, allowing the separation to be made. As shown in Fig. 6.17, the crude column overhead product is fed to the azeotropic distillation column where cyclohexane is used to form an azeotrope with water. Ethanol is recovered as the bottom product. The overhead azeotropic cyclohexane/water mixture is condensed where water-rich and cyclohexane-rich liquid phases are formed. Residual cyclohexane is removed from the water-rich phase in a stripper.

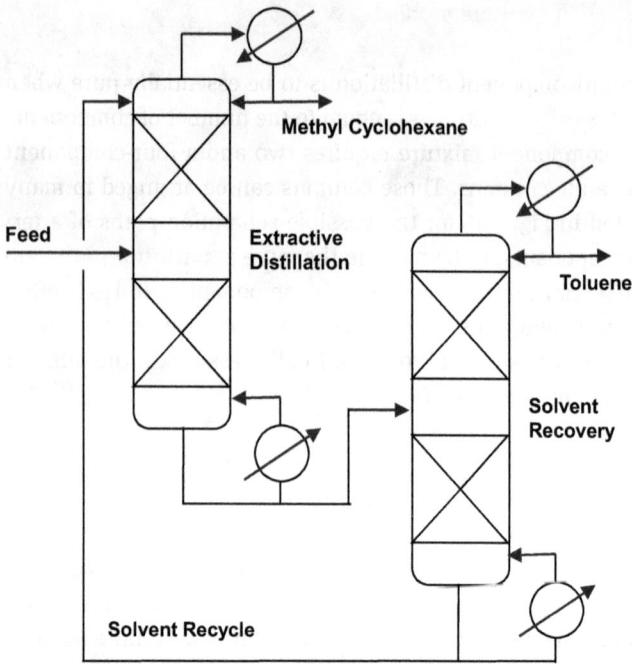

Fig. 6.16: Extractive distillation separation of toluene from methylcyclohexane (MCH) using solvents such as *N*-methylpyrrolidone (NMP).

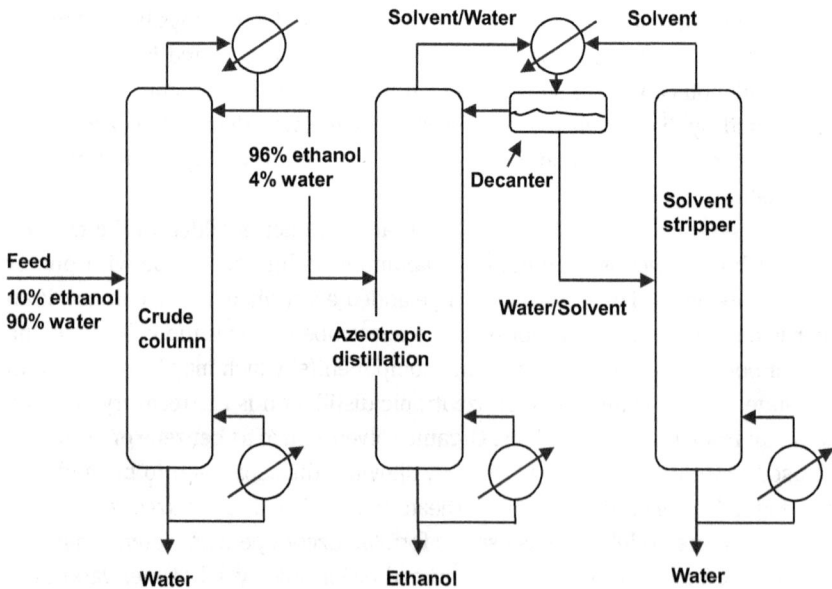

Fig. 6.17: Dehydration of alcohol by azeotropic distillation with cyclohexane as entrainer.

It is well known that many minimum boiling azeotrope containing mixtures allow the position of the azeotrope to be shifted a change in system pressure. This effect can be exploited to separate a binary azeotrope containing mixture when appreciably changes (>5 mol%) in azeotropic composition can be achieved over a moderate pressure range. Pressure swing distillation uses a sequence of two columns operated at different pressures for the separation of pressure-sensitive azeotropes. This is illustrated in Fig. 6.18 where the effect of pressure on the temperature and composition of a minimum boiling azeotrope is given. The binary azeotrope can be crossed by first separating the component boiling higher than the azeotrope at low pressure. The composition of the overhead should be as close as possible to that of the azeotrope at this pressure. As the pressure is increased the azeotropic composition moves toward a higher percentage of A and component B can be separated from the azeotrope as the bottom product in the second column. The overhead of the second column is returned to the first column.

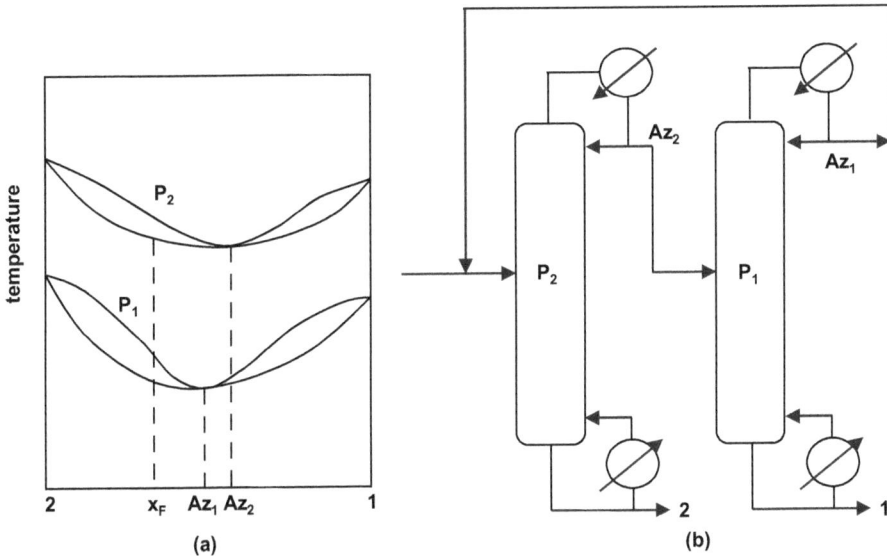

Fig. 6.18: Pressure-swing distillation: (a) T–y–x curves for minimum boiling azeotrope at pressures P_1 and P_2 and (b) distillation sequence for minimum boiling azeotrope.

6.3 Distillation equipment

6.3.1 Basic functions

Distillation is usually conducted in vertical cylindrical vessels that provide intimate contact between the rising vapor and the descending liquid. The distillation column normally contains internal devices for effective vapor–liquid contact that provides the opportunity for the two streams to achieve some approach to thermodynamic equilibrium. Their basic function is to provide efficient mass transfer between a two-phase vapor–liquid systems. The requirements for efficient mass transfer across the vapor–liquid interface can be deduced from

$$\Phi = k_{OV} \, A \, (y^* - y) \tag{6.26}$$

The maximum mass transfer rate is obtained when all three terms on the right-hand side of the above equation are as large as possible:
- The mass transfer coefficient k_{OV} increases proportionally with the relative velocities between the liquid and vapor phases and is also improved by constant regeneration of the contact area between the phases.
- A large contact area A is desirable for mass transfer and mainly determined by the used column internals. Depending upon the type of internal devices used, the contacting may occur in discrete steps, called plates or trays, or in a continuous differential manner on the surface of a packing material. Tray columns and packed columns are most often used for distillation since they guarantee excellent counter-current flow and permit a large overall height. The internals provide a large mass transfer area, which is constantly renewed, especially in tray columns.
- The largest concentration driving force (y^*-y) is achieved when the overall flow pattern allows countercurrent contact between equal streams of gas and liquid without significant remixing. An important condition is that both phases are distributed uniformly over the entire flow area.

6.3.2 Tray columns

Figure 6.19 shows the most important features of a tray column. The gas flows upwards within the column through perforations in horizontal trays, and the condensed liquid flows counter currently downward. However, as indicated in Fig. 6.20, the two phases exhibit cross-flow to each other on the individual trays. The liquid enters the cross-flow tray from the bottom of the downcomer belonging to the tray above and flows across the perforated active or bubbling area. The ascending gas from the tray below passes through the perforations and aerates the liquid to form a large interfacial area between the two phases. It is in this zone where the main vapor–liquid mass transfer occurs. The vapor subsequently disengages from the aerated mass on the tray and rises to the tray above. The aerated liquid flows over the exit weir into the downcomer where most of the trapped vapor escapes from the liquid and flows back to the interplate vapor space. Some of the liquid accumulates in each downcomer to compensate for the pressure drop caused by the gas as it passes through the tray. The liquid then leaves the tray by flowing through the downcomer outlet onto

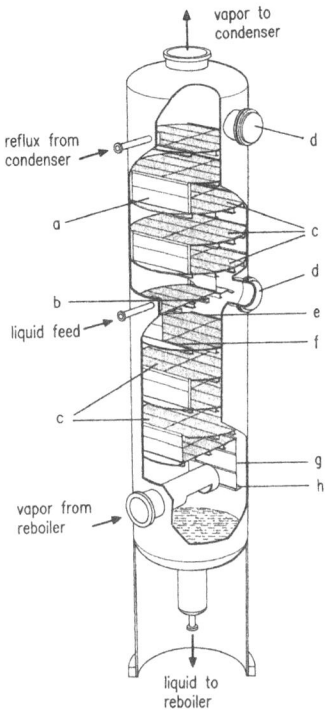

Fig. 6.19: Cutaway section of a tray column (reproduced with permission from [76]).

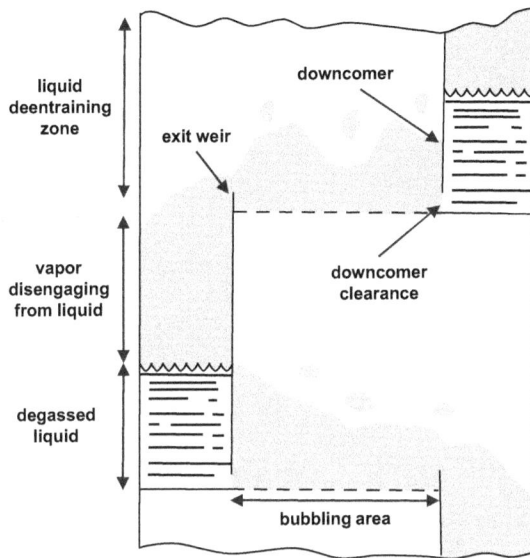

Fig. 6.20: Schematic of flow pattern in a cross-flow tray distillation column.

the tray below. In large diameter cross-flow trays, multiple liquid-flow-path trays with multiple downcomers are used to prevent that the hydraulic gradient of the liquid flowing across the tray becomes excessive.

Three principal vapor–liquid contacting devices are used for cross-flow tray design (see Fig. 6.21):

- Sieve trays have become very important because they are simple, inexpensive, have high separation efficiency and produce a low pressure drop across the tray. Conventional sieve trays contain typically 1–12 mm holes and exhibit ratios of open area to active area ranging from 1:20 to 1:7. If the open area is too small the pressure drop across the tray is excessive, while if the open area is too large the liquid weeps or dumps through the holes.
- Valve trays are a relatively new development, that represent a variation of the sieve tray with liftable valve units such as those shown in Fig. 6.22 fitted in the holes. The liftable valves prevent the liquid from leaking at low gas loads and to avoid excessive pressure increase at high gas loads. The main advantage is the ability to maintain efficient operation while being able to vary the gas load up

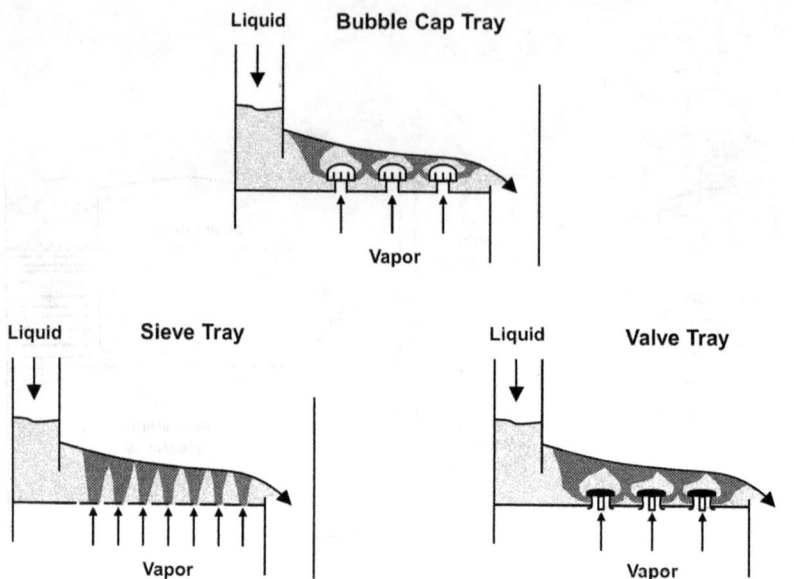

Fig. 6.21: Schematic of a sieve, valve and bubble cap tray.

to a factor of 4–5. This capability gives valve trays a much larger operational flexibility than any other tray design.

– Bubble cap trays have been used almost exclusively in the chemical industry until the early 1950s. As shown in Fig. 6.21, their design prevents liquid from leaking downward through the tray. The vapor flows through a hole in the tray floor, through the riser, reverses direction in the dome of the cap, flows downward and exits through the slots in the cap. However, the complex bubble caps are relatively expensive and have a higher pressure drop than other designs. This limits their usage in newer installations to low liquid flow rate applications or to those cases where the widest possible operating range is desired.

Although the column requirements are calculated in terms of theoretical or equilibrium stages, the real design must specify the actual number of trays. This requires the determination of the performance of an actual tray to approaching equilibrium. This is often done through the overall tray efficiency, which is defined by the number of theoretical stages divided by the required number of actual trays:

$$E_o = \frac{N}{N_{act}} \times 100\% \tag{6.27}$$

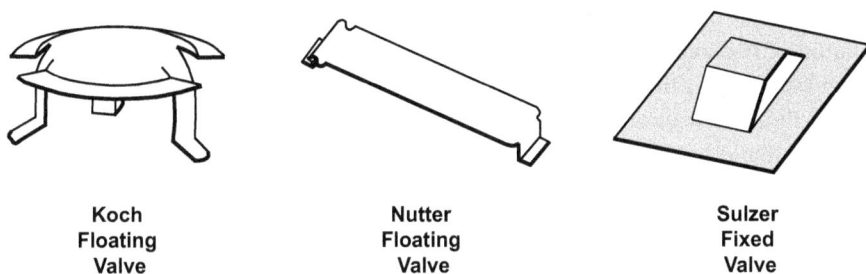

| Koch Floating Valve | Nutter Floating Valve | Sulzer Fixed Valve |

Fig. 6.22: Examples of valves uses in valve trays.

As a result the overall efficiency is an average of all the individual trays. Most hydrocarbon distillation systems in commercial columns achieve overall tray efficiencies of 60–80%. In absorption processes the range is 10–50%.

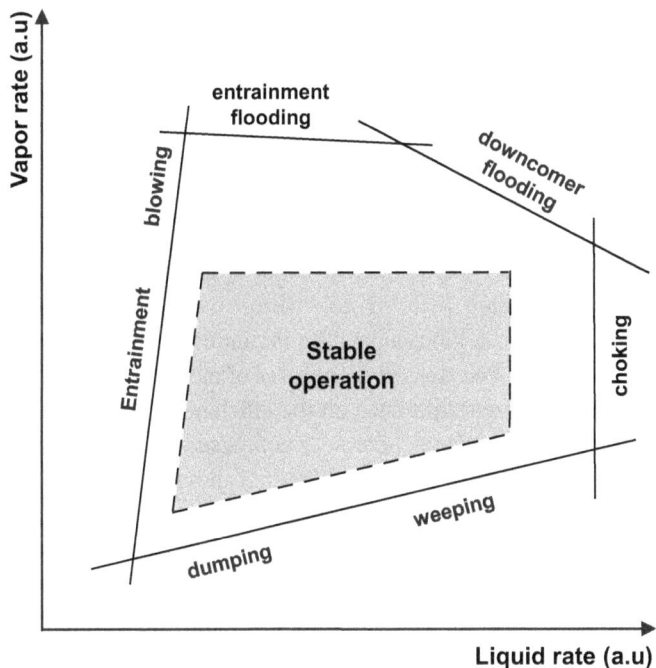

Fig. 6.23: Operating diagram for cross-flow trays.

More detailed column dimensioning requires determination of the entire operating region bound by a range of liquid and vapor flow rates as shown in Fig. 6.23. The weeping/dumping line represents the minimum operable vapor flow rate at various liquid flow rates. Below the line, the vapor rate is too low to maintain the liquid on the tray and the liquid weeps or dumps through the tray orifices. At high vapor rates the entrainment line represents the boundary where the gas blows the liquid of the tray in the form of fine droplets. The liquid then no longer flows counter currently to the gas, and proper column operation ends. The same occurs when the liquid flow through the column becomes larger than the downcomer capacity and the liquid holdup becomes excessive. Flooding conditions can occur at high liquid rates or by excessive pressure drop across the trays restricting the liquid flow rate through the downcomer. Due to the increased holdup part of the liquid recycled back to the previous tray with the rising vapor and proper countercurrent column operation breaks down. Although trays can be operated at very low liquid loads, extremely low liquid loads cause an uneven liquid distribution across the tray that decreases the mass transfer efficiency.

6.3.3 Packed columns

The most important features of a packed column are shown in Fig. 6.24. The vapor enters the bottom of the column and flows upward through the free cross-sectional area of the internals in countercurrent contact with the down flowing liquid. To promote mass transfer the packing should have a large surface area per unit of volume and be wetted by the liquid as completely as possible. Because a countercurrent flow exists throughout the column, packed columns are in principle more effective for mass transfer than tray columns. However, the countercurrent flow of gas and liquid in a packed column is not perfect since the liquid flow is not uniform over the cross section. Well-known mechanisms causing liquid maldistribution are liquid channeling and wall flow. The internals of the column must offer minimum resistance to gas flow. Modern packings have a relative free cross-sectional area of more than 90%.

The type of packing used has a great influence on the efficiency of the column. Most industrial packed columns use random packings composed of a large number of specially formed particles. Presently, some 50 different types of random packings are offered on the market. Some examples of the more important types are shown in Fig. 6.25. The oldest packing element is the Rashig ring, with its characteristic feature that the length of the ring is equal to its diameter. This feature makes the particles form quite a homogeneous bed structure during pouring into the column. Rashig rings have a rather high pressure drop since the walls of those rings in horizontal position block the gas flow. The Pall ring avoids this disadvantage since parts of the wall are punched out and deflected into the inner part of the ring. A Pall ring has the same porosity and the same volumetric area as a Rashig ring but a considerably lower pressure drop. The feasible packing size depends on column diameter but should not exceed 1/10 to 1/30 of the column diameter. Most packings are made of metal or ceramics, but plastics are increasingly used.

A certain degree of inhomogeneity is unavoidable in any random packing. These inhomogeneities, that cause liquid maldistribution, are avoided by using

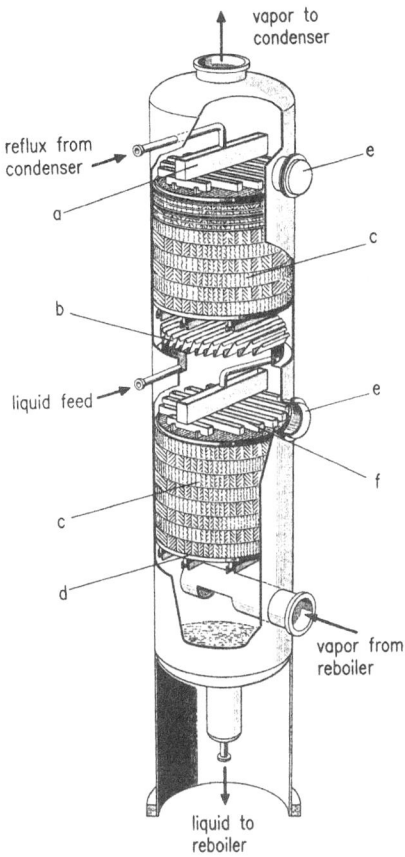

vapor to
condenser

reflux from
condenser

e

a

c

b

e

liquid feed

c

f

c

d

vapor from
reboiler

liquid to
reboiler

Fig. 6.24: Cutaway section of a packed column with
a structured packing (reproduced with permission
from [76]).

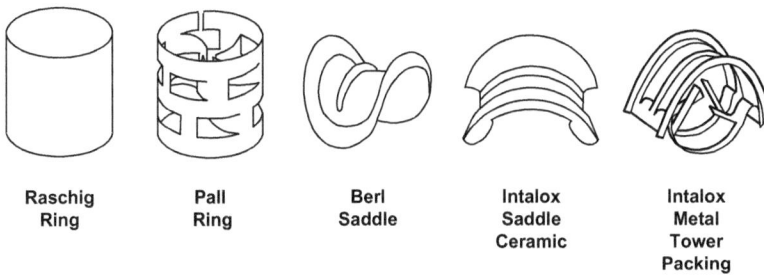

| Raschig Ring | Pall Ring | Berl Saddle | Intalox Saddle Ceramic | Intalox Metal Tower Packing |

Fig. 6.25: Common random column packings.

ordered packing structures such as the corrugated sheet structure that was developed by Sulzer in the mid-1960s. As Fig. 6.26 shows the corrugated sheets are assembled parallel in vertical direction with alternating inclinations of the corrugations of neighboring sheets. Since the packing does not fit perfectly into the cylindrical column shell, additional tightening strips have to be installed between packing and column wall. These structured packings provide a homogeneous bed structure, and a low pressure drop due to the vertical orientation of the sheets. It is therefore not surprising that their first applications were found in vacuum distillation. At present structured packings are available in different types (gauze, sheet) and in a variety of materials (metals, plastics, ceramics, carbon). For most applications structured sheet metal packings offer a more attractive performance to cost ratio than structured gauze packings because the cost of structured sheet is about one-third that of gauze while the efficiencies are about the same. Disadvantages of structured packings are high costs relative to trays, criticality of initial liquid and vapor distributions and the associated hardware required. Typically the installed cost of structured sheet metal packing plus associated hardware is about three to four times that for conventional trays.

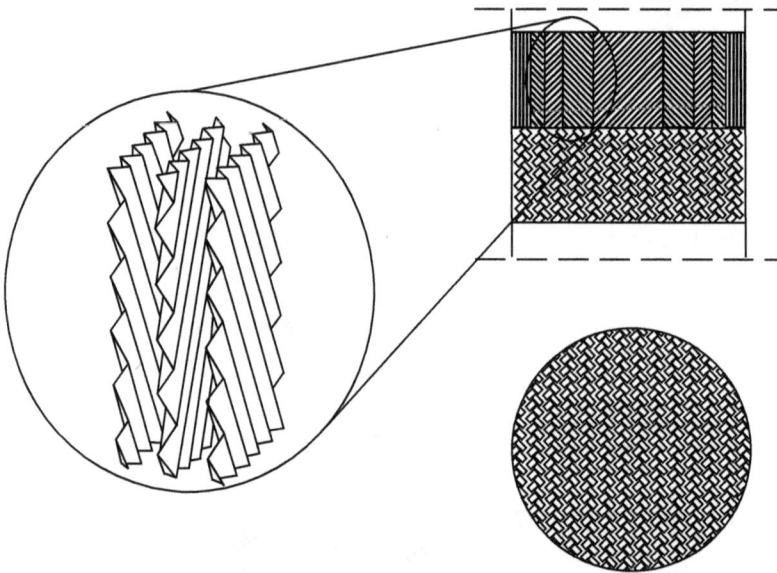

Fig. 6.26: Schematic of corrugated structured packing (adapted from [76]).

Obviously, the most important element of packed column internals is the packing itself. However, some supplementary elements, schematically depicted in Fig. 6.24, are necessary for proper column operation. Good contacting efficiency is only obtained with a uniform liquid distribution over the entire cross-sectional area. For this purpose, liquid distributors and redistributors are used. Redistributors are necessary to avoid the built up of a high degree of liquid maldistribution when the bed exceeds a height of 6 m. Other important supplementary internals are:

- Liquid collectors are installed for the withdrawal of side stream products, pump arounds and the collection of liquid before each liquid distributor
- Wall wiper to direct the liquid flowing along the wall back into the packing
- Support grid is installed to support the packing and the liquid holdup of the packing.
- Hold down plate has the primary function to prevent expansion of the packed bed as well as to maintain a horizontal bed level.
- Gas distributor is used to obtain a uniform gas flow across the column cross section.

A good packing design should have high capacity, high separation efficiency and large flexibility to gas and liquid throughput. Figure 6.27 shows a rough comparison of capacity and separation efficiency of several metal packings. Structured packings are generally superior to random packings in both capacity and separation efficiency. Modern random packings perform better than the standard Rashig and Pall ring packings. In packed columns the height equivalent to a theoretical plate (HETP) or height equivalent to a theoretical stage (HETS) concept is commonly used to represent the mass transfer efficiency and to allow comparison of tray and packed columns. For packed columns, the HETP is defined as the ratio of the height of the mass transfer zone containing packing, H_P, and the number of equilibrium stages N:

$$\text{HETP} = \frac{\text{total packed height}}{\text{number of equilibrium stages}} = \frac{H_P}{N} \tag{6.28}$$

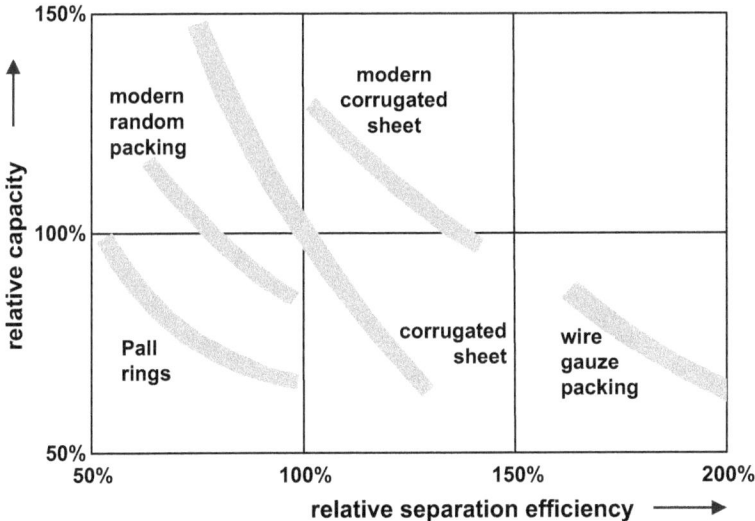

Fig. 6.27: Comparison of relative capacity and relative separation efficiency of various random and structured packings. (adapted from [76]).

For tray columns, the mass transfer zone is equal to the tray spacing H_{Tray} times the number of trays N_{act}, giving

$$\text{HETP} = \frac{H_{\text{Tray}} \times N_{\text{act}}}{N} = \frac{H_{\text{Tray}}}{E_o} \tag{6.29}$$

The feasible operating region of packed columns differs considerably from tray columns and is limited by flooding and wetting. As illustrated in Fig. 6.28, flooding sets the upper capacity limit. At the flood point the pressure drop of the gas flow through the bed increases so drastically that the liquid is no longer able to flow downward against the gas flow. Hence, the countercurrent flow of gas and liquid breaks down and the separation efficiency decreases dramatically. In contrast to tray columns, packed columns require a minimum liquid load to ensure sufficient mass transfer. Below this minimum value, only a very small part of the packing surface is wetted and liquid and gas are no longer in intimate contact. This results in a considerable drop of separation efficiency. The lower capacity limit of liquid load depends on the type of the packing, the quality of the packing supplements and the physical properties of the liquid.

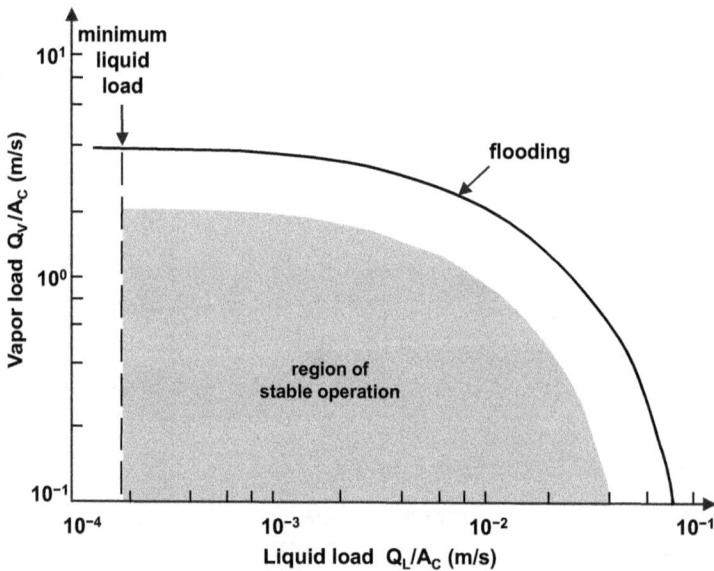

Fig. 6.28: Operating region of a packed column.

Proper choice of equipment is very important for effective and economical distillation. Tray columns are generally employed in large diameter (>1 m) towers. The gas load must be kept within a relatively narrow range, only valve trays allow greater operational flexibility. The liquid load can be varied over a wide range even down to very low liquid loads. For vacuum operation their relatively high pressure drop (typically 7 mbar per equilibrium stage) is a disadvantage. Tray columns have a relatively high liquid holdup compensating for fluctuations in feed compositions. However, the resulting high liquid residence time in the column may lead to the decomposition of thermally unstable substances. Tray columns are also relatively insensitive to impurities in the liquid.

Packed columns are used almost exclusively in small diameter (<1 m) towers. Only structured packings permit the use of packed columns with very large diameters. They are extremely flexible with regard to gas load, but require a minimum liquid load for packing wetting. Structured packings with an extremely small pressure loss, typically under 0.5 mbar per equilibrium stage, are a tremendous advantage in vacuum operation. The danger of decomposition of thermally unstable substances is also less in packed columns because of lower liquid holdup.

6.4 Polymer devolatilization

6.4.1 Introduction

In the production of most polymeric materials the removal of solvents, residual monomers and other volatile side products is an important operation that can contribute to manufacturing costs as much as the actual polymerization step or monomer costs. These low-molecular-weight components are often collectively referred to as volatiles and their presence in the final polymer product is usually undesired. Separating them from the polymer may be performed for several reasons:
– Improvement of polymer properties
– Monomer/solvent recovery
– Comply with health and environmental regulations
– Elimination of odors
– Increase the extend of polymerizations (polycondensation)

Since in most cases the contaminants are far more volatile than the polymer, they are removed by evaporation into a continuous gas phase. The process by which volatiles are separated from the bulk polymer is called devolatilization. Though devolatilization can be carried with the polymer in the solid state, it is more commonly performed in the molten phase with the polymer above its glass transition temperature or above the melting temperature for crystalline polymers. Devolatilization in the solid state behaves more like a drying process and is therefore not discussed in this chapter.

Melt devolatilization is often the intermediate step between polymerization and product finishing in which solvents and residual monomers are recovered. Separation is affected by pressure reduction, applying vacuum or by adding inert substances such as nitrogen gas or steam. Difficulties arise from the very viscous nature of the materials being handled and the severe heat and mass transfer limitations. Examples are the bulk polymerization process for polystyrene and solution polymerization processes for polyethylene and EPDM. To enhance end-product properties, devolatilization is also carried out in secondary processes such as compounding of plastics, or processes involving polymer modification. In these operations extrusion-type machinery fitted with venting capability are most commonly used.

6.4.2 Basic mechanisms

Devolatilization of molten polymers is a thermodynamically driven, mass transfer limited separation process. One of the basic parameters of interest is the maximum degree of separation that may be obtained. This is determined by the thermodynamic equilibrium relating the composition of the polymer to the partial pressure of the volatile solvent in the gas phase:

$$P_S = P_S^o \gamma_S \phi_S \tag{6.30}$$

where the solvent volume fraction ϕ_S is used for composition, P_{SO} is the saturated vapor pressure of the pure volatile solvent and γ_S is the solvent activity coefficient that comes from the Flory–Huggins theory:

$$\ln \gamma_S = (1 - \phi_S) + \chi (1 - \phi_S)^2 \tag{6.31}$$

In eq. (6.31), χ represents the Flory–Huggins binary interaction parameter. An example of the resulting pressure–composition diagram is shown in Fig. 6.29. For sufficiently small contaminant concentrations, these equations take the form of Henry's law:

$$P_S = K_S w_S \tag{6.32}$$

where

$$K_S = P_S^o \left(\frac{\rho_P}{\rho_S} \right) \exp(1 + \chi) \text{ and } \phi_S = \left(\frac{\rho_P}{\rho_S} \right) w_S$$

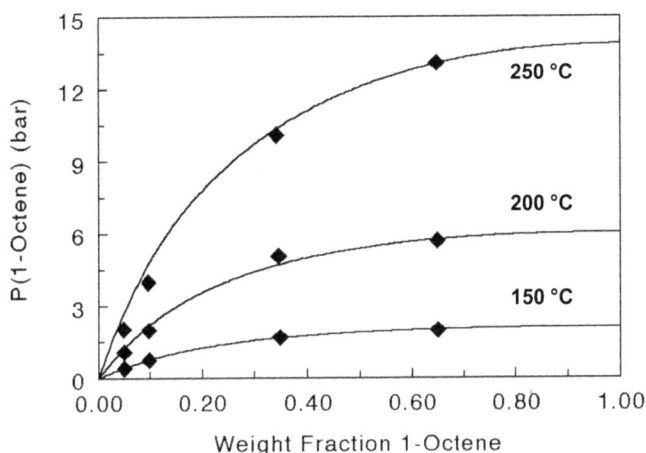

Fig. 6.29: Pressure–composition diagram for the LLDPE/1-octene system.

In actual devolatilization processes with finite residence times equilibrium is never attained. The overall mechanism is generally described by a combination of volatile transport to a polymer-vapor interface, evaporation of volatiles at the interface and their subsequent removal by a vacuum system. For most polymer-volatile systems, the time it takes to achieve a certain degree of separation is strongly dependent on the volatile migration rate through the polymer. This migration rate is relatively high in flash evaporation, where the molten feed stream contains large amounts of solvent. Application of a reduction in pressure such that the pressure is less than the vapor pressure of the solvent to the superheated polymer solution initiates boiling or foaming. As shown schematically in Fig. 6.30, these bubbles may then grow, coalesce and finally rupture at the polymer-vapor interface, where they release their volatile contents to the vapor phase. The bubble growth is determined by the diffusion rate of the volatile material from the polymer bulk to the bubble surface, and the resistance of the viscous polymer melts to displacement by the growing bubble. During flash evaporation the evaporation rate is frequently controlled by the rate of heat transfer through equipment walls to the polymer. This is because of the large amount of latent heat required and the notoriously poor polymer heat transfer coefficient.

At low volatile concentrations large amounts of latent heat are no longer required and the devolatilization rate becomes completely diffusion controlled. This is true for the migration of volatiles directly to the surface as well as to bubble nucleation sites and vapor bubbles that grow within the polymer melt. The diffusion in concentrated polymer solutions is several orders of magnitude smaller than in low-molecular-weight liquids. It can be strongly enhanced by increasing the temperature of the system. An additional reason for devolatizing at high temperatures is the increase in the vapor pressure of the volatile component.

Reduced pressure P < P$_s$

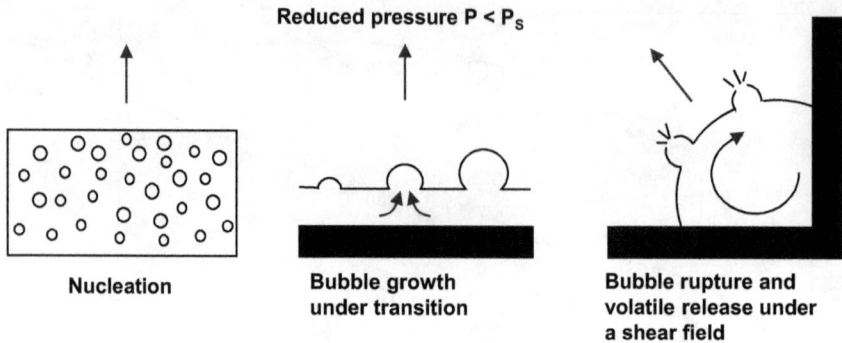

| Nucleation | Bubble growth under transition | Bubble rupture and volatile release under a shear field |

Fig. 6.30: Sequential process steps during bubble-assisted devolatilization (adapted from [78]).

6.4.3 Multistage operation and devolatilization aids

In the design of a separation system, one often has the knowledge of feed composition and the required residual volatile level in the exiting stream. Two extreme situations that require special consideration are large separation loads and/or stringent requirements for residues in the end product. When the total separation requirements exceed one order of magnitude, it is often beneficial to use several separation stages that operate at different absolute pressures, to achieve the desired separation. At the higher pressures the bulk of the volatiles are removed, thereby strongly reducing the demand on the high vacuum system. Finally the residual traces of the volatile component can then be removed at high vacuum with minimal load. The devolatilization efficiency in those final stages is commonly increased by the intentional addition of small amounts of a devolatilizing aid (stripping agent) to the polymer. Introduction of an inert substance enhances devolatilization in several ways:
- It reduces the partial pressure of the volatile component and therefore increases the concentration gradient for devolatilization
- The additivity of vapor pressures induces boiling/foaming in the polymer solution at a lower temperature and volatile concentration
- Boiling of the inert substance creates bubbles that increase the area available for mass transfer from the polymer to the vapor phase

Both staging and stripping agent addition are often combined in devolatizing extruders as shown in Fig. 6.31. In the backward and first forward devolatilization zones the polymer melt is degassed only by increased vacuum. The final two devolatilization zones use the addition of a stripping agent for enhanced efficiency. The stripping agent (water, nitrogen, carbon dioxide) should be of low cost, inert, easily recoverable and tolerable in trace quantities in the final product. Additional advantages of staging are the reduction of polymer entrainment arising from the high vapor velocities in the initial stages and opportunity to reheat the polymer between the stages to compensate the evaporative cooling.

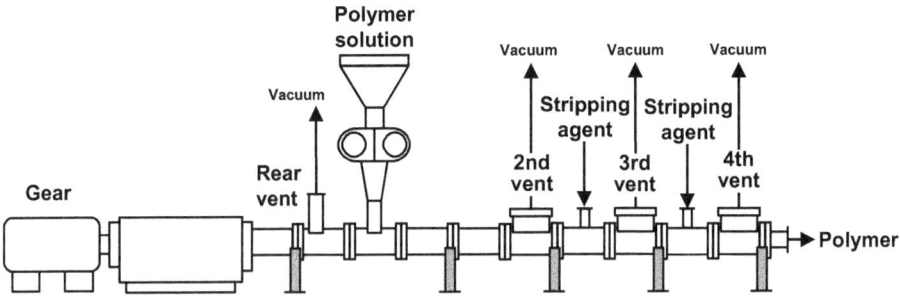

Fig. 6.31: Multistage devolatizing extruder installation.

6.4.4 Devolatilization equipment

A variety of commercial equipment is available for the industrial separation of volatile components from the base polymer matrix. The equipment may be broadly classified as nonrotating and rotating equipment. Some of the main units in each category are:

Nonrotating devolatizers
–   Flash evaporators
–   Falling-film devolatizers
–   Falling-strand devolatizers

Rotating devolatizers
–   Thin-film evaporators
–   Single-screw extruders
–   Double-screw extruders
–   Kneaders

In nonrotating equipment gravitational forces transport the polymer through the devolatilization zone. Thermal energy is usually introduced by heating the polymer melt under pressure in an external heat exchanger. In rotating equipment the melt is conveyed by its contact with moving elements and devolatilization is significantly enhanced because of mechanical agitation. Thermal energy is introduced into molten polymers by direct heat transfer of heat through equipment walls and transformation of mechanical energy (shaft work) into thermal energy via viscous dissipation. The mechanical introduction of heat is significant only when the polymer melt viscosity is sufficiently high. Accordingly, the viscosities that nonrotating devolatizers may handle are much lower than those processed in rotating equipment.

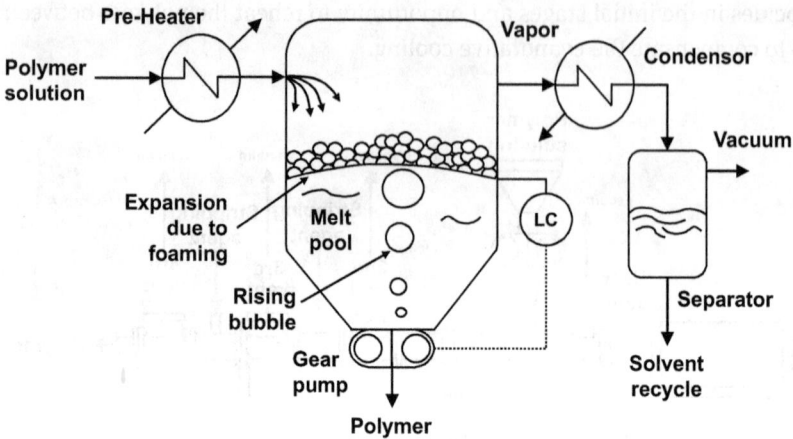

Fig. 6.32: Schematic of a flash evaporation devolatizer.

For polymer solutions with a high volatile concentration and relative low viscosity simple flash evaporation is the preferred process. In flash evaporators, schematic in Fig. 6.32, the liquid to be devolatilized is first heated in an external heat exchanger. The superheated solution under pressure is then flashed into an expansion tank where the pressure is relieved with an overhead vent for vapor removal. Because the sudden pressure reduction promotes foaming and the low solution viscosities, heat and mass transfer limitations are small and extensive surface renewal is not necessary. The heat of evaporation is obtained at the expense of the sensible heat of the feed solution. The devolatilized polymer melt is usually not very viscous and can be discharged from the evaporating chamber using a gear or screw pump. A specific configuration of a flash chamber is the falling strand devolatizer, shown in Fig. 6.33, where the material is extruded into a multitude of strands and allowed to fall by gravity, while flashing, into a melt pool. The free fall of strands allows easy disengagement of the already grown bubbles.

The equipment used to handle more viscous polymer solutions usually contains rotating parts that not only aid forward movement of the material but also impart surface renewal for heat and mass transfer. Thin-film evaporators are a modification of the falling-film evaporator. In the latter, a thin liquid film flows by gravity down the inner wall of a vertical cylinder and contacts a vapor stream that passes through the center of the cylinder. For polymer devolatilization, the equipment is modified to accommodate the higher viscosities of the liquid phase. Figure 6.34 shows a cross section of a vertical thin-film evaporator. The apparatus comprises of a vertical heated drum within which a mechanical agitator rotates. The incoming liquid is first distributed evenly along the inside of the heated body and flows downward by gravity. The revolving rotor comprises several blades that spread the downflowing solution into thin films. Thus, a film is deposited on the wall, which provides

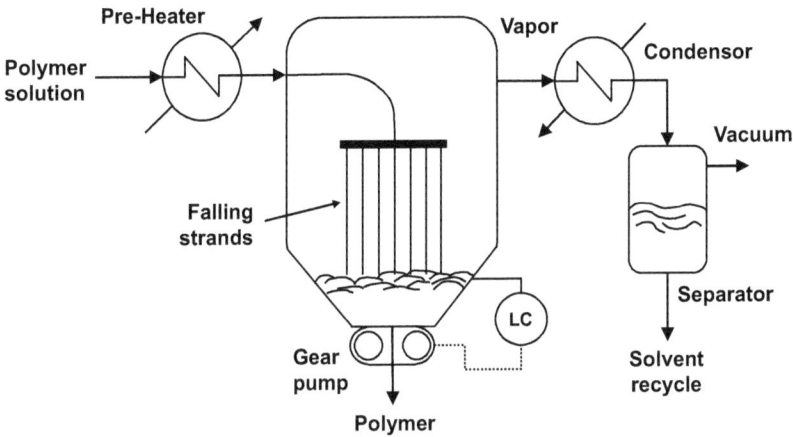

Fig. 6.33: Schematic of a falling strand devolatizer.

a large surface area for the evaporation of the contaminant. Evaporated vapors leave the solution through the core of the evaporator, either in a co-current fashion (high volatile content) or counter-current fashion (residual volatiles removal). For high viscosity fluids, the downward flow is aided by pitched instead of straight rotor blades that provide a downward motion to assist polymer flow.

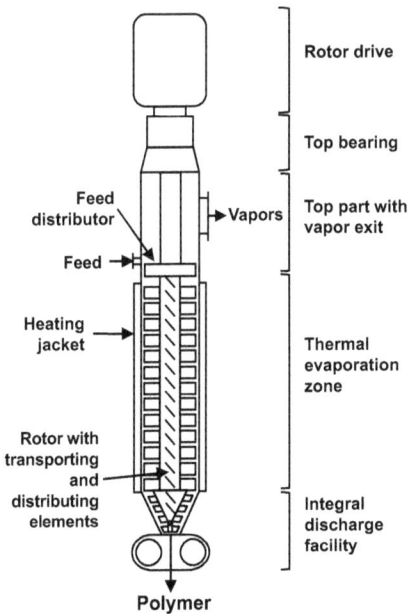

Fig. 6.34: Schematic of a thin-film evaporator.

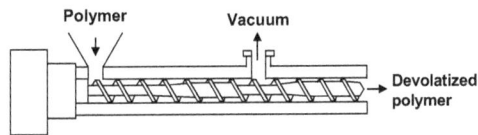

Fig. 6.35: Schematic of a devolatilization extruder.

For highly viscous, difficult to pump polymers vented, single-screw or multi-screw extruders are the commonly used devolatizers (Fig. 6.35). Extruders can efficiently handle fluids with viscosities varying over several orders of magnitude on the same shaft. They can be used stand-alone or in combination with other polymer processing functions. In the extruder the rotating screw conveys the polymer melt from the hopper to the exit by a combination of drag and pressure flow. Heat is introduced by heating through the barrel wall and generated by viscous dissipation. The vent zone of a screw extruder has deep channels to provide a pressure-free zone in which the polymer is exposed to lower pressures or vacuum. Devolatilization is considered to occur through two supporting parallel evaporation processes. This is illustrated in Fig. 6.36 that shows a schematic of this process. Firstly, the vented section of the screw channel is partially filled with rotating polymer melt. Secondly, the clearance between the barrel and the screw flight deposits a thin layer of polymer on the barrel as the screw flight rotates. As a result evaporation occurs either at the surface of the rotating bulk or at the surface of the deposited films. Both mechanisms are subject to surface renewal. The polymer melt pool through cross-channel, drag-induced flow, and the film via periodic introduction into the mixed pool. It is therefore no surprise that the separation efficiency is strongly dependent upon screw speed.

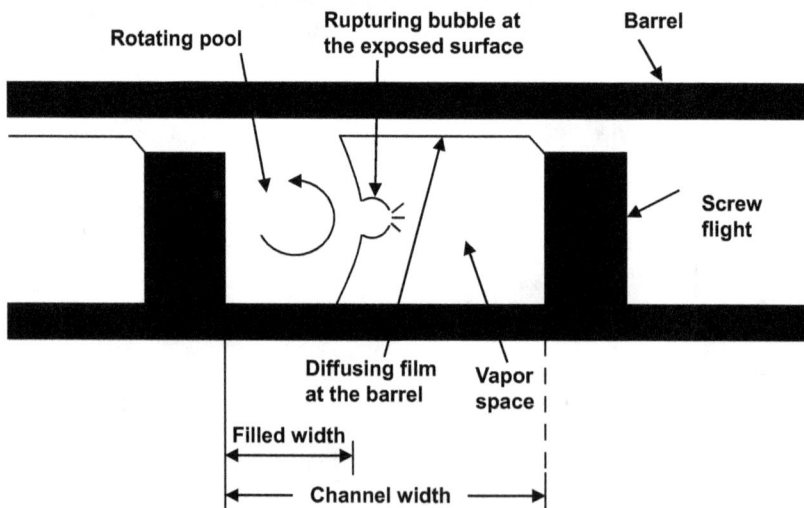

Fig. 6.36: Schematic of surface evaporation in a partially filled vented extruder channel (adapted from [78]).

Nomenclature

A	Interfacial area	$[m^2]$
B	Bottom product flow	$[mol/s]$
C_P	Specific heat capacity	$[J/kg\ K]$
D	Top product flow	$[mol/s]$
E_o	Overall plate efficiency	$[-]$
F	Feed flow	$[mol/s]$
H	Height	$[m]$
HETP	Height equivalent of a theoretical plate	$[-]$
HETS	Height equivalent of a theoretical stage	$[-]$
ΔH_{vap}	Molar heat of vaporization	$[J/mol]$
K_i	Distribution coefficient of component i	$[-]$
k	Mass transfer coefficient	$[m/s]$
L	Liquid flow (in rectifying section L', in stripping section L'')	$[mol/s]$
m	Mass rate	$[kg/s]$
N	Number of (theoretical) stages	$[-]$
p	Partial pressure	$[N/m^2]$
p^0	Saturation pressure	$[N/m^2]$
P_{tot}	Total pressure	$[N/m^2]$
Q	Heat flow	$[J/s^1]$
R	Reflux ratio L'/D	$[-]$
T	Temperature	$[K]$
V	Vapor flow (in rectifying section V', in stripping section V'')	$[mol/s]$
w	Weight fraction	$[-]$
x, y, z	Mole fraction (liquid, vapor, feed)	$[-]$
α_{ij}	Selectivity, relative volatility or equilibrium constant	$[-]$
χ	Flory–Huggins binary interaction parameter	$[-]$
ϕ	Volume fraction	$[-]$
φ	Mass transfer rate	$[kg/s]$
γ	Activity coefficient	$[-]$
ρ	Density	$[kg/m]$

Indices

1,2,*i,j*	Components
act	Actual
av	Average
B	Bottom
C	Condenser
D	Distillate
F	Feed
L	Liquid
lk	Light key
hk	Heavy key
MIN	Minimum
OV	Overall based on vapor phase
P	Packing
P	Polymer
R	Reboiler
S	Solvent
V	Vapor

7 Extraction and leaching

7.1 Liquid-liquid extraction

7.1.1 Introduction

Since the introduction of industrial liquid-liquid extraction processes, a large number of applications have been proposed and developed. An overview of some important industrial example applications is presented in Tab. 7.1. The first and until now largest volume industrial application of solvent extraction is in the petrochemical industry. Extraction processes are well suited to the petroleum industry because of the need to separate heat-sensitive liquid feeds according to chemical type (e.g. aliphatic, aromatic, naphthenic) rather than by molecular weight or vapor pressure. Other major applications include the purification of antibiotics and the recovery of vegetable oils from natural substrates. In metals processing the recovery of metals such as copper from acidic leach liquors and the refining of uranium, plutonium and other radioactive isotopes from spent fuel elements. Recently extraction is gaining increasing importance as a separation technique in biotechnology.

Tab. 7.1: Industrial liquid-liquid extraction processes.

Solute	Carrier	Solvents
Acetic acid	Water	Ethyl acetate, isopropyl acetate
Aromatics	Paraffins	Diethylene glycol, furfural, sulfolane, NMP, DMSO
Caprolactam	Aqueous ammonium sulfate	Benzene, toluene, chloroform
Benzoic acid	Water	Benzene
Formaldehyde	Water	Isopropyl ether
Phenol	Water	Benzene
Penicillin	Broth	Butyl acetate
Vanilla	Oxidized liquors	Toluene
Vitamins A, D	Fish liver oils	Liquid propane
Vitamin E	Vegetable oils	Liquid propane
Copper	Acidic leach liquors	Chelating agents in kerosene
Uranium	Acidic leach liquors	Tertiary amines in kerosene

Liquid-liquid extraction is based on the partial miscibility of liquids and used to separate a dissolved component from its solvent (carrier) by transfer to a second solvent. The principle of liquid-liquid extraction and its special terminology is illustrated in Fig. 7.1. In the simplest case the feed solution consists of the carrier solvent containing the desired solute. This liquid feed is contacted with a solvent, which is immiscible or only partly miscible with the liquid feed. In general the solvent has a higher affinity for the solute than the carrier and the solute is extracted into the

https://doi.org/10.1515/9783110712445-007

Feed flow, F		Solvent flow, S		Raffinate flow, R	Extract flow, E
Carrier A + Solute C	+	Solvent B	⇒	Carrier A + Solute C	Solvent B + Solute C

Phases in equilibrium, one ideal stage

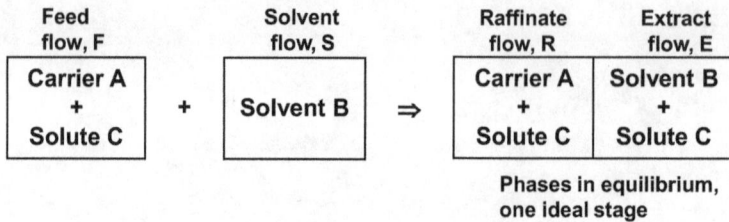

Fig. 7.1: Principle of a single extraction contacting stage.

solvent phase. For efficient contact a large interfacial area must be created across which the solute can transfer until equilibrium is closely approached. This is achieved by bringing the feed mixture and the solvent into intimate contact. When equilibrium is reached, the stage is defined as an ideal or theoretical stage and the equilibrium conditions can be expressed in terms of the extraction factor E for the solute:

$$E = \frac{\text{amount of solute in solvent phase}}{\text{amount of solute in carrier phase}} = K \frac{S}{F} \qquad (7.1)$$

where S and F are the solute-free solvent and feed streams, respectively, and K is the distribution coefficient of the solute between the solvent and the carrier. The larger the value of E, the greater the extent to which the solute is extracted. Large values of E result from large values of the distribution coefficient, K, or large solvent to carrier ratios. The resulting solute loaded solvent phase is called the extract, while the other liquid phase is designated the raffinate.

After extraction at least one distillation column (or other separation process) is required to separate the solvent from the extract and recycle the solvent. If the solvent is partially miscible with the feed, a second separation process (normally distillation) is required to recover solvent from raffinate. Figure 7.2 illustrates such an extraction system where two distillation columns are needed. This example is based on a high-boiling solvent, which is recovered as the bottom product of the distillation column. A low-boiling solvent would be recovered as the top product. Solvent recovery is an important factor in the economics of industrial extraction processes. Especially if the solvent and solute have close boiling points, the distillation column may require many trays and a high reflux ratio, resulting in a costly process.

The main disadvantage of extraction is that the necessity of a solvent increases the complexity and thereby costs of the process. Situations where extraction is preferred to distillation include the following application areas:

- Dissolved or complexed inorganic substances in organic or aqueous solutions
- Removal or recovery of components present in small concentrations
- When a high-boiling component is present in relatively small quantities in a waste steam

–   Recovery of heat-sensitive materials and low to moderate processing tempera-
    tures are needed
–   Separation of a mixture according to chemical type rather than relative volatility
–   Separation of close-melting or close-boiling liquids, when solubility differences
    can be exploited
–   Mixtures that form azeotropes or exhibit low relative volatilities and distillation
    cannot be used

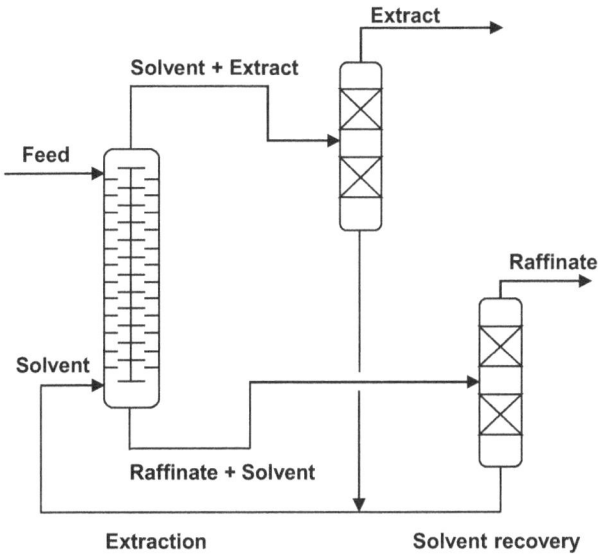

Fig. 7.2: Schematic of a liquid-liquid extraction process.

7.1.2 Liquid-liquid equilibria

Addition of a solvent to a binary mixture of the solute in a carrier can lead to the
formation of several mixture types. The simplest ternary system is the type I system
for one immiscible-liquid pair shown in Fig. 7.3. In such a system, the carrier and
the solvent are essentially immiscible, while the solute is miscible with the carrier as
well as the solvent. Such ternary systems are commonly represented in a triangular
diagram, showing the two-phase envelope that encloses the region of overall compositions in which two phases exist in equilibrium. The equilibrium compositions of the
extract and raffinate phases are connected by tie-lines that can be used to determine
distribution coefficients and selectivities. For each component the distribution coefficient is given by the ratio of the concentrations in the two phases as

$$K_i = \frac{x_{i,E}}{x_{i,R}} \tag{7.2}$$

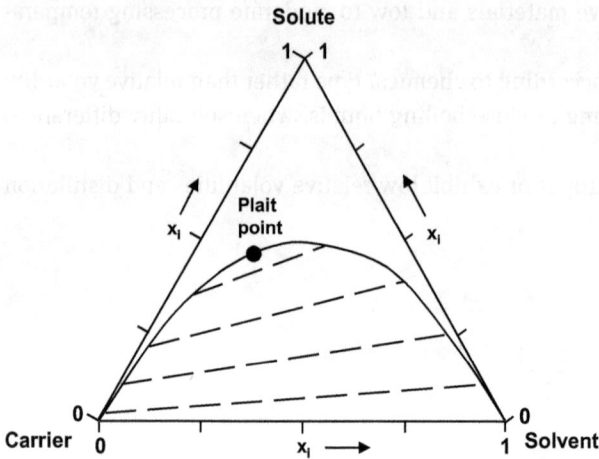

Fig. 7.3: Triangular diagram for a type 1 system with the tie-lines represented by the dashed lines.

where $x_{i,E}$ is the mole fraction of component i in the extract phase and $x_{i,R}$ is the mole fraction of component i in the raffinate phase. The selectivity is, equivalent to the relative volatility in distillation, defined as the ratio of K values:

$$\beta_{12} = \frac{K_1}{K_2} = \frac{\left(x_1/x_2\right)_{\text{EXTRACT}}}{\left(x_1/x_2\right)_{\text{RAFFINATE}}} \tag{7.3}$$

At equilibrium, thermodynamics requires the activity of each component to be the same in the two liquid phases. However, due to the strong thermodynamic nonideal behavior of the two liquid phases, the large difference in activity coefficients causes significant differences in equilibrium compositions:

$$\gamma_{i,E}\, x_{i,E} = \gamma_{i,R}\, x_{i,R} \tag{7.4}$$

Replacing the composition ratio with the ratio of activity coefficients gives for the selectivity:

$$\beta_{12} = \frac{\left(\gamma_2/\gamma_1\right)_{\text{EXTRACT}}}{\left(\gamma_2/\gamma_1\right)_{\text{RAFFINATE}}} \tag{7.5}$$

The required activity coefficients or distribution coefficients are generally calculated from thermodynamic models such as NRTL, UNIQUAC or predicted from group contribution methods like UNIFAC. In Fig. 7.3 it can be seen that as the concentration of the solute is increased, the tie-lines become shorter because of the increased mutual miscibility of the two phases. At the plait point the raffinate-phase and extract-phase boundary curves intersect and the selectivity becomes equal to one. An important additional use of the triangular diagram is the graphical solution of material balance problems, such as the calculation of the relative amounts of equilibrium phase obtained from a given overall mixture composition. Liquid-liquid equilibria having more than three components cannot, as a rule, be presented in a two-dimensional diagram.

7.1.3 Solvent selection

The key to an effective extraction process is the selection of the right solvent. A solvent or extractant may be a pure chemical compound or a mixture of chemical compounds. Proposing a solvent requires the knowledge of the physical and chemical properties of the solvent, solute, carrier and other constituents of the extraction system. While pure component properties are easily found in the literature, mixture physical properties are available in the literature for very few systems. The following criteria are important to consider during the selection of a solvent:

- Distribution coefficient – A high value of distribution coefficient indicates high affinity of the solvent for the solute and permits lower solvent/feed ratios
- Selectivity – A high value of the selectivity reduces the required number of equilibrium stages. If the feed is a complex mixture where multiple components need to be extracted from, selectivities between groups of components become important.
- Density – Higher density differences between extract and raffinate phases permits higher capacities in extraction devices using gravity for phase separation. The density difference must at least be large enough to ease the settling of the liquid phases.
- Viscosity – High viscosities lead to difficulties in pumping, dispersion, and reduce rate of mass transfer. Low viscosities benefit rapid settling and capacity.
- Solvent recoverability – Recovery of the solvent should be easy. A solvent which boils much higher than the solute generally leads to better results, though solvents boiling lower than the solute are also used commercially.
- Solubility of solvent – Mutual solubilities of carrier and solvent should be low to avoid an additional separation step for recovering solvent from the raffinate.
- Interfacial tension – Low interfacial tension facilitates the phase dispersion but may require large volumes for phase separation due to slow coalescence. High interfacial tension permits a rapid settling due to an easier coalescence, allowing higher capacities. A too low interfacial tension leads to emulsification.
- Availability and cost – Solvent cost may represent a large initial expense for charging the system, as well as a heavy continuing expense for replacing solvent losses. Therefore, one should make sure that the solvent of interest is commercially available and relatively inexpensive.
- Toxicity, compatibility and flammability – These criteria are important occupational health and safety considerations a suitable solvent has to meet.

Especially for food and pharmaceutical products only nontoxic solvents will be taken into consideration. In general, any hazard associated with the solvent will require extra safety measures and increases costs.

- Thermal and chemical stability – It is important that the solvent should be thermally and chemically stable because it is recycled. In particular, the solvent should resist breakdown during recovery in for example a distillation column.
- Corrosivity – Corrosive solvents can lead to increased equipment cost but might also require expensive pre- and posttreatment of streams.
- Environmental impact – The solvent should not only be compatible with the process, but also with the environment (minimal losses due to evaporation, solubility and entrainment).

In addition to being nontoxic, inexpensive and easily recoverable, a good solvent should be relatively immiscible with feed components other than the solute and have a different density from the feed to facilitate phase separation. Also, it must have a very high affinity for the solute, from which it should be easily separated by distillation, crystallization or other means. Obviously no solvent will be best from all of these viewpoints, and the selection of a desirable solvent involves compromises between the various criteria. In this selection process some of the criteria are essential for the separation while others are desirable properties that will improve the separation and/or make it more economical. The most important compromise is always between solute solubility and selectivity that usually behave exactly opposite. When a high selectivity is obtained, solubilities are usually low and the other way around. In most cases, the selectivity is the most important parameter, since this determines whether a certain separation can be accomplished.

7.1.4 Extraction schemes

In the simplest extraction scheme a feed and solvent are contacted as shown in Fig. 7.1. In the case that the solvent and the carrier are completely immiscible and the solvent contains no solute, the material balance over the solute becomes

$$x_F F = x_E E + x_R R \tag{7.6}$$

which can be rearranged by elimination of $x_E = K x_R$ to give

$$\frac{x_R}{x_F} = \frac{F}{F + KS} = \frac{1}{1 + E} \tag{7.7}$$

The main disadvantage of single stage contacting is that residual solute is left behind in the carrier. Therefore often a series of contacting stages is arranged in a cascade to accomplish a separation that cannot be achieved in a single stage and/or reduce the required amount of the mass-separating agent. Multistage liquid-liquid extraction cascades can be arranged in a cocurrent, cross-current and countercurrent arrangement shown in Fig. 7.4. When the stages are ideal, cocurrent stage wise contact is not necessary because equilibrium is reached between the streams after the first stage.

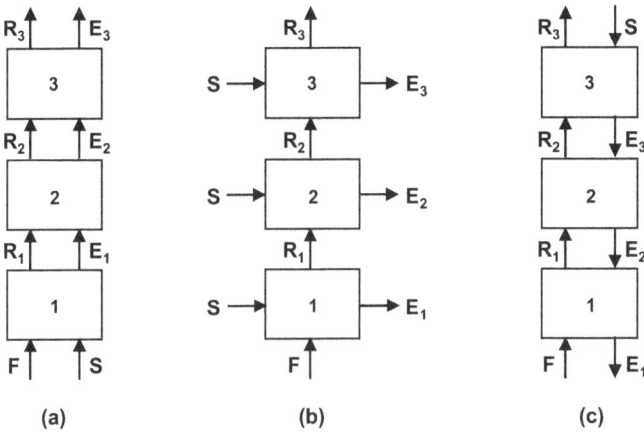

Fig. 7.4: Schematic of cocurrent (a), cross-current (b) and countercurrent (c) extraction.

A cross-current cascade in which the fresh solvent distributed over multiple stages gives an improvement over the separation obtainable in a single stage for a given ratio of solvent to feed. Since the total amount of solvent is distributed over N stages, the extraction factor for each stage becomes E/N, giving for the raffinate concentration after N stages:

$$\frac{x_N}{x_F} = \frac{1}{(1 + E/N)^N} \tag{7.8}$$

Thus, unlike the cocurrent cascade, the value of x_N decreases in each successive stage. The best compromise between the objectives of high extract concentration and a high degree of extraction of the solute is the countercurrent arrangement. The feed entering stage 1 is brought into contact with a solute-rich stream which has already passed through the other stages, while the raffinate leaving the last stage has been in contact with fresh solvent. Because of economic advantages, continuous countercurrent extraction is normally preferred for commercial scale operations. For a countercurrent operation the raffinate concentration after N stages is given by

$$\frac{x_N}{x_F} = \frac{1}{\sum\limits_{n=0}^{N} E^n} \tag{7.9}$$

As with the cross-current arrangement, the value of x_N decreases in each successive stage. The decrease for the countercurrent arrangement is larger than for the cross-current arrangement, and the difference increases exponentially with increasing extraction factor, E. Therefore the countercurrent cascade is the most efficient of the three linear cascades. In the case that the solvents are substantially immiscible and the distribution coefficient remains constant, the number of theoretical stages, N_S, for countercurrent contact can be calculated from the Kremser equation:

$$N_S = \frac{\ln \left[\left(\dfrac{x_F - x_E/K}{x_N - x_E/K} \right) \left(1 - \dfrac{1}{E} \right) + \dfrac{1}{E} \right]}{\ln E} \tag{7.10}$$

7.2 Industrial liquid-liquid extractors

Extractors are usually classified according to the methods applied for interdispersing the phases and producing the countercurrent flow pattern. To maximize the dispersion of one phase in the other, and to minimize back-mixing, extractors are equipped with trays, packings or mechanical moving internals. The location of the principal interface depends upon which phase is dispersed. When the light phase is dispersed, the interface is located at the top of the extractor. When the heavy phase is dispersed, the interface is located at the bottom. The solvent can be the heavy or light phase, and dispersed or continuous. Usually the phase that is fed at the highest rate is the dispersed phase.

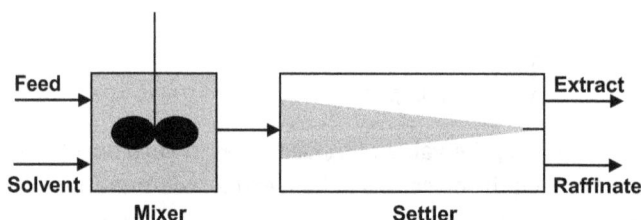

Fig. 7.5: Mixer settler.

7.2.1 Mixer settlers

As shown in Fig. 7.5, mixer-settler systems involve a mixing vessel for phase dispersion that is followed by a settling vessel for phase separation. They are widely used in the chemical process industry because of reliability, flexibility and high capacity. Although any number of mixer-settler units may be connected together, these extractors are particularly economical for operations that require high throughput and few stages. Dispersion can be achieved by pump circulation, air agitation or mechanical stirring. Intense agitation in the dispersion vessel leads to high rates of mass transfer and close approach to equilibrium. However, because the resulting dispersion can be difficult to separate, designs of mixer-settler systems must be carefully balanced between dispersion intensity and time of settling. Scale-up and design of mixer settlers is relatively reliable because they are practically free of interstage back-mixing and stage efficiencies are high. The main disadvantages of mixer settlers are high capital cost per stage and large inventory of material in the vessels. In large industrial units, the settlers usually represent at least 75% of the total volume.

7.2.2 Mechanically agitated columns

Many modern differential contactors employ rotary agitation for phase dispersion. The best known commercial rotary agitated contactors are shown in Fig. 7.6. The Scheibel column is designed to simulate a series of mixer-settler extraction units. An impeller agitates every alternate compartment with self-contained mesh-type coalescers between each contacting stage. The rotating-disk contactor (RDC) uses the shearing action of a rapidly rotating disk to interdisperse the phases. RDC's have been used widely throughout the world for propane deasphalting, sulfolane extraction for aromatics/aliphatics separation and caprolactam purification. The Oldshue–Rushton column consists essentially of a number of compartments separated by horizontal stator-ring baffles. Kuhni contactors are similar to the Scheibel columns and have gained considerable commercial application. A baffled turbine impeller promotes radial discharge within a compartment. The principal features are the use of a shrouded impeller to promote radial discharge within the compartments.

A more energy-efficient way to obtain phase dispersion in a column are reciprocating or vibrating of plates in a column. Reciprocating of plates requires less energy and has the same effect in terms of mixing patterns and uniform dispersion. The difference between the different reciprocating-plate columns that have been built for industrial use lies in the plate design. The open-type Karr column is the best known. This type of column has gained increasing industrial application in the petrochemical, pharmaceutical and hydrometallurgical industries.

7.2.3 Unagitated and pulsed columns

Despite their low efficiency, unagitated columns are widely used in industry because of their simplicity and low cost. They are particularly suited for processes requiring few theoretical stages and for corrosive systems where the absence of mechanical moving parts is advantageous. The three main types of unagitated column extractors are shown in Fig. 7.7. Spray columns are the simplest in construction but suffer from

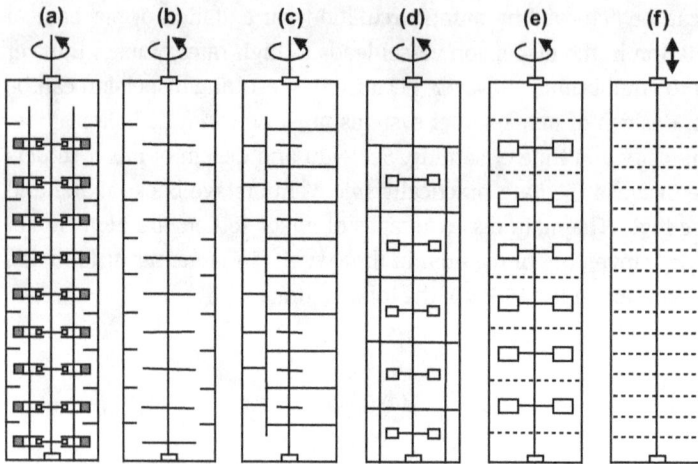

Fig. 7.6: Mechanically agitated columns: Scheibel column (a), rotating disk contactor (b), asymmetric rotating disk contactor (c), Oldshue–Rushton multiple mixer column (d), Kuhni column (e) and Karr reciprocating plate column (f).

very low efficiency because of poor phase contacting and excessive back-mixing in the continuous phase. They generally provide no more than the equivalent of one or two equilibrium stages and are typically used for basic operations, such as washing and neutralization.

Packed columns have better efficiency because of improved contacting and reduced back-mixing. The used packings in extraction are similar to the ones used in distillation and include random and structured packings. It is important that the packing material should be wetted by the continuous phase to avoid coalescence of the dispersed phase. The main functions of the packing elements are to reduce back-mixing in the continuous phase and promote mass transfer due to jostling and breakup of dispersed phase drops. Because the packing elements reduce the cross-sectional area for flow and decrease the velocity of the dispersed phase, the column diameter for a given rate will always be greater than for a spray tower. However, a packed column is commonly preferred over a spray column because the reduced flow capacity is less important than the improved mass transfer.

The sieve tray extractor resembles sieve tray distillation. If the light phase is dispersed, the light liquid flows through the perforations of each plate and is dispersed into drops, which rise through the continuous phase. The continuous phase flows horizontally across each plate and passes to the plate beneath through a downcomer. If the heavy phase is dispersed, the column is reversed and upcomers are used for the continuous phase. Perforated plate columns are operated semi-stage-wise and are reasonably flexible and efficient.

An increased efficiency of sieve-plate and packed extraction columns is obtained by applying a sinusoidal pulsation to the contents of the column. The well-

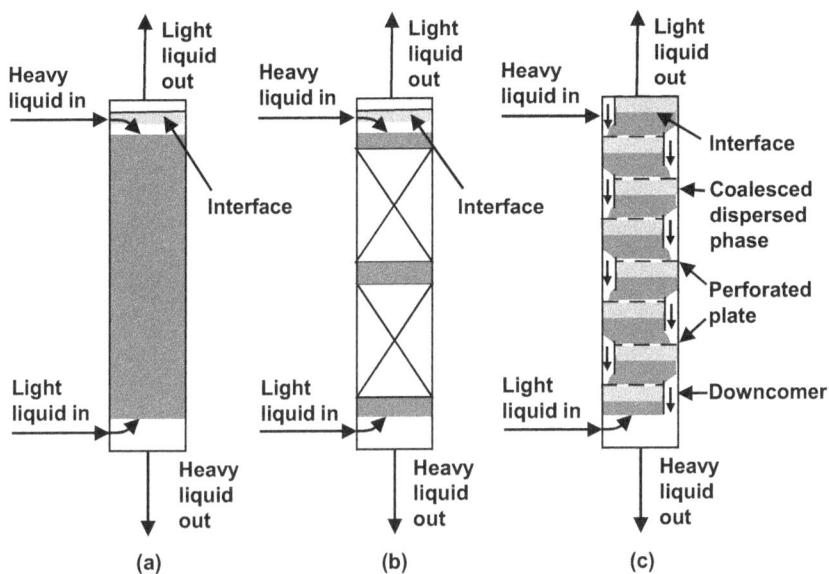

Fig. 7.7: Unagitated column extractors: spray column (a), packed column (b) and perforated plate column (c).

distributed turbulence promotes dispersion and mass transfer while tending to reduce axial dispersion in comparison with the unpulsed column. A pulsed-plate column is fitted with horizontal perforated plates that occupy the entire cross section of the column. The total free area of the plate is about 20–25%. Pulsed-packed columns (Fig. 7.8) contain random or structured packing. The light and dense liquids passing countercurrently through the columns are acted on by pulsations transmitted hydraulically to form a dispersion of drops. The pulsation device is connected to the side of the column, usually at the base, through a pulse leg. Typical operating conditions are frequencies of 1.5–4 Hz with amplitudes of 0.6–2.5 cm.

7.2.4 Centrifugal extractors

In centrifugal extractors centrifugal forces are applied to reduce the contact time between the phases and accelerate phase separation. The units are compact and a relatively high throughput per unit volume can be achieved. Centrifugal extractors are particularly useful for chemically unstable systems such as the extraction of antibiotics or for systems in which the phases are slow to settle. Advantages include short contact time for unstable materials, low space requirement and easy handling of emulsifed materials or fluids with small density differences. The disadvantages are complexity

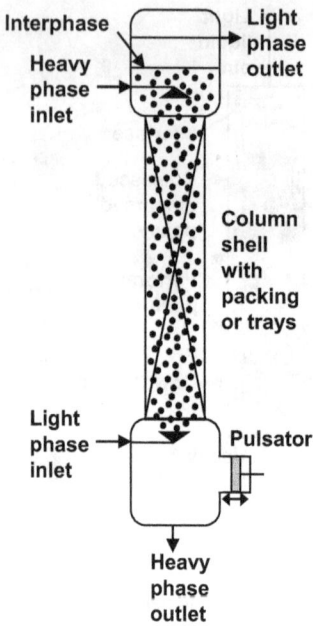

Fig. 7.8: Pulsed packed column.

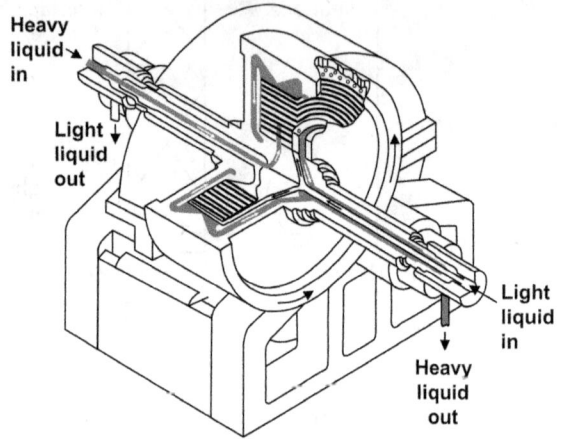

Fig. 7.9: Podbielniak centrifugal extractor.

and high capital and operating costs. Centrifugal extractors have been widely used in the pharmaceutical industry and are increasingly used in other fields.

Figure 7.9 shows a schematic of the first differential centrifugal extractor used in industry. The Podbielniak extractor can be regarded as a perforated-plate column wrapped around a rotor shaft. This extractor consists of a drum rotating around a shaft equipped with annular passages at each end for feed and raffinate. The light phase is introduced under pressure through the shaft and then routed to the periphery of the drum. The heavy phase is also fed through the shaft but is channeled to the center of the drum. Centrifugal forces acting on the phase-density difference causes dispersion as the phases are forced through the perforations. Other manufacturers of centrifugal extractors are Robatel, Westfalia, Alfa-Laval and Cinc.

7.2.5 Selection of an extractor

The large number of extractors available and the number of design variables complicate selection of a contactor for a specific application. Some important criteria that should be taken into consideration during contactor selection are:
- Stability and residence time
- Settling characteristics of the solvent system
- Number of stages required

- Capital cost and maintenance
- Available space and building height
- Throughput
- Experience with the type of extractor

The preliminary choice of an extractor for a specific process is primarily based on consideration of the system properties and number of stages required for the extraction. Choosing a contactor is still both an art and a science. Although cost ought to be a major balancing consideration, in many actual cases experience and practice are the deciding factors. Mechanically agitated devices show some efficiency advantages but whether these advantages come at the expense of higher cost must be evaluated by designing for the specific system at hand.

Tab. 7.2: Some industrial solid-liquid extraction processes.

Solute	Solid	Solvents
Soluble coffee	Roasted beans	Water
Caffeine	Green beans	Water, methylene chloride
Fish oil	Fish scraps	Hexane, butanol
Vegetable oil	Seeds, beans	Hexane, acetone
Spice extracts	Leaves, etc.	Ethanol
Sugar	Sugar beets	Water
Quinine	Rind of quinine tree	Toluene
Gold	Ore	Aqueous sodium cyanide
Copper	Ore	Aqueous sulfuric acid or ammoniacal solutions

7.3 Leaching

Many natural substances occur in a mixture of components dispersed through a solid material. The separation of a desired soluble constituent from its solid by means of a solvent is referred to as leaching. Originally the term leaching was used to describe the process of percolating a liquid through a solid material. During those days, leaching was carried out mainly as a batch process but nowadays many continuous plants have been developed.

Liquid-solid extraction covers a variety of processes such as the extraction of oil from seeds with hexane and sugar from beet or coffee from ground roasted coffee beans with hot water. Leaching is often accompanied by a chemical reaction, which converts the extract into soluble form. Examples are the hydrolysis of wood and many leaching processes used in metallurgy and the pigment industries. A list of industrial leaching operations is given in Tab. 7.2.

7.3.1 Mechanism and process of leaching

If the solute is uniformly dispersed in the solid, the material close to the surface dissolves first, leaving a porous structure in the solid residue. The solvent will then have to penetrate this outer porous region before it can reach further solute. The process becomes progressively more difficult and the rate of extraction will decrease. If the solute forms a large proportion of the volume of the original particle, its removal can destroy the structure of the particle that may crumble away. In such cases the solvent easily accesses further solute and the extraction rate does not fall as rapidly. Sometimes the soluble material is distributed in small isolated pockets in a material that is impermeable to the solvent. An example is gold dispersed in rock that is crushed to expose all the soluble material to the solvent. The limiting extraction rate is influenced by four important factors:

- Solid particle size – A smaller particle size increases the relative interfacial area between the solid and liquid and the extraction rate. However, smaller particle sizes tend to lead to lower drainage rates from the solid residue and can create problems in the solids flow through countercurrent extraction equipment.
- Solvent – The liquid chosen should be selective and its viscosity should be sufficiently low to allow free circulation.
- Temperature – In general the solute solubility and rate of extraction increase with temperature.
- Agitation of fluid – Agitation of the solvent improves the mass transfer of material form the surface of the particles to the bulk of the solution.

Traditionally that part of the solvent retained by the solid is referred to as the underflow or holdup, whereas the solid-free solute-rich solvent separated from the solid after extraction is called the overflow. The solid can be contacted with the solvent in a number of different ways. Countercurrent extraction offers the most economical use of solvent, permitting high concentrations in the final extract and high recovery from the initial solid. In a multistage operation fresh solid enters the first stage and fresh solvent enters the final stage. The solvent is gradually enriched in solute until it leaves the extraction battery as overflow of from the first stage. The operation is usually discontinuous in that the solvent is pumped from one vessel to the next intermittently and allowed to remain until equilibrium extraction is approached. A complete solid-liquid extraction process consists of the following stages:

- Preparation of the feed material in such a way that the solvent can dissolve the extract quickly. This is achieved by crushing, grinding or flaking.
- Contact of liquid solvent with the solid to effect transfer of solute to the solvent
- Separation of resulting solution from the residual solid
- Recovery of the extracted solute from the solvent
- Separation of solvent residues from the extracted solid

In leaching processes special attention is needed for solvent regeneration and for solute recovery. Solvent recovery is often energy intensive and a full process energy analysis is recommended to reduce costs. Recovery of organic solvents from the exhausted solids is also important and can be more troublesome than recovery from a liquid.

7.3.2 Solid-liquid extractors

Extractors often contribute substantially to the capital and operating costs of a plant. They use either percolation or agitation to ensure intimate contact between the solids and solvent. In a percolation system extraction rate needs to be high, as the solvent residence time is often relatively short. A percolation process can be carried out either in stage wise or in a differential contactor. For an immersion process stage wise contact is often more practicable, especially when a low extraction rate requires a long residence time or multiple contact with the solvent.

7.3.2.1 Batch extractors

Batch extraction is based on the principle of displacement or enrichment. In the displacement method a given quantity of fresh solvent is added to the feed material. After thorough mixing, the extraction residue is allowed to settle and the loaded solvent (extract) is drawn off. The procedure is repeated until the required degree of extraction has been achieved. This system, known as cross-flow extraction, is normally carried out in agitated vessels, as shown in Fig. 7.10. Filling and emptying take place through the top of the vessel. After extraction, the solvent is distilled of from the residue by direct heating of the extractor.

In the enrichment method several batch extractors are connected to form a battery operating on a cyclic basis. The vessels are charged with the solids to be extracted and the solvent is passed countercurrently through the extraction system. The fresh solvent flows into the extractor containing the fully extracted material, and becomes increasingly enriched with extract as it flows through the following stages. In the cyclic operation, the most exhausted diffuser is bypassed and emptied, and an empty one is charged with fresh solids. This plant layout, schematically shown in Fig. 7.11, is used for the extraction of coffee solubles using hot water.

7.3.2.2 Continuous extractors

Continuous extractors are available in a variety of forms. The main difference between them is the way by which the solids are transported through the equipment. Only a few characteristic types are described here. Moving bed percolation systems are used for extraction from many types of cellular particles such as seeds, beans and peanuts. An example is the Bollman extractor shown in Fig. 7.12. The solids are

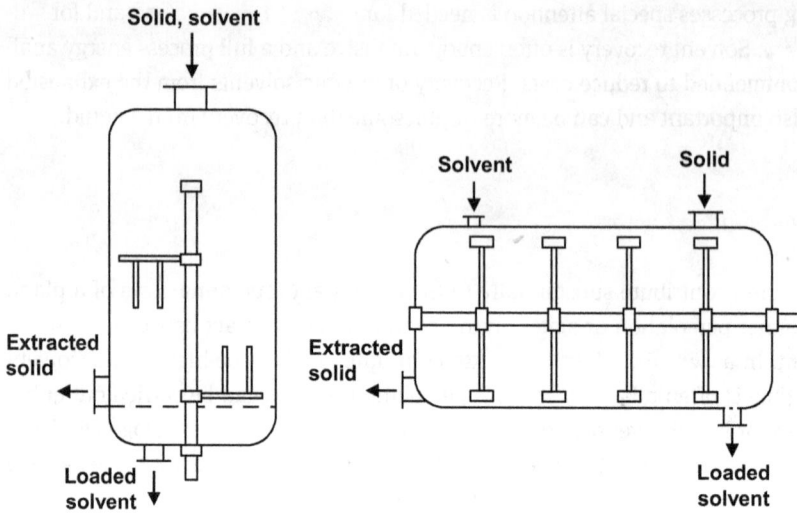

Fig. 7.10: Vertical and horizontal agitated extraction vessels.

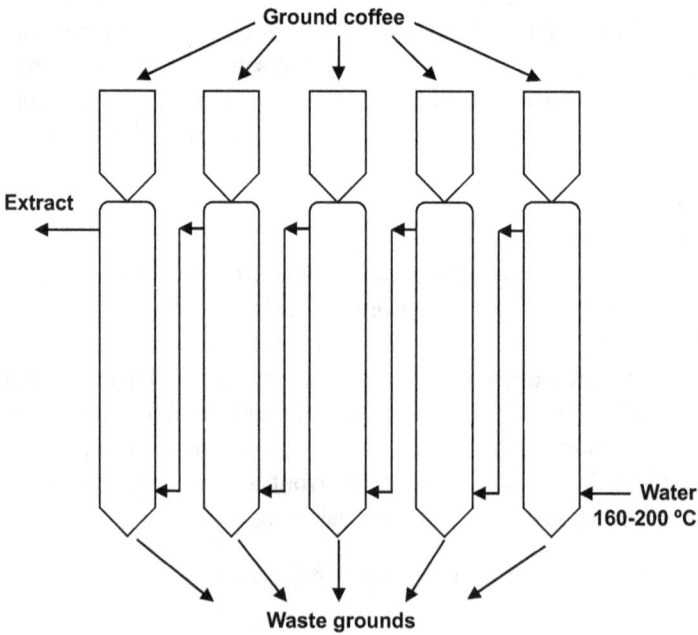

Fig. 7.11: Batch percolation battery for soluble coffee extraction.

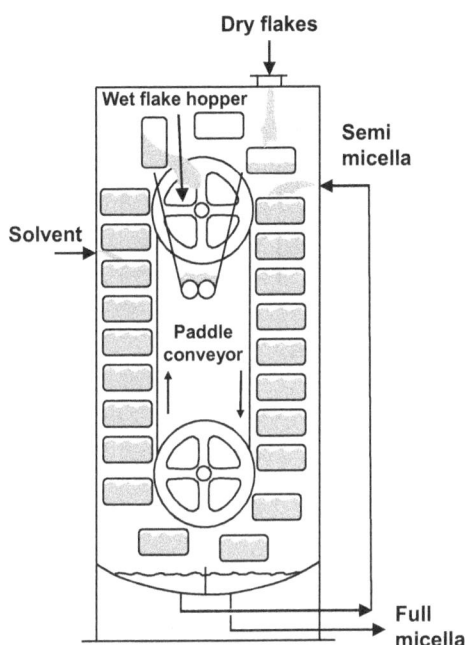

Fig. 7.12: The Bollman extractor.

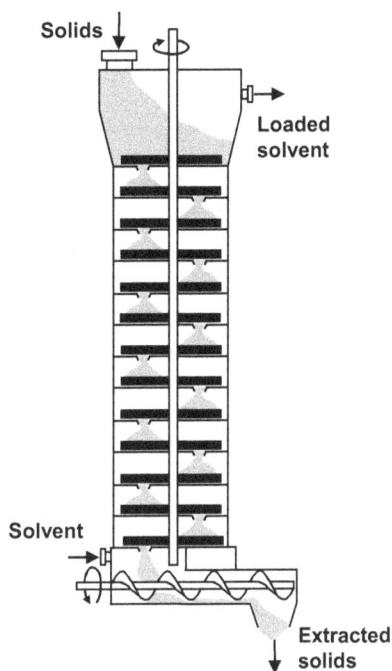

Fig. 7.13: The Bonotto extractor.

loaded into perforated baskets fixed to a chain conveyor in a closed vessel as in a bucket elevator. Solid is fed to the top basket on the downward side of the conveyor and is discharged from the top basket on the upward side. Fresh solvent is sprayed on the solid about to be discharged and passes downward through the baskets to effect a countercurrent flow. An alternative tower design, the Bonotto extractor, consists of a tall cylindrical vessel with a series of slowly rotating horizontal trays (Fig. 7.13). The solid is fed continuously on the top tray and a stationary scraper moves it toward the center of the tray. The solid then falls through an opening onto the tray beneath, where another scraper moves the solid across the tray in the opposite direction. The solvent is fed to the bottom of the vessel and flows upward to give a flow countercurrent to the solids flow direction.

Endless belt percolation extractors are similar to a belt filter. Figure 7.14 shows that they are fitted with a slow-moving perforated belt on which the solid is fed from a hopper. Fresh solvent is sprayed onto the bed close to the discharge end of the belt. The collected solvent is sprayed back onto the bed at a point closer to the solids feed end of the belt. This process is repeated to achieve extraction with a countercurrent flow. The extraction and percolation rates determine the belt speed, amount of drainage area, and hence the length of the belt required.

Immersion extraction systems are useful in handling finely ground material or when the percolation rate through the material to be extracted is too rapid to allow

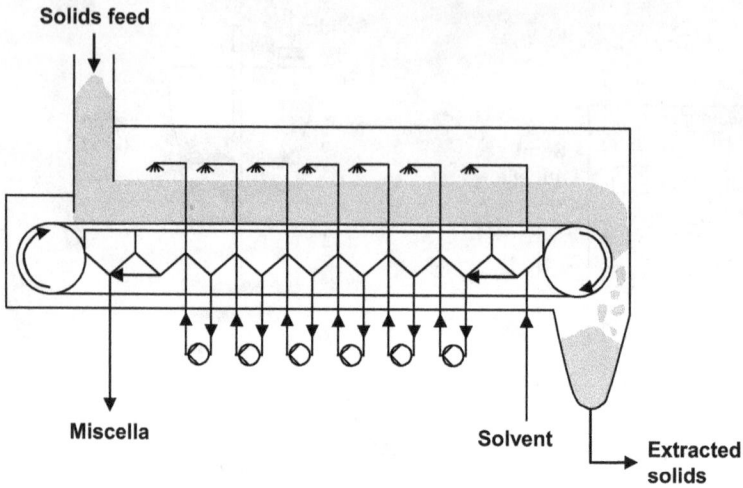

Fig. 7.14: Schematic of the belt-type extractor.

effective extraction from the solids. They have been made continuous through the in-clusion of screw conveyors to transport the solids. A relative simple version of these machines is the De Danske Sukkerfabriker diffuser extractor shown in Fig. 7.15. A double screw is used to transport the solids up the gradient of the shell, while the solvent flows countercurrently down the gradient.

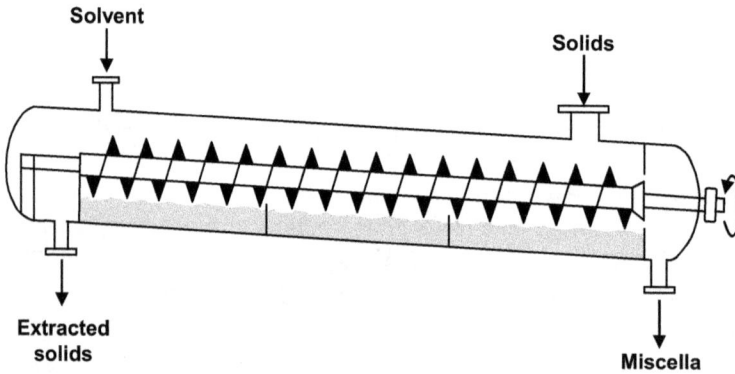

Fig. 7.15: The "De Danske Sukkerfabriker" extractor.

7.4 Supercritical extraction

7.4.1 Introduction

Supercritical extraction is a separation process that applies a supercritical fluid as a separating agent in the same way as for instance, liquid solvents are used in extraction or absorption. The supercritical region of a pure fluid is defined as the area above the critical pressure P_c and critical temperature T_c, as shown by the cross-hatched area in Fig. 7.16. In this area, the operating temperature is above the critical temperature and the operating pressure above the critical pressure of the solvent. The region in the immediate vicinity of the critical point is called the critical fluid region and the region just under the critical point is the near-critical region.

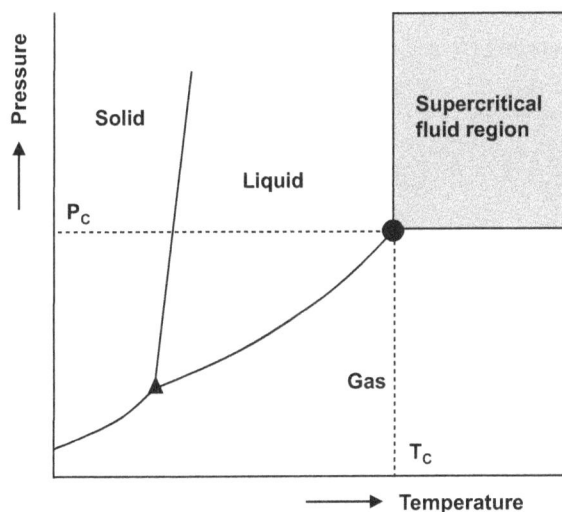

Fig. 7.16: Schematic pressure–temperature diagram for a pure substance showing the supercritical fluid region.

Supercritical fluids may be used in the same way as other ordinary solvents, although their somewhat different behavior can be advantageous to take into account. The main difference is that above the critical temperature the discontinuity between liquid and vapor disappears, enabling a continuous transition by isothermal compression form a dilute phase to a dense phase, without passing through a two-phase region.

7.4.2 Properties of supercritical fluids

The solvent strength of a given supercritical compressed fluid is related directly to the fluid density and extremely sensitive to pressure and temperature near the critical point. This effect of pressure on the solubility in the vicinity of the critical point can be explained by considering the density behavior of a pure solvent, shown in Fig. 7.17. For reduced temperatures (T/T_c) ranging from 1.0 to 1.2, small changes in reduced pressure (P/P_c) can change the reduced density (ρ/ρ_c) of a solvent from about 0.1, a gas-like density, to about 2.5, a liquid-like density. This means that by adjusting the pressure and/or temperature the properties of a supercritical solvent can be changed from a gas, with little solvating power, to a liquid, with good solvating power. At higher reduced temperatures, $T_r > 1.5$, the same variation in density can only be achieved by a much larger increase in pressure. Similarly a significant increase in density can be obtained with only a small decrease in temperature for a reduced pressure P_r close to unity, while for $P_r = 7$ the effect of temperature on the density is much smaller.

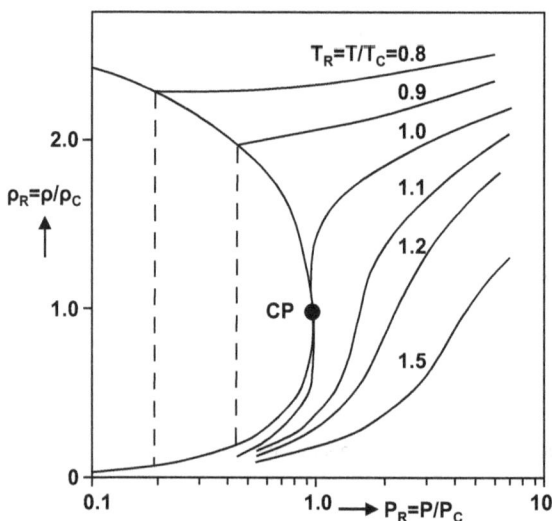

Fig. 7.17: Variation of the reduced density of a pure component in the vicinity of its critical point (CP).

Thus, solvent strength may be manipulated over a wide range by making small changes in temperature and pressure. Based on this principle, there are three possible operation modes of supercritical extraction processes. These operation modes are illustrated in Fig. 7.18, showing the solubility of naphthalene in supercritical carbon dioxide as a function of pressure and temperature. Depending upon the pressure level, the temperature is seen to affect the naphthalene solubility quite differently. For example, at 300 atm an increase in the temperature increases the solubility of naphthalene in carbon dioxide, whereas at a low pressure of 80 or 90 atm an increase in temperature decreases the naphthalene solubility. Most commonly used is the combination of extraction at elevated pressure with separation of solvent and extract by pressure reduction (a). At very high operating pressures this configuration becomes uneconomical due to the large recompression costs. In that case isobaric cooling of the solvent (b) will separate the extract because the solubility depends mainly on the vapor pressures of the solutes. An opposite effect of temperature on the solubility occurs when operating close to the critical pressure of the solvent. Isobaric heating of the solvent (c) will reduce the solubility as a result of a large decrease in solvent density.

Tab. 7.3: Comparison of some physical properties for a gas, supercritical solvent and a liquid.

	ρ (kg/m^3)	η (mPa.s)	D (m^2/s)
Gas (1 bar, 20 °C)	0.6–2.0	0.01–0.03	$1–4 \times 10^{-5}$
Supercritical solvent			
T_C, P_C	200–500	0.01–0.03	$7\ 10^{-7}$
$T_C, 4P_C$	400–900	0.03–0.09	$2\ 10^{-7}$
Liquid (20 °C)	600–1,600	0.2–3.0	$0.2–2 \times 10^{-9}$

Table 7.3 compares some physical properties of a gas, a supercritical solvent and a liquid. It is seen that although a supercritical solvent has a liquid like density, the viscosity is more like that of a gas, resulting in diffusion coefficients that are much higher than diffusion coefficients in liquids. This makes the supercritical fluid a very mobile phase, capable of faster and better penetration into a solid matrix containing substances to be extracted.

A list of the critical temperatures, pressures and densities of some gases and liquids is given in Tab. 7.4. The solvents of interest for most of the practical applications have moderate critical temperatures ranging from 0 to 100 °C. For large-scale applications propane, ethane and carbon dioxide are the most suitable solvents because they are cheap, readily available and nontoxic. It is clear that carbon dioxide is the most attractive solvent because it is nonflammable and therefore easy to handle. Water has an unusually high critical temperature owing to its polarity. At supercritical conditions water can dissolve gases such as oxygen and non-polar organic compounds.

7.4.3 Processes and applications

A generalized process scheme for the production of an extract from a starting mixture with supercritical CO_2 is shown in Fig. 7.19. The supercritical fluid extracts the desired components in the extraction vessel. The fluid phase is expanded in a pressure-reducing valve, and because of the decrease in the density of the CO_2, the extract separates from the fluid phase and is collected in the separator. CO_2 is recompressed

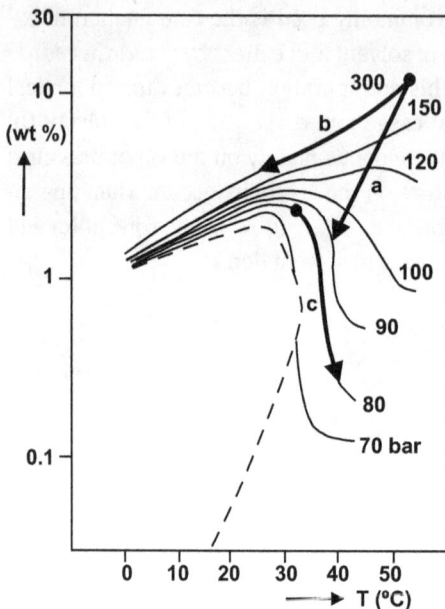

Fig. 7.18: Schematic representation of the three basic process operations in a solubility diagram for naphthalene in carbon dioxide.

Tab. 7.4: Critical properties for common supercritical solvents.

Solvent	T_C (°C)	P_C (bar)	ρ_C (kg/m³)
Ethylene	10	50	220
Carbon dioxide	31	73	470
Ethane	32	48	200
Nitrous oxide	37	72	450
Propane	97	42	220
Ammonia	133	111	240
Butane	152	38	230
Methanol	240	81	270
Toluene	319	41	290
Water	374	220	320

after being liquefied in the condenser and finally heated up to the desired extraction temperature before reentering the extraction vessel.

Already for some decades several commercial processes have used the unique dissolving properties of solvents near their critical point. In the propane deasphalting process lube oils are refined with near-critical propane as shown in the schematic diagram in Fig. 7.20. In the asphalt settler saturated, liquid propane at approximately 50 °C dissolves all the constituents of a lube-oil feedstock except for the asphalt.

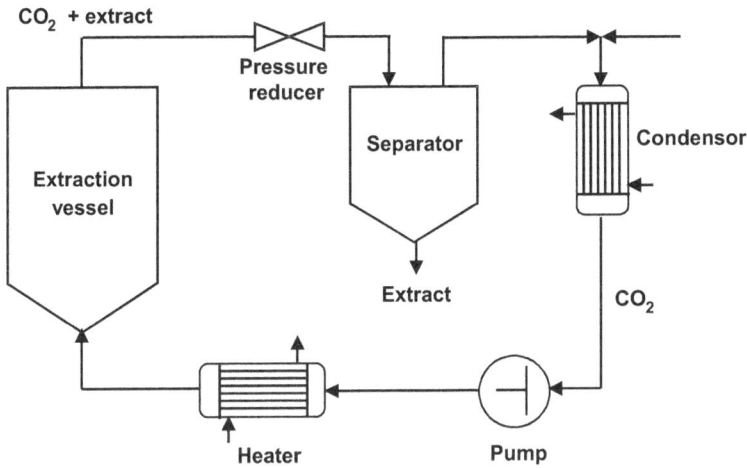

Fig. 7.19: Schematic representation of the supercritical extraction process.

Heating the remaining propane–oil mixture to temperatures near 100 °C decreases the solvent power of liquid propane, resulting in the sequential precipitation of the resins, thus leaving only the lightest paraffins in solution.

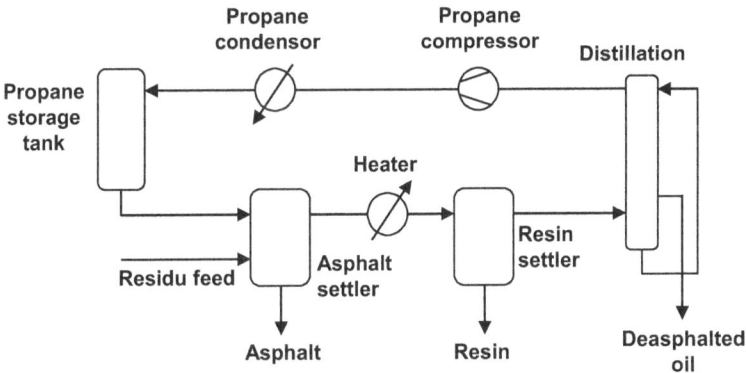

Fig. 7.20: Schematic diagram of the propane-lube oil refining process.

Another important application, the ROSE process, employs near critical butane or pentane for the separation of heavy components such as asphaltenes, resins and oils from oil residues. In the process (Fig. 7.21), residuum and pentane are mixed and the soluble resins and oils recovered in the supercritical phase. By stepwise isobaric temperature increases, which decrease the solvent density, the resin and oil fractions are precipitated sequentially.

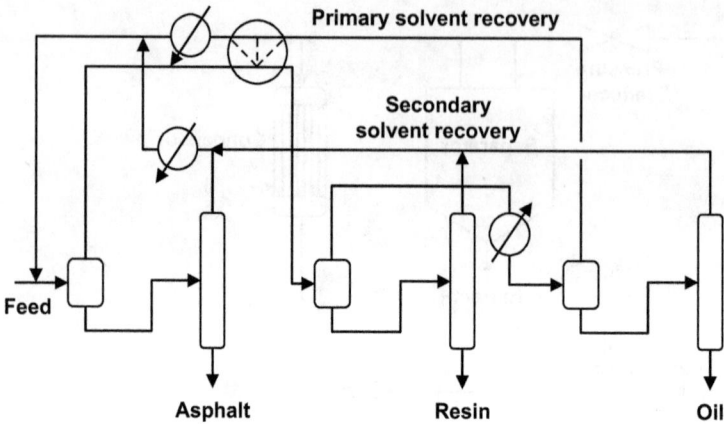

Fig. 7.21: Schematic diagram of the ROSE process.

Supercritical carbon dioxide is well known for its ability to extract thermally labile food components at near-ambient temperature without contaminating the food product. In this area decaffeination of coffee by supercritical CO_2 has become the major industrial application. By using water-saturated CO_2 as a solvent at a temperature around 60 °C and pressure around 300 bar, the caffeine content of coffee beans can be reduced from 3 to nearly 0.2 wt%. Figure 7.22 shows a part of the process for green coffee beans decaffeination. Saturation of both the green (unroasted) coffee beans and the CO_2 with water has been found to improve the caffeine extraction rates. Also, increasing the temperature and pressure improves the partitioning of caffeine into the supercritical phase. A large excess of carbon dioxide is required because caffeine does not dissolve to its solubility limit during the extraction of coffee beans.

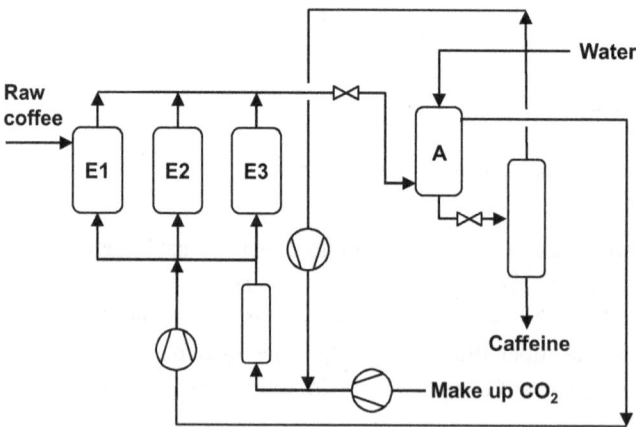

Fig. 7.22: Semicontinuous countercurrent supercritical coffee decaffeination.

In principle the separation of caffeine from the carbon dioxide can be achieved by reducing pressure. However, pressure reduction must be substantial and is only of use if the loading of caffeine in carbon dioxide is high. For the decaffeination of green coffee beans and tea leaves this is not the case, which would lead to high costs for recycling the CO_2. Therefore alternative techniques are used to separate the caffeine isobaric from the supercritical CO_2. One option is to adsorb caffeine on activated carbon, by which high loadings are achieved. The advantage of adsorption over the classical pressure reduction is that energy consumption is reduced. Alternatively absorption in water is feasible since the equilibrium concentration of caffeine in water is much higher than that in the gaseous phase. Absorption in water has the additional advantage over adsorption that the extracted caffeine can be recovered and sold as an additional product. Both techniques, absorption in water and adsorption on activated carbon, are currently used in commercial processes for decaffeination.

In supercritical decaffeination processes multistage countercurrent contacting is the most effective mode. It reduces the amount of solvent and enables the continuous production of extract. Real countercurrent contact is not easily established for solids, since special effort is necessary for moving the solid, with increased difficulties at elevated pressures. Therefore, it is easier not to move the solid material and to achieve countercurrent contact by other measures. For the purposes of extraction from solids with supercritical solvents, several fixed beds in countercurrent contact with the solvent are the best configuration.

In the food industry, carbon dioxide is also used to extract α-acids from hops. These acids impart a characteristic bitter taste to beer. Although the yields are similar to those using methylene chloride, the extract's color, composition, odor and texture are better controllable. Many other flavors, spices and fragrances have been extracted. Supercritical solvents can also be used to fractionate liquid mixtures such as glycerides, vegetable oils, waxes and polymers into numerous components. In that case the extraction is typically carried out in countercurrent columns as illustrated schematically in Fig. 7.23. The main part of the scheme is the column where the countercurrent contact of the liquid and supercritical solvent takes place. The column is usually equipped with internals such as packing or trays to intensify the contact between both phases.

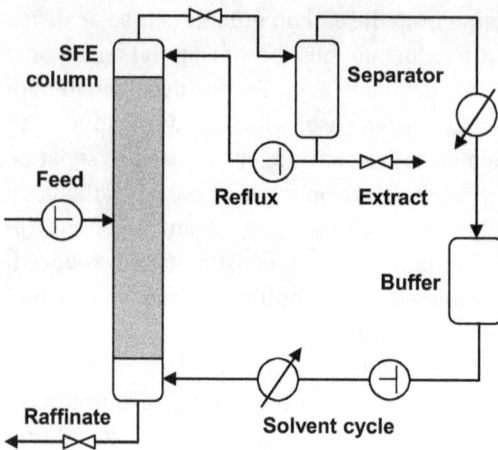

Fig. 7.23: Schematic of an installation for countercurrent extraction of liquids with supercritical fluids.

Nomenclature

A, B, C	Components	
CP	Critical point	[–]
D	Diffusion coefficient	[m^2/s]
E	Extraction factor	[–]
E, F, R, S	Extract, feed, raffinate, solvent flow	[amount/s]
K_i	Distribution coefficient of component i	[–]
N_s	Number of theoretical stages	[–]
N	Number of stages	[–]
P	Pressure	[bar]
T	Temperature	[K]
x	Mole or weight fraction	[–]
β_{12}	Selectivity of component 1 over component 2	[–]
ρ	Density	[kg/m]
η	Viscosity	[mPa s]
γ	Activity coefficient	[–]

Indices

1,2, l	Components
C	Critical
E, F, R, S	Extract, feed, raffinate, solvent flow
R	Reduced

8 Absorption and stripping

8.1 Introduction

In absorption (also called gas absorption, gas scrubbing and gas washing) a gas mixture is contacted with a liquid (the absorbent or solvent) to selectively dissolve one or more components by transfer from the gas to the liquid. The components transferred to the liquid are referred to as solutes or absorbate. The operation of absorption can be categorized on the basis of the nature of the interaction between absorbent and absorbate into the following three general types:

1. Physical solution. In this case, the component being absorbed is more soluble in the liquid absorbent than are the other gases with which it is mixed but does not react chemically with the absorbent. As a result, the equilibrium concentration in the liquid phase is primarily a function of partial pressure in the gas phase and temperature.
2. Reversible reaction. This type of absorption is characterized by the occurrence of a chemical reaction between the gaseous component being absorbed and a component in the liquid phase to form a compound that exerts a significant vapor pressure of the absorbed component.
3. Irreversible reaction. In this case, a reaction occurs between the component being absorbed and a component in the liquid phase which is essentially irreversible.

The use of physical absorption processes is usually preferred whenever feed gases are present in large amounts at high pressure and the amount of the component to be absorbed is relatively large. On the other hand, absorption processes with simultaneous chemical reaction are always preferred when the components to be separated from feed gases are present in small concentrations and at low partial pressures.

Gas absorption is usually carried out in vertical countercurrent columns as shown in Fig. 8.1. The solvent is fed at the top whereas the gas mixture enters from the bottom. The absorbed substance is washed out by the lean solvent and leaves the absorber at the bottom as a liquid solution. Usually, the absorption column operates at a pressure higher than atmospheric, taking advantage of the fact that gas solubility increases with pressure. The loaded solvent is often recovered in a subsequent stripping or desorption operation. After preheating the rich solvent is transported to the top of a desorption column that usually operates under lower pressure than in the absorption column. This second step is essentially the reverse of absorption in which the absorbate is removed from the solvent. Desorption can be achieved through a combination of methods:

https://doi.org/10.1515/9783110712445-008

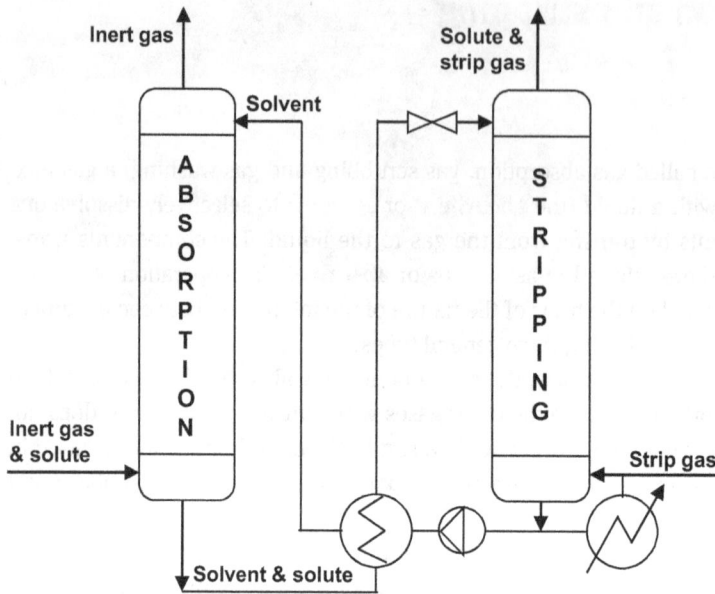

Fig. 8.1: Basic scheme of an absorption installation with stripping for regeneration.

1. Flashing the solvent to lower the partial pressure of the dissolved components
2. Reboiling the solvent, to generate stripping vapor by evaporation of part of the solvent
3. Stripping with an inert gas or steam

Wide use is made of desorption by pressure reduction because the energy require-ments are low. After depressurization, stripping with an inert gas is often more eco-nomic than thermal regeneration. However, because stripping is usually not perfect, the absorbent recycled to the absorber contains species present in the vapor entering the absorber. The desired purity of the gas determines the final costs of desorption. The necessary difference between the partial pressure of the absorbed key component over the regenerated solution and the purified gas serves as a criterion for determining the dimensions of the absorption and desorption equipment. The lean solvent, devoid of gas, flows through a heat exchanger, where part of the heat needed for heating the rich solvent is recovered, and then through a second heat exchanger, where it is cooled down to the desired temperature before it is returned to the absorption column. Usually a small amount of fresh solvent should be added to the column to replenish the solvent which was partly evaporated in the desorption column or underwent irre-versible chemical reactions which take place in the whole system.

Tab. 8.1: Typical applications of absorption for gas purification.

Impurity	Process	Absorbent
Ammonia	Indirect process (coke oven gas)	Water
Carbon dioxide and hydrogen sulfide	Ethanolamine	Mono- or diethanolamine in water
	Benfield	Potassium carbonate and activator in water
	Selexol	Polyethylene glycol dimethyl ether
Carbon monoxide	Copper ammonium salt	Cuprous ammonium carbonate and formate in water
Hydrogen chloride	Water wash	Water
Toluene	Toluene scrubber	Toluene
Cyclohexane	Scrubber	Cyclohexane

Tab. 8.2: Typical applications of absorption for product recovery.

Product	Process	Absorbent
Acetylene	Steam cracking of hydrocarbons (naphtha)	Dimethylformamide
Acrylonitrile	Ammoxidation of propylene	Water
Maleic anhydride	Butane oxidation	Water
Melamine	Urea decomposition	Water
Nitric acid	Ammonia oxidation (NO_x absorption)	Water
Sulfuric acid	Contact process (SO_3 absorption)	Water
Urea	Synthesis (CO_2 and NH_3 absorption)	Ammonium carbamate solution

8.2 The aim of absorption

Generally the commercial purpose of absorption processes can be divided into gas purification or product recovery, depending on whether the absorbed or the unabsorbed portion of the feed gas has the greater value. Typical gas purification applications are listed in Tab. 8.1. The removal CO_2 from synthesis gas in ammonia production and the removal of acid gases (CO_2, H_2S) from natural gas are some of the most widespread applications that are being improved continuously by the development of new solvents, process configurations and design techniques. In both applications stripping is used for absorbent regeneration.

Examples of absorption processes for product recovery are listed in Tab. 8.2. The absorption of SO_3 and NO_x in water to make concentrated sulfuric respectively nitric acid are probably the most widely used product recovery applications of absorption. Other frequently encountered examples are the recovery of various products from a gaseous product stream by inert absorbents such as water. In some cases the absorber is used as a reactor where the desired chemical compound is obtained by a liquid phase reaction of the absorbed gases. An illustration of such a process is the production of urea from CO_2 and ammonia.

8.3 General design approach

Design or analysis of an absorber (or stripper) requires consideration of a number of factors, including:
1. Entering gas (liquid) flow rate, composition, temperature and pressure
2. Desired degree of recovery of one or more solutes
3. Choice of absorbent (stripping agent)
4. Operating pressure and temperature, and allowable gas pressure drop
5. Minimum absorbent (stripping agent) flow rate and actual absorbent (stripping agent) flow rate as a multiple of the minimum rate needed to make the separation
6. Number of equilibrium stages
7. Heat effects and need for cooling (heating)
8. Type of absorber (stripper) equipment
9. Height of the absorber (stripper)
10. Diameter of the absorber (stripper)

The initial step in the design of the absorption system is selection of the absorbent and overall process to be employed. There is no simple analytical method for accomplishing this step. In most cases, the process requirements can be met by more than one solvent, and the only satisfactory approach is an economic evaluation, which may involve the complete but preliminary design and cost estimate for more than one alternative. The ideal absorbent should
– have a high solubility for the solute(s) to minimize the need for absorbent;
– have a low volatility to reduce the loss of absorbent and facilitate separation of solute(s);
– be stable to maximize absorbent life and reduce absorbent makeup requirement;
– be noncorrosive to permit use of common materials of construction;
– have a low viscosity to provide low pressure drop and high mass and heat transfer rates;
– be nonfoaming when contacted with the gas;
– be nontoxic and nonflammable to facilitate its safe use;
– be available, if possible within the process, or be inexpensive.

The most widely used absorbents are water, hydrocarbon oils, and aqueous solutions of acids and bases. For stripping the most common agents are water vapor, air, inert gases and hydrocarbon gases. Once an absorbent is selected, the design of the absorber requires the determination of basic physical property data such as density, viscosity, surface tension and heat capacity. The fundamental physical principles underlying the process of gas absorption are the solubility and heat of solution of the absorbed gas and the rate of mass transfer. Information on both must be available when sizing equipment for a given application. In addition to the fundamental design concepts based on solubility and mass transfer, many practical details have to be considered during actual plant design.

The second step is the selection of the operating conditions and the type of contactor. In general, operating pressure should be high and temperature low for an absorber, to minimize stage requirements and/or absorbent flow rate. In contrast operating pressure should be low and temperature high for a stripper to minimize stage requirements or stripping agent flow rate. However, because maintenance of a vacuum is expensive, strippers are commonly operated at a pressure just above ambient. A high temperature can be used, but it should not be so high as to cause undesirable chemical reactions. Choice of the contactor may be done on the basis of system requirements and experience factors such as those discussed in paragraph 8.5. Following these decisions it is necessary to calculate material and heat balance calculations around the contactor, define the mass transfer requirements, determine the height of packing or number of trays and calculate contactor size to accommodate the liquid and gas flow rates with the selected column internals.

8.3.1 Gas solubilities

The most important physical property data required for the design of absorbers and strippers are gas–liquid equilibria. Since equilibrium represents the limiting condition for any gas–liquid contact, such data are needed to define the maximum gas purity and rich solution concentration attainable in absorbers, and the maximum lean solution purity attainable in strippers. Equilibrium data are also needed to establish the mass transfer driving force, which can be defined simply as the difference between the actual and equilibrium conditions at any point in a contactor.

At equilibrium, a component of a gas in contact with a liquid has identical fugacities in both the gas and liquid phase. For ideal solutions, Raoult's law applies:

$$y_A = \frac{P_A^0}{P_{tot}} x_A \tag{8.1}$$

where y_A is the mole fraction of A in the gas phase, P_{tot} is the total pressure, $P_A 0$ is the vapor pressure of pure A and x_A is the mole fraction of A in the liquid. For nonideal mixtures, Raoult's law modifies into

$$y_A = \frac{y_A^\infty P_A^0}{P_{tot}} x_A \tag{8.2}$$

where $\gamma_A{}^\infty$ is the activity coefficient of solute A in the absorbent at infinite dilution. A more general way of expressing solubilities is through the *vapor–liquid equilibrium constant K defined by*

$$y_A = K x_A \tag{8.3}$$

The value of the equilibrium constant K is widely employed to represent hydrocarbon vapor-liquid equilibria in absorption and distillation calculations. Correlations and experimental information for the equilibrium K values of hydrocarbons are available from various sources.

For moderately soluble gases with relatively little interaction between the gas and liquid molecules equilibrium data are usually represented by Henry's law:

$$p_A = P_{tot}\, y_A = H\, x_A \tag{8.4}$$

where p_A is the partial pressure of component A in the gas phase. H is Henry's constant, which has the units of pressure per composition. Usually H is dependent upon temperature, but relatively independent of pressure at moderate levels. In general, for moderate temperatures, gas solubilities decrease with an increase in temperature. Henry's constants for many gases and solvents are tabulated in various literature sources. Examples of Henry's constants for a number of gases in pure water are given in Fig. 8.2.

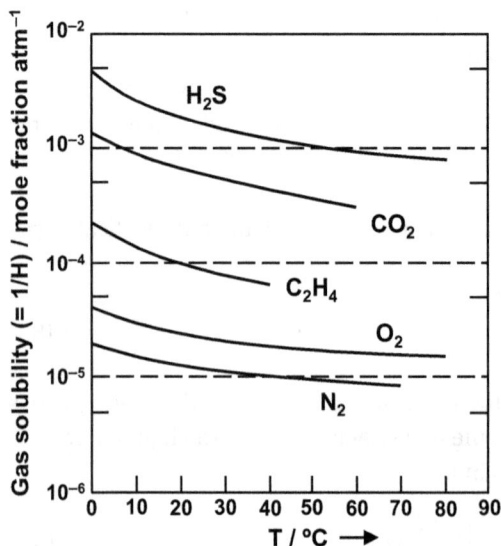

Fig. 8.2: Solubilities of various gases in water expressed as the reciprocal of the Henry's law constant (adapted from [9]).

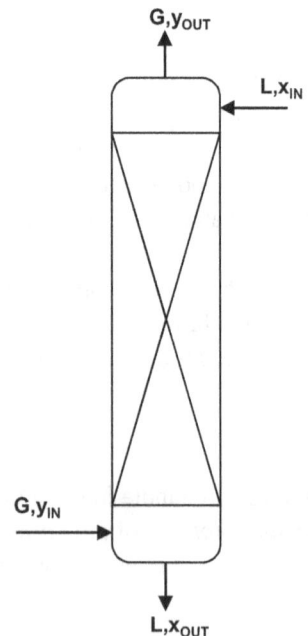

Fig. 8.3: Continuous, steady-state operation in a countercurrent absorber.

8.3.2 Minimum absorbent flow

For each feed gas flow rate, absorbent composition, extent of solute absorption, operating pressure and operating temperature, a minimum absorbent flow rate exists that
corresponds to an infinite number of countercurrent equilibrium contacts between the
gas and liquid phases. As a result a tradeoff exists in every design problem between
the number of equilibrium stages and the absorbent flow rates at rates greater than
the minimum value. This minimum absorbent flow rate L_{MIN} is obtained from a
mass balance over the whole absorber, assuming equilibrium is obtained between in
incoming gas and outgoing absorbent liquid in the bottom of the column. An overall
mass balance over the column illustrated in Fig. 8.3 gives the following result:

$$G\,y_{IN} + L\,x_{IN} = G\,y_{OUT} + L\,x_{OUT} \tag{8.5}$$

Introduction of the assumption that the outgoing liquid is in equilibrium with in
incoming gas ($x_{OUT} = K\,y_{IN}$) gives the minimum absorbent flow:

$$L_{MIN} = G\left(\frac{y_{IN} - y_{OUT}}{K\,y_{IN} - x_{IN}}\right) \tag{8.6}$$

A similar derivation of the minimum stripping gas flow rate G_{MIN} for a stripper
results in an analogous expression:

$$G_{MIN} = L\left(\frac{x_{IN} - x_{OUT}}{K\,x_{IN} - y_{IN}}\right) \tag{8.7}$$

8.3.3 Number of equilibrium stages

If the gas and liquid flow rates in absorbers and strippers are relatively constant
and temperature effects can be neglected, the number of equilibrium stages N may
be calculated with the Kremser method:

$$\text{Fraction of a solute, i, absorbed} = \frac{A_i^{N+1} - A}{A_i^{N+1} - 1} \tag{8.8}$$

and

$$\text{Fraction of a solute, i, stripped} = \frac{S_i^{N+1} - S}{S_i^{N+1} - 1} \tag{8.9}$$

Where the solute absorption factor A and stripping factor S are given by

$$A_i = \frac{L}{K_i\,G} \qquad \text{respectively} \qquad S_i = \frac{K_i\,G}{L}$$

Values of L and G in mole per unit time may be taken as the entering values. Values of K_i depend mainly on temperature, pressure and liquid phase composition.

8.4 Basic characteristics of absorbers

The main purpose of various industrial absorbers is to ensure large gas–liquid mass transfer area and to create such conditions that a high intensity of mass transfer is achieved. Although small-scale processes sometimes use batch-wise operation where the liquid is placed in the equipment and only gas is flowing, continuous absorption is most commonly applied in large-scale industrial processes. There are various criteria for classifying absorbers. It seems that the best one is a widely used criterion that takes into account which of the phases (gas or liquid) is in a continuous or disperse form. Using this criterion, absorbers can be classified into the following groups:
1. Absorbers in which both phases are continuous
    - Packed columns
    - Wetted wall columns
    - Contactors with flat surface
    - Laminar jet
    - Disk (sphere) columns
2. Absorbers with a disperse gas phase and a continuous liquid phase
    - Tray columns
    - Tray columns with packing
    - Bubble columns
    - Packed bubble columns
    - Mechanically agitated absorbers
    - Jet absorbers
3. Absorbers with a dispersed liquid phase and a continuous gas phase
    - Spray columns
    - Venturi scrubbers

In each apparatus, due to different hydrodynamic conditions, various values of the mass transfer coefficients occur in both phases. Therefore, when choosing a given type of absorber, the following criteria should be taken into account:
1. The required method of absorption (continuous, semicontinuous)
2. The flow rate of the gas and the liquid entering the absorber (e.g., high gas flow rate and low liquid flow rate need different types of equipment than in the case where the two flow rates are of the same size)
3. The required liquid holdup (large or small liquid holdup is needed)
4. Which phase controls mass transfer in the absorber (gas phase or liquid phase)?
5. Is it necessary to remove heat from the absorber?

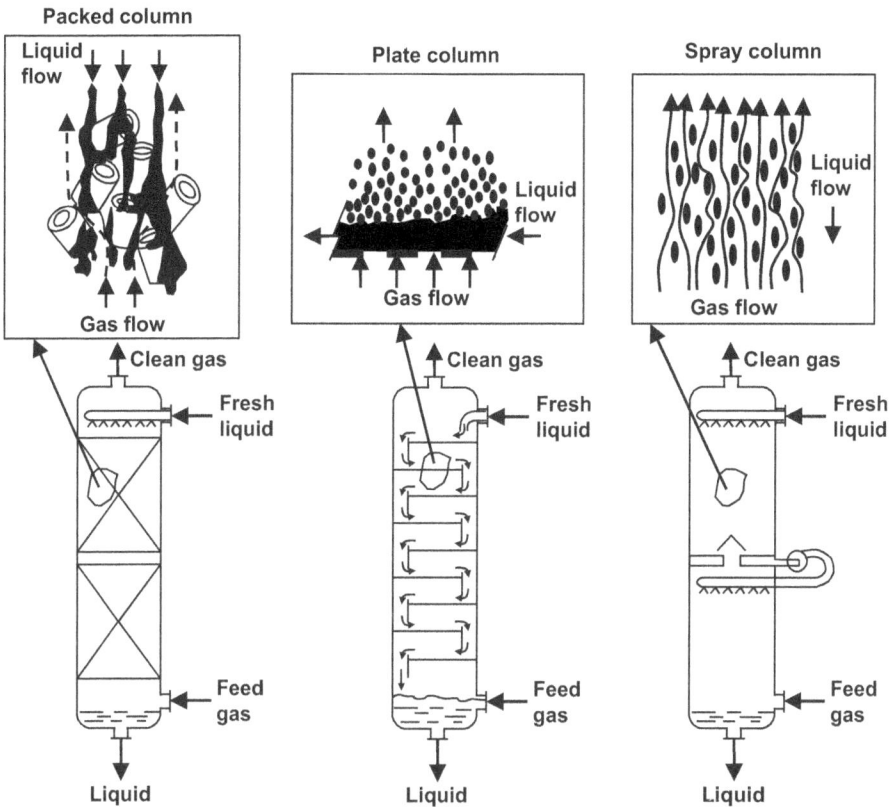

Fig. 8.4: Schematic and operating principles of packed, tray and spray towers (adapted from [8]).

6. The corrosiveness of the absorption systems
7. The required size of the interface (large or small mass transfer area)
8. The physicochemical properties of the gas and the liquid (particularly viscosity and surface tension)

Apart from these factors there are many others which influence the selection of equipment. For instance, gas impurities or deposits formed during the absorption process require the application of a given absorber.

8.5 Industrial contactors

The industrially most frequently used absorbers are packed columns, tray towers, spray columns and bubble columns.

8.5.1 Packed columns

Packed columns are the units most often used in absorption operations (Fig. 8.4). Usually, they are cylindrical columns up to several meters in diameter and over 10 meters high. A large gas–liquid interface is achieved by introducing various packings into the column. The packing is placed on a support whose free cross section should be at least equal to the packing porosity. Liquid is fed in at the top of the column and distributed over the packing through which it flows downwards. To guarantee a uniform liquid distribution over the cross section of the column, a liquid distributor is employed. The gas flows upward countercurrent to the falling liquid, which absorbs the soluble species from the gas. The gas, which is not absorbed, exits from the top of the column, usually through a mist eliminator. The mist eliminator separates liquid drops entrained by the gas from the packing. The separator may be a layer of the packing, mesh or it may be specially designed.

Packed columns are preferable to tray columns for small installations, corrosive service, liquids with a tendency to foam, very high liquid-to-gas ratios and low pressure drop applications. In the handling of corrosive gases packing, but not trays, can be made from ceramic or plastic materials. Packed columns are also advantageous in vacuum applications because the pressure drop, especially for regularly structured packings, is usually less than in tray columns. In addition, packed columns offer greater flexibility because the packing can be changed with relative ease to modify column-operating characteristics.

The main virtues of a good packing are a large specific surface area, high void fraction (porosity), chemical resistance to liquid and gas, high mass transfer rates in both phases, a small pressure drop and low cost. Two of the features mentioned above characterize the packing particularly well. These are the specific surface of the packing and the void fraction. The specific surface of the packing is the total surface area of all packing elements contained in a unit volume. Usually, the specific surface area of industrial packings ranges from 100 to 1,000 m^2/m^3. The void fraction is the ratio of the free volume to the bed volume. For ceramic packings the void fraction is usually about 0.7, while for metal packing it is about 0.95 and more.

Packings may be divided into two main groups: random and structured. Structured packings are manufactured for a given column diameter and column height. Usually the same packing cannot be used in columns of different diameter. Random packings may be relocated to various columns.

The most popular random packings are rings. Raschig rings have a large specific packing surface and high porosity. They are hollow cylinders with an external diameter equal to the ring height. They can be made of ceramic, metal, graphite or plastic. Since the time when Raschig introduced his rings in industry, they have been continuously modified. Figure 8.5 shows a Raschig ring and its main modified forms. The objective in modifying the rings is to increase the mass transfer area, the porosity of the packing, and to guarantee a low-pressure drop and good conditions for mixing the gas and liquid phases. From experimental investigations it follows that the most efficient of all the types of ring are Pall rings.

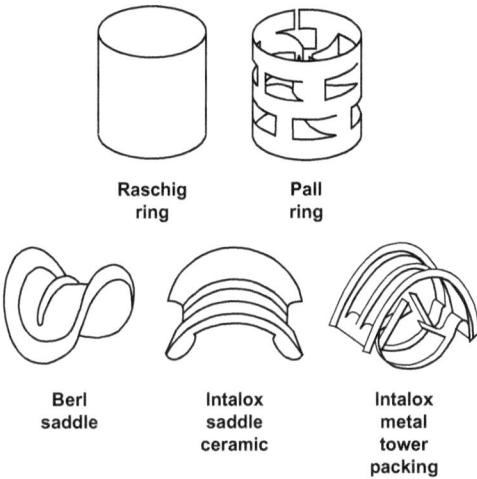

Fig. 8.5: Schematic drawing of various random packings.

Fig. 8.6: Schematic of structured Sulzer BX packing.

Recently, the application of structured packing in absorption has increased rapidly owing to the fact that it has a larger mass transfer area than random packing. Another advantage is that by using this packing it is possible, in contrast to random packing, to obtain the same values of mass transfer coefficient in the entire column. Furthermore, it is also easier and cheaper to install columns already with the structured packing inside rather than assemble a column and then fill it with random packing on site. One of the first structured packings is the Sulzer BX packing. A classical form of the Sulzer BX packing is a grid strip bent alternately at an angle of about 300 to the column axis. Bent strips form bunches located in the column one on top of the other. The liquid and the gas flow in channels which change their flow direction while passing through the bunches. Figure 8.6 presents schematically the Sulzer BX packing. Several modifications have been recently introduced to this packing.

8.5.2 Tray columns

Another basic type of equipment widely applied in absorption processes is a tray column (Fig. 8.4). The diameter and height of the column can reach 10 and 50 m, respectively, but usually they are much smaller. In tray columns various tray constructions guarantee good contact of the gas with the liquid. Taking into account the whole column, the flow of the gas with the liquid has a countercurrent character.

Liquid is supplied to the highest tray, flows along it horizontally and, after reaching a weir, flows through a downcomer to a tray below. Gas is supplied below the lowest tray, then it flows through perforations in the tray and bubbles through a liquid layer. The application of such a flow pattern is aimed at ensuring the maximum mass transfer area and high turbulence of the gas and liquid phases, which results in obtaining high mass transfer coefficients in both phases. The distances between trays should be such that liquid droplets entrained by the gas are separated from the liquid and the gas is separated from the liquid in the downcomer. Usually, the tray-to-tray distance ranges from 0.2 to 0.6 m and depends mainly on the column diameter and the liquid load of the tray.

Tray columns are particularly well suited for large installations and low-to-medium liquid flow rate applications. In general, they offer a wider operating window for gas and liquid flow than a countercurrent packed column. That is, they can handle high gas flow rates and low liquid flow rates that would cause flooding in a packed column. Tray columns are also preferred in applications having large heat effects since cooling coils are more easily installed in tray towers and liquid can be withdrawn more easily from trays than from packings for external cooling. Furthermore, they are advantageous for separations that require tall columns with a large number of transfer units because they are not subject to channeling of vapor and liquid streams which can cause problems in tall packed columns. The main disadvantages of tray columns are their high capital cost, especially when bubble-cap trays or special proprietary design are used, and their sensitivity to foaming.

As mentioned above, the most widely applied in apparatus mass transfer processes are trays with a cross-flow of phases. There are a variety of tray designs for this flow. On an industrial scale, three tray types are most important (Fig. 8.7):
– bubble cap tray
– sieve tray
– valve tray

The bubble cap tray was used in absorption and distillation for many years until the early 1950s. Due to high production costs the classical form of the bubble cup trays is not recommended. Only under conditions where the column load of gas and liquid is changing over a wide range, the use of these trays is justified. On classical bubble cup trays gas flows through risers under the caps and then in the form of bubbles passes through a liquid layer. In the most recent design solutions flat caps or caps without risers are applied. Because of an appreciable decrease in risers a decrease in the cost of tray production and gas pressure drop on the tray are achieved while maintaining high efficiency of the tray.

A characteristic feature of valve trays is that they can operate efficiently over a very wide range of column loading with gas and liquid. Perforations of 35–50 mm diameter made in the trays are equipped with valves which, depending on the gas and liquid flow rates, open to a greater or lesser extent. There is always, irrespective of gas

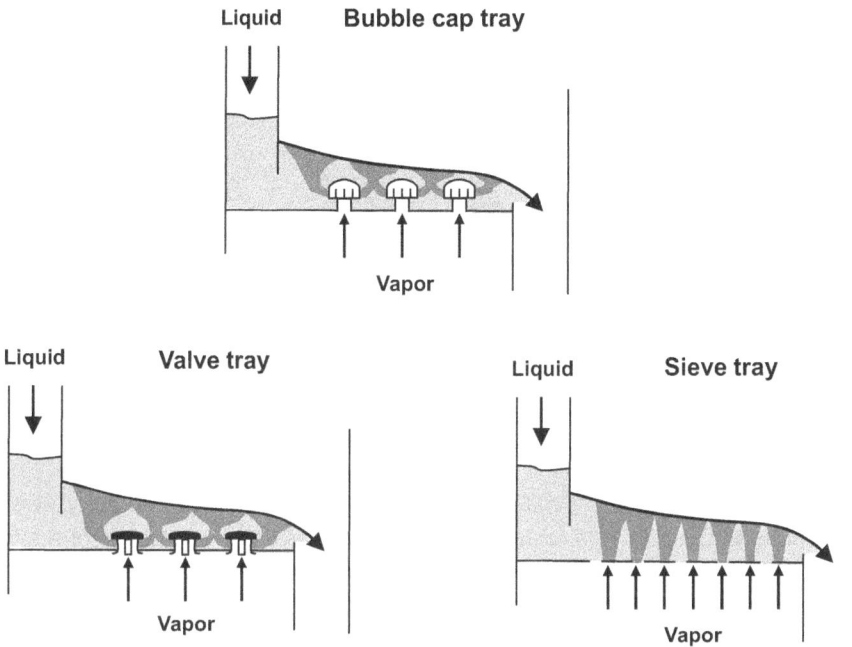

Fig. 8.7: Drawing of the most important industrial tray types.

flow velocity, some minimum opening of the valve. With an increase in the gas velocity the valve floats up, opening to the maximum position.

Taking into account low costs, the most important and most frequently used trays now are sieve trays. The steel sheets are perforated in such a way that the perforation forms equilateral triangles. Initially, the diameter of the holes in the sieve trays was 3–6 mm. Now, trays with holes reaching 25 mm are used. However, the most frequently applied trays have holes of about 5 mm in diameter.

8.5.3 Spray towers

In spray towers liquid is sprayed as fine droplets, which make contact with a cocurrently or countercurrently flowing gas (Fig. 8.4). The gas and liquid flows are similar to plug flow. Spray contactors are used almost entirely for applications where pressure drop is critical, such as flue gas scrubbing. They are also useful for slurries that might plug packings or trays. Other important applications include particles removal and hot gas quenching. When used for absorption, spray devices are not applicable to difficult separations because they are limited to only a few equilibrium stages even with countercurrent spray column designs. The low efficiency of spray columns originates from the entrainment of droplets in the gas and back mixing of the gas induced by the sprays. The high energy consumed for atomizing liquid and liquid entrainment in the gas outlet stream are two additional important disadvantages.

8.5.4 Bubble columns

Bubble columns (Fig. 8.8) are finding increasing application in processes when absorption is accompanied by a chemical reaction. They are also widely used as chemical reactors in processes where gas, liquid and solid phases are involved. In bubble columns the liquid is a continuous phase while the gas flows through it in the form of bubbles, The character of the gas flow is well represented by plug flow, while that of the liquid is between plug and ideally mixed flow. These absorbers can operate in cocurrent and countercurrent phase flow.

Bubble columns are particularly well suited for applications when significant liquid holdup and long liquid residence time are required. An advantage of these columns is their relatively low investment cost, a large mass transfer area and high mass transfer coefficients in both phases. Disadvantages of bubble columns include a high-pressure drop of the gas and significant back mixing of the liquid phase. The latter disadvantage can be reduced by introducing an inert packing of high porosity to the bubble column. Such a packing eliminates to a large extent the effects of

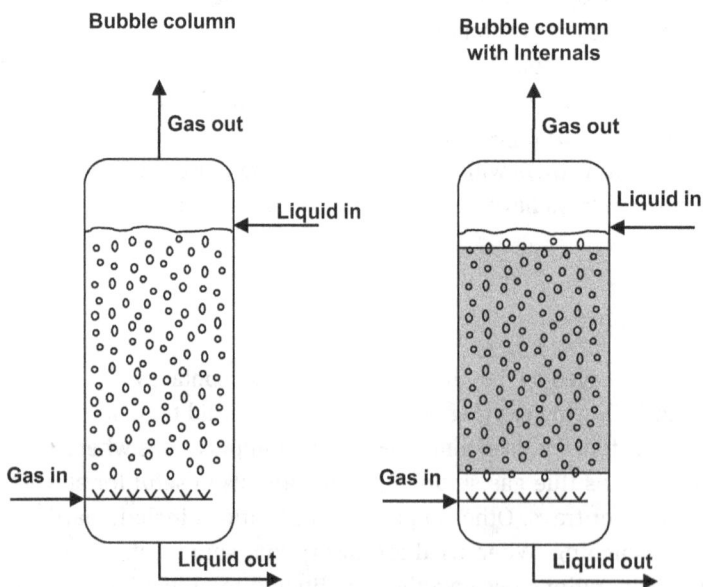

Fig. 8.8: Schematic of a bubble column absorber, without and with internal packing.

liquid phase mixing along the column height. In addition the packing may also cause an increase in the mass transfer surface area with relation to the bubble column at the same flows of both phases. The advantages of bubble columns appreciably exceed their disadvantages and therefore an increasing number of these columns are applied in industry. The column diameter sometimes exceeds 5 m and its height reaches over 10 m. Beside the modified columns mentioned above there are many design solutions which are particularly useful in cases where in addition to the gas and liquid phases there is also a solid phase present.

Nomenclature

A	$L/K \cdot G$, absorption factor	[–]
f	Fraction absorbed or stripped	[–]
G	Gas flow	[mol/s]
H	Henry coefficient	[–]
K	Equilibrium constant	[–]
K_A	Distribution coefficient	[–]
L	Liquid flow	[mol/s]
N	Number of (theoretical) stages	[–]
p	Partial pressure	[N/m^2]
p^0	Saturation pressure	[N/m^2]
P_{tot}	Total pressure	[N/m^2]
S	$K \cdot G/L$, stripping factor	[–]
x, y	Mole fraction	[–]
γ	Activity coefficient	[–]

Indices

A	Component

9 Adsorption and ion exchange

9.1 Introduction

Adsorption and ion exchange are sorption operations, in which certain components of a fluid phase are selectively transferred to insoluble particles. Adsorption processes use the natural tendency of liquid or gas components to collect at the surface of a solid material. As a result, selective concentration, adsorption, of one or more components, adsorbates, occurs at the surface of a (micro)porous solid, adsorbent. An adsorbent is an example of a mass separating agent, which is used to facilitate the separation. In most cases, the attractive forces binding the adsorbate are weaker than those of chemical bonds, allowing the adsorption to be reversed by either raising the temperature of the adsorbent or by reducing the concentration or partial pressure of the adsorbate. This combination of selective adsorption followed by regeneration, shown in Fig. 9.1, is the basis for a separation when more of one component is removed from a gas or liquid mixture than of the other components. In an overall process, the desorption or regeneration step is very important. It allows recovery of the adsorbates when they are valuable and permits reuse of the adsorbent for further cycles. The downside of the need to regenerate the adsorbent is that the overall process is necessarily cyclic in time. Only in a few cases, desorption is not practical, and the adsorbate must be removed by thermal destruction, another chemical reaction or the adsorbent is simply discarded.

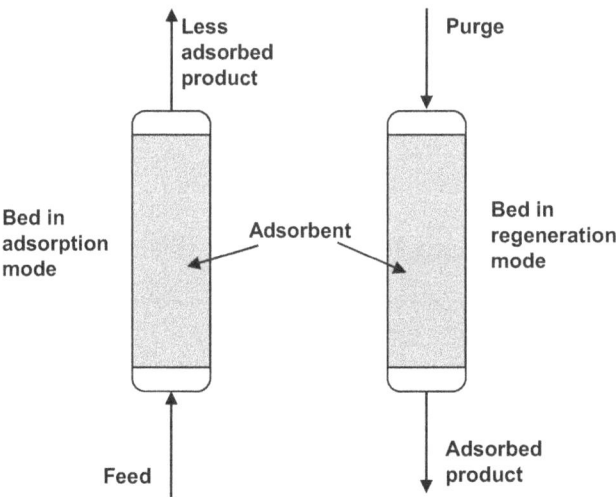

Fig. 9.1: Schematic of an adsorption/desorption process.

https://doi.org/10.1515/9783110712445-009

In ion exchange processes, positive (cations) or negative (anions) ions from an aqueous solution are exchanged with cations or anions on a solid ion exchanger. Water softening by ion exchange involves a cation exchanger, in which calcium ions are removed by the following reaction:

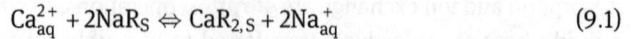

$$Ca^{2+}_{aq} + 2NaR_S \Leftrightarrow CaR_{2,S} + 2Na^+_{aq} \tag{9.1}$$

where R is the polymeric backbone of the ion exchanger. The exchange of ions is reversible, which allows extended use of the ion exchange resin before replacement is necessary. Regeneration is usually accomplished with concentrated acid, base or salt solutions. In demineralization or deionization, the ion exchange concept is extended to complete removal of inorganic salts from water by a two-step process. In the first step, a cation resin is used to exchange hydrogen ions for cations such as calcium, magnesium and sodium. In the second step, an anion resin exchanges hydroxyl ions for sulfate, nitrate and chloride anions. The hydrogen and hydroxyl ions produce water.

Adsorption processes may be classified as purification or bulk separations, depending on the concentration in the feed fluid of the components to be adsorbed. Nowadays, it is the most widely used nonvapor–liquid technique for molecular separations throughout a wide range of industries. Application should be considered when separation by distillation becomes difficult or expensive and a suitable adsorbent exists. Suitable means that the adsorbent shows proper selectivity and capacity can be easily regenerated and causes no damage to the products. Selectivity between key components should in general be greater than 2. A representative list of industrial applications is given in Tab. 9.1. Industrial applications of ion exchange range from the purification of low-cost commodities such as water to the purification and treatment of high-cost pharmaceutical derivatives as well as precious metals. Table 9.1 illustrates that the largest single application is water treatment. Other major industrial applications are the processing and decolorization of sugar solutions and the recovery of uranium from relatively low-grade mineral acid leach solutions.

Tab. 9.1: Examples of industrial adsorptive and ion exchange separations.

Separation	Application	Adsorbent
Gas bulk separations	Normal paraffins, isoparaffins, aromatics	Zeolite
	N_2/O_2	Zeolite
	O_2/N_2	Carbon molecular sieve
	CO, CH_4, CO_2, N_2, NH_3/H_2	Zeolite, activated carbon
	Hydrocarbons from vent streams	Activated carbon
Gas purification	H_2O removal from cracking gas, natural gas, air, etc.	Silica, alumina, zeolite
	CO_2 from C_2H_4, natural gas, etc.	Zeolite

Tab. 9.1 (continued)

Separation	Application	Adsorbent
	Organics from vent streams	Activated carbon, silicalite
	Sulfur compounds from organics	Zeolite
	Solvents and odors from air	Activated carbon, silicalite
Liquid bulk separations	Normal paraffins, isoparaffins, aromatics	Zeolite
	p-Xylene/o-xylene, m-xylene	Zeolite
	Fructose/glucose	Zeolite, ion exchange resins
Liquid purifications	H$_2$O removal from organic solvents	Silica, alumina, zeolite
	Organics from water	Activated carbon, silicalite
	Odor and taste components from drinking water	Activated carbon
	Product decolorizing	Activated carbon
	Fermentation product recovery	Activated carbon, affinity agents
Ion exchange	Water softening	Polymeric ion exchange resins
	Water demineralization	
	Water dealkalization	
	Decolorization of sugar solutions	
	Recovery of uranium from leach solutions	
	Recovery of vitamins from fermentation broths	

9.2 Adsorption fundamentals

9.2.1 Industrial adsorbents

The role of the adsorbent is to provide the selectivity and capacity required for the separation of a mixture. In the majority of adsorptive separation processes, the selectivity is provided by the physical adsorption equilibrium. Because adsorption forces depend on the nature of the adsorbing molecule as well as on the nature of the surface, different substances are adsorbed with different affinities. In Tab. 9.2, a simple classification scheme is given where equilibrium-controlled adsorbents are primarily divided into hydrophilic and hydrophobic surfaces. It is clear that adsorbents with hydrophilic surfaces will preferentially attract polar molecules, in particular water, and adsorbents with hydrophobic surfaces nonpolar molecules.

Besides high selectivity, high capacity is also desirable since the adsorption capacity determines the size and therefore the cost of the adsorbent bed. To achieve a high adsorptive capacity, commercial adsorbents are made from microporous materials with a high specific area. The most important properties of various microporous industrial adsorbents are listed in Tab. 9.3. According to the definition of the IUPAC, the pores in adsorbents fall into three categories:

1. Micropores <2 nm
2. Mesopores 2–50 nm
3. Macropores >50 nm

Tab. 9.2: Classification of commercial adsorbents.

Equilibrium selective		Kinetically selective	
Hydrophilic	Hydrophobic	Amorphous	Crystalline
Activated alumina	Activated carbon	Carbon molecular sieves	Small-pore zeolites
Silica gel Al-rich zeolites	Microporous silica Silicalite, dealuminated mordenite and other silica-rich zeolites		
Polymeric resins containing –OH groups or cations	Other polymeric resins		

In a micropore, the guest molecule never escapes from the force field of the solid surface, while in mesopores and macropores, the molecules in the central region are essentially free from the surface force field. Macropores provide very little surface area relative to the pore volume and hardly contribute to the adsorptive capacity. Their main role is to provide a network of super highways to facilitate rapid penetration of molecules into the interior of the adsorbent particle.

In industrial processes, only four types of adsorbents dominate in usage: activated carbon, silica gel, activated alumina and molecular sieve zeolites. Among these four types, activated carbon is most often used for removing hydrophobic organic species from both gas and aqueous liquid streams. Activated carbon is produced in many different forms that differ mainly in pore size distribution and surface polarity. For liquid-phase adsorption, a relatively large pore size is required, while the activated carbons used in gas adsorption have much smaller pores. Commercial carbons are used in the form of hard pellets, granules, cylinders, spheres, flakes or powders. Silica gels represent an intermediate between highly hydrophilic and highly hydrophobic surfaces. Most often, these adsorbents are used for removing water from various gases, but hydrocarbon separations are also sometimes feasible. Activated alumina is essentially a microporous (amorphous) form of Al_2O_3 that has quite a high affinity for water and is often used in drying applications for various gases. Like with silica gels, the water bond with the alumina surface is not as strong as with zeolites, so that regeneration of aluminas can often be accomplished at somewhat milder temperatures. As can be seen in Fig. 9.2, activated alumina can have a higher ultimate water capacity, but the zeolites have a higher capacity at low water partial pressures. Thus, zeolites are typically chosen when very high water removal is necessary, while activated alumina is preferred if adsorbent capacity is more important.

Tab. 9.3: Properties of industrial adsorbents.

Adsorbent	Pore diameter (nm)			Particle density	Specific area
	Micropores $d < 2$ nm	Mesopores	Macropores $d > 50$ nm	g/cm^3	m^2/g
γ-Alumina		3–6	200–600	1.2–1.3	200–400
Silica gel	2	25		0.6–1.1	200–900
Activated carbon		2–4	800	0.5–0.9	400–1,200
Zeolite 5A	1		30–1,000	0.7	
Carbon molecular sieve	0.4–0.5		10–100	0.9–1.0	100–300
Polymeric adsorbents		25–40		0.4–0.8	100–700

Polymer-based adsorbents are presently used in a few operations removing organic constituents from vent gas streams, such as the removal of acetone from air. As illustrated by Fig. 9.3, these materials are usually styrene–divinylbenzene copolymers, which in some cases have been derivatized to give the desired adsorption properties.

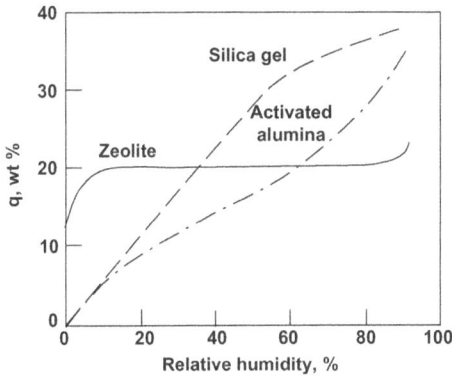

Fig. 9.2: Comparison of water adsorption on various adsorbents.

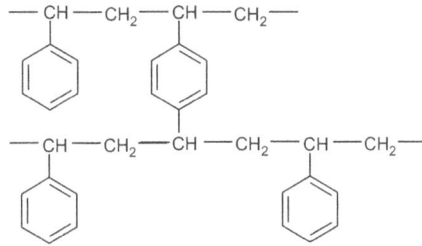

Fig. 9.3: Structure of styrene–divinylbenzene copolymer adsorbents.

In addition to equilibrium adsorption, selectivity of adsorbents may also originate from two other important separating mechanisms. Exclusion of certain molecules in the feed because they are too large to fit into the pores of the adsorbent (molecular sieving) and differences in the diffusion of different adsorbing species in the pores of the adsorbent (kinetic effect). Significant kinetic selectivity is a characteristic feature of molecular sieve adsorbents such as carbon molecular sieves and zeolites. Molecular sieve zeolites are highly crystalline aluminosilicate structures with highly regular channels and cages as displayed in Fig. 9.4. Zeolites are selective for polar, hydrophilic species, and very strong bonds are created with water, carbon dioxide and hydrogen sulfide while weaker bonds are formed with organic species. Molecular sieving is possible when a channel size is of molecular dimensions, restricting the diffusion sterically. As a result, small differences in molecular size or shape can lead to very large differences in diffusivity. In the extreme, certain molecules or a whole class of compounds may be completely excluded from the micropores. This is illustrated in Fig. 9.5, where a range of molecular sizes is compared with the channel diameters of various zeolites. The most important examples of such a processes are the separation of linear hydrocarbons from their branched and cyclic isomers using a 5A zeolite adsorbent and air separation over carbon molecular sieve or 4A zeolite, in which oxygen, the faster diffusing component, is preferentially adsorbed.

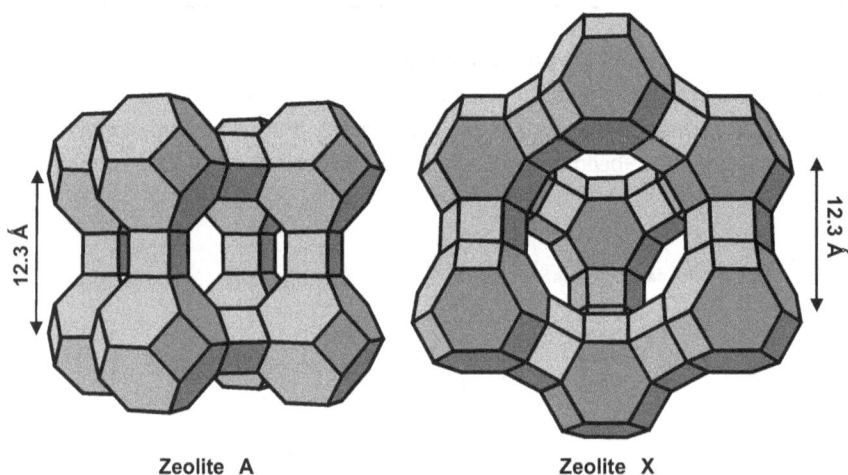

Zeolite A   Zeolite X

Fig. 9.4: Schematic diagrams of structures of two common molecular sieve zeolites.

9.2.2 Adsorption equilibrium

In adsorption, a dynamic equilibrium is established for the distribution of the solute between the fluid and the solid surface. This equilibrium is commonly expressed in the form of an isotherm, which is a diagram showing the variation of the equilibrium adsorbed-phase concentration or loading with the fluid-phase concentration or partial pressure at a fixed temperature. For pure gases, experimental physical adsorption isotherms have shapes that are classified into five types by Brunauer, as shown in Fig. 9.6. Both Types I and II are desirable isotherms, exhibiting strong adsorption.

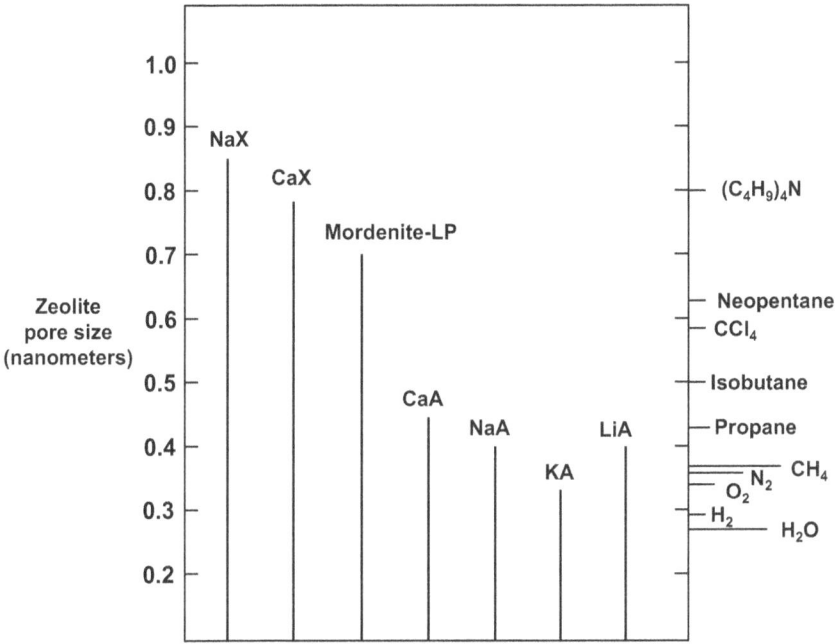

Fig. 9.5: Comparison of molecular dimensions and zeolite pore sizes.

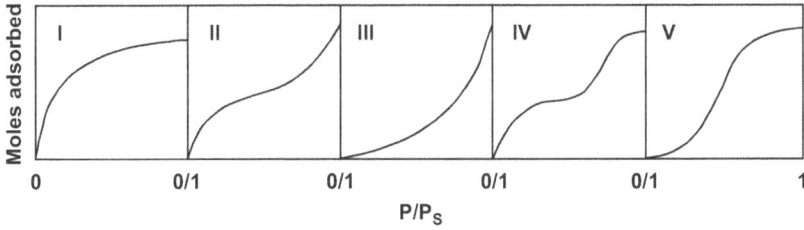

Fig. 9.6: The Brunauer classification of adsorption isotherms.

When the amount adsorbed is low, the isotherm should approach a linear form and the following form of Henry's law, called linear isotherm, is obeyed:

$$q = bp \quad \text{or} \quad q = bc \tag{9.2}$$

where q is the equilibrium loading or amount adsorbed per unit mass of adsorbent, p is the partial pressure of a gas and c is the concentration of the solute. b is a temperature-dependent empirical constant that is related to the heat of adsorption through an Arrhenius-type equation:

$$b = b_0 e^{-\Delta H_{ADS}/RT} \tag{9.3}$$

For an exothermic process, ΔH_{ads} is negative, and the Henry constant decreases with increasing temperature. At higher concentrations, competition for surface sites and significant interaction between adsorbed molecules starts to play a role. Under these conditions, many industrial systems exhibit the type I behavior, which is commonly represented by the ideal Langmuir model:

$$\frac{q}{q_S} = \frac{bc}{1+bc} \tag{9.4}$$

In the Langmuir model q_S is the saturation limit and b is again an equilibrium constant that is directly related to the Henry constant. This expression is the correct form to represent a type I isotherm since at low pressure, it approaches Henry's law, while at high pressure, it tends asymptotically to the saturation limit. Although only few systems behave exactly to the Langmuir model, it provides a simple qualitative representation of the behavior of many systems and it is therefore widely used. The Langmuir model can also be simply extended to multicomponent systems, reflecting the competition between species for the adsorption sites:

$$\frac{q_a}{q_S} = \frac{b_a c_a}{1+b_a c_a + b_b c_b + \cdots} \tag{9.5}$$

In equilibrium-based separations, the selectivity of the adsorbent is determined by the separation factor α' that for the multicomponent Langmuir model simply corresponds to the ratio of the equilibrium constants:

$$\alpha'_{AB} = \frac{q_A/c_A}{q_B/c_B} = \frac{b_A}{b_B} \tag{9.6}$$

Because this selectivity is independent of composition, the ideal Langmuir model is often referred to as the constant separation factor model.

9.2.3 Adsorption kinetics

As illustrated in Fig. 9.7, the adsorption of a solute onto the porous surface of an adsorbent requires the following steps:
1. External mass transfer of the solute from the bulk fluid through a thin film or boundary layer to the outer solid surface of the adsorbent.
2. Internal mass transfer of the solute by pore diffusion from the outer surface of the adsorbent to the inner surface of the internal pore structure.
3. Surface diffusion along the porous surface.
4. Adsorption of the solute onto the porous surface.

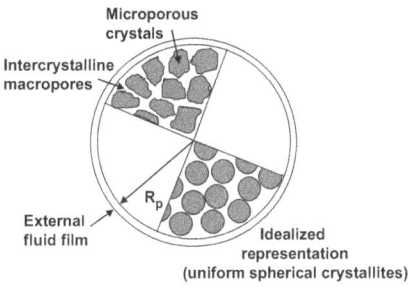

Fig. 9.7: Resistances to mass transfer in a composite adsorbent pellet (adapted from [97]).

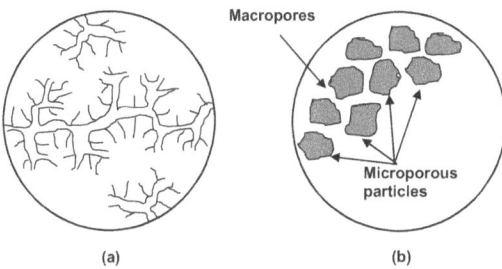

Fig. 9.8: Homogeneous and composite microporous adsorbent particle (adapted from [97]).

During regeneration of the adsorbent, the reverse of the four steps occurs. In general, the rate of physical adsorption is controlled by diffusional limitations rather than by the actual rate of equilibration at a surface. A laminar fluid film or boundary layer through which the solute must diffuse always surrounds the outside of an immersed particle. This rate of mass transfer is normally expressed in terms of a simple linear rate expression:

$$\frac{\partial q}{\partial t} = k_f a (c - c^*) \tag{9.7}$$

However, under most practical conditions, the external film resistance is seldom rate limiting, so that internal mass transfer generally controls the adsorption/ desorption rate. From this perspective, adsorbents may be divided into two broad classes: homogeneous and composite. Figure 9.8 illustrates that in homogeneous adsorbents (activated alumina, silica gel), the pore structure persists throughout the particle, while the composite adsorbent particles (zeolites, carbon molecular sieves) are formed by aggregation of small microporous microparticles. As a result, the pore size distribution in composite particles has a well-defined bimodal character with micropores within the microparticles connected through macropores within the pellet. Transport in a macropore can occur by bulk molecular diffusion, Knudsen diffusion, surface diffusion and Poiseuille flow. In liquid systems bulk molecular diffusion is generally dominant, but in the vapor phase, the contributions from Knudsen and surface diffusion may be large or even dominant. Knudsen diffusion becomes dominant when collisions with the pore wall occur more frequently than collisions with diffusing molecules. In micropores, the diffusing molecule never escapes from the force field of the pore wall and the Knudsen mechanism no longer applies. Diffusion occurs by jumps from site to site, just as in surface diffusion, and the diffusivity becomes strongly dependent on both temperature and concentration. The selectivity in a kinetically controlled adsorption process depends on both kinetic and equilibrium effects. This kinetic selectivity can be approximated when two species (A and B) diffuse independently and their isotherms are also independent. Under these conditions, the ratio of their uptakes at any time will be given by

$$\alpha_{AB} = \frac{q_A/p_A}{q_B/p_B} = \frac{b_A}{b_B} \sqrt{\frac{D_A}{D_B}} \tag{9.8}$$

A typical example is shown in Fig. 9.9, where the equilibrium adsorption and loading curves of oxygen and nitrogen on carbon molecular sieves are shown. The dramatic difference between the uptake rates is caused by the fact that the pore diameter of the carbon is just very slightly larger than the diameters of the two gases. As a result, nitrogen has much more difficulty traversing a pore than oxygen. Thus, even though there is virtually no equilibrium selectivity, the operation in the kinetic region results in very high selectivities for oxygen/nitrogen separation.

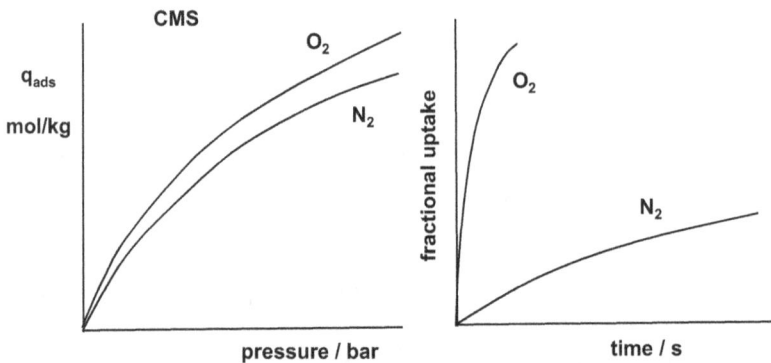

Fig. 9.9: Equilibrium adsorption and uptake rates for carbon molecular sieves used in air separation.

9.2.4 Fixed bed adsorption

In a fixed bed, a nearly solute-free liquid or gas effluent can be obtained until the adsorbent in the bed approaches saturation. Under the conditions that external and internal mass transfer resistances are very small, plug flow is achieved, axial dispersion is negligible, the adsorbent is initially free of adsorbate and the adsorption isotherm begins at the origin where equilibrium between the fluid and the adsorbent is achieved instantaneously. Under these ideal fixed bed adsorption conditions, the instantaneous equilibrium results in a shock-like wave, the stoichiometric front (Fig. 9.10), moving as a sharp concentration step through the bed. Upstream of the front, the adsorbent is saturated with adsorbate and the concentration of solute in the fluid is that of the feed. In the upstream region, the adsorbent is spent. Downstream of the stoichiometric front and in the exit fluid, the concentration of the solute in the fluid is zero, and the adsorbent is still adsorbate free. After a period of time, called the stoichiometric time, the stoichiometric wave front reaches the end of the bed and the concentration of the solute in the fluid rises abruptly to the inlet value. Because no further adsorption is possible, this point is referred to as the breakthrough point.

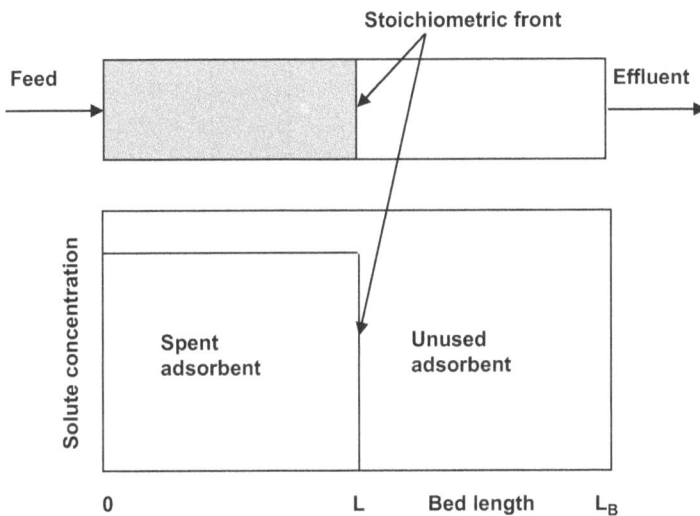

Fig. 9.10: Stoichiometric equilibrium concentration front for ideal fixed bed adsorption.

For ideal fixed bed adsorption, the location of the concentration wave front L as a function of time can be obtained from a material balance under adsorption equilibrium considerations. The adsorbent loading in equilibrium with the feed is designated by the adsorption isotherm. A material balance over the adsorbate gives the position of the front in the adsorbent bed:

$$\text{Adsorbate in} = \text{Adsorbate accumulation}$$

$$\phi_{\text{FEED}} c t_{\text{ideal}} = q_{\text{FEED}} m_{\text{adsorbent}} \frac{L}{L_{\text{BED}}} \tag{9.9}$$

In real fixed bed adsorption, internal and external transport resistance are finite and axial dispersion can be significant. These factors contribute to the development of broader S-shaped concentration fronts like those in Fig. 9.11. The S-shaped curve is called the breakthrough curve. The steepness of the breakthrough curve determines the extent to which the capacity of an adsorbent bed can be utilized. Another complication is the fact that the steepness of the concentration profiles increases or decreases with time, depending on the shape of the adsorption isotherm. If the adsorption isotherm is curved, regions of the front at a higher concentration move at a velocity different from regions at a lower concentration. Thus, for a linear isotherm, the shape remains constant. For a favorable isotherm of the Langmuir type, high-concentration regions move faster than low-concentration regions, and the wave front steepens with time until a constant pattern front is developed.

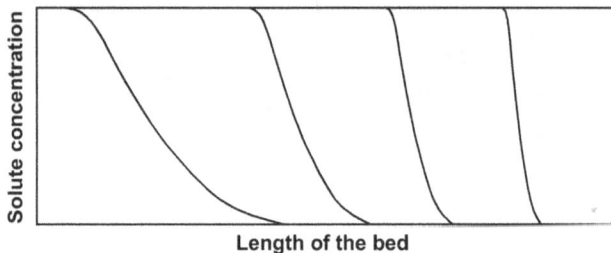

Fig. 9.11: Self-sharpening wave front caused by a favorable adsorption isotherm.

9.3 Basic adsorption cycles

Commercial adsorptions can be divided into bulk separations, in which about 10 wt% or more of a stream must be adsorbed, and purifications, in which usually considerably less than 10 wt% of a stream must be adsorbed. Such differentiation is desirable because, in general, different process cycles are used for the two categories. As schematically illustrated in Fig. 9.12, there are three basic ways to influence the loading of an adsorbate on an adsorbent:
1. Changing temperature
2. Changing (partial) pressure in a gas or concentration in a liquid
3. Adding a component which competitively adsorbs with the adsorbate of interest

A major difference in processes based on temperature change and those based on pressure or concentration change is the time required to change the bed from an adsorbing to a desorbing or regenerating condition. In short, pressure and concentration can be changed much more rapidly than temperature. The consequence is that temperature swing processes are limited almost exclusively to low adsorbate concentrations in the feed, that is, purifications. For higher adsorbate feed concentrations, the adsorption time would become very short compared to the regeneration time. In contrast,

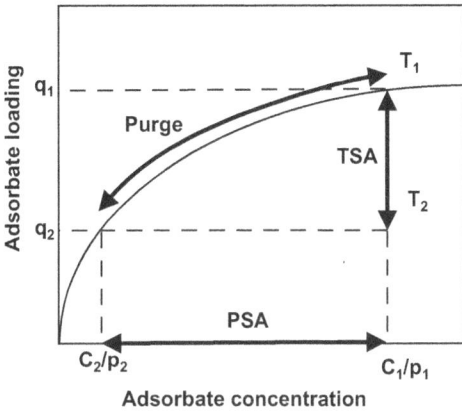

Fig. 9.12: Schematic illustration of adsorption/desorption cycles.

pressure swing or concentration swing processes are much more suitable for bulk separations because they can accept feed, change to regeneration and be back to the feed condition within a reasonable fraction of the cycle time.

9.3.1 Temperature swing

A temperature swing or thermal swing adsorption (TSA) cycle is one in which desorption takes place at a temperature much higher than adsorption. In this cycle, shown in Fig. 9.13, a stream containing a small amount of an adsorbate at concentration c_1 is passed through the adsorbent bed at temperature T_1. After equilibrium between adsorbate in the feed and on the adsorbent is reached, the bed temperature is raised to T_2, and desorption occurs by a continued feed flow through the bed. In general, the regeneration is more efficient at higher temperatures. Heating, desorbing and cooling of the bed usually takes in the range of a few hours to over a day. During this long regeneration time, the bed is not effectively separating feed. Therefore, temperature swing processes are used almost exclusively to remove small concentrations of adsorbates in order to maintain the on-stream time a significant fraction of the total process cycle time.

The most common application of TSA is drying. Zeolites, silica gel and activated aluminas are widely used in the natural gas, chemical and cryogenics industries to dry streams. Of these adsorbents, silica gel requires the lowest temperatures for regeneration. Other TSA applications range from CO_2 removal to hydrocarbon separations, and include the removal of pollutants, odors and contaminants with activated carbon.

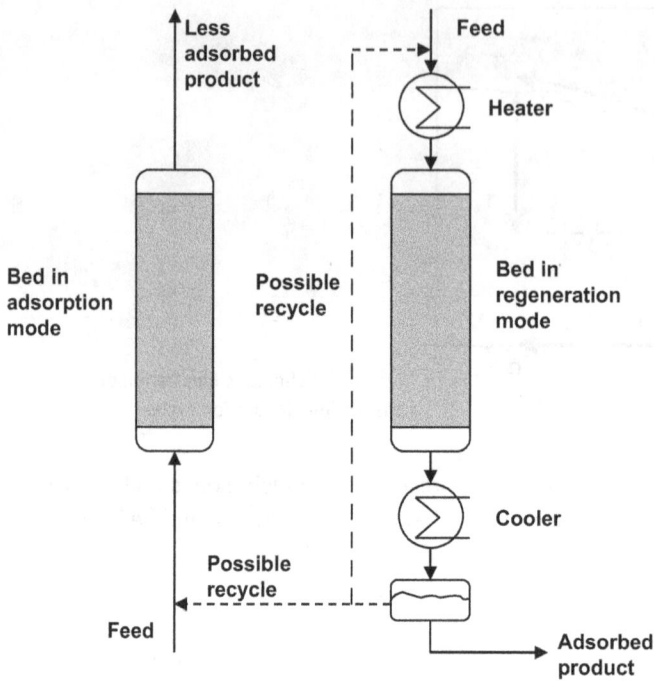

Fig. 9.13: Temperature swing adsorption (TSA).

9.3.2 Pressure swing

In a pressure swing adsorption (PSA) cycle, desorption is effected at a pressure much lower than adsorption. The most common processing scheme has two, illustrated in Fig. 9.14, or three fixed bed adsorbers alternating between the adsorption step and the desorption step. A complete cycle consists of at least three steps: adsorption, blow down and repressurization. During adsorption, the less selectively adsorbed components are recovered as product. The selectively adsorbed components are removed from the adsorbent by adequate reduction in partial pressure. During this countercurrent depressurization (blow down) step, the strongly adsorbed species are desorbed and recovered at the adsorption inlet of the bed. The repressurization step returns the adsorber to feed pressure and can be carried out with product or feed. PSA cycles operate at nearly constant temperature and require no heating or cooling steps. They utilize the exothermic heat of adsorption remaining in the adsorbent to supply the endothermic heat of desorption.

The principle application of PSA is for bulk separations where contaminants are present at high concentration. The major purification applications for PSA are for hydrogen, methane and drying. Air separation, methane enrichment and iso/normal alkane separations are the principal bulk separations for PSA. Fortunately, packed beds of adsorbent respond rapidly to changes in pressure. In most applications, equilibrium adsorption is used to obtain the desired separation. However, in air separation PSA processes, the adsorptive separation is based on kinetically limited systems where oxygen is preferentially adsorbed on 4A zeolite when the equilibrium selectivity favors N_2 adsorption.

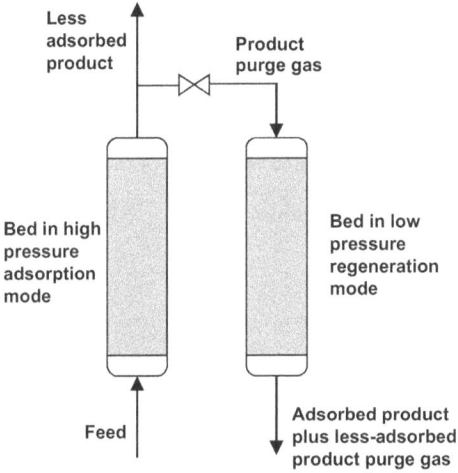

Fig. 9.14: Pressure-swing adsorption (PSA). Fig. 9.15: Inert purge cycle.

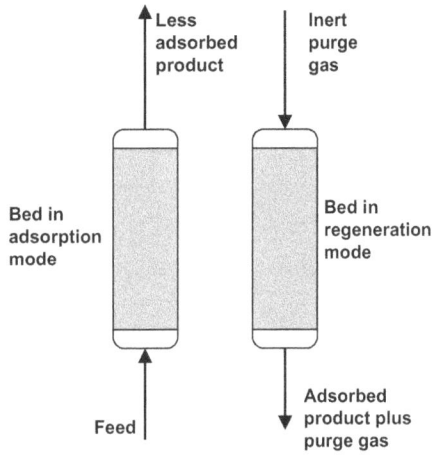

9.3.3 Inert and displacement purge cycles

In a purge swing adsorption cycle, the adsorbate is removed by passing a nonadsorbing gas or liquid, containing very little to no adsorbate, through the bed. Desorption occurs because the partial pressure or concentration of the adsorbate around the particles is lowered. If enough purge gas or liquid is passed through the bed, the adsorbate will be completely removed. Most purge swing applications use two fixed bed adsorbers to provide a continuous flow of feed and product (Fig. 9.15). Its major application is for bulk separations when contaminants are present at high concentration. Applications include the separation of normal from branched and cyclic hydrocarbons, gasoline vapor recovery and bulk drying of organics.

The displacement purge cycle differs from the inert purge cycle in that a gas or liquid that adsorbs about as strongly as the adsorbate is used to remove the adsorbate. Desorption is thus facilitated by adsorbate partial pressure or concentration reduction in the fluid around the particles in combination with competitive adsorption of the displacement medium. The use of an adsorbing displacement purge fluid causes a major difference in the process. Since it is actually absorbed, it will be present when the adsorption cycle begins and therefore contaminates the less-adsorbed product. This means that the displacement purge fluid must be recovered from both product streams, as illustrated by Fig. 9.16.

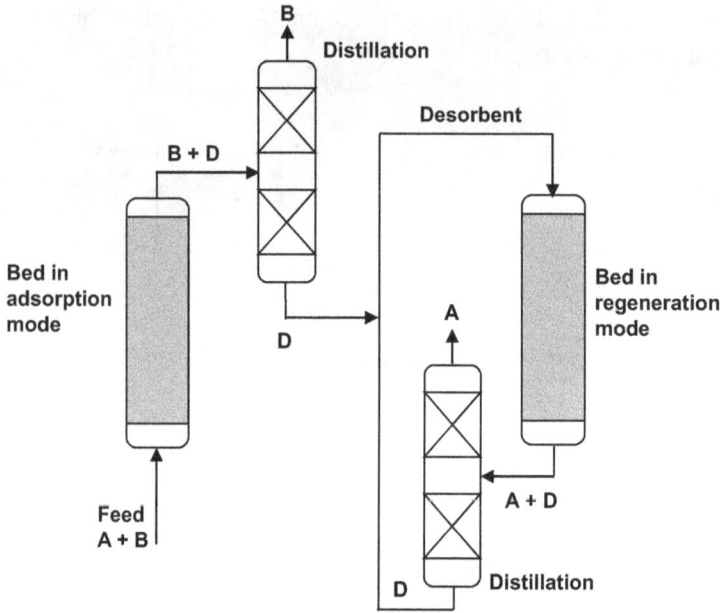

Fig. 9.16: Displacement purge cycle.

9.4 Principles of ion exchange

9.4.1 Ion exchange resins

Naturally occurring inorganic aluminosilicates (zeolites) were the first ion exchangers used in water softening. Today, synthetic organic polymer resins based on styrene or acrylic acid-type monomers are the most widely used ion exchangers. These resin particles consist of a three-dimensional polymeric network with attached ionic functional groups to the polymer backbone. Ion exchange resins are categorized by the nature of functional groups attached to a polymeric matrix, by the chemistry of the particular polymer in the matrix. Strong acid and strong base resins are based on the copolymerization of styrene and a crosslinking agent, divinylbenzene, to produce the three dimensional cross-linked structure as shown in Fig. 9.17. In strong acid cation exchange resins, sulfonic acid groups in the hydrogen form exchange hydrogen ions for the other cations present in the liquid phase:

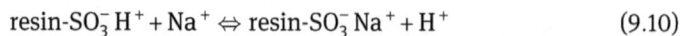

$$\text{resin-SO}_3^- \, \text{H}^+ + \text{Na}^+ \Leftrightarrow \text{resin-SO}_3^- \, \text{Na}^+ + \text{H}^+ \tag{9.10}$$

It is not always necessary for the resin to be in the hydrogen form for adsorption of cations. Softening of water is accomplished by displacing sodium ions from the resin by calcium ions, for which the resin has a greater affinity:

$$2\text{resin-SO}_3^- \text{Na}^+ + \text{Ca}^{2+} \Leftrightarrow (\text{resin-SO}_3^-)_2\text{Ca}^{2+} + 2\text{Na}^+ \tag{9.11}$$

In strong base anion exchange resins, quaternary ammonium groups are used as the functional exchange sites. They are used most often in the hydroxide form to reduce acidity:

$$\text{resin-N}^+(\text{CH}_3)_3\text{OH}^- + \text{H}^+ + \text{Cl}^- \Leftrightarrow \text{resin-N}^+(\text{CH}_3)_3\text{Cl}^- + \text{H}_2\text{O} \tag{9.12}$$

During the exchange, the resin releases hydroxide ions as anions are adsorbed from the liquid phase. The effect is elimination of acidity in the liquid and conversion of the resin to a salt form. Ion exchange reactions are reversible. After complete loading, a regeneration procedure is used to restore the resin to its original ionic form. For strong cation and anion resins, this is typically done with dilute (up to 4%) solutions of hydrochloric acid, sulfuric acid, sodium hydroxide or sodium chloride.

Weak acid cation exchanger resins have carboxylic acid groups attached to the polymeric matrix derived from the copolymerization of acrylic acid and methacrylic acid. Weak base anion exchanger resins may have primary, secondary or tertiary amines as functional groups. The tertiary amine is most common. Weak base resins are frequently preferred over strong base resins for removal of strong acids in order to take advantage of the greater ease of regeneration.

9.4.2 Equilibria and selectivity

Ion exchange differs from adsorption in that one adsorbate (a counterion) is exchanged for a solute ion, and the exchange is governed by a reversible, stoichiometric chemical reaction equation. The exchange equilibria depend largely on the type of functional group and the degree of crosslinking in the resin. The quantity of ions, acids or bases that can be adsorbed or exchanged by the resin is called the operating capacity. Operating capacities vary from one installation to another as a result of differences in composition of the stream to be treated.

In ion exchange, significant exchange does not occur unless the functional group of the resin has a greater selectivity for the ions in solution than for ions occupying the functional groups, or unless there is excess in concentration as in regeneration. This is pictured in the following reaction where the cation exchange resin removes cation B^+ from solution in exchange for A^+ on the resin:

$$\text{resin-SO}_3^- A^+ + B^+ \Leftrightarrow \text{resin-SO}_3^- B^+ + A^+ \tag{9.13}$$

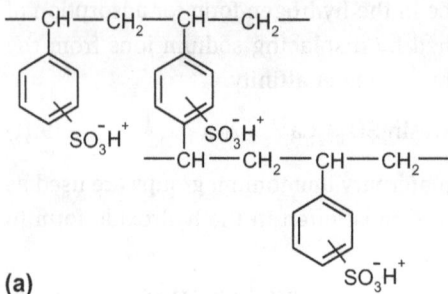

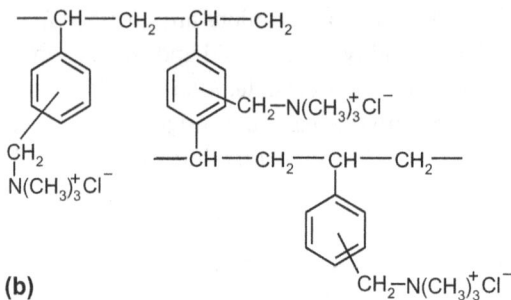

Fig. 9.17: Styrene and divinylbenzene based ion exchange resins: (a) sulfonated cation exchanger and (b) aminated anion exchanger.

For this case, we can define a conventional chemical equilibrium constant, determining the selectivity of B over A:

$$K_A^B = \frac{q_{B,\,resin}\ m_{A,\,liquid}}{q_{A,\,resin}\ m_{B,\,liquid}} \tag{9.14}$$

where m and q are the concentrations of the ions in the solution and the resin phase, respectively. A typical plot of K^B_A for univalent exchange is given in Fig. 9.18. B is preferred by the exchanger if $K^B_A > 1$, while for $K^B_A < 1$, B is less preferred and the ion exchanger preferentially absorbs species A. Selectivity for ions of the same charge usually increases with atomic weight (Li < Na < K < Rb < Cs) and selectivities for divalent ions are greater than for monovalent ions. For practical applications, it is rarely necessary to know the selectivity precisely. However, knowledge of relative differences is important when deciding if the reaction is favorable or not. Selectivity differences are marginally influenced by the degree of cross-linking of a resin. The main factor is the structure of the functional groups.

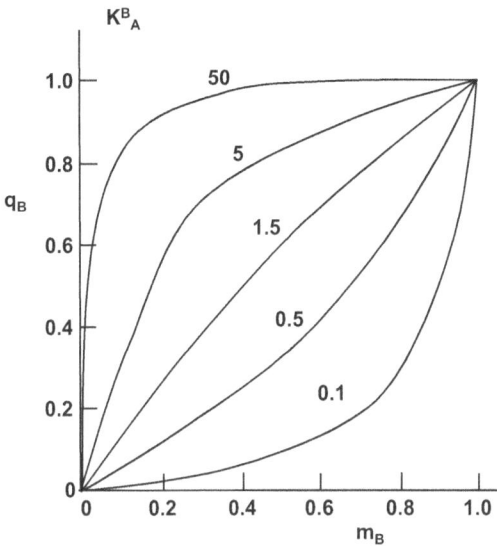

Fig. 9.18: Equilibrium plot for univalent–univalent ion exchange.

9.5 Ion exchange processes

Ion exchange cycles consist of two principal steps, adsorption and regeneration. During adsorption, impurities are removed or valuable constituents are recovered. Regeneration is, in general, of much shorter duration than the adsorption step. Industrial systems may be batch, semicontinuous or continuous. They vary from simple one-column units to numerous arrays of cation and anion exchangers, depending upon the required application and quality of the effluent. A single-column installation is satisfactory if the unit can be shut down for regeneration. However, for the processing of continuous streams, two or more columns of the same resin must be installed in parallel. Most ion exchanger columns are operated concurrently with the process stream and regeneration solution flowing down through the resin bed to avoid possible fluidization of the resin particles.

Deionization processes require two columns, one containing a cation exchanger and one an anion exchanger. The cation exchanger unit must be a strong acid-type resin and it must precede the anion exchange unit. Placing the anion exchanger first generally causes problems with precipitates of metal hydroxides. A column containing a mixture of cation and anion exchanger resins is called a mixed bed. The majority of the used resins consist of strong acid cation and strong base anion exchangers. Mixed bed systems yield higher quality deionized water or process streams than when the same resins are used in separate columns. However, regeneration of a mixed bed is more complicated than of a two-bed system.

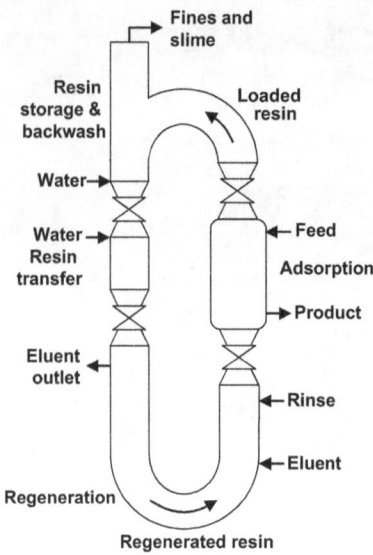

Fig. 9.19: Higgins moving bed contactor (adapted from [9]).

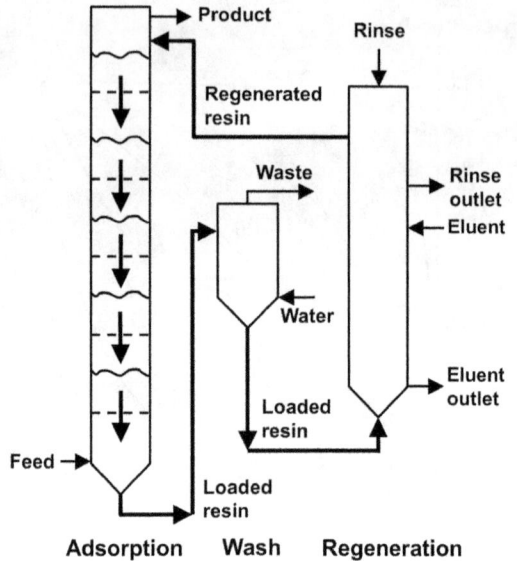

Fig. 9.20: Himsley fluidized bed contactor (adapted from [9]).

To increase resin utilization and achieve high ion exchange reaction efficiency, numerous efforts have been made to develop continuous ion exchange contactors. Two examples of them are shown in Figs. 9.19 and 9.20. The Higgins loop contactor operates as a moving, packed bed by using intermittent hydraulic pulses to move incremental portions of the bed from the contacting section to the regeneration section. In other continuous systems, such as the Himsley contactor, columns with perforated plates are used. The liquid is pumped into the bottom and flows upward through the column to fluidize the resin beads on each plate. Periodically, the flow is reversed to move incremental amounts of resin to the stage below. The resin at the bottom is lifted via the wash column to the regeneration column. A more recent approach involves the placement of a number of columns on a carousel that rotates constantly at an adjustable speed.

Nomenclature

a	Particle volume/surface area ratio (= $d_{part}/6$ for spheres)	$[m^3/m^2]$
b	Adsorption constant	$[m^3\,mol, Pa^{-1}]$
c	Concentration	$[mol/m^3]$
D	Diffusivity	$[m^2/s]$
k_f	External mass transfer coefficient	$[m/s]$
K	Equilibrium constant	$[-]$

L	Column/bed length	[m]
m	Mass	[kg]
m	Ion concentration in liquid	[mol/m^3]
p	Partial pressure of a gas	[Pa (bar)]
q	Amount adsorbed	[mol/kg (mol/m^3)]
q	Concentration ions in resin	[mol/m^3]
R	Gas constant	[8.314 J/mol K]
t	Time	[s]
T	Temperature	[K]
a	Selectivity	[−]
ϕ	Flow rate	[m^3/s]

Indices

a, b	Components
ADS	Adsorption
ideal	For ideal fixed bed adsorption
S	Maximum (monolayer) capacity
*	At equilibrium

10 Solid-liquid separation

10.1 Introduction

Solid liquid and liquid-liquid separation processes may be classified according to the involved principles. A perfect solid-liquid separation would result in a stream of liquid going one way and dry solids going another. Unfortunately, none of the separation devices works perfectly. Typically there may be some fine solids leaving in the liquid stream, and some of the liquid may leave with the bulk of the solids. This imperfection is characterized by the mass fraction of the solids recovered and the residual moisture content of the solids. Solid-liquid separation is used in many processes with the aim of recovering:
- valuable solids, the liquid being discarded
- liquid, the solids being discarded
- both the solids and the liquid
- neither but for example prevent water pollution

If the particles or droplets can move freely within the continuous liquid phase we deal with sedimentation and settling. Sedimentation exploits the density difference between the solids and the liquid for the partial separation or concentration of suspended solid particles from a liquid by gravity or centrifugal settling. These two main groups of sedimentation technology are further subdivided into the operations and equipment listed in Fig. 10.1. Additionally sedimentation may be divided into the functional operations thickening and clarification. The primary purpose of thickening is to increase the concentration of suspended solids in a feed stream, while that of clarification is to remove a relatively small quantity of suspended particles and produce a clear effluent. Separation of two liquid phases, immiscible or partially miscible liquids, is a common requirement in the process industries. The simplest form of equipment used to separate liquid phases is the gravity settling tank. Their design is largely analogous to the design of gravity settling tanks for solid-liquid separation. For difficult to separate liquid-liquid mixtures, such as emulsions, centrifugal separators are used. In addition to solid-liquid separation, it is often desirable to remove either the coarse or the fine particles from the product. This process is referred to as classification or solid-solid separation and can be achieved in many types of solid-liquid separation equipment.

   If the particles are constrained by a medium, which allows the liquid or a gas to flow through, we deal with separation by filtration. In this chapter we focus on the industrially important separation of solid particles from liquids by a porous filtration medium contained in a housing with flow of liquid in and out (Fig. 10.2). Flow

https://doi.org/10.1515/9783110712445-010

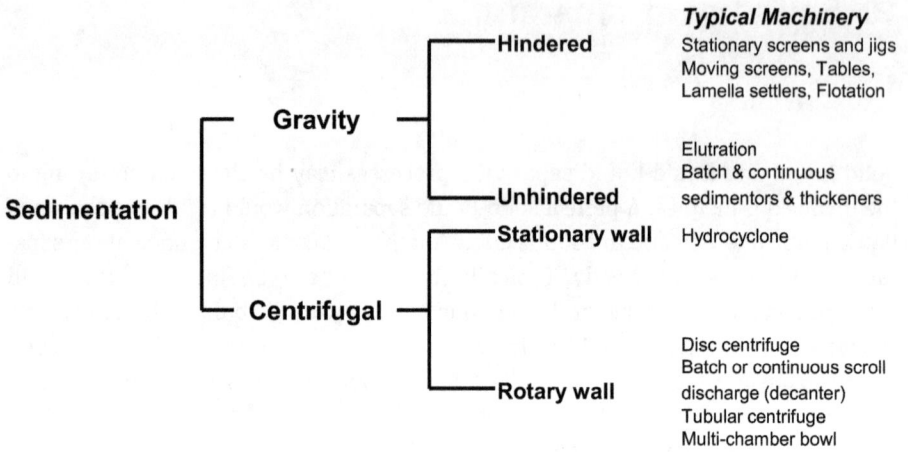

Fig. 10.1: General classification of sedimentation equipment.

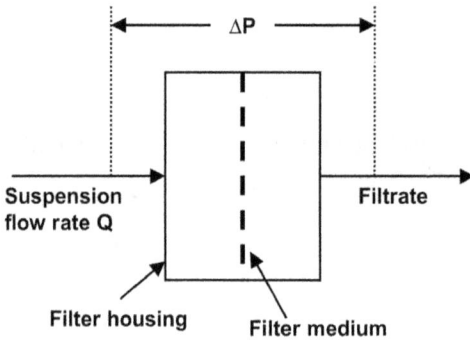

Fig. 10.2: Schematic of a filtration system.

through the filter medium is usually achieved by applying a static pressure differ-ence as the driving force.

The two most often used types of filtration are cake, or surface filtration and deep bed filtration. In cake filtration (Fig. 10.3a) separation is to be achieved on the upstream side of a relatively thin filter medium. The particles to be separated must be larger than the pores of the medium or able to form bridges to cover the pores. In the latter case, the initial breakthrough of particles will stop as soon as the bridges are formed. On top of this first particle layer, successive layers of solids de-posit and form a cake. Bridging over the pores and porosity of the cake are usually improved with relatively high feed solids concentration. The main disadvantage of dead-end filtration is the declining rate over time due to the increasing pressure drop across the growing cake. This is circumvented in cross-flow filtration systems

(Fig. 10.3b), where the slurry moves tangentially to the filter medium so that the cake is continuously sheared off and little or no cake is allowed to form on the medium.

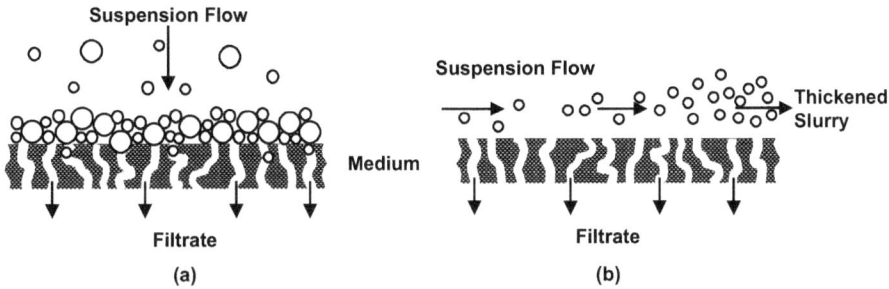

Fig. 10.3: Schematic of the cake (a) and cross-flow and (b) filtration mechanisms.

Cake filters are used in clarification of liquids, recovery of solids, dewatering of solids, thickening of slurries and washing of solids. The filtration medium may be fitted into various forms of equipment that can be operated in several modes. Vacuum operation is widely used in the industry and the laboratory. In the filters a low pressure is maintained downstream the filter medium by vacuum pumps. Pressure filters operate at pressure levels above atmospheric. The pressure differential created across the medium causes flow of fluid through the equipment. In centrifugal filtration, flow of liquid is created by the centrifugal force resulting from spinning the suspension. Such centrifugal filters are found in many applications in the food, beverage and pharmaceutical industries. The large range of machinery shown in Fig. 10.4 reflects the uncertainty involved in the processing of solids, particularly those in small particle size ranges.

An important first step in the rationalization of solid-liquid separation problems is to choose between sedimentation, filtration or their combination. Filtration is typically used when it is desired to obtain relatively dry solids or when the application of sedimentation techniques becomes impractical due to small density differences, small particle diameters or low solids concentrations. Combination with sedimentation may have the advantage of a pre-thickened suspension. As with other unit operations, filtration is never complete. Some of the solids may leave in the liquid stream, and some liquid will be entrained with the separated cake. Separation of solids is measured by the fraction recovered while the separation of liquid is usually characterized through the slurry moisture content or solids concentration. Although filter media such as a woven wire mesh easily retain coarse materials, "screens" with smaller openings or pores such as nonwoven cloths or membranes are required as the size of the particulates decreases. When the particulates are small compared to the medium pores, deposition may occur in the depths of the filter medium. This so-called deep bed filtration is

		Discharge	Typical Machinery
Filtration	**Gravity**	**Discontinuous**	Strainer or Nutsche Sand-Charcoal Filter
		Continuous	Grids, Sieve bends Rotary screen Vibratory screen
	Vacuum	**Discontinuous**	Nutsche filter Candle and cartridge
		Continuous	Table or pan filter Rotary drum or disc filter Horizontal belt
	Pressure	**Discontinuous**	Pressure Nutsche Plate and frame filter Tube, candle and leaf filter Cartridge filter
		Continuous	Belt press Screw press
	Centrifugal	**Discontinuous**	Basket centrifuge Peeler centrifuge
		Continuous	Pusher, vibratory and tumbler Centrifuge Helical conveyor Conical & cylindrical basket centrifuge

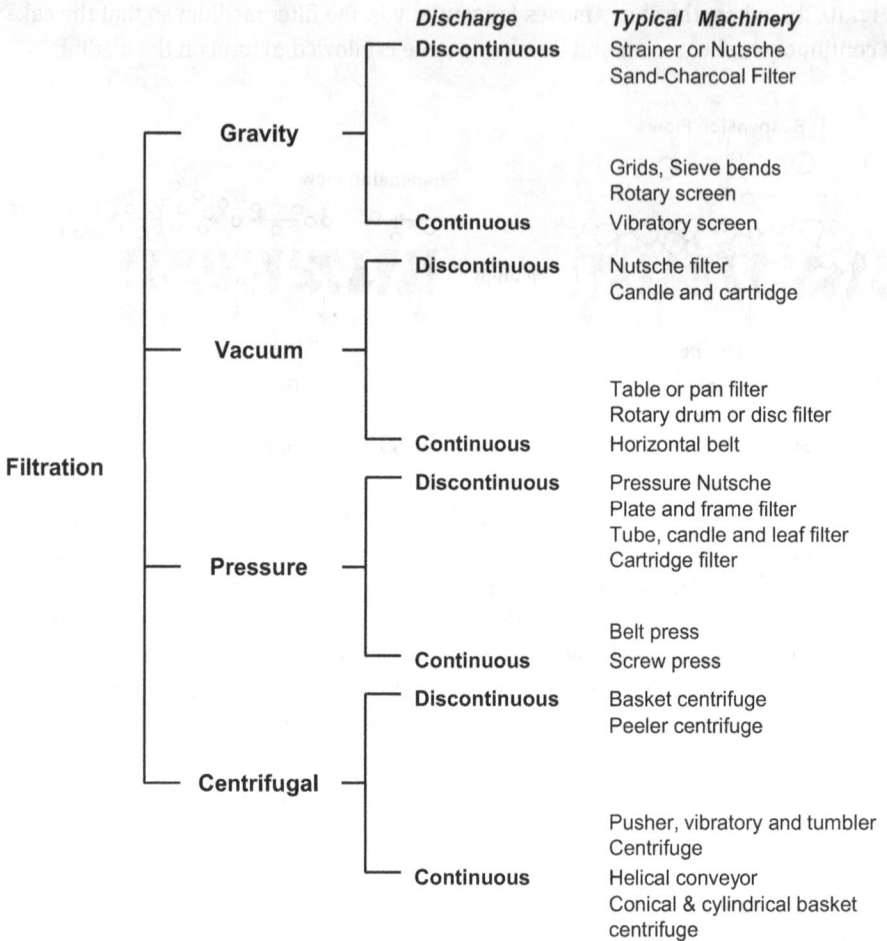

Fig. 10.4: General classification of filtration equipment.

fundamentally different from cake filtration in principle and application. The filter medium is a deep bed with pore sizes much greater than the particles it is meant to remove. Particles penetrate into the medium where they separate due to gravity settling, diffusion and inertial forces. Deep bed filters were developed for potable water treatment as the final polishing process. They find increasing application in the treatment of industrial wastewater.

Filtration is frequently combined with several other processes to enhance its performance. Washing is applied to replace the mother liquid in the solids stream with a clean wash liquid. Especially for high purity products, washing may represent a dominant portion of the total filtration cost. The three most common techniques are washing by displacement, re-slurrying and successive dilution. Dewatering is used to reduce the moisture content of filter cakes either by mechanical compression of the filter cake or by displacement with air. It is enhanced by addition of dewatering aids to the suspension in the form of surfactants that reduce surface tension. Another important aspect to improve the performance of a filtration step may be conditioning or pretreatment of the feed suspension. Common techniques are coagulation, flocculation and the addition of inert filter aids. Coagulation and flocculation increase the effective particle size with the accompanying benefits of higher settling rates, higher permeability of filter cakes and better particle retention in deep bed filters. Coagulation brings particles into contact to form agglomerates. Flocculation agents such as natural or synthetic polyelectrolytes interconnect colloidal particles into giant flocs up to 10 mm in size.

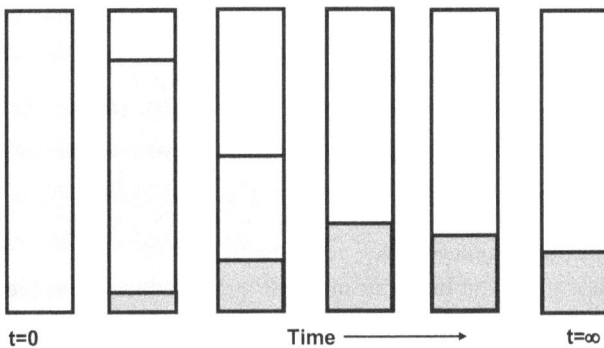

t=0   Time ━━━━━━▶   t=∞

Fig. 10.5: Schematic of a batch sedimentation experiment.

10.2 Gravity sedimentation

Gravity sedimentation is a process of solid-liquid separation under the effect of gravity. A slurry feed is separated into an underflow slurry of higher solids concentration and an overflow of substantially clear liquid. Difference in density between the solids and the suspending liquid, as well as sufficient particle size (to prevent particle dispersion by Brownian motion), are necessary prerequisites. Flocculation agents are often used to enhance settling.

Sedimentation is used in industry for solid-liquid separation and solid-solid separation. In solid-liquid separation, the solids are removed from the liquid either because the solids or the liquid are valuable or because these have to be separated before disposal. If the clarity of the overflow is of primary importance, the process is called clarification and the feed slurry is usually dilute. If the primary purpose is the production of concentrated slurry, the process is called thickening and the feed slurry is usually more concentrated. In solid-solid separation, the solids are separated into fractions according to size, density, shape or other particle properties. Sedimentation is also used for size separation of solids. One of the simplest ways to remove the coarse or dense solids from a feed suspension is by sedimentation. Successive decantation produces closely controlled size fractions of the product. In all cases, the assessment of the sedimentation behavior of the solids within the fluids will allow the correct size of vessel to be determined. It is therefore important to know about the way in which settling solids behave during sedimentation.

10.2.1 Sedimentation mechanisms

Gravity sedimentation is the separation of particles from fluids under the effect of gravity. The particle sedimentation rates are dependent upon particle properties such as size, size distribution, shape and density. Materials with particle diameters of the order of a few microns settle too slow for most practical operations, and particles of one micron or smaller do not sediment at all because of dispersion caused by Brownian motion. Wherever possible, such particles are agglomerated or flocculated into relatively large clumps called flocs that settle out more rapidly. Spherical or near-spherical particles and agglomerates settle faster than plate or needle-like particles of similar weights. Other readily recognized factors to be considered are the density and viscosity of the surrounding medium.

The main factors that determine the behavior of a settling suspension are the concentration of the particulate solids and the state of aggregation of the particles. Their effect on the characteristics of sedimentation is best understood through analyzing a batch settling experiment, as in Fig. 10.5. Solid particles without the tendency to cohere with each other generally settle at a steady rate (depending on local solids concentration) and are described as discrete particles. At low solids concentrations, the individual particles are generally able to settle as individuals, while the fluid which is displaced flows upward between them. Regardless of their properties, the particles are sufficiently far apart to settle freely in this dilute

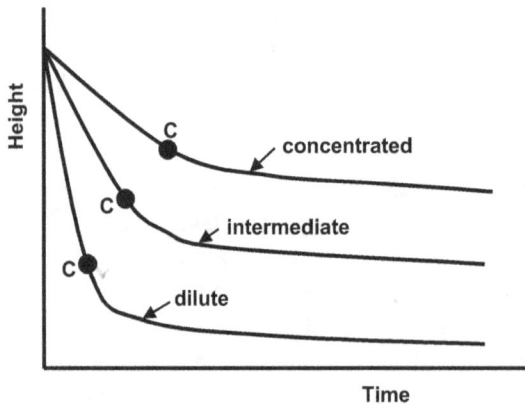

Fig. 10.6: **Effect of concentration on sedimentation.**

sedimentation region. Contact between two flocculent particles may result in cohesion, resulting in an increase in size and hence a more rapidly sedimenting particle.

Increasing the particle concentration in a fluid decreases the settling rate of each individual particle. This phenomenon, known as hindered settling, is readily appreciated if it is considered that the settling of each particle is accompanied by a return flow of the sedimentation liquid. Since the fluid is unable to pass through the particles, its velocity must increase to compensate for this partial blocking of the flow channel. The relative velocity effect does not account completely for the slowing of sedimentation with increases in solids concentration. Hydrodynamic interactions between particles become stronger at higher particle concentration. Particle interference by collision and coagulation are other factors. For flocculent particles, the effect of initial solids concentration on sedimentation behavior has been observed to exhibit three quite distinct modes. In dilute suspensions, the individual particles or flocs again behave as discrete particles. At intermediate concentrations the flocs, which are in loose mutual contact, settle by channeling. The channels are about the same size as the particles and develop during an induction period in which an increasing quantity of fluid forces its way through the mass.

Particles near the bottom of the cylinder pile up into a concentrated sludge, whose height increases as more particles settle. This continues until the suspension zone disappears and all the solids are contained in the sediment. This condition, illustrated in Fig. 10.6, is known as the critical sedimentation point. Until this point is reached, the solid-liquid interface follows an approximately linear relationship with time. In such concentrated suspensions, however, fluid flow is only possible through minute voids between the primary particles. The resistance of the touching particles below drastically reduces the sedimentation rate to a relatively low compaction rate. In this compression regime the rate of sedimentation is a function of both the solids concentration and the depth of settled material in the tank. Particles closer to the base will be compressed by the mass of solids above, resulting in more concentrated sediment by slowly expelling the liquid, which accompanies the flocs into the deposit. This continues until equilibrium is established between the weight of the flocs and their mechanical strength. The discussed effects of particle coherence and concentration of the settling characteristics of a feed suspension is summarized in Fig. 10.7. It is important to realize that although the feed stream may start in one regime, it may pass through all regimes during clarification or thickening.

10.2.2 Rate of sedimentation

Newton's second law of motion describes the movement of a solid particle through a viscous liquid under the influence of gravity. For a free-falling particle, the nonstationary force balance reads

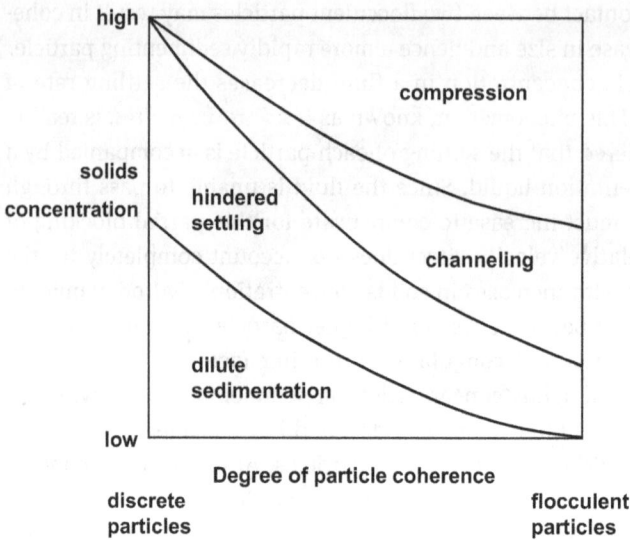

Fig. 10.7: Effect of particle coherence and solids concentration on the settling characteristics of a suspension.

$$\text{acceleration force} = \text{gravitational force} + \text{buoyant force} + \text{drag force} \qquad (10.1)$$

According to Archimedes' principle, the buoyant force is opposite to the gravitational force and is equal to the weight of the liquid displaced by the particle. The drag force is opposite to the velocity v of the particle. An important parameter that characterizes the nature of liquid flow around the particles is the particle Reynolds number Re_p.

For small spherical particles (*typically* 0.1 mm or less) that sediment under laminar flow conditions, *Stokes' law* can be used to describe the drag force. Together, the force balance reads

$$\frac{\pi d_p^3}{6} \rho_s \frac{dv}{dt} = \frac{\pi d_p^3}{6}(\rho_s - \rho)g - 3\pi\eta d_p v \qquad (10.2)$$

After a short initial acceleration period, the particle will attain a constant velocity that is calculated by setting the acceleration term to zero. This gives the so-called terminal settling velocity v_t of a single small sphere in a viscous liquid:

$$v_t = \frac{d_p^2}{18\eta}(\rho_s - \rho)g \qquad Re_p = \frac{\rho v_t d_p}{\eta} < 0.2 \qquad (10.3)$$

For larger spherical particles (*typically* >0.1 mm), the Stokes drag force in eq. (10.2) must be replaced by a suitable correlation for the drag force applicable to the transition from laminar to turbulent flow conditions, such as the empirical Schiller and Naumann relation:

$$F_{drag} = \begin{cases} -3\pi\eta d_p v_t(1 + 0.15\text{Re}_p^{0.687}) & 0.2 < \text{Re}_p < 1{,}000 \\ -0.44\left(\frac{\pi}{8}\rho d_p^2 v_t^2\right) & \text{Re}_p > 1{,}000 \end{cases} \tag{10.4}$$

Note that for $0.2 < \text{Re}_p < 1{,}000$, the terminal settling velocity v_t, can only be obtained iteratively.

Equations (10.3) and (10.4) show that the sedimentation of a particle is determined by the physical characteristics of the particle and the continuous phase. Increased sedimentation rates are obtained for larger particle diameters, greater density difference between the particle and the continuous phase and lower viscosity of the continuous phase. However, as the concentration of the suspension increases, the particles get closer together and no longer settle as individual particles. Hydrodynamic interactions between particles become stronger and the fluid must move in the opposite direction, leading to a reduction in the rate of sedimentation in a swarm of particles, compared to that of a single particle. A well-known and simple empirical relation between the hindered settling velocity v_s of the swarm of particles and the (single particle) terminal settling velocity v_t is the Richardson–Zaki relation:

$$v_s = v_t \varepsilon^n \tag{10.5}$$

where the void fraction ε is the volume fraction of the fluid defined by

$$\varepsilon = \frac{V_l}{V_l + V_p} \tag{10.6}$$

The exponent n in eq. (10.5) is dependent on the flow regime. For very small particles in the laminar flow regime ($\text{Re}_p \ll 1$) the exponent n tends to 4.7, while for relatively large particles (for which $\text{Re}_p \gg 1$) the exponent n drops down to 2.35. To good approximation, n can be related to the particle's Reynolds number (evaluated for the terminal velocity of a single particle) through the implicit relation

$$\frac{4.7 - n}{n - 2.35} = 0.175\text{Re}_p^{0.75} \tag{10.7}$$

Note that the Richardson–Zaki relation is known to be inaccurate for very dense suspensions, when the voidage drops below approximately 0.5. Moreover, these correlations apply only to cases where flocculation is absent. Suspensions of fine particles often flocculate and therefore show different behavior.

Inlet

H

$\rightarrow$V$_{LIQ}$

Outlet

v$_t$

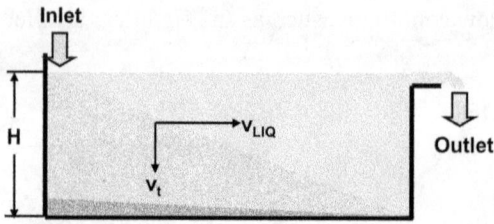

Fig. 10.8: Tank for continuous removal of solids particles from a process liquid.

10.2.3 Design of continuous sedimentation tanks

Figure 10.8 illustrates a tank for the continuous removal of solids particles from a process liquid. The liquid is introduced at one end of the tank and flows toward the outlet at the other end. The dispersed particles are separated out and fall to the bottom with their settling velocity v_s. These processes are described most simply by the ideal continuous sedimentation tank model, which equates the required settling time of the suspension and the residence time derived from the horizontal flow of the liquid. The residence time t of the liquid in the tank is obtained by dividing the volume V of the tank by the volumetric flow rate of liquid Q:

$$t = \frac{V}{Q} = \frac{AH}{Q} \tag{10.8}$$

where H is the tank depth and A its area. During the same time t, the suspension must have time to fall to the bottom of the tank. Note that in reality, because of sedimentation, the solid concentration will be higher closer to the bottom of the tank, but in the ideal model this is neglected and the initial suspension voidage is used everywhere to estimate v_s. Thus

$$t = \frac{H}{v_s} \tag{10.9}$$

Since the time is equal in both cases, the two expressions may be equated and the depth of the tank is eliminated from the equations:

$$A = \frac{Q}{v_s} \tag{10.10}$$

Two important conclusions may be drawn from eq. (10.10). The first is that the height H of the tank does not influence the throughput. The second is that the throughput of this type of tank is directly proportional to the area that can be utilized for separation. Accordingly an increased throughput Q is obtained when the tank area A is increased through fitting in a number of horizontal plates as illustrated in Fig. 10.9. This increases the number of separation channels to N, giving for the total throughput of the tank:

$$Q = v_s NA \tag{10.11}$$

It is therefore the total area NA that determines the throughput. In the case of continuous separation, horizontal channels will eventually become clogged with sediment and separation will cease. If inclined plates are used, as shown in Fig. 10.10, the sediment slides down the plates under the influence of gravity and collects at the bottom of the tank.

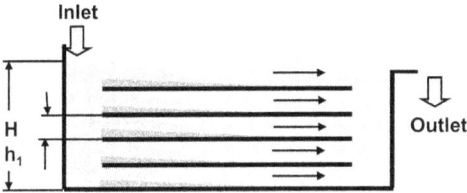

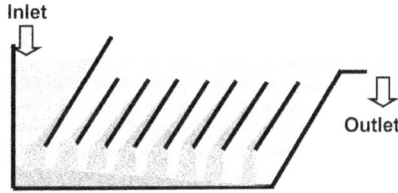

Fig. 10.9: Tank with horizontal plates.   Fig. 10.10: Tank with inclined plates.

10.2.4 Gravity sedimentation equipment

Two distinct forms of sedimentation vessels are in common usage. The clarifier is used for the clarification of a dilute suspension to obtain an overflow containing minimal suspended solids. In a thickener, the suspension is concentrated to obtain an underflow with a high solids content while also producing a clarified overflow. Sedimentation equipment can be divided into batch settling tanks and continuous thickeners of clarifiers. Most commercial equipment is built for continuous sedimentation in relatively simple settling tanks.

The largest user of clarifiers is probably the water-treatment industry. The conventional one-pass clarifier shown in Fig. 10.11 uses horizontal flow in circular or rectangular vessels with the feed at one end and overflow at the other. The feed is pre-flocculated in a paddle flocculator. Settled solids are pushed to a discharge trench by paddles or blades on a chain mechanism or suspended from a travelling bridge. Circular basin clarifiers are most commonly fed through a centrally located feed well. The overflow is led into a trough around the periphery of the basin. The bottom gently slopes to the center and the settled solids are pushed down by a number of scraper blades. The conventional one-pass clarifier is designed for the lowest specific overflow rate, which is typically 1–3 m/h depending on the degree of flocculation.

The most common thickener is the circular basin type shown in Fig. 10.12. After treatment with flocculant, the feed stream enters the central feed well, which dissipates the stream's kinetic energy and disperses it gently into the thickener. In an operating thickener, the downward increasing solids concentration gives stability to the process. The settling solids and some liquid move downward. Most of the liquid

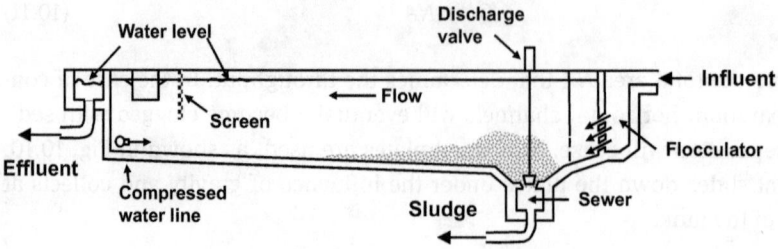

Fig. 10.11: A schematic diagram of a clarifier with a paddle flocculator.

flows upward and into the overflow, which is collected in a trough around the periphery of the basin. Thickeners are widely used, particularly in the mineral processing industry and in wastewater treatment. Stacking of sedimentation units in vertical arrangements increases the capacity per unit area. A development in this category is the lamella thickener (Fig. 10.13), which consists of a number of inclined plates stacked closely together. The feed enters the stack from the side feed box. The flow moves upward between the plates while the solids settle onto the plate surfaces and slide down into the sludge hopper underneath.

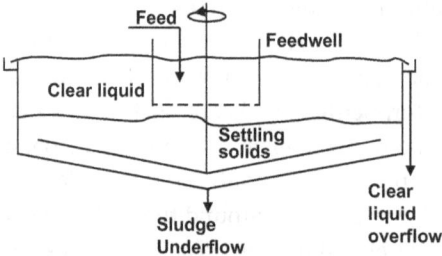

Fig. 10.12: The circular basin continuous thickener.

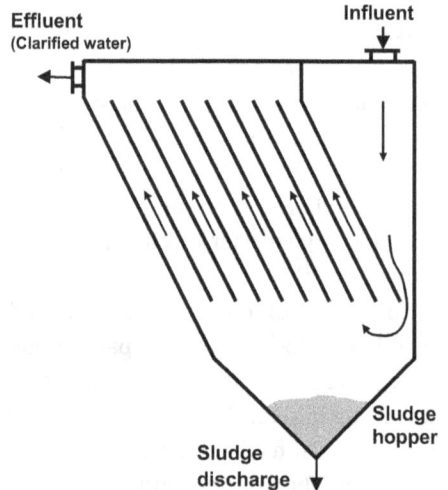

Fig. 10.13: A lamella thickener.

10.3 Centrifugal sedimentation

Centrifugal sedimentation increases the force on particles over that provided by gravity and extends sedimentation to finer particle sizes and to emulsions that are normally stable (i.e., dispersed by Brownian motion) in the gravity field. Centrifugation equipment is divided into and rotating wall (sedimenting centrifuges) and fixed-wall (hydrocyclones) devices.

10.3.1 Particle velocity in a centrifugal field

In a liquid-filled rotating vessel, a particle that is corotating with the liquid at angular velocity w at a distance r from the axis of rotation, feels a centrifugal acceleration given by

$$a = rw^2 \tag{10.12}$$

which acts away from the axis of rotation. As a consequence of this force, particles in suspension move radially outward. Replacing the gravitational acceleration g in eq. (10.3) by the acceleration in a centrifugal field provides the sedimentation velocity of small particles at low solids concentration in a rotating vessel. However, unlike in gravity settling where the acceleration is constant, the radial particle velocity in a centrifugal field depends on the radial distance from the center because the radius r appears in the acceleration expression. Hence, to derive the design equations for centrifugal sedimenting machines, the sedimentation velocity must be written in the differential form:

$$v_t = \frac{dr}{dt} = \frac{d_p^2}{18\eta}(\rho_s - \rho)rw^2 \quad (\text{Re}_P < 1) \tag{10.13}$$

Comparison of eq. (10.13) with eq. (10.3) provides the commonly quoted g-factor:

$$g - \text{factor} = \frac{rw^2}{g} \tag{10.14}$$

10.3.2 Sedimenting centrifuges

Industrial centrifuges are commonly divided into batch, continuous and semi-continuous. The common laboratory centrifuge is a simple batch bottle centrifuge designed to handle small batches of material for laboratory separations. The basic structure is usually a motor-driven vertical spindle supporting various heads or rotors. There are three types of rotors: swinging bucket, fixed-angle head or small perforate or imperforate baskets for larger quantities of material. The bottle centrifuge is often used for preliminary testing to provide the basis for the design of commercial centrifuges.

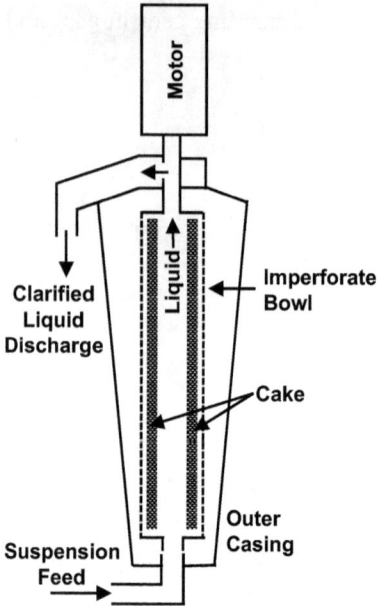

Fig. 10.14: A tubular centrifuge.

Process centrifuges are more complex and available in a variety of sizes and types. Tubular centrifuges are used to separate liquid-liquid mixtures or to clarify liquid-solid mixtures with less than 1% solids content and fine particles. The solids are collected at the bowl wall and removed manually when sufficient bowl cake has accumulated. Liquid is discharged continuously. As the name suggests, these machines (Fig. 10.14) have long tubular bowls that rotate around their vertical axis. Feed material is introduced at the base of the rotor. The longer the feed material spends in the bowl, the longer the centrifugal force is allowed to act on the particles, resulting in a progressively clarified feed stream as it flows up the length of the tubular bowl. Multichamber centrifuges, Fig. 10.15, utilize a closed bowl that is subdivided into a number of concentric vertical cylindrical compartments through which the suspension flows in series. Their efficiency is high because of the reduced travelling distance to the collecting surface. Cleaning of multichamber centrifuges is more difficult and takes longer than for the tubular type.

Centrifuges that channel feed through a large number of conical disks combine high flow rates with high theoretical capacity factors. The basic idea of increasing the settling capacity by using a number of layers in parallel is the same as the lamella principle in gravity sedimentation. Both liquid-liquid and liquid-solid separations are

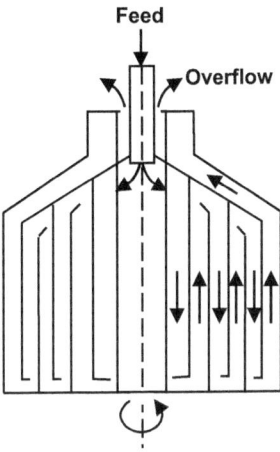

Fig. 10.15: Schematic of a multichamber bowl.

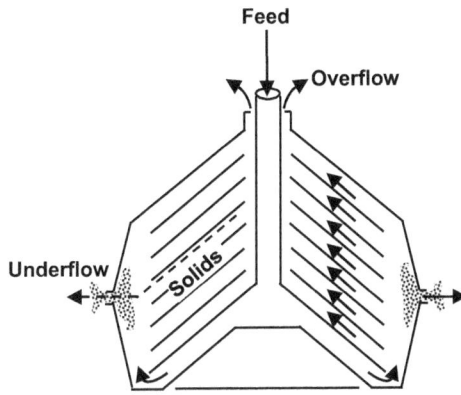

Fig. 10.16: Clarifying disk stack centrifuge.

performed for slurries with solids concentrations below 15% and small particle sizes. The general flow patterns in a disk stack centrifuge are illustrated in Fig. 10.16. Feed enters near the center of the bowl from either the top or the bottom. The clarified medium is discharged at a relatively small radius, generally at the top of the bowl. Solids are collected at the underside of the disks, slide outward along the surfaces and finally move from the outer edges of the disks to the bowl wall by free settling. Continuous solids discharge is achieved by sloping the inner walls of the bowl toward the discharge point. Generally disk centrifuges have the best ability to collect fine particles at a high rate.

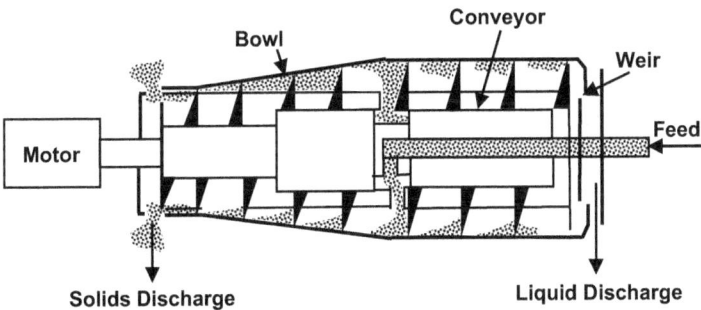

Fig. 10.17: Continuous-scroll discharge decanter.

Scroll centrifuges or decanters discharge solids continuously and usually drier than disk and imperforate bowl centrifuges. The feed is introduced through a stationary axial tube. Solids are collected on the bowl wall by sedimentation and continuously moved up a sloping beach by a helical screw conveyor operating at a differential speed with respect to the bowl. As shown in Fig. 10.17, the solids discharge is usually at a radius smaller than that of the liquid. Centrifugal fields are lower than in disk or tubular centrifuges because of the conveyor and its associated mechanism. For clarification, this type of centrifuge recovers medium and coarse particles from feeds at high or low solids concentration.

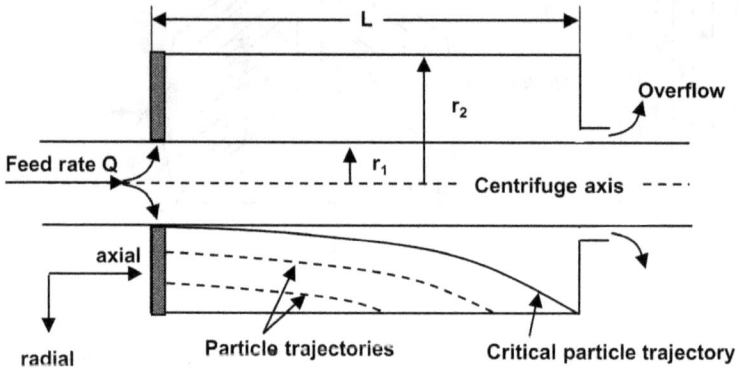

Fig. 10.18: Critical particle trajectory for 100% capture efficiency.

10.3.3 Bowl centrifuge separation capability

The separation capability of cylindrical bowl centrifuges can be analyzed by equating the time required for settling of a spherical particle to the bowl wall to the time required for the feed liquid to travel to the discharge. In the model derivation plug flow is assumed to apply for the liquid flowing down the machine axis. The particles are considered to settle at low solid fraction under Stokes conditions and reach their terminal settling velocity the moment they enter the centrifuge pond. The particles that reach the wall of the centrifuge are removed from the system while particles that do not reach the wall will be swept out of the machine in the overflow. Under these conditions the particle which is just captured within the centrifuge will travel along the critical particle trajectory shown in Fig. 10.18. The critical particle trajectory reflects the particle diameter for which the trajectory goes from the top surface (inner radius) of the centrifuge bowl to the bottom surface (outer radius) in the residence time within the machine. Particles of a similar diameter entering the machine at a radial position between r_1 and r_2 will not present a problem as they will follow a parallel trajectory and intercept the wall before the end of the machine. The residence time of the particle in the axial direction equals

$$t = \frac{V_C}{Q}$$

(10.15)

where V_C is the volume of the centrifuge and Q is the volume flow rate of material fed to the machine. The radial velocity of the sedimenting particles is given by eq. (10.13). Integrating this equation using the limits of $r = r_1$ at $t = 0$ and $r = r_2$ at $t = t$, and rearranging for the sedimentation time gives

$$t = \frac{18\eta \, \ln(r_2/r_1)}{d_p^2 \, (\rho_S - \rho) \, \omega^2} \tag{10.16}$$

Equating the equation for the sedimentation time to that for the residence time yields the bowl centrifuge separation capability:

$$\frac{Q}{V_C} = \frac{d_p^2 \, (\rho_S - \rho) \, \omega^2}{18\eta \, \ln(r_2/r_1)} \tag{10.17}$$

10.3.4 Hydrocyclones

Hydrocyclones offer one of the least expensive means of solid-liquid separation from both an operating and an investment viewpoint. They are cheap, compact, versatile and similar in operation to a centrifuge, but with much larger values of g-force. This force is, however, applied over a much shorter residence time. The most significant difference with a centrifuge is that centrifugal forces are generated without the need for mechanically moving parts other than a pump. The energy needed for the rotation of the liquid is delivered by the velocity of that liquid. Cyclones have been employed to remove solids and liquids from gases and solids from liquids and are operated at temperatures as high as 1,000 °C and pressures up to 500 bar.

The principal features and flow patterns are shown schematically in Fig. 10.19. In a hydrocyclone, the liquid path involves a double vortex with the liquid spiraling downward at the outside and upward at the inside. The primary vortex at the outside carries suspended material down the axis of the hydrocyclone. In the inside, material is carried up the axis and into the overflow vortex finder by the secondary vortex. The overflow usually consists of a dilute suspension of fine solids, while the underflow is a concentrated suspension of more coarse solids. Depending on their design, either the thickening or classifying action of the hydrocyclone is enhanced. The long cone shown in Fig. 10.20 provides thicker underflow concentrations, but poorer sharpness of separation than the long cylinder. The vortex finder is important in reducing the loss of unclassified material.

As a first approximation, a hydrocyclone can be considered as a rolled up settling chamber of area A in which gravitational acceleration is replaced by centrifugal acceleration. Equation (10.13) shows that the sedimentation velocity v_t depends on the particle diameter d_p, leading to a separation efficiency that is expected to scale with d_p^2 up to a critical diameter d_0 determined by the settler criterion:

Fine particles
Overflow

Vortex finder

Slurry rotation develops
high rotational forces
throughout cyclone

Pressurized slurry
enters tangentially

Fine particles move
inward and upward
as spiraling vortex

Coarse particles driven
toward wall and downward
in accelerating spiral

Apex

Underflow
Coarse particles

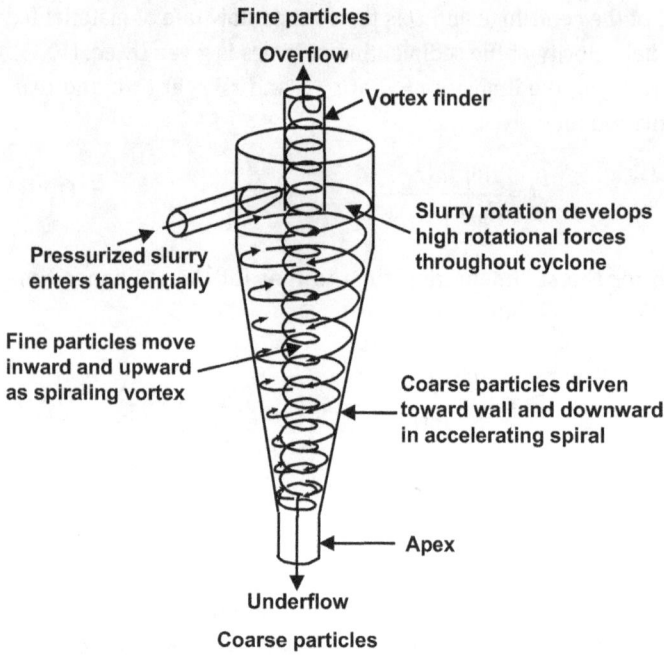

Fig. 10.19: Principal features and flows inside a hydrocyclone.

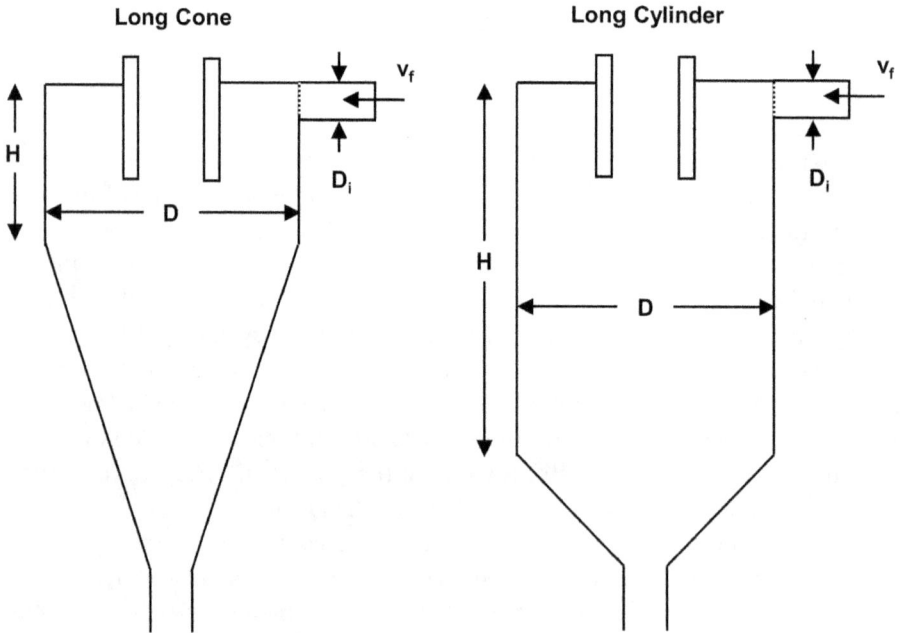

Long Cone

Long Cylinder

v_f

v_f

H

D_i

D

D_i

H

D

Fig. 10.20: Basic hydrocyclone designs.

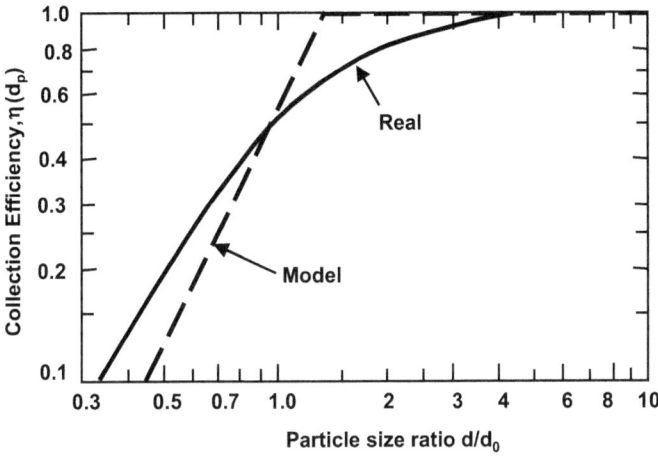

Fig. 10.21: Comparison between theory (line) and practice for the separation efficiency of a hydrocyclone.

$$\frac{v_t A}{Q} > 1 \qquad (10.18)$$

All particles with a diameter beyond this d_0 are expected to be separated. Although Fig. 10.21 illustrates that this criterion gives a reasonable indication where the separation can be expected, the distinction between particles that are and are not separated by hydrocyclones is not that sharp in reality. The poor sharpness of separation can be overcome by employing hydrocyclones in series, as illustrated in Fig. 10.22. High retention efficiency of solids is obtained in hydrocyclones of low diameter. This imposes limitations on the throughput of a single device. Therefore it is common to install packages with multiple hydrocyclone units operating in parallel.

10.4 Filtration fundamentals

10.4.1 Flow through packed beds

During its flow through a packed bed of solids, the liquid passes through the open space between the particles. The part of the total bed volume available for fluid flow is called the porosity:

$$\varepsilon = \frac{\text{volume of voids}}{\text{total bed volume}} \qquad (10.19)$$

As the liquid flows through the bed, friction with the surface of the solid packing leads to a pressure drop. In filtration, generally very small particles are used,

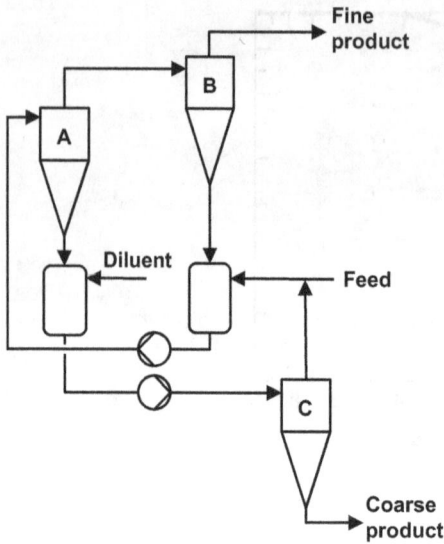

Fig. 10.22: Multicyclone arrangement for sharper separation.

leading to a dominance of the first, laminar, term in the Ergun equation (4.1). In this Darcy flow limit, the filtrate velocity v_F is proportional to the imposed pressure difference ΔP and inversely proportional to the viscosity of the flowing fluid η and the sum of the cake resistance R_C and the resistance of the filter medium R_M. This is schematically shown in Fig. 10.23 and may be written as follows:

$$v_F = \frac{\Delta P}{\eta \left(R_M + R_C \right)} \tag{10.20}$$

Implicitly it is assumed that the resistance of the filter medium does not change during the process. Although true for the medium, filter cake resistances can vary over a wide range, depending on porosity, particle shape and packing, particle size and distribution, cake formation rate, slurry concentration. This is the reason that the relations derived for the resistance of idealized particle beds such as the Ergun equation should only be used as guidelines and are not applicable to practical filtration operations. In filtration, it is common practice to deduce an empirical permeability from simple laboratory tests or existing operating data. Finally the filtrate velocity is replaced by the amount of filtrate dV collected in period dt divided by the filter cross-sectional area A:

$$v_F = \frac{1}{A} \frac{dV}{dt} \tag{10.21}$$

superficial filtrate velocity

$$V_F = \frac{1}{A}\frac{dV}{dt}$$

porous cake

filter medium

V_F

pressure drop ΔP

slope: $\eta \cdot R_{tot}$

superficial filtrate velocity v_F

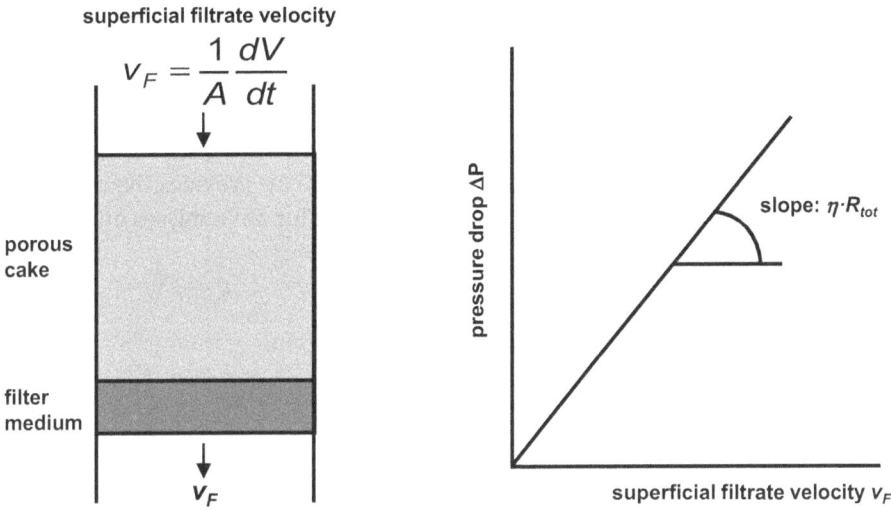

Fig. 10.23: Schematic of flow through a porous medium.

10.4.2 Cake filtration

The mathematical description of the cake filtration process shown in Fig. 10.3a starts by combining both relations for the filtrate velocity into Darcy's law to relate the filtrate flow rate and pressure drop:

$$\Delta P = \frac{\eta (R_M + R_C)}{A}\frac{dV}{dt} \tag{10.22}$$

During filtration, the deposition of solids increases the cake height, which is accompanied by an increase in the total cake resistance. A material that exhibits a linear increase in cake resistance with the cake height shown in Fig. 10.24 is known as an incompressible cake. In an incompressible cake filtration process, the solids concentration or porosity of the filter cake remains constant and the cake volume increases by a constant amount for each unit volume of suspension. However, when filtering at constant pressure the rate of filtration and solids deposition declines as shown in Fig. 10.25, because each new element of filter cake increases the total resistance to filtrate flow through the deposited cake. The total resistance of the cake can be represented through the proportionality constant α and the deposited mass of dry solids per unit area of the filter w:

$$R_C = \alpha w \tag{10.23}$$

The proportionality constant α is known as the specific cake resistance and has the units m/kg. The amount of deposited dry solids per unit area can be obtained from the dry solids concentration c in the suspension and the total amount of filtrate V divided by the filter area:

$$w = \frac{cV}{A} \qquad (10.24)$$

Introduction of the specific cake resistance into eq. (10.22) provides the two resistances in series form of Darcy's law we will use in the further analyses of filtration processes:

$$\Delta P = \frac{\eta c \alpha}{A^2} V \frac{dV}{dt} + \frac{\eta R_M}{A} \frac{dV}{dt} \qquad (10.25)$$

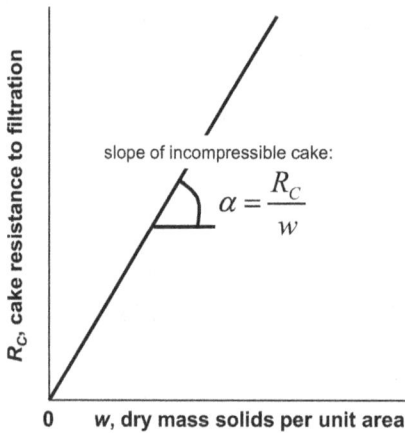

Fig. 10.24: Specific resistance of an incompressible cake.

slope of incompressible cake:

$$\alpha = \frac{R_C}{w}$$

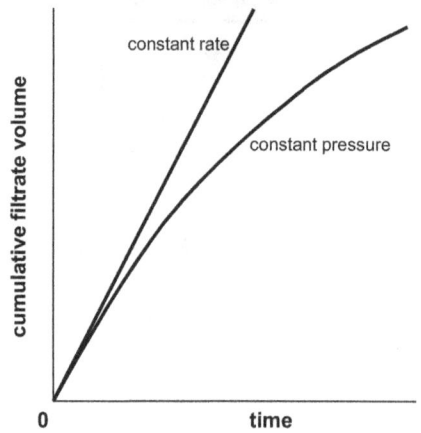

Fig. 10.25: Declining filtrate rate during constant pressure filtration.

10.4.3 Constant pressure and constant rate filtration

For an incompressible cake this modified Darcy's law equation contains three variables and five constants. The equation can be solved analytically only if one of the three variables (time, filtrate volume, and pressure) is held constant. This reflects the two main operation modes of industrial filters. Vacuum filtration tends to be under constant pressure and pressure filtration is often under constant rate. Under constant differential pressure, eq. (10.25) can be rearranged and integrated using the boundary condition of zero filtrate at zero time. After integration the following equation, known as the linearized parabolic rate law for constant pressure filtration, is obtained:

$$\frac{t}{V} = \frac{\eta c \alpha}{2A^2 \Delta P} V + \frac{\eta R_M}{A \Delta P} \qquad (10.26)$$

This is a straight line, where t/V is the dependent and V is the independent variable. Thus a graph of the experimental data points of t/V against V permits determination of the slope and the intercept from which the specific cake resistance and resistance of the filter medium can be calculated. Constant rate filtration is encountered when a positive displacement pump feeds a pressure filter. Due to the increasing cake resistance, the pressure delivered by the pump must increase during the filtration process to maintain a constant filtration rate. Constant rate filtration is easily observed on a plot of filtrate volume against time, as illustrated in Fig. 10.26. Because under these circumstances:

$$\frac{dV}{dt} = \frac{V}{t} = \text{constant} \tag{10.27}$$

Equation (11.9) can be rearranged to give

$$\Delta P = \left(\frac{\eta c \alpha}{A^2}\frac{V}{t}\right) V + \left(\frac{\eta R_M}{A}\frac{V}{t}\right) \tag{10.28}$$

which is again a straight line when the filtration pressure is plotted against the filtrate volume. In the same manner as with constant pressure filtration, the slope and the intercept taken from the graph provides the values for the specific cake and filter medium resistance.

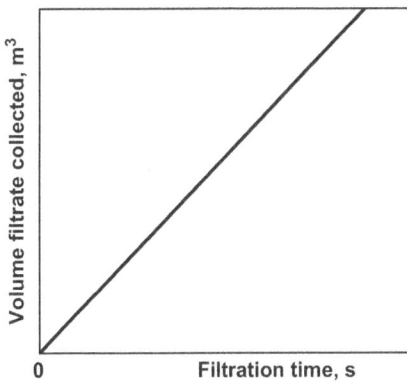

Fig. 10.26: Constant rate filtration.

10.5 Filtration equipment

An important factor in the optimization of particular processes is the thickness of the filter cake. Too thick filter cakes lead to an uneconomic lengthening of the filter cycle due to low filtering, dewatering and washing rates. On the other hand, thin filter cakes may be difficult to remove from the equipment, again increasing the filter cycle time. The productivity of all filters is related to the time required to complete a full filtration cycle and may be described as

$$\text{Productivity} = \frac{\text{Volume of filtrate per cycle}}{\text{Cycle time}} = \frac{cV}{t_C} \qquad (10.29)$$

Besides filtration, additional time is required for dewatering, cake washing and finally discharge of the filter cake and cleaning/reassembly/filling of the filter. It is usual to lump the latter two and other not mentioned operations into a total downtime period t_{DW}. The overall cycle time is then given by

$$t_C = t_F + t_D + t_W + t_{DW} \qquad (10.30)$$

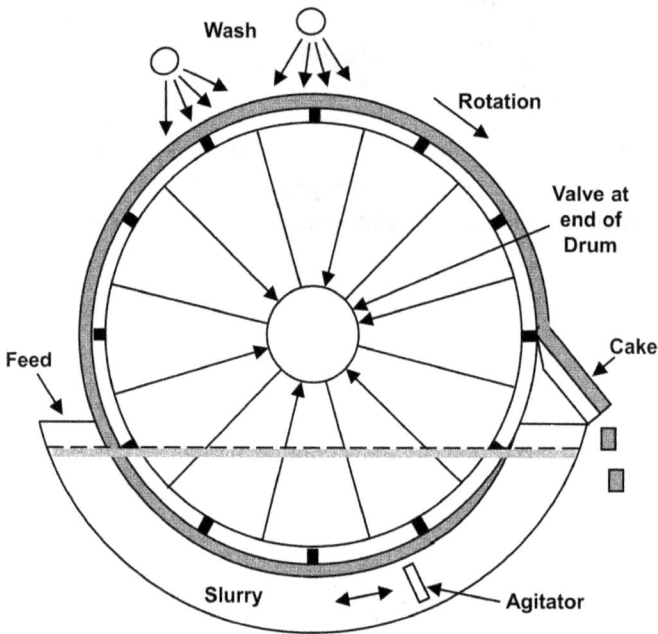

Fig. 10.27: Schematic of a rotary vacuum drum filter.

10.5.1 Continuous large-scale vacuum filters

Vacuum filters are the only truly continuous filters that can provide for washing, drying and other process requirements on a large scale. Examples include rotary drum, rotary disc, horizontal tilting pan and horizontal belt filters. Many of them use a horizontal filtering surface with the cake forming on top. The suspension is delivered to the filter at atmospheric pressure, and vacuum is applied on the filtrate side of the medium to create the driving force for filtration. The rotary vacuum drum filter is the most popular vacuum filter. As depicted in Fig. 10.27, the filtering surface of rotating drum filters is usually situated on the outer face of a cylindrical drum that rotates slowly about its horizontal axis and is partially submerged in a slurry reservoir. The drum surface is covered with a cloth filter medium and divided into independent longitudinal sections that are connected to the vacuum source by a circular connector or rotary valve. Filtration takes place when a section is submerged in the feed slurry. After cake formation, dewatering and washing by vacuum displacement takes place, followed by cake discharge at the end of the rotational cycle. In some applications, compression rolls or belts are used to close possible cracks or to further dewater the cake by mechanical compression. Cake discharge can be affected by knife, belt or string and roller discharge. Compared to drum filters, significant savings in required floor space and costs is possible with vertical disc filters. Rotary disk filters use a number of disks mounted vertically on a horizontal shaft and suspended in a slurry reservoir. The feed suspension is supplied continuously into troughs, in which the suspension flow is arranged in the same direction as the rotating discs. A particular disadvantage of disk filters is the inefficient cake washing and difficult cloth washing.

Horizontal filters largely circumvent the geometric constraints inherent in the design of rotary drum and vertical disc units. They have the advantage that gravity settling can take place before the vacuum is applied, and are ideal for cake washing, dewatering and other process operations such as leaching. Of the available industrial units in this category, the horizontal belt filter is the most popular. A schematic diagram of a typical unit is shown in Fig. 10.28. The feed slurry is supplied to the upper surface of the horizontal filter cloth that is supported by an endless belt situated above the vacuum box filtrate receivers. The top strand of the endless belt is used for filtration, cake washing and drying. There is appreciable flexibility in the relative areas allocated to each cycle step. Efficient cake discharge can be affected by separation of the belt from the cloth and directing the latter over a set of discharge rollers. Here the produced cake is cracked and discharged by a sharp turn in the cloth over a small diameter roller. In module-type horizontal belt filters, stainless steel or plastic trays replace the endless rubber belt. Horizontal belt filters are well suited to either fast or slowly draining solids, especially where washing requirements are critical. The primary advantages of this filter are its simple design and low maintenance costs. The main disadvantage is the difficulty of handling very fast filtering materials on a large scale.

10.5.2 Batch vacuum filters

The best-known batch vacuum filter units include Nutsche, horizontal table and vertical leaf filters. The Nutsche filter (Fig. 10.29) is a scaled-up version of the simple Buchner funnel consisting of a tank divided into two compartments by a horizontal

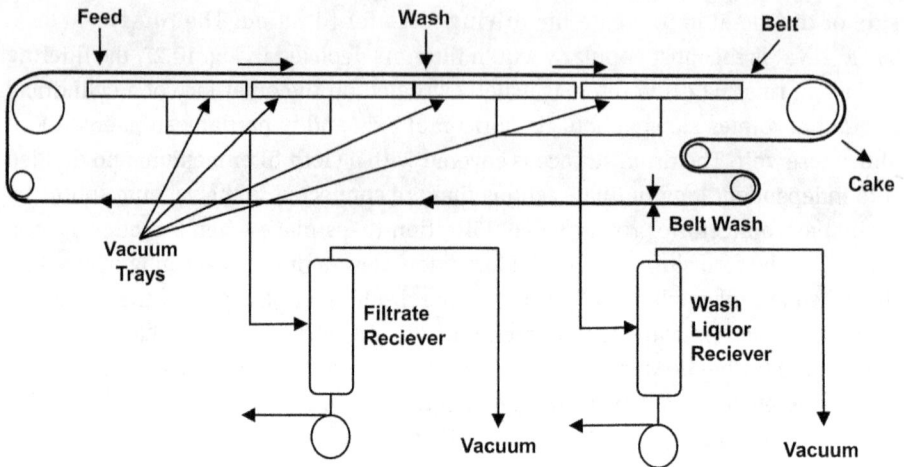

Fig. 10.28: Schematic of a horizontal belt filter (adapted from [7]).

filter medium supported by a filter plate. Vacuum is applied to the lower compartment where the filtrate is collected. These filters are particularly advantageous for separations where it is necessary to keep batches separated and when rigorous washing is required. They are simple in design but laborious and prone to high wear in cake discharge. This problem is circumvented with the mechanized Nutsche filter shown in Fig. 10.30, which is provided with a shaft carrying a stirrer passing through the cover. The stirrer can sweep the whole filter area and can be lowered or raised vertically as required. The agitators are fitted to facilitate slurry agitation, cake smoothening prior to washing and cake removal. A discharge door is provided at the edge and the rotor moves the cake toward the door. Enclosed agitated filters are useful when volatile solvents are in use or when the solvent gives off toxic vapor or fume. Another advantage is that their operation does not require any manual labor. The horizontal table filter has overall features similar to the other horizontal vacuum filters, except that during filtration, washing and dewatering, the filter element is stationary. This facilitates optimization of filter cake thickness, wash times, etc. The totally closed system is opened to allow band movement for cake discharge.

10.5.3 Pressure filters

In circumstances where large separating areas may be necessary because of slow settling characteristics, poor filterability, high solids content or other factors, the use of pressure filtration is beneficial. Pressure filters can be operated at constant pressure differential or at constant flow rate. At constant pressure differential, the liquid flow rate will decrease with time while at constant flow rate the pressure differential needs

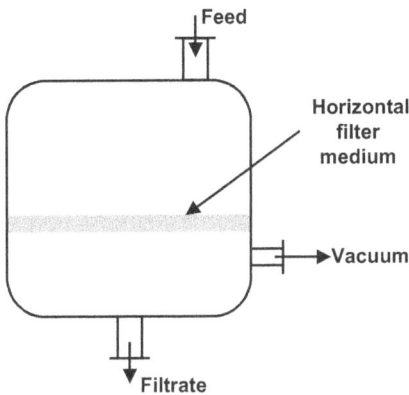

Fig. 10.29: Nutsche filter.

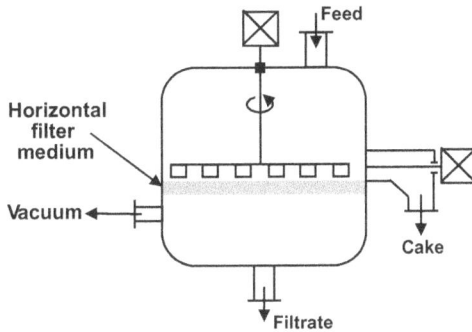

Fig. 10.30: Mechanized Nutsche filter.

to increase to compensate for increasing cake resistance. Within the extremely large variety of pressure filters three main groups may be distinguished:
1. Plate and frame filter presses
2. Pressure vessels containing tubular or flat filter elements
3. Variable-chamber presses

Various types of plate-and-frame filter presses, designed for cake formation and squeezing, are available for use within the process industry. The conventional plate-and-frame filter press shown in Fig. 10.31 contains a sequence of perforated square, or rectangular, plates mounted on suitable supports alternating with hollow frames and pressed together with hydraulic screw-driven rams. The plates are covered with a filter cloth, which also forms the sealing gasket. Most units are operated batch wise. Aftering frames with slurry the filtrate is drained through the plates and the machine is disassembled or opened for cake discharge. Washing is performed by introducing the wash liquid either through the main feed port or through a separate port behind the filter cloth. Plate-and-frame filter presses are most versatile since their effective area is easily varied by blanking off some of the plates and cake holding capacity altered by changing the frame thickness. In deciding the overall economics of the process the time taken in discharging the cake and refitting the filter is of great importance.

All pressure vessel filter units consist of a multitude of leaf, candle or cartridge filter elements mounted horizontally or vertically in a pressure vessels housing (Fig. 10.32). Vertical vessels with vertical leaf or candle filters are the cheapest of the pressure vessel filters and have the lowest volume-to-area ratio. In order to avoid filter cake bridging between the elements, serious attention has to be given to candle or leaf spacing. Deposition on the outer surface is advantageous in view of the increase in area with cake growth. Horizontal leaf filters consist of a stack of

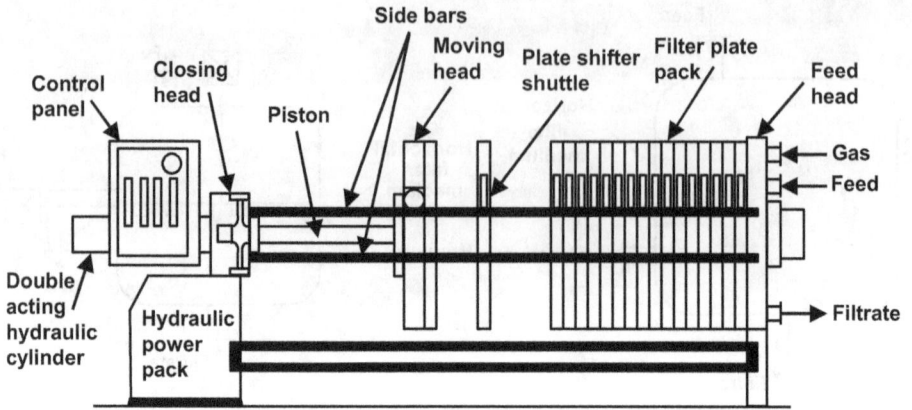

Fig. 10.31: Typical arrangement of a plate-and-frame filter press.

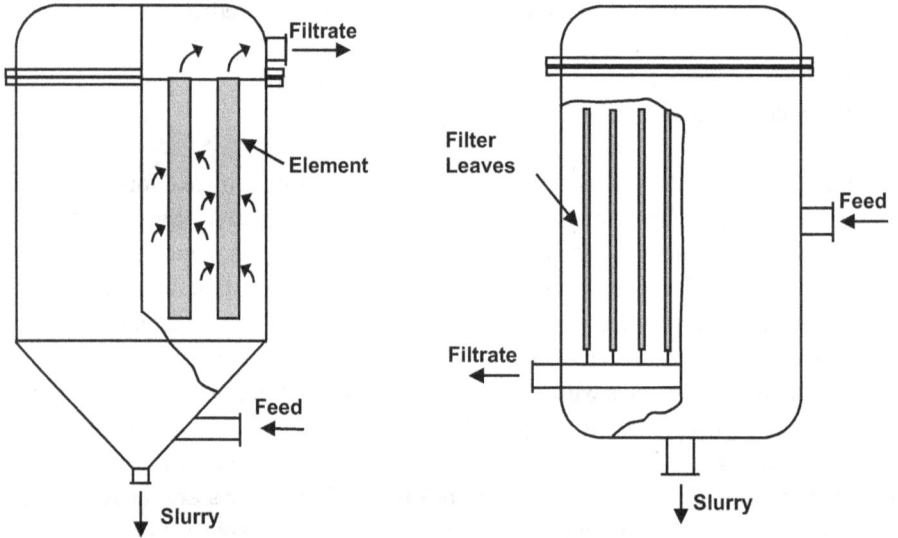

Fig. 10.32: A vertical candle (a) and vertical leaf (b) pressure vessel filter.

rectangular horizontal trays mounted inside the vessel that can be withdrawn for cake discharge. They have the disadvantage that half the filtration area is lost because the underside of the leaf cannot be used for filtration. Cartridge filtration is limited to liquid polishing or clarification in order to keep the frequency of cartridge replacements down.

Most continuous pressure filters have their roots in vacuum filtration technology. They have been adapted to pressure by enclosing in a pressure cover. As such one finds continuous disk, drum and belt pressure filters. Special designs of continuous pressure filters are belt presses and screw presses. Belt presses combine gravity drainage with mechanical squeezing of the cake between two running belts. In screw presses a screw mounted inside a perforated cage conveys the material along the barrel (Fig. 10.33). The available volume for transport diminishes continuously along the length of the screw in order to compress the filter cake. Washing liquid can be injected at points along the length of the cage. Screw presses are only suitable for the dewatering of high solids containing pastes or sludges, because no filtration stage is included.

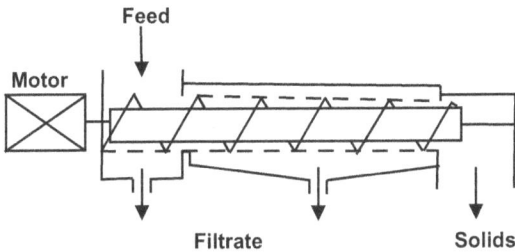

Fig. 10.33: Schematic of a screw press.

10.6 Filter media

The filter medium is the heart of any filtration step, and the importance of a careful selection cannot be overstated. Many industrial process difficulties relate to the interaction between impinging particles and the pores of the filter medium. Selection of the filter medium, however, depends largely on experience in small and large-scale operations. The performance of the filter medium may be characterized by the resistance to filtrate flow, clarity of filtrate, and durability. Resistance to flow depends to a great extent on the porosity or free area of the medium, which itself is dependent on the material used and the type of manufacturing. Clarity of the filtrate depends on the pore size of the filter medium, the particle size and the ability of the particles to form bridges to cover the pores. The ideal circumstance, where all particles are retained on the surface of the filter is often not realized. Particle penetration into cloth or membrane pores leads to an increase in resistance of the medium to the flow of filtrate.

A wide variety of filter media are available to the user. Table 10.1 gives an overview of some important filter media and their characteristics. For most applications the medium of particular interest will be the one that is readily installed in the filter to be used in the process. Thus, woven and non-woven fabrics, constructed from natural (cotton, silk, wool) or synthetic fibers, are probably the most common industrial filter medium in pressure, vacuum and centrifugal filters. Wire cloths and meshes, produced by weaving monofilaments of ferrous or non-ferrous metals, are also widely used in industrial filtrations. At the small scale filter papers are common. The same materials and also rigid porous media (porous ceramics, sintered metals, woven wires) can be incorporated into cartridge and candle filters. These filter elements are usually constructed in the form of a cylinder. Cartridge filters are either depth or surface type and widely used throughout the whole process industry for the clarification of liquids. They are particularly useful at low solid contaminant concentration and particle sizes smaller than 40 μm.

Tab. 10.1: Overview of filter media.

Filter medium	Examples	% Free area	Minimum trapped particle size (μm)
Woven fabrics	Cloths, natural and synthetic fibers	20–40	10
Rigid porous media	Ceramics, sintered metal, glass	50–70	1
			3
Metal sheets	Perforated plates, woven wire	5–20	100
			5
Non-woven sheets	Felts paper, mats	60–80	10
			5
Loose solids	Sand, diatomaceous earth, perlite, carbon	60–95	

If the specific cake resistance is too high or the cake is too compressible, the addition of filter aids can improve the filtration rate considerably. Filter aids are rigid, porous and highly permeable powders that are applied as a pre-coat which then acts as a filter medium on a coarse support, or mixed with the feed suspension as body feed to increase the permeability of the resulting cake. A pre-coat of filter aids allows filtration of very fine or compressible solids from suspensions of 5% or lower solids concentration. In fact, we speak here about deep bed filtration. In body feed filtration, the filter aid serves as a body builder to obtain high cake porosities in order to maintain high flow rates. Materials suitable as filter aids include diatomaceous earth, expanded perilitic rock, cellulose, non-activated carbon, ashes, ground chalk, or mixtures of those materials.

10.7 Centrifugal filtration

In a centrifugal filter the suspension is directed on the inner surface of a perforated rotating bowl to effect the separation of liquid from the solids. Centrifugal filters essentially consist of a rotating basket equipped with an appropriate filter medium

(Fig. 10.34). The driving force for filtration is the centrifugal force acting on the fluid. During cake formation, the filtrate passes radially outward through the filter medium and the bowl. Because the centrifugal forces tend to pull out the liquid from the cake, filtering centrifuges are excellent for dewatering applications. Solids removal may take place continuously or batch-wise. In comparison to other filtration equipment, centrifuges show good separation performance at the expense of relatively high costs. This is due to required special foundations to absorb vibrations combined with high manufacturing and maintenance costs associated with the parts rotating at high speed. Centrifuges are generally only applicable to the coarser particles, typically 10 µm to 10 mm.

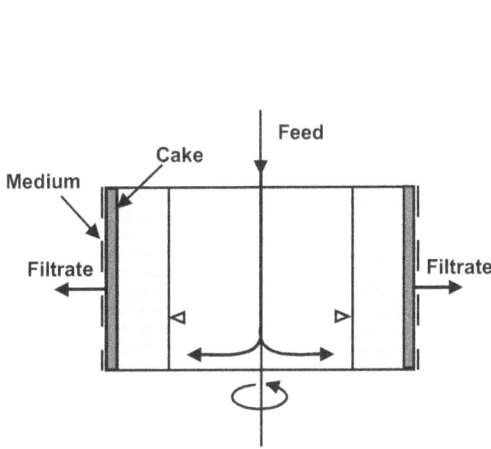

Fig. 10.34: Perforated basket centrifuge.

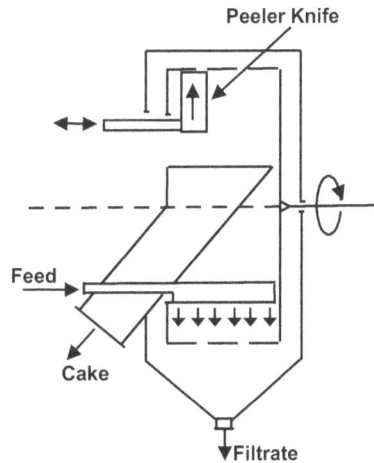

Fig. 10.35: Peeler centrifuge.

The simplest of the fixed bed centrifuge is the perforated basket centrifuge, which has a vertical axis, a closed bottom and an overflow restriction at the top end. The slurry is fed through a pipe or rotating feed cone into the basket. The cake is discharged manually or by a scraper that moves into the cake after the basket slows down to a few revolutions per minute. The plough directs the solids toward a discharge opening provided at the bottom of the basket. Some residual cake remains since the plough cannot be allowed to reach too close to the cloth. The basket centrifuge has a wide range of applications in the filtration of slow draining products that require long feed, rinse and draining times. It can be applied to the finest suspensions of all filtering centrifuges because filter cloths may be of pore sizes down to 1 µm. Making the axis horizontal may eliminate the non-uniformity due to gravity with a vertical basket. This is known as the peeler centrifuge shown in Fig. 10.35, which is designed to operate at constant speed to eliminate nonproductive periods. The cake is also discharged at full speed, by means of a sturdy knife, which peels off the cake into a screw conveyor or a chute in the center of the basket. The peeler centrifuge is particularly attractive where filtration and dewatering times are short, such as high output duties with non-fragile crystalline materials.

Continuously operating, moving bed centrifuges use either conical or cylindrical screens. The conical centrifuge in Fig. 10.36 has a conical basket rotating on either a vertical or a horizontal axis. The feed suspension is fed into the narrow end of the cone. If the cone angle is sufficiently large for the cake to overcome its friction on the screen, the centrifuge is self-discharging. Pusher-type centrifuges have a cylindrical basket with a horizontal axis. The feed is introduced through a distribution cone at the closed end of the basket (Fig. 10.37) and the cake is pushed along the basket by means of a reciprocating piston that rotates with the basket. Pusher centrifuges can be made with multistage screens consisting of several steps of increasing diameter.

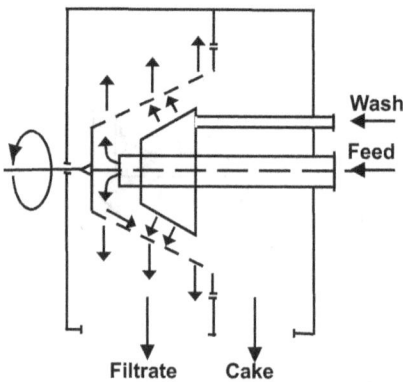

Fig. 10.36: Conical centrifuge.

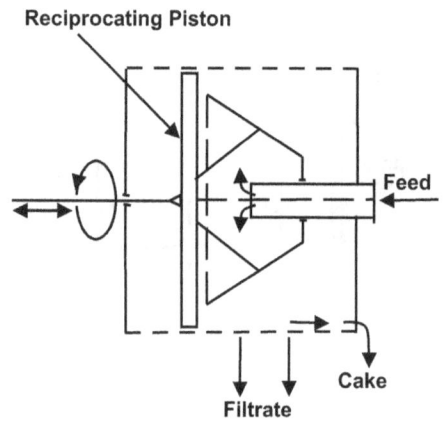

Fig. 10.37: Pusher centrifuge.

10.8 Deep bed filtration

Deep bed filtration is typically used for the removal of very low concentrations of suspended submicrometer solids that do not settle or filter readily. Their relatively high efficiency in removing fine particles present in low concentration is used extensively for drinking water filtration and the final polishing of effluents before discharge or process liquids prior to further processing. The high capacity available in modern deep bed filters offers the possibility of use in cases where conventional equipment could not produce an economical separation.

Commercial deep bed filters are of simple construction, consisting of a holding tank for the granular bed of solids through which the liquid to be filtered is passed.

Flow may be directed downward or upward, with typical liquid velocities of 40 and 15 m/h. The granular bed is usually made up of sand, gravel, anthracite or a variety of other materials having a particle size of 0.5–5 mm. Although the diameter of the pores is 100–10,000 times larger than the diameter of the suspended particles, a mixture of mechanisms are responsible for their deposition in the bed. In addition to mechanical interception, particle deposition due to electrostatic and London/van der Waals attractive forces takes place. The deposited particles cause blockage and an increased pressure drop of the fluid flowing through the bed. A common design to improve the effectiveness of the deep bed filter is to use multiple solid layers on top of each other. This is illustrated in Fig. 10.38, where a dual layer of anthracite/sand is contained in a vessel. The anthracite particles, being coarser than the sand, serve to prevent the formation of surface deposits on the sand surface. During filtration, the bed is contaminated with the particles from the filtered liquid. At regular times, the filter is back-washed with water in counter flow to rinse the bed clean. To this end, the filter bed is fluidized, quickly releasing the particles deposited during the filtration stage.

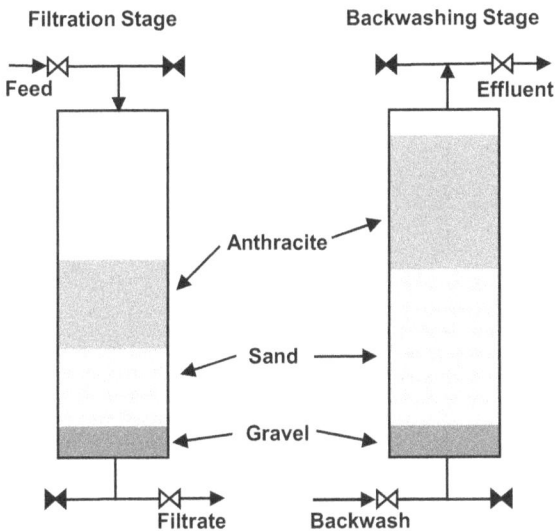

Fig. 10.38: Dual-layer deep bed filter.

Nomenclature

A	Area	$[m^2]$
a	Centrifugal acceleration	$[m/s^2]$
c	Mass of dry solid per unit volume suspension	$[kg/m^3]$
d, D	Diameter	$[m]$

g	Gravitational acceleration	$[m/s^2]$
H	Height	$[m]$
N	Number of separation channels	$[-]$
t	Time	$[s]$
P	Pressure	$[N/m^2]$
Q	Volume flow rate	$[m^3/s]$
r	Radius	$[m]$
Re	Reynolds number	$[-]$
R_M, R_C	Resistance of filter medium and cake, respectively	$[m^{-1}]$
v	Velocity	$[m/s]$
V	Volume	$[m^3]$
w	Mass of dry solids per unit area	$[kg/m^2]$
a	Specific cake resistance	$[m/kg]$
ε	Porosity or fluid void fraction	$[-]$
η	Fluid viscosity	$[kg/s\ m = N \cdot s/m^2 = Pa \cdot s]$
ρ, ρ_s	Fluid density, solid density	$[kg/m^3]$
ω	Angular velocity	$[rad/s]$

11 Particle removal from gases

11.1 Introduction

In many industrial processes, solid (dust) or liquid (droplets, mist) particles are at some stage generated and entrained by a carrier gas. As a rule, the particles must be separated from the transporting gas for a variety of process reasons. Examples are the recovery of a valuable product from the gas stream or the separation of solid and gas at the end of a pneumatic conveying line. Important other reasons are the purification of gaseous feeds and protection of the environment. A single-stage separator can be schematically drawn as in Fig. 11.1. As the efficiency of separation is very often particle size dependent, the feed flow is separated into a coarse fraction containing the separated particles and the cleaned gas that still contains some residual fine material. In order to meet the required limits set by process demands or environmental regulations, collection efficiencies greater than 99.99% are often necessary. This illustrates the high importance of highly efficient and well-designed gas–particle separation equipment. One difficulty is that the size distribution of the particles to be collected generally extends over a wide range, from less than 0.1 µm to more than 100 µm.

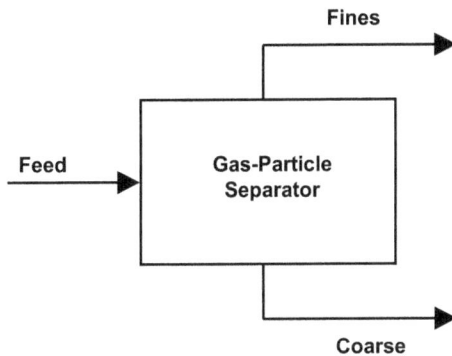

Fig. 11.1: Schematic of a gas–particle separator.

11.1.1 Separation mechanisms

The removal of particles (liquids, solids or mixtures) from a gas stream requires their deposition or attachment to a surface. The surface may be continuous, such as the wall of a cyclone or the collecting plates of an electrostatic precipitator, or it may be discontinuous such as spray droplets in a scrubbing tower or fibers of a filter. Once deposited on a surface, the collected particles must be removed at intervals

https://doi.org/10.1515/9783110712445-011

without appreciable re-entrainment in the gas stream. Figure 11.2 illustrates that there are four major mechanisms used to separate particles from gas steams:

1. Gravitation
2. Centrifugal
3. Electrostatic
4. Interception

Although this list is not exhaustive, it represents the mechanisms for the majority of gas-solid separation process equipment currently in operation. The first category includes simple gravity settling chambers, where the particles are fed to a large box-like vessel and allowed to settle under their own weight. Gravity sedimentation is efficient only for sufficiently large ($d_p > 50$ μm) particles. The second category includes cyclones and swirl separators, in which the applied centrifugal forces extend the applicability of sedimentation to considerably smaller particle sizes. The third category encompasses process equipment where electrical charge is used to separate the solids from the gas. While their use is not widespread in the chemical industry, electrostatic precipitators find wide application in the power industry. Electrostatic forces are the strongest forces available and able to separate even very fine particles loosely defined as ≤ 2–3 μm. The last category encompasses a variety of dust collectors such as filters and wet scrubbers where interception of the particles by a medium (filter bag, plates, droplets) is used to separate the particles. These mechanisms are effective for the separation of particles down to 2–3 μm.

All of these collectors separate particles from gas by a common principle. Forces acting on the particles cause them to enter regions from which the gas cannot transport them away. These regions may be the inner wall of a cyclone, the droplet surface in a scrubber, the fiber or grain surface in a filter, or the collecting electrode in a precipitator.

11.2 Collecting efficiency

The evaluation of any particulate separation process requires the determination of the collecting efficiency. A first start is to use the overall efficiency of particle collection defined on a mass basis:

$$\eta_M = \frac{\text{mass of particles entering} - \text{mass of particles leaving}}{\text{mass of particles entering}} \tag{11.1}$$

to describe the effectiveness of the separation. Besides the mass basis, the collecting efficiency may also be expressed in terms of the number of particles or area of particles entering and leaving the collector. A major complication is that the overall efficiency depends not only on the operating conditions but also on the particle size distribution of the feed. It is therefore impossible to characterize the performance of

Typical Machinery

Gravity ———————————— Settling Chambers

Centrifugal ———————————— Cyclone
Swirl Flow Separators

Particle
Separation ⎯ Electrostatic ———————————— Electrostatic Precipitator

Inertial

Wave-Plate Separators
Wire Filters
Interception ———————— Flow-Line Packed Bed Separators
Deep-Bed Filters
Surface Filters
Wet Scrubbers

Diffusion

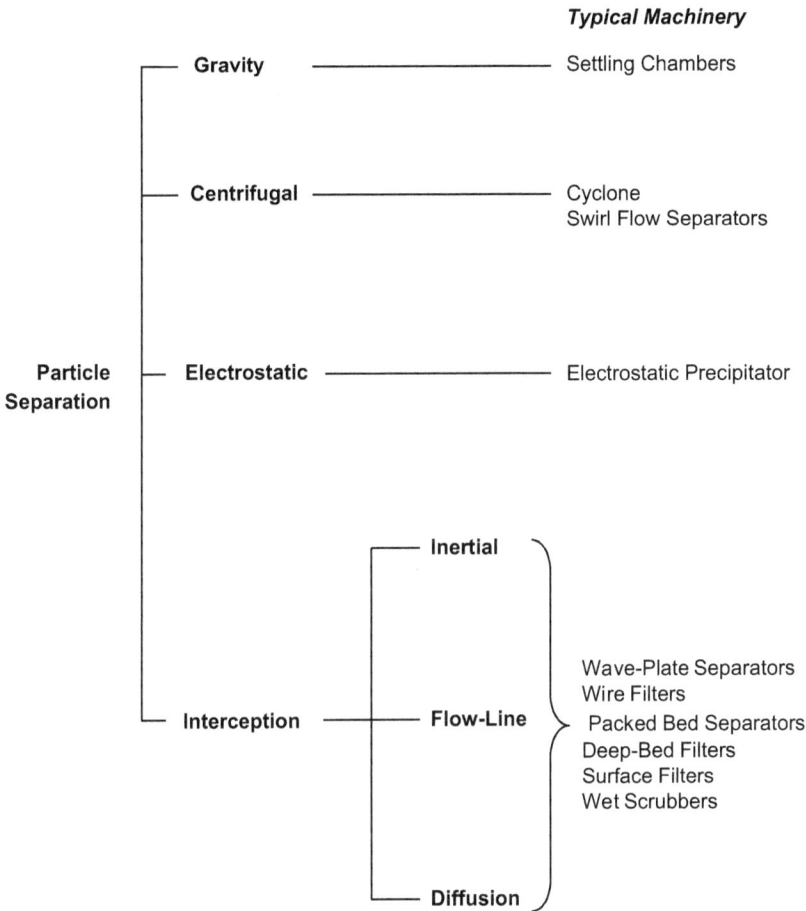

Fig. 11.2: Overview of most important gas–particle separation mechanisms and equipment.

gas-solid separation equipment without proper quantification of particle size and size distribution. This requires the identification of the most suitable definition of particle size and the appropriate quantification of the distribution (number, surface or volume). The distribution of particle sizes is normally plotted as either a differential or a cumulative distribution (see Fig. 11.3). The cumulative distribution can be represented as either oversize or undersize distribution. In an oversize cumulative distribution plot, the y-axis represents the fraction of population with particle sizes greater than the selected size (on the x-axis). The plots also allow interpolation of size fractions for particle sizes that have not been experimentally determined.

The dependence of collection efficiency on particle size is represented by the grade efficiency function. It is most common to define the grade efficiency curve as the collecting efficiency obtained for each particle size d_p:

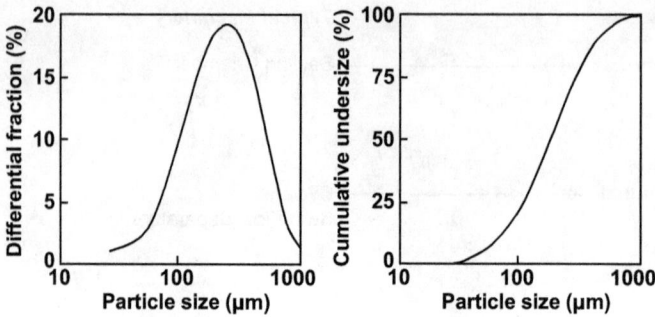

Fig. 11.3: Differential (a) and cumulative (b) particle size distribution curves.

$$G(d_p) = \frac{\text{mass of size } d_p \text{ in the coarse}}{\text{mass of size } d_p \text{ in the feed}} \tag{11.2}$$

For an ideal separation the vertical line in Fig. 11.4 would describe the grade efficiency curve. In practice, however, this situation is rarely realized and consequently the s-shape curve represents a typical grade efficiency curve. There are several key aspects to the grade efficiency curve. The size for which the particle has equal probability to be collected in the coarse or the fines steams is called the cut size. Another useful concept is the sharpness of cut. The sharpness of cut defines the amount of material that is misclassified into the two product fractions. One of the most common definitions is the ratio of the sizes associated with two different efficiencies:

$$S_C = \frac{d_{80}}{d_{20}} \tag{11.3}$$

where d_{80} and d_{20} are the particle sizes associated with 80% and 20% collecting efficiencies on the grade efficiency curve.

11.3 Gravitational separators

The simplest way to separate particles from gases is by gravity. The gravity settling chamber is one of the oldest forms of gas–particle separation. It may be nothing more than a large room where the well-distributed gas enters at one end and leaves at the other (Fig. 11.5). In the settling chamber, the gas stream is slowed down sufficiently to allow particles to settle under their own weight. In theory, a very large settling chamber would give sufficient time for even very small particles to be collected, down to micron-sized particles (generally, submicron particles remain suspended by the Brownian motion). Practical size limitations, however, restrict the applicability of these chambers to the collection of coarse particles. In the horizontal-type chamber an average gas velocity can be assumed to represent piston flow

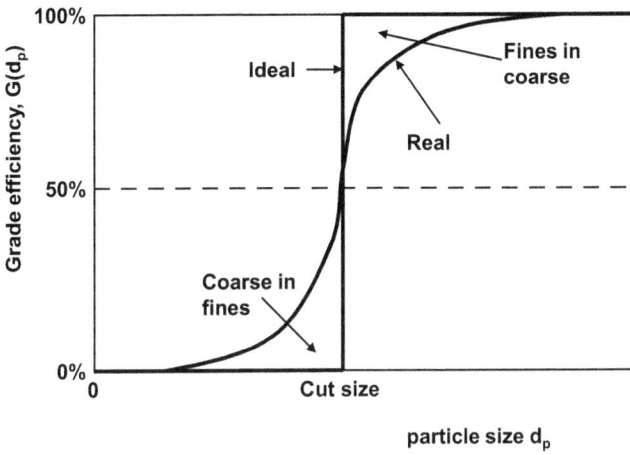

Fig. 11.4: Ideal and real grade collecting efficiency curves.

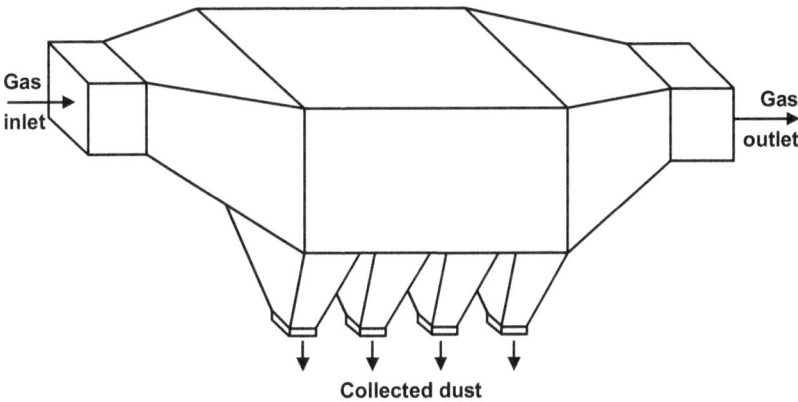

Fig. 11.5: Horizontal flow settling chamber with square cross section.

through the chamber. This can be simply derived from the gas flow rate Q and the height H and width W of the chamber:

$$v_g = \frac{Q}{HW} \tag{11.4}$$

From the length L of the chamber, the residence time t of the gas can be calculated:

$$t = \frac{L}{v_g} = \frac{LHW}{Q} = \frac{V}{Q} \tag{11.5}$$

During this residence time a particle of size d_p will settle a distance h equal to the product of the terminal settling velocity v_t of the particle and the residence time t:

$$h = v_t\, t \tag{11.6}$$

where the steady-state (terminal) settling velocity of spherical particles in the Stokes regime is given by

$$v_t = \frac{d_p^2(\rho_p - \rho)g}{18\eta} \qquad Re_p < 1 \tag{11.7}$$

In general, nonspherical particles settle slower than spherical ones of the same size. If a particle settles a distance h, then h/H represents the fraction of particles of this size that will be collected. If h is equal to or greater than H, all particles of that size or larger will be collected in the settling chamber. A curve of h/H ratios for the different sizes of a material provides the fractional or grade efficiency curve for the settling chamber:

$$G(d_p) = \frac{h(d_p)}{H} \tag{11.8}$$

Gravity settlers have largely disappeared because of bulky size and low collection efficiency. They are generally impractical for particles smaller than 50 μm. Improved settling chamber efficiencies can be achieved by decreasing the height a particle has to fall before being collected. This has been applied in the Howard settling chamber (Fig. 11.6) where a number of collecting trays have been inserted in the chamber. Because gravity settling is technically feasible only for coarse particles, these settlers have low importance in gas–particle separations.

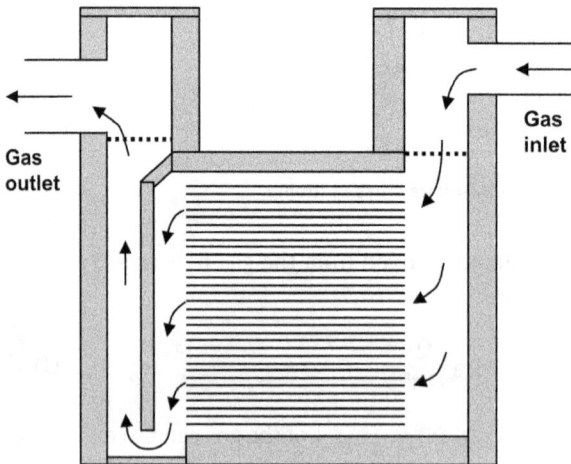

Fig. 11.6: Multitray settling chamber.

11.4 Cyclones

Already for decades it is known that collection by means of centrifugal forces is much more efficient than simple gravity separation. Because of their simple design, reliable operation, small space requirement and low cost, cyclones are widely used in many branches of industry. They are frequently employed in material recovery from gas recycle systems, pneumatic conveying and other areas. Because cyclones can be operated at temperatures over 1,000 °C, they currently provide the only method of particle collection at high temperature for use on an industrial scale. The advantages of cyclones over other collectors are offset by the fact that they are less efficient collectors for particle sizes below 1–5 μm. Therefore, the use of cyclones is restricted and they are often used as first-stage collectors.

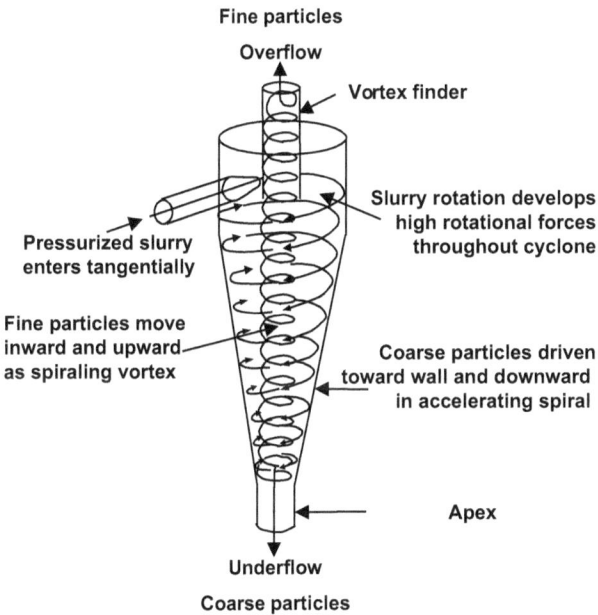

Fig. 11.7: Schematic diagram of a Reverse-flow cyclone.

Figure 11.7 shows the most common type cyclone with flow reversal. The gas path is designed to impart a twist to the particle-laden gas entering the rotationally symmetrical apparatus. A turbulent, three-dimensional, rotational flow is produces. In the separating chamber spiral, downward flow takes place at larger radii. Flow is then reversed and the gas spirals upward at smaller radii, and reaches the exit duct. The particles in the revolving gas experience centrifugal forces hundreds to thousands of times greater than the force of gravity. Thus, the larger particles

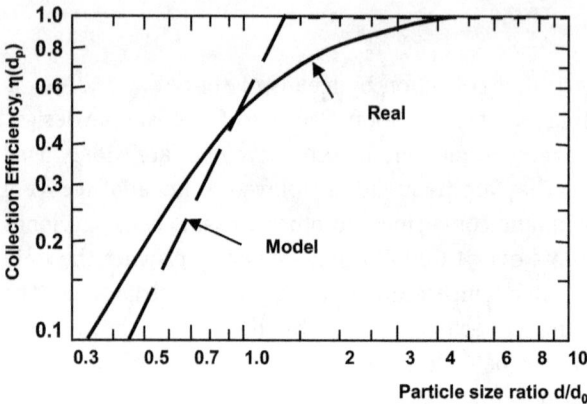

Fig. 11.8: Comparison of theoretical and real collection efficiency of a cyclone. d_0 is the 50% cut diameter.

migrate outward to the cyclone wall, where they collect and are carried downward and out of the separating chamber.

The collection performance of a cyclone depends mainly on its geometry, gas throughput, inlet gas concentration and properties of the material to be separated. An increase in gas viscosity decreases performance. Collection efficiency can be improved by altering the cyclone geometry or by increasing the volume flow rate through the cyclone. In both cases the tangential velocity increases so that a greater centrifugal force acts on the particles. Most cyclone manufacturers provide grade-efficiency curves to predict overall collection efficiency of a dust stream in a particular cyclone. A problem with the development of generalized grade efficiency curve equations is that a cyclone's efficiency is affected by its geometric design. The following equation was proposed by Lapple to calculate the cut diameter d_0, which is the particle size collected with 50% efficiency in a cyclone:

$$d_0 = \left(\frac{9\eta W_i}{\pi N_e v_i (\rho_p - \rho)} \right)^{1/2}$$

(11.9)

where W_i is the width of the cyclone inlet channel, v_i is the mean gas velocity in the inlet channel, and N_e is the effective number of spirals the gas makes in the cyclone. For smaller particles, the theory indicates that efficiency decreases according to the dotted line of Fig. 11.8. Experimental data show that eq. (11.9) tends to overstate the collection efficiency for moderately coarse particles and understate efficiency for the finer fraction.

An alternative to classical cyclones is the so-called swirl tube separator shown in Fig. 11.9b. It essentially consists of a swirl generator, which comprises guide vanes and a central body. The guide vanes impart a rotational motion to the passing gas so that the entrained particles are flung against the wall. These swirl tubes differ from axial cyclones in that there is no inversion of the gas flow. Occasionally, they are referred to as straight through cyclones. For the most part, swirl separators are installed in vertical gas outlets and stacks in order to prevent particle discharge. Cleaning a gas with a given flow rate usually requires more cyclones or swirl tubes in parallel. Systems composed of many small cyclones are often called multicyclones such as the one shown in Fig. 11.9c.

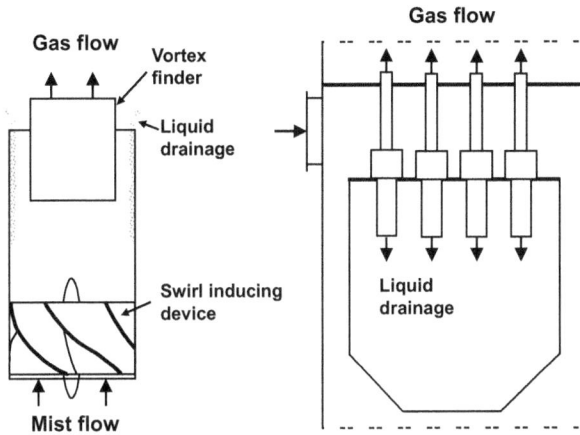

Fig. 11.9: Schematic diagrams of a double-inlet cyclone (a), swirl tube separator (b) and multicyclone package (c).

11.5 Electrostatic precipitation

11.5.1 Principles

In electrostatic precipitation, electrostatic forces are used to remove the solid or liquid particles suspended in a gas stream. An efficient separation is achieved by creating an electrostatic charge on the particles through the generation of ions in the gas, which unload their charge on the particles by collision or diffusion. In electrostatic precipitators, ions are created in the gas by exposure to corona discharge in high-density electrostatic fields. This is accomplished by applying high voltages (30–100 kV) to an assembly of negatively charged wires and positively grounded collection plates. At the wires, the corona discharge occurs when positive ions in the gas impinge on the negatively charged wires and release electrons that ionize the gas into positive ions and electrons. While the positive ions produce further ionization, the produced electrons associate with gas molecules to form negative ions traveling across the gas flow to the positively charged collecting electrodes. This continuous flow of negative ions effectively fills the space between the electrodes with negative ions that charge the particles passing through the space with the gas by two mechanisms: bombardment charging and diffusion charging. Bombardment charging occurs by ions that move toward the positive electrode with high velocity and impact on the particles. In diffusion charging the Brownian motion of the ions in the gas leads to particle-ion collisions. The charging continues until the so-called equilibrium charge on the particle repels further ions.

As soon as the dust particles acquire some negative charge, they will migrate toward the positive collector electrodes away from the negative discharge electrode. Although the overall picture is very complex, the calculation of the particle drift velocity is usually based on a relatively simple model with the following assumptions:

1. The particle is considered fully charged during the whole of its residence in the precipitating field
2. Uniform distribution of particles through the precipitator cross-section
3. Particles moving toward the electrode normal to the gas stream encounter fluid resistance in the viscous flow regime and move at their terminal velocity according to Stokes law
4. Electric repulsion effects between the particles are neglected
5. There are no hindered settling effects in the concentrated dust near the wall
6. The effect of repulsion between electric wires and ions, sometimes called ion wind, is neglected
7. The gas velocity through the precipitator does not affect the migration velocity of the ions

Although it can hardly be realistic, the migration velocity based on these assumptions has been found to give reasonable estimates of the cross-stream drift velocity. For a particle with saturation charge q moving in a field of constant intensity E the terminal migration velocity v_t toward the collecting electrode may be determined from an equilibrium between the electrostatic force and the drag force:

$$Eq = 3\pi\eta d_p v_t \tag{11.10}$$

Rewriting gives for the migration velocity of the particle:

$$v_t = \frac{Eq}{3\pi\eta d_p} \tag{11.11}$$

On arrival at the earthed collector, the particles adhere and are discharged to earth potential. The strong electrostatic field and adhesive properties of the particulate inhibit re-entrainment. When a layer of particles has formed, these are shaken off by rapping and fall into a hopper.

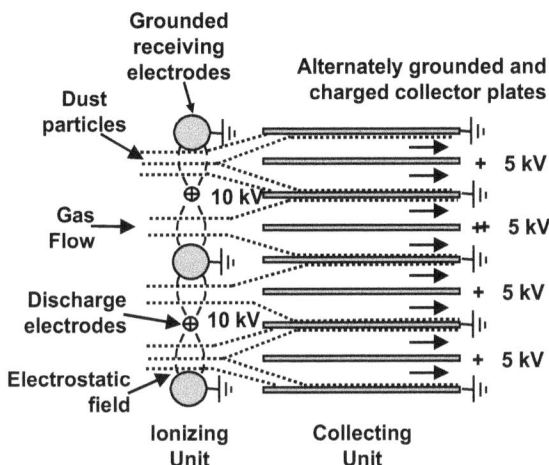

Fig. 11.10: Schematic of a two-stage electrostatic precipitating unit.

11.5.2 Equipment and collecting efficiency

Electrostatic precipitators can be classed as single- or two-stage. In the two-stage designs, used for air purification, air conditioning or ventilation, particle charging and collection are carried out in two separate stages. As illustrated by Fig. 11.10, the dust particles are first charged in a separate charging section, providing only a fraction of a second residence time to avoid collection. Particle collection follows in the second stage, which consists of alternately charged parallel plates. As laminar flow conditions usually prevail in two-stage precipitators, the particle collecting efficiency η_F can be expressed in direct analogy with sedimentation tanks:

$$\eta_F = \frac{v_t A}{Q} \tag{11.12}$$

where v_t is the effective migration velocity of the dust particles, A is the collecting area and Q is the gas flowrate. Equation (11.12) can be rewritten in terms of the gas velocity v_g, the electrode spacing W and the length of the collector electrodes L:

$$\eta_F = \frac{v_t L}{v_g W} \tag{11.13}$$

Application of this equation is limited to the two-stage precipitator which is the only type built for laminar flow. In most industrial applications single-stage units are used that are designed for operation in the turbulent flow regime causing turbulent diffusion and re-entrainment. It was found experimentally that the efficiency of a precipitator was an exponential function of the gas stream residence time in the precipitator field. Under the conditions that the dust is uniformly distributed in the beginning, the uncollected dust remains uniformly distributed and the migration velocity is effectively constant, Deutsch derived the following equation for the collecting efficiency:

$$\eta_F = 1 - \exp\left(-\frac{v_t A}{Q}\right) = 1 - \exp\left(-\frac{v_t L}{v_g W}\right) \qquad (11.14)$$

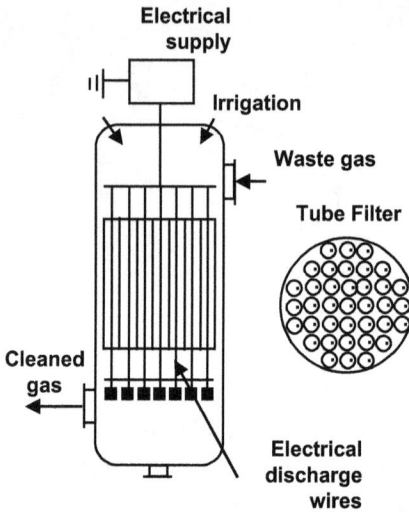

Fig. 11.11: Diagram of a tube-type single-stage electrostatic precipitator.

In single-stage units, the particles are charged and collected in the same electrical field, thus making the design simpler. For a round wire axially suspended in a tube, a radial evenly distributed electrostatic field is obtained. To accommodate larger gas flows, banks of tubes are nested together vertically to facilitate the removal of the deposited dust by gravity (Fig. 11.11). Wire-in-tube separators are employed for small quantities of gas. The wire-in-tube precipitator was the earliest type used but has now been almost entirely replaced by the plate type. Wire-and-plate or simply plate precipitators consist of vertical parallel plates with vertical wires arranged at intervals of 0.1–0.2 m along the centerline between the plates (Fig. 11.12). The wires have negative charge whilst the plates are earthed. A plate precipitator may have 10–15 m tall parallel flat plates with 0.2–0.4 m horizontal spacing and discharge electrodes that are suspended midway between the plates.

Both the discharge and collecting electrodes must be rapped at preset time intervals because clean electrodes are essential for the generation of an effective corona discharge and electrostatic field. Particles are removed by vibrating the collection plates, thereby dislodging the particles, which drop into collection hoppers. Swing hammers or camshaft and coil springs are most commonly used for mechanical rapping. Electromagnetic or pneumatic devices allow easier adjustment of the timing and intensity of the rapping than the mechanical devices. In the wet two-stage tubular precipitator, the deposited dust is removed from the collection electrodes by a flowing water film. This device consists of a short ionizing section followed by a relatively long collection system. The discharge electrode is in the form of a rod or a tube, with a section of sharp discharge points at the end, centered in the collection tube.

Electrostatic precipitators are applied wherever very large volumes of gases have to be cleaned with high collection efficiency on fine particles and there is no explosion risk. The plants are invariably used for fly ash collection in large (800–1,200 MW) coal-fired electric power stations and for the collection of dusts in the cement industry. Precipitators are also employed for large-scale fume collection systems in the metallurgical industry.

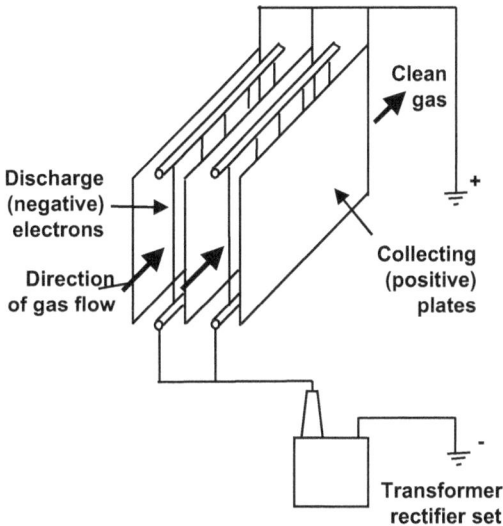

Fig. 11.12: Schematic of a single-stage electrostatic plate separator.

11.6 Particle interception mechanisms

Impingement separators require the transport of particles or droplets to a surface on which they are deposited by collision. Deposition requires a body for impact to be placed in the gas stream. This can be a wire, a fiber or a large droplet of washing liquid around which the gas can flow on all sides. However, as in the case of wave plate separators, it can also be an impact surface which forces the gas flow to change its direction. If the particle contacts the surface of the body, it can be regarded as collected. The three important mechanisms for particle collection, schematically drawn for the case of a collecting fiber in Fig. 11.13, are: inertial interception, Brownian diffusion and flow-line interception. Inertial interception occurs when particles approach targets such as a baffle, impaction element, fiber or droplet with sufficient velocity to cause a collision with the target by inertia of the particle. As a result of inertia, particle trajectories deviate from the streamline of the gas so that the particles can strike the collecting surface of the target. This applies to all particles of large enough diameter present in the oncoming gas. Because sufficiently large particles will tend to move in a straight line, their collection efficiency will be close to unity. Brownian motion causes particles to diffuse randomly across flow streamlines, making them more susceptible to capture by a target. Brownian motion becomes appreciable for smaller particles of 1 μm diameter and becomes totally dominant for very small particles of 0.1 μm or less. Because most gas-solid separation processes involve particles larger than 1 μm, Brownian motion is often of secondary importance. Flow-line interception is the striking of a target by a particle that passes the target in a streamline within one particle radius. While inertial interception separation efficiency increases with droplet diameter, diffusional deposition efficiency increases with decreasing droplet size. This results in a separation minimum at particle or droplet sizes between 0.2 and 0.7 μm. It is mainly this minimum range where flow-line interception can be predominant. The resulting fractional collection efficiency of droplets captured by a single fiber is shown in Fig. 11.14. Similar results apply to particles.

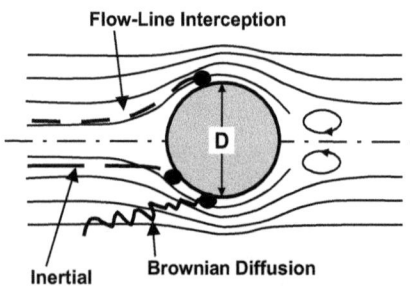

Fig. 11.13: From top to bottom: inertial interception, flow-line interception and Brownian diffusion particle separation mechanisms on a fiber.

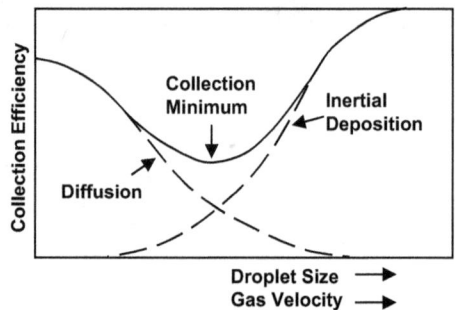

Fig. 11.14: Collection efficiency of a droplet by a single fiber. Similar results apply to particles.

For interception separators, the collection efficiency is usually estimated in terms of the target efficiency of the single baffle. The target or collection efficiency is described as a function of the separation number defined as follows:

$$N_S = \frac{v_t V_0}{g D_b} \tag{11.15}$$

where g is the gravitational acceleration, D_b a characteristic baffle dimension, v_t the terminal settling velocity and V_0 the approach gas velocity. Figure 11.15 shows that the efficiency increases with increasing approach gas velocity because of the particle inertia. Likewise, efficiency increases with increasing terminal velocity. Somewhat counterintuitive is the decrease in efficiency with larger characteristic dimension. This is a result of the increased distance that the gas is deflected by larger baffles.

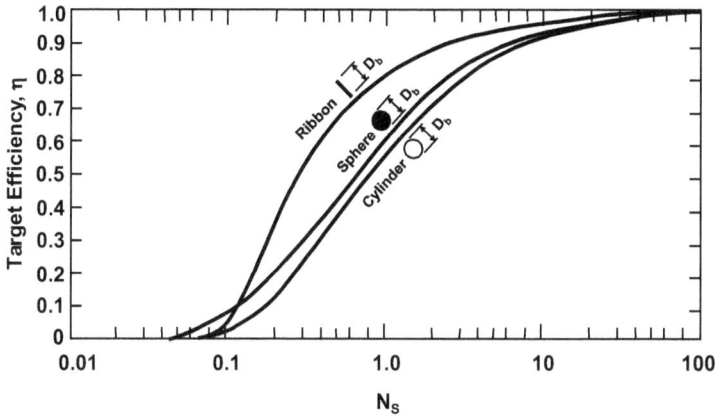

Fig. 11.15: Target efficiency of single spheres, cylinders and ribbons as a function of separation number.

11.7 Dry impingement separators

Baffle bundles, also known as impingement separators, are widely employed in industry for removal of droplets and particles from gas flows. In plate separators rows of baffles can be arranged in series to increase the overall efficiency of impingement. In this manner, the rather modest collection efficiencies that are possible for a single row can result in industrially practical collection efficiencies for a multiple row unit. A series of parallel impact plates is commonly named a wave plate separator. Multiple body contacting is also used in fibrous filters, which are especially important in modern particle collection technology when highly efficient collection in the finest particle size range is required. Of all collectors, these have the widest spectrum of application and thus a large share of the market. With regard to application, design and operation, filtration devices for particle collection can be classified as deep bed filters or surface filters.

11.7.1 Deep bed filtration

Deep bed filters consist of a relatively loose fiber mat with a porosity often >99%. The term "filter" denotes here a packing consisting of wires or fibers through which a gas containing dust particles will flow. The filter can be arranged either horizontally or vertically and can consist of uniformly or randomly arranged wires or fibers. Particle collection takes place in the interior of the layer where the dust particles adhere to the fibers and accumulate. As the gas flows through the fibrous layer, it must flow past a large number of cylindrical elements. Hereby, the particles in the gas continue their path until they strike a wire or fiber element by inertial or flow-line interception. Brownian diffusion contributes mainly for submicrometer particles. An important concept for the collecting efficiency of a filter is its overall on-flow area. This follows from the total length of all the fibers from which the filter is made up. Based on the filter cross-sectional area A, the overall relative on-flow area is given by

$$p = f\frac{LD}{A} = f\frac{4\,G\,H}{\pi\rho_F\,D}$$ (11.16)

where L is the total wire length of the filter, D is the wire diameter, G the effective filter density (filter mass mass/filter volume), H the thickness of the packing and ρ_F the density of the wire material. The fact that the fibers are not at right angles to the direction of flow is accounted for through the correction factor f. With this relative on-flow area concept it is possible to transfer the collecting efficiency of a single wire to the wire filter. For a single layer of wire the collection efficiency with relative on-flow area p_1 and single wire collection efficiency $\eta_F(0)$ becomes

$$\eta_F(1) = \eta_F(0)\,p_1$$ (11.17)

Extension to n consecutive layers provides the collection efficiency of a filter as the product of the single layer probabilities:

$$\eta_F(n) = 1 - (1 - \eta_F(0)\,p_1)^n \approx 1 - \exp(-\eta_F(0)\,p)$$ (11.18)

with

$$p = p_1 \times n$$ (11.19)

With eq. (11.18) and a known value of the relative on-flow area p of a wire or a fiber filter, its collection efficiency can be calculated from that of a single cylinder. The number of layers in the filter is not explicitly required and the indicator n of the collection efficiency can henceforth be omitted.

For the collection of liquid droplets, fogs and mists, in-depth fiber bed filters made out of horizontal pads of knitted metal wire are used (Fig. 11.16). Collection from the up-flowing gas is mainly by inertial interception. Thus efficiency will be low at low superficial velocities and for fine droplets or particles. For collection of

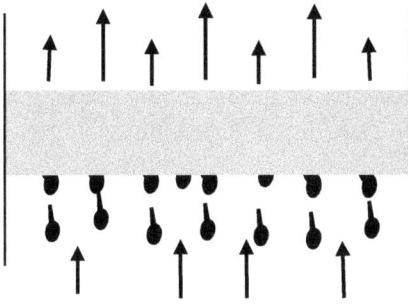

Fig. 11.16: Schematic of a wire mesh separator.

fine mist particles, the use of randomly oriented fiber beds is preferred. Fine particle removal by filtration through a bed of granular solids has appeal because of corrosion and temperature resistance. Several types of aggregate bed filters are
available which provide in-depth filtration. Both gravel and particle bed filters have
been developed for removal of dry particulates. Important parameters for the collection efficiency in granular beds are bed thickness, gravel size and air velocity.

11.7.2 Surface filters

Surface filters are used widely because of their excellent collection performance,
even for the finest particles. With porosities in the range of 70–90%, separation of
particles from the carrying gas is not a sieving or simple filtration process since the
filter fabric pore size is much larger than the particles collected. Until a small dust
layer is formed on the filter surface, the collecting efficiency is quite low. The dust
bed (filter cake) is the true filter medium and highly efficient. Because the pressure
drop increases as the cake grows, these filters must be cleaned periodically.

The three most important basic surface-filter designs are round bag, envelope
and cartridge. Envelope filters are not as widely used as round bag filters because of their low dust-handling capacity. They must be discarded or cleaned when
the pressure drop becomes too large. The cloth or bag dust filter is the oldest and
often the most reliable of the many methods for removing dusts from an air stream.
Among their advantages are high (often 99 +%) collection efficiency, moderate
pressure drop and the recovery of the dust in a dry reusable form. The maintenance
for bag replacement, their large size and incompatibility with liquid particle laden
gases are the main disadvantages. Bag filters may be woven or felted, an envelope
supported with an internal wire cage. The classical design is a multi-chamber filter with shaker cleaning. Particle-laden gas is fed into the bags from below, flows
through the filter medium and is directed to the clean gas duct at the top of the
device. Particle collection takes place on the outside of the bags. For cleaning, the
gas supply is cut off and the dust is removed by shaking or rapping the bag support.

Newer systems use reverse flow cleaning (Fig. 11.17) where a compressed air pulse is given to dislodge the dust from the outside of the bag. No bag fabric can withstand truly high temperatures.

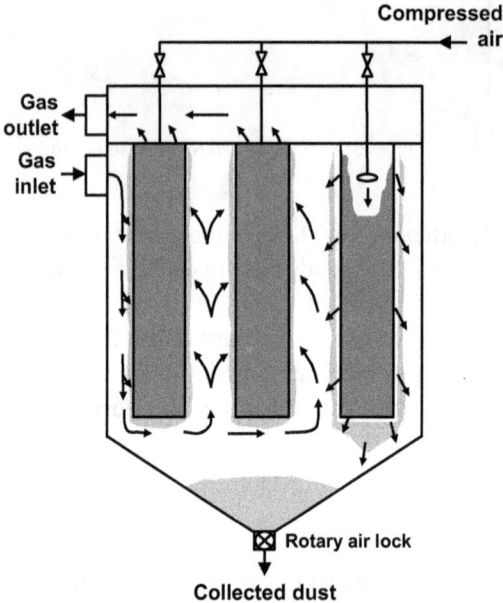

Fig. 11.17: Multicompartment bag filter.

In cartridge filters, the filter medium is folded to pack more filtration area into a unit volume. They are frequently employed for the purification of gases containing particles. Filter candles consist of ceramic or glass sintered elements and are resistant to heat and chemicals. Figure 11.18 shows the construction of such a separator. The gas to be cleaned flows through the porous stone material from the exterior to the interior. The particles are collected on the outside surface and need to be removed frequently.

11.7.3 Lamellar plate separators

Lamellar plate separators are often used because of their simple construction and low pressure drop. These devices consist of parallel channels with multiple deflections and retaining grooves (Fig. 11.19), that can capture particles or droplets. Here we will follow the example of droplets. The droplet-laden gas stream undergoes many changes of direction in the channels. When the gas is diverted, entrained droplets that are sufficiently large will continue their motion until they strike the collecting surface. On the other hand, very small droplets behave almost as gas molecules and follow the flow lines of the gas. Droplets intermediate in size can follow the curvatures of the gas flow only partially. Depending on which flow line they are starting from, these droplets will hit or miss the impact body. Hence only a partial separation is possible for these droplets. It follows that for impingement separators, the collection efficiency varies from zero for sufficiently small particles/droplets to 1 (=100% collection) for large particles/droplets. The plates can have various profiles. Like all inertial collectors, lamellar separators become more efficient as flow velocity increases. On account of their simple and compact construction, wave plate separators are relatively cost effective and space saving.

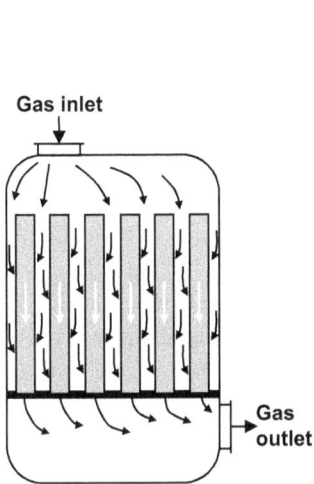

Fig. 11.18: Filter candle separator.

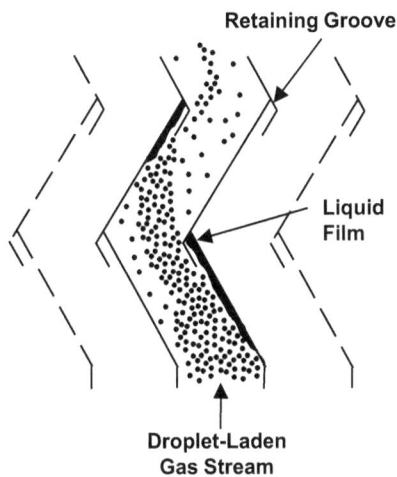

Fig. 11.19: Operating principle of a wave plate separator.

11.8 Wet scrubbers

The effectiveness of rain in removing airborne dust from the atmosphere has been used as inspiration in industry to develop a variety of liquid scrubbing equipment. In scrubbers, the gas to be purified must pass through a dense cloud of drops formed by the scrubbing liquid. The gas must flow past the scrubbing drops that act as targets for the particles. Scrubbers operate by collecting dust particles on the scrubbing droplets and binding them to it. Wettability of the particles does not play a crucial role in binding. Decisive in separation are the collecting mechanisms that are the same as in impingement separators and filters: inertial impaction, interception and diffusion. Although inertial interception predominates in most scrubbers diffusion becomes important for particle sizes in the submicron range. Wet scrubbers are able to collect particles and droplets up to very small diameter (<0.1 μm).

Development and sizing of scrubbers are often purely empirical because the processes that occur in these complicated pieces of equipment are complex and difficult to predict with theoretical models. However, some models have been developed that are quite helpful because they allow identification of important parameters. In these models the scrubbing liquid is assumed to be in droplet form, and inertial forces are considered to be the dominant collecting mechanism of dust particles on the droplet. Like fiber filters, the collecting efficiency of a wet scrubber is the result of the collective cleaning capacities of all the droplets present. The separation zone can be thought of as a dense curtain of washing droplets which must be crossed by the gas flow. Then, in principle, one can apply the theory for filters to estimate the collecting efficiency of scrubbers:

$$\eta_F = 1 - \exp\left(-p' \times \eta_F(0)\right) \tag{11.20}$$

where $\eta_F(0)$ denotes the collection efficiency of a single droplet and p' the total relative on-flow area of all n washing droplets in the scrubber with cross-sectional area A. Hence p' is

$$p' = n\frac{\pi}{4}\frac{d^2}{A} \tag{11.21}$$

The collection efficiency of a scrubber thus increases with water loading and the diameter of the sprayed droplets. However, the single drop collecting efficiency is also a strong function of the diameter of the particles to be captured. The optimum spray droplet size for maximum collecting efficiency can be obtained from Fig. 11.20.

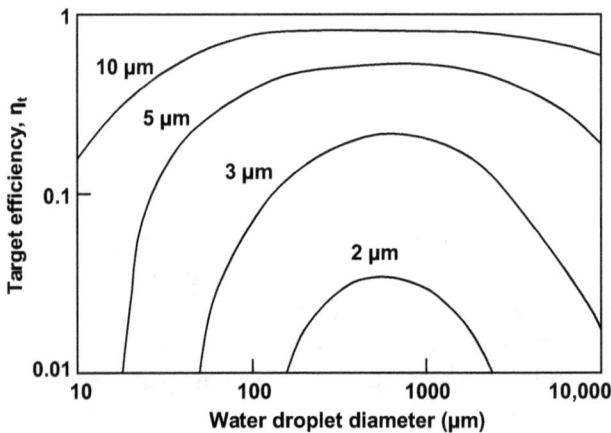

Fig. 11.20: Collecting efficiency of a wet scrubber.

Scrubbers are relatively easy to adapt to various operating conditions. They are employed mainly when products are to be recovered from a wet phase or when the risk of dust explosion rules out dry collection. The spray tower is the simplest and oldest type of scrubber. Liquid droplets are produced by spray nozzles mounted at different levels and allowed to fall downward through a rising stream of dust-laden gas. In order not be entrained by the gas, the droplets must be sufficiently large to have a terminal falling velocity greater than the upward velocity of the gas stream. Because the droplets used are in the order of 0.1–1 mm diameter, the dust particles collected are comparatively large and the predominant collection mechanisms are inertial impaction and interception. The dust-laden liquid leaves the bottom of the scrubber. A typical spray tower is shown in Fig. 11.21. Packings are sometimes used to improve collection efficiency. If the dust-laden gas stream encounters a series of liquid films or impinges on a succession of pools, a better collection efficiency is achieved than for the simple scrubbers. These ideas have been incorporated in packed bed scrubbers that contain packings such as Raschig rings or saddles. The efficiency of particle collection can be improved further by increasing the relative velocity between the droplets and the gas stream. This is achieved by using the centrifugal force of a spinning gas stream rather than the gravitational force in the simple tower. In rotational scrubbers the washing fluid is dispersed radially in the gas by rotating spray nozzles. The gas spirals upward and the dust-laden water is removed from the bottom of the scrubber. In commercial designs, the spinning motion is imparted to the gas stream by tangential entry with velocities between 15 and 60 m/s. The liquid is directed outward from sprays set in a central pipe.

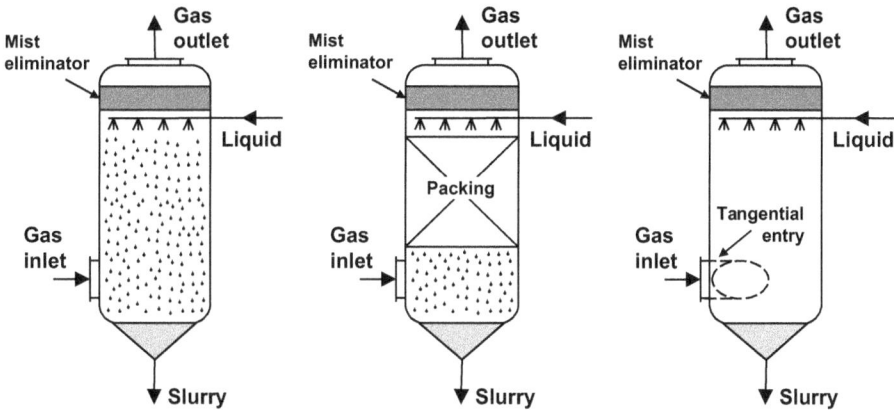

Fig. 11.21: Drawing of a spray, packed and rotational scrubber.

In many scrubber designs, the gas stream breaking through a sheet of liquid or impinging on a pool forms the scrubbing droplets. An example such a self-induced spray scrubber is shown in Fig. 11.22, where the gas stream breaks through a pool of liquid, and then creates a liquid curtain because of the specially designed orifices. These plants are extensively used in the metallurgical industry for dusts and

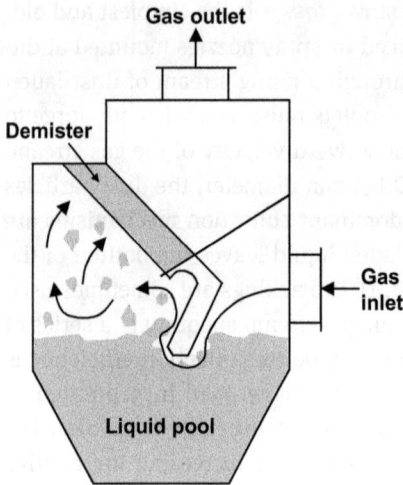

Fig. 11.22: Self-induced spray scrubber.

sticky materials. Entrained dust-laden liquid collects inside the scrubber in the collection tank. The resulting slurry must be removed regularly.

Venturi scrubbers (Fig. 11.23) have the highest collection efficiency. The gas is accelerated up to 100 m/s in the venturi throat (orifice). Washing fluid enters the middle of the throat just in front of its narrowest point or is fed radially into the throat. The liquid is atomized into very small droplets by the high speed of the inflowing gas. Aerosol particles are collected and captured by the accelerating liquid

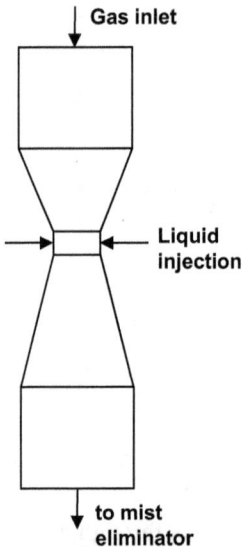

Fig. 11.23: Venturi scrubber.

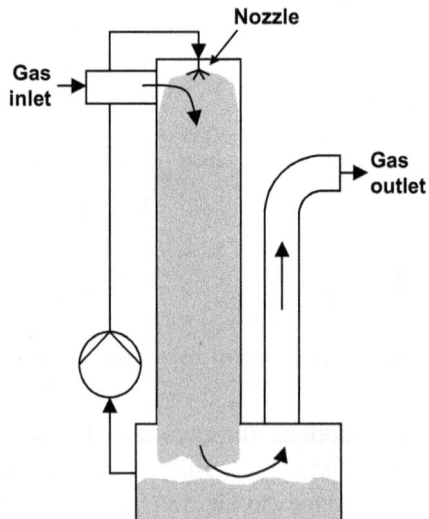

Fig. 11.24: Ejector venturi scrubber.

droplets. A cyclone or wire demister is needed to separate the dust-laden water droplets from the gas.

The ejector venturi scrubber (Fig. 11.24) is basically a water-jet pump. Water is sprayed into the scrubber and provides a draft for moving the gas. The velocity of the gas causes disintegration of the liquid. Two successive stages are often used to reach the desired collection efficiency.

Nomenclature

A	Area	$[m^2]$
d, D	Diameter	$[m]$
D_b	Characteristic baffle dimension	$[m]$
E	Electric field strength	$[V/m = N/C = m\ kg/s^3\ A]$
f	Correction factor for wire alignment	$[–]$
g	Gravitational acceleration	$[m/s^2]$
G	Effective filter density	$[kg/m^3]$
$G(d_p)$	Collection efficiency	$[–]$
h, H	Height	$[m]$
L	Length	$[m]$
n	Number of layers/droplets	$[–]$
N_e	Effective number of spirals in cyclone	$[–]$
N_S	Separation number	$[–]$
p	Relative on-flow area	$[m^2]$
q	Particle charge	$[C = A\ s]$
Q	Volume flow rate	$[m^3/s]$
Re	Reynolds number	$[–]$
S_c	Sharpness of cut	$[–]$
t	Time	$[s]$
v	Velocity	$[m/s]$
v_i	Gas velocity in cyclone inlet channel	$[m/s]$
V	Volume	$[m^3]$
W	Chamber width	$[m]$
W	Electrode spacing	$[m]$
W_i	Width of cyclone inlet channel	$[m]$
η	Gas viscosity	$[kg/s\ m = N \cdot s/m^2 = Pa \cdot s]$
η_M	Efficiency of particle collection	$[–]$
η_F	Collection efficiency	$[–]$
$\eta_F(n)$	Collection efficiency of n wires	$[–]$
$\rho_o\ \rho_s$	Gas density, solid particle density	$[kg/m^3]$
ρ_F	Density of filter wire material	$[kg/m^3]$

12 Membrane separations

12.1 Introduction

As a method of separation membrane processes are rather new. Although membranes were already applied for the filtration of drinking water samples at the end of the Second World War, the breakthrough as an industrial separation process took off only after the discovery of the Loeb–Sourirajan process for making defect-free, high flux, asymmetric reverse osmosis membranes in the early 1960s. As illustrated in Fig. 12.1, the membrane splits a feed stream into a permeate stream, enriched in particles capable of permeating the membrane and a retentate or concentrate stream that leaves the module without passing through the membrane. Although there are many membrane processes, based on different separation principles or mechanisms, they are all characterized by the use of a membrane to accomplish a particular separation. Every application exploits the ability of a membrane to control the permeation of a chemical species in contact with it. Separation is achieved because the membrane transports one component from the feed mixture more readily than any other component or components. Transport through the membrane takes place as result of a driving force acting on the components in the feed.

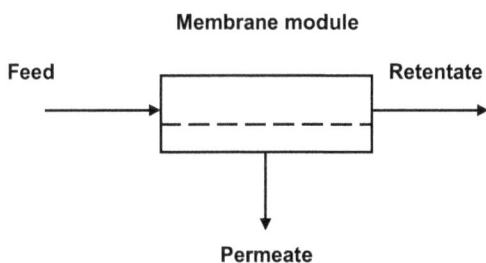

Fig. 12.1: Basic diagram of a membrane process.

This is schematically represented in Fig. 12.2. Phase 1 is usually considered as the feed or upstream phase while phase 2 is considered as the permeate or downstream side. In many cases the permeation rate (N) through the membrane is proportional to the driving force across the membrane (ΔX):

$$N = -k\,\Delta X \tag{12.1}$$

where k is the proportionality coefficient. Driving forces can be gradients in pressure, concentration, electrical potential or temperature.

https://doi.org/10.1515/9783110712445-012

Phase 1 Membrane Phase 2

Feed Permeate

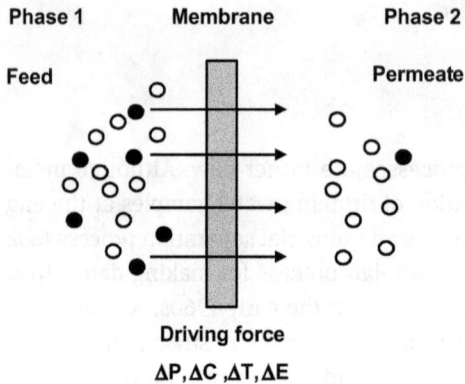

Driving force

$\Delta P, \Delta C, \Delta T, \Delta E$

Fig. 12.2: Schematic representation of selective membrane permeation.

Tab. 12.1: Most important industrial membrane processes.

Membrane process	Driving force	Pressure range (bar)	Flux range (L/m$_2$.h bar)
Microfiltration	ΔP	0.1–3.0	> 50
Ultrafiltration	ΔP	1.0–10	10–50
Nanofiltration	ΔP	10–30	1.4–12
Reverse osmosis	ΔP	10–200	0.05–1.4
Gas separation	ΔP	Feed side up to 140 bar or vacuum permeate side	
Vapor permeation	ΔP	Feed atmospheric, vacuum or low partial	
Pervaporation	ΔP	pressure at permeate side	
Electrodialysis	ΔE		

An overview of the most important membrane processes and driving forces is given in Table 12.1. Well-established industrial membrane separation processes are microfiltration, ultrafiltration, nanofiltration and reverse osmosis. The range of application of these four pressure-driven membrane water separation processes is summarized in Fig. 12.3. Both, ultrafiltration and microfiltration employ sieving through increasingly fine pores as the mode of separation. Microfiltration membranes filter colloidal particles and bacteria from 0.1 to 10 μm in diameter. Ultrafiltration membranes can be used to filter dissolved macromolecules, such as proteins, from solutions. The mechanism of separation by reverse osmosis membranes is quite different. In reverse osmosis the membrane pores are so small that they are within the range of thermal motion of the polymer chains that form the membrane. Nanofiltration membranes fall into a transition region between pure reverse osmosis and pure ultrafiltration. Both techniques are used when low-molecular-weight solutes such as inorganic salts or small organic molecules have to be separated from a solvent. The most important application of reverse osmosis is the desalination of brackish groundwater or seawater.

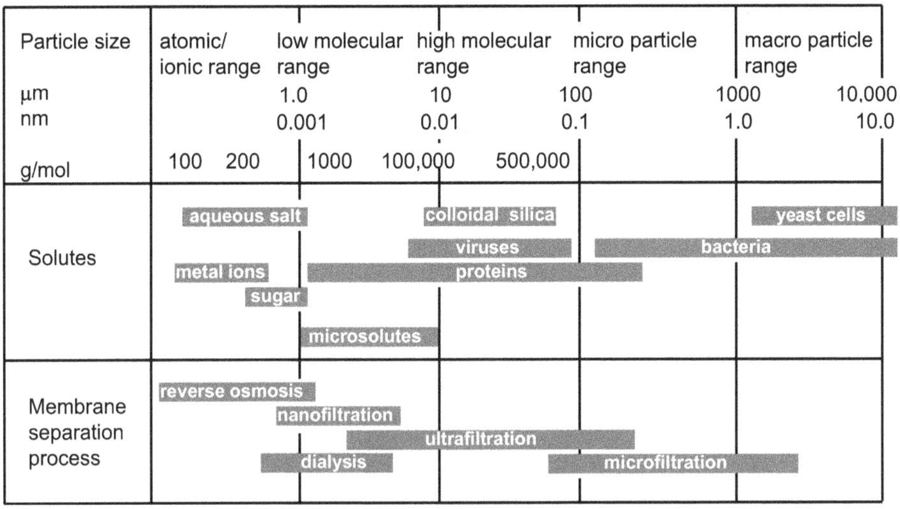

Particle size	atomic/ ionic range	low molecular range	high molecular range	micro particle range	macro particle range	
μm		1.0	10	100	1000	10,000
nm		0.001	0.01	0.1	1.0	10.0
g/mol	100 200	1000 100,000 500,000				
Solutes	aqueous salt / metal ions / sugar	microsolutes	colloidal silica / viruses / proteins		bacteria	yeast cells
Membrane separation process	reverse osmosis / nanofiltration	dialysis	ultrafiltration		microfiltration	

Fig. 12.3: Application range of various membrane processes.

In gas separation and vapor permeation, a gas mixture is passed across the surface of a membrane that is selectively permeable to one component of the feed mixture. Major current applications of gas permeation include the separation of hydrogen from nitrogen, argon and methane in ammonia plants, the production of nitrogen from air and the separation of carbon dioxide from methane in natural gas applications. Pervaporation is the only membrane process where a phase transition occurs. A liquid mixture contacts one side of a membrane, and the permeate is removed as a vapor from the other side. The driving force for the process is the low vapor pressure on the permeate side of the membrane. Currently the main industrial application of pervaporation is in the dehydration of organic solvents. Electrodialysis applies charged membranes to separate ions from aqueous solutions under the driving force of an electrical potential difference. The principal application is the desalting of brackish groundwater. However, other industrial applications such as in pollution control are rapidly growing.

Membrane processes possess a number of distinct advantages compared to alternative separation methods, such as distillation, absorption and extraction. The main advantage is that most membrane separations consume less energy because generally no phase transition is involved. Additional advantages are that the separation can be carried out under mild conditions and scale-up is easily accomplished by simply increasing the membrane area. However, scaling up by applying more modules in parallel makes the investment costs of a membrane separation process increase almost linearly with scale, which becomes a disadvantage at higher capacities. Additional limitations of membrane processes are limited amount of stages requiring a high selectivity for a given separation and the chemical/thermal stability of the membrane material. Also fouling of membranes may restrict permeability and/or lifetime of membranes and make them even unsuited for some types of feed streams.

12.2 Principles

12.2.1 Membranes

In essence a membrane is a discrete, thin interface that should be semipermeable to at least one component of the mixture to be separated. A wide variety of synthetic membranes exists that differ in chemical composition, physical structure and the way they operate. They are commonly classified by their morphology because the membrane structure determines the separation mechanism and hence the type of application. The principal types of membranes are shown schematically in Fig. 12.4.

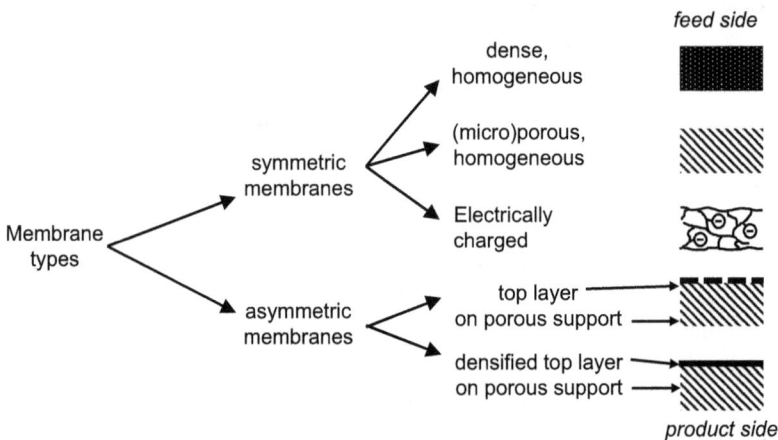

Fig. 12.4: Principal types of solid membrane structures.

An homogeneous microporous membrane is very similar in its structure and function to a conventional filter. It has a rigid, highly porous structure with randomly distributed interconnected pores. However, compared to a conventional filter these pores are extremely small (0.01–10 µm). All particles larger than the largest pores are completely rejected by the membrane by size exclusion. Partial rejection occurs for particles smaller than the largest pores, but larger than the smallest pores. Thus separation of solutes by micro-porous microfiltration and ultrafiltration membranes depends mainly on size of the solute/particle and pore size distribution of the membrane. In general effective separations are possible only for molecules that differ considerably in size. Homogeneous dense membranes consist of a dense film through which permeants are transported by diffusion under the driving force of a pressure, concentration or electrical potential gradient. The separation of various components of a solution is related directly to their relative transport rate within the membrane, which is determined by their diffusivity and solubility in the membrane material. Most gas separation, pervaporation and reverse osmosis membranes use dense membranes to perform the separation. However, the thickness (10 to 200 µm) required for mechanical stability makes transport through homogeneous dense membranes uneconomically low. Because the transport rate is inversely proportional to the membrane thickness, high transport rates can be achieved by making the membrane as thin as possible. A breakthrough for many industrial applications was the development of asymmetric membrane structures. Asymmetric membranes consist of a very dense top layer with a thickness of 0.1–0.5 µm on a porous sublayer with a thickness of about 50–150 µm that acts as the mechanical support. These membranes combine the high selectivity of a dense membrane with the high permeation rate of a very thin membrane. The separation properties and permeation rates are determined largely or completely by the thin top layer. The advantage of higher fluxes is so great that almost all commercial dense membranes have an asymmetric structure. Electrically charged membranes are most commonly microporous with positively or negatively charged ions fixed to the pore walls. They are used for processing electrolyte solutions in electrodialysis. A membrane with positively charged ions is referred to as anion-exchange membrane because it binds anions in the surrounding fluid. Similarly, a membrane containing negatively charged ions is called a cation-exchange membrane. Separation with charged membranes is achieved mainly by exclusion of ions with similar charge as the ion-exchange membrane and is affected by the charge and concentration of the ions in solution. Monovalent ions are excluded less efficiently than divalent ions and selectivity decreases in solutions of high ionic strength.

Membrane materials can be polymeric, inorganic or even metallic. In principle all polymers can be used for the preparation of organic (polymeric) membranes, but the chemical and physical properties differ so much that only a limited number are used in practice. Often a distinction is made between the open porous membranes and dense nonporous membranes. For the porous microfiltration/ultrafiltration membranes the choice of the material is mainly determined by the processing requirements, fouling tendency and chemical/thermal stability of the membrane. Hydrophobic materials such as polytertrafluorethylene (PTFE), polyvinylidenefluoride (PVDF) and isotactic polypropylene (PP) are often used for microfiltration membranes. The best-known hydrophilic polymers are cellulose and its ester derivatives such as cellulose acetate and cellulose triacetate. Besides in microfiltration and ultrafiltration they are also used in reverse osmosis, gas separation and pervaporation. Other important hydrophilic polymers are aromatic and aliphatic polyamides that both show outstanding performance and membrane stability. Because ultrafiltration membranes require a different preparation method than microfiltration membranes, often polysulfones (PS) and polyethersulfones (PES) are a very important class for ultrafiltration membranes. Other polymers of importance for the manufacturing of ultrafiltration membranes are polyimides and polyacrylonitrile. In nonporous composite or asymmetric membranes the porous support layer is often made of the polymers already mentioned. The choice of material for the permselective skin is determined by the required performance and type of application. The applied types of polymer range from elastomers to glassy materials. Inorganic materials generally possess superior chemical and thermal stability relative to polymeric materials. Types that can be distinguished are ceramic, glass and zeolite membranes. Ceramic, glass and metallic membranes are usually prepared by sintering or by sol-gel processes. Zeolites are crystals with a defined, very narrow pore size that can be sintered into a membrane.

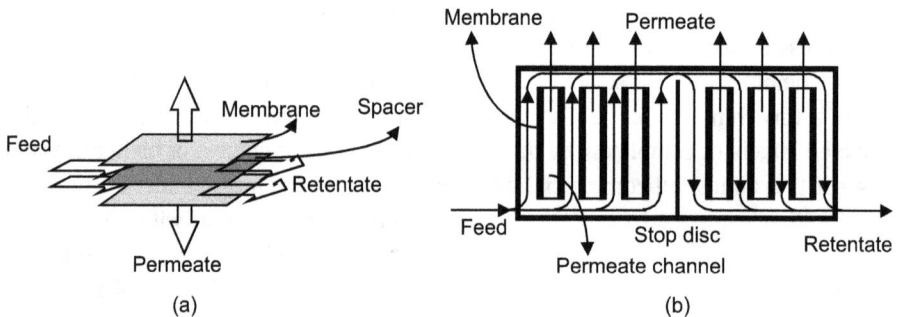

Fig. 12.5: Schematic drawing of construction (a) and flow path (b) in a plate-and-frame module.

12.2.2 Modules

Industrial membrane plants often require hundreds to thousands of square meters of membrane area to perform the separation required on a useful scale. In these plants a bare membrane cannot be applied without the proper connections to direct a feed stream onto the membrane surface and to collect the permeate and the raffinate. The smallest unit into which the membrane area is packed and provides the necessary connections is called a membrane module. A number of module designs, based on flat or tubular membranes, are possible. Plate-and-frame and spiral wound modules involve flat membranes, whereas tubular, capillary and hollow fiber modules are based on tubular membranes.

Plate-and-frame modules were one of the earliest types of membrane systems. A schematic drawing is given in Fig. 12.5. Membrane, feed spacers and product spacers are layered together between two end plates. The feed mixture is forced across the surface of the membrane. A portion passes through the membrane, enters the permeate channel, and makes its way to a central permeate collection manifold. The packing density of such modules is about 100 to 400 m^2/m^3. The spiral wound module can be considered as a plate-and-frame system wrapped around a central collection pipe. The packing density of this module (300–1,000 m^2/m^3) is greater than of the plate-and-frame module. Membrane and permeate side spacer material are glued along three edges to build a membrane envelope. As shown in Fig. 12.6 this membrane envelope is wound around a perforated central collection tube and placed inside a tubular pressure vessel. Feed passes axially down the module across the membrane envelope where the feed-side spacer also acts as a turbulence promoter. Part of the feed permeates into the membrane envelope, where it spirals toward the center and exits through the collection tube.

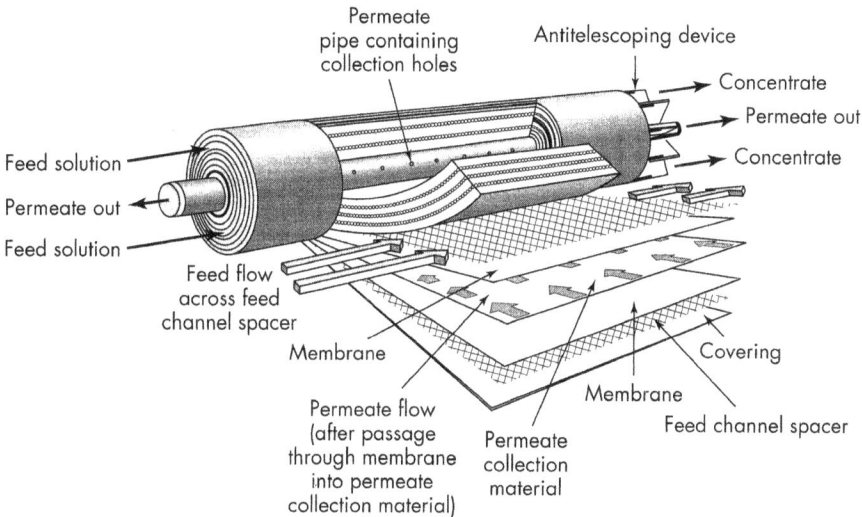

Fig. 12.6: Spiral wound module (reproduced with permission from [9]).

Tubular modules are now generally limited to ultrafiltration applications because their packing density of less than 300 m^2/m^3 is rather low. In general tubular membranes are not self-supporting. They consist of a porous tubular stainless steel, ceramic or plastic support with the membrane placed on the inside of the tubes. As illustrated schematically in Fig. 12.7, multiple tubes (4–18) with diameters larger than 10 mm are put together in a module. The feed solution always flows through the center of the tubes while the permeate flows through the porous supporting tube into the module housing. Capillary and hollow fiber modules consist of a large number of self-supporting membrane capillaries assembled together in a module, as shown schematically in Fig. 12.7. The free ends of the fibers are potted with agents such as epoxy resins, polyurethanes or silicone rubber. In the bore-side feed type the fibers are open at both ends and the feed fluid is circulated through the bore of the fibers with the permeate collected on the outside of the capillaries in the module housing. A packing density of about 600–1,200 m^2/m^3 is obtained with modules containing capillaries. The difference between the capillary module and the hollow fiber module is simply a matter of membrane fiber dimensions. The hollow fiber module is the configuration with the highest packing density, which can attain values up to 30.000 m^2/m^3. The second type of capillary module is the shell-side feed type. In this module a closed bundle of fibers is contained in a pressure vessel. The system is pressurized from the shell side, the permeate passes through the fiber wall and exits through the open fiber ends.

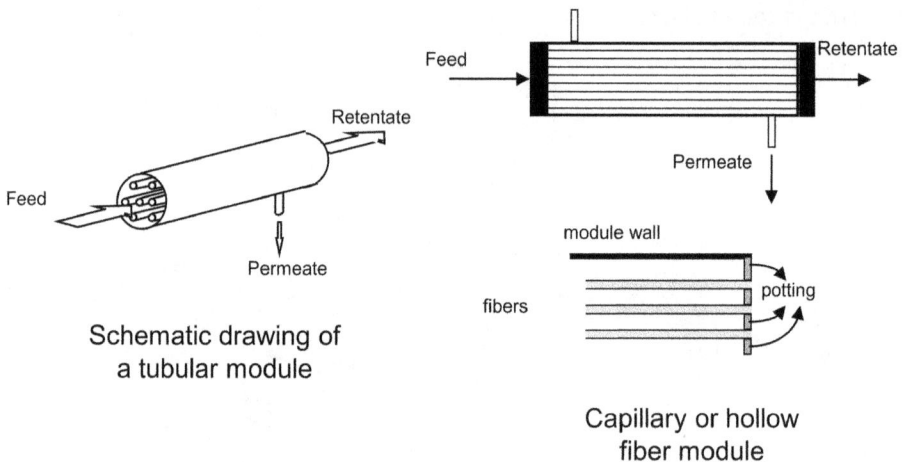

Fig. 12.7: Diagram of a tubular and bore-side feed capillary or hollow fiber module.

12.2.3 Flux, permeability and selectivity

The performance or efficiency of a given membrane is determined by two parameters: its selectivity and its flow through the membrane. The latter, often denoted as the flux or permeation rate, is defined as the volume flowing through the membrane per unit area and time. In porous membranes the most common transport mechanism is pressure-driven convective flow of the solvent and certain solutes flow through the pores. Separation occurs because certain larger solutes are excluded from some or all of the pores in the membrane through which the other permeants move. The convective flux through the membrane can be written as follows:

$$\text{flux} = \frac{\text{driving force}}{\text{viscosity} \times \text{total resistance}} \tag{12.2}$$

which in the case of pressure-driven processes such as microfiltration, ultrafiltration, nanofiltration and reverse osmosis becomes

$$N_i = \frac{\Delta P}{\eta \, R_{tot}} = \left(\frac{P_m}{\delta_m}\right) \Delta P \qquad (12.3)$$

where the membrane resistance R_{tot} is related to the membrane thickness δ_m by the permeability$_m$ that contains structural factors such as the porosity and pore size as well as the viscosity of the permeating liquid.

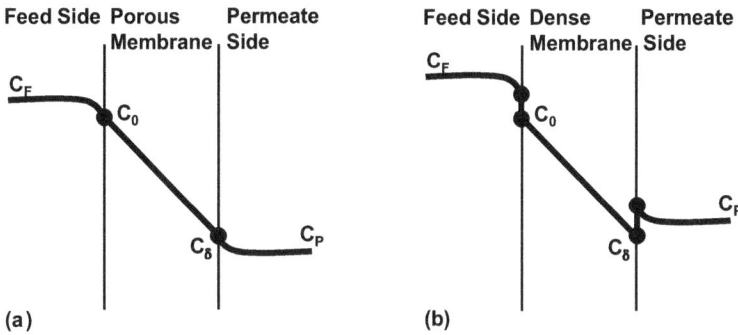

Fig. 12.8: Concentration profiles for solute transport through porous (a) and dense (b) membranes.

For porous membranes the concentration profile is continuous from the bulk feed liquid to the bulk permeate liquid because liquid is continuously present from one side to the other (Fig. 12.8a). This is not the case for the dense membrane in Fig. 12.8b, where permeants dissolve in the membrane material and then diffuse through the membrane down a concentration gradient. As a result the liquid-phase concentration gradient ΔC should be multiplied by the thermodynamic equilibrium distribution coefficient K_i to obtain the true solute concentration gradient over the membrane:

$$N_i = \left(\frac{K_i \, D_{e,i}}{\delta_m}\right) \Delta C_i \qquad (12.4)$$

For gases the solubility can be related to the partial pressure by Henry's law, giving

$$N_i = \left(\frac{H_i \, D_{e,i}}{\delta_m}\right) \Delta P_i \qquad (12.5)$$

In both equations D_e is the effective diffusion coefficient of the permeant through the membrane. They illustrate that for transport of a gas, vapor or liquid through a dense, nonporous membrane the permeability is always described in terms of

$$\text{Permeability } (P) = \text{Solubility } (S) \times \text{Diffusivity } (D) \qquad (12.6)$$

where the solubility provides a thermodynamic measure of the amount of permeant sorbed by the membrane under equilibrium conditions and the diffusivity depends on the geometry of the permant. Separation is obtained because of the differences in solubilities of the permeants in the membrane and the differences in the rates at which the permeants diffuse through the membrane. In general, convective pressure-driven membrane fluxes are high compared with those obtained by simple diffusion.

The selectivity of a membrane toward a mixture is generally expressed by the retention (R) or the separation factor (SF). For dilute aqueous mixtures it is more convenient to express the selectivity in terms of the retention toward the solute. The solute is partly or completely retained while the solvent (water) molecules pass freely through the membrane. The retention is given by

$$R = \frac{C_F - C_P}{C_F}$$

(12.7)

where C_F is the solute concentration in the feed and C_P is the solute concentration in the permeate. The value of R varies between 100% for complete retention of the solute and 0% when solute and solvent pass freely through the membrane. Membrane selectivity toward gas and liquid mixtures is usually expressed in terms of the separation factor. For a mixture consisting of two components the separation factor is given by

$$SF_{AB} = \frac{y_A/y_B}{x_A/x_B}$$

(12.8)

where y is the concentration of each component in the permeate and x the concentration of each component in the feed. The separation factor is usually calculated in such a way that its value is greater than unity. For the separation of a binary gas or liquid mixture in the absence of boundary layer or film mass-transfer resistances, the separation factor equation can be combined with the transport flux equations given in equations (12.4) and (12.5) and rearranged to give the ideal separation factor:

$$SF^*_{AB} = \frac{P_{M,A}}{P_{M,B}} = \frac{K_A \, D_A}{K_B \, D_B} = \frac{H_A \, D_A}{H_B \, D_B}$$

(12.9)

Thus a high separation factor can be achieved from a high solubility ratio, a high diffusivity ratio or both.

12.2.4 Concentration polarization and fouling

In membrane separation processes a separation is achieved because certain components are transported more readily than other. Because the feed mixture components permeate at different rates, concentration gradients form in the fluids on both sides of the membrane. This phenomenon, illustrated in Fig. 12.9, is called concentration

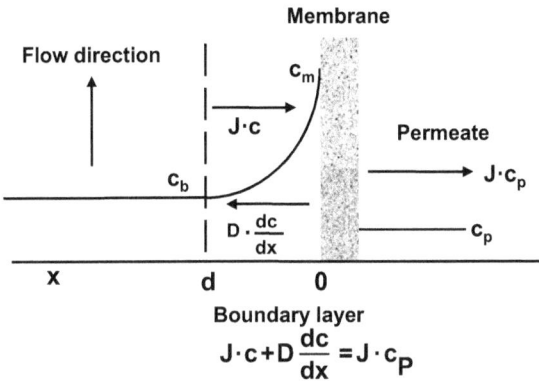

Fig. 12.9: Concentration polarization under steady-state conditions.

polarization. In membrane filtration processes, such as reverse osmosis, nanofiltration, ultrafiltration and microfiltration, certain solutes are retained by the membrane. As a result the concentration of retained solutes in the permeate is lower than the concentration in the bulk. The retained molecules accumulate at the membrane surface where their concentration will gradually increase. Such a concentration build-up will generate a diffusive flow back to the bulk of the feed. At steady state the convective transport of solute to the membrane surface is balanced by the sum of its flux through the membrane plus the diffusive back transport of retained solute from the membrane surface to the bulk. In mathematical terms this balance can be expressed as

$$J c + D \frac{dc}{dx} = J c_p \tag{12.10}$$

where D is the diffusion coefficient of the molecule in the boundary layer. The main effect of concentration polarization is a reduced flux at a finite time because of the counteracting retained solute concentration gradient. Especially in microfiltration and ultrafiltration the process flux may often be less than 5% of the pure water flux. The problem is less severe in gas separation and pervaporation. Other consequences of concentration polarization can be a lower retention because of the increased solute concentration at the membrane surface or a higher retention when higher molecular weight solutes form a kind of second membrane. In principle concentration polarization is a reversible process, meaning that when steady-state conditions have been attained the no further flux decrease should be observed. In practice, however, often a continuous flux decline can be observed due to membrane fouling. Fouling occurs mainly in microfiltration/ultrafiltration where the membrane may be blocked by some solutes or particles that penetrate into the pores. Additionally adsorption phenomena may also play an important role in fouling and hence it is important to select an appropriate membrane material. In general hydrophobic materials have a larger tendency to foul than hydrophilic materials.

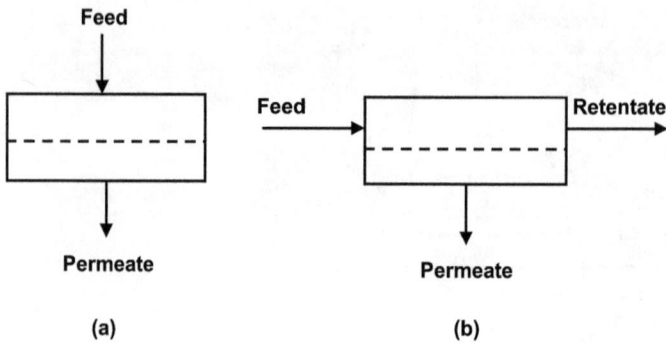

Fig. 12.10: Schematic of a dead-end (a) and cross-flow (b) module operation.

12.2.5 System design and cascades

A single membrane module or a number of such modules arranged in parallel or in series without recycle constitutes a single-stage membrane separation process. The simplest design is the dead-end operation shown in Fig. 12.10a. For most industrial applications a cross-flow operation (Fig. 12.10b) is preferred because of the lower fouling tendency. Although counter-current flow gives the best performance in cross-flow operation, most systems are designed with perfect permeate mixing. In a single-pass system the feed solution passes only once through the modules and the volume of the feed decreases with path length. Multi-stage single-pass designs compensate for this loss of volume by arranging the modules in a tapered design. As shown in Fig. 12.11, the cross-flow velocity through the systems remains virtually constant by gradually reducing the number of modules in parallel.

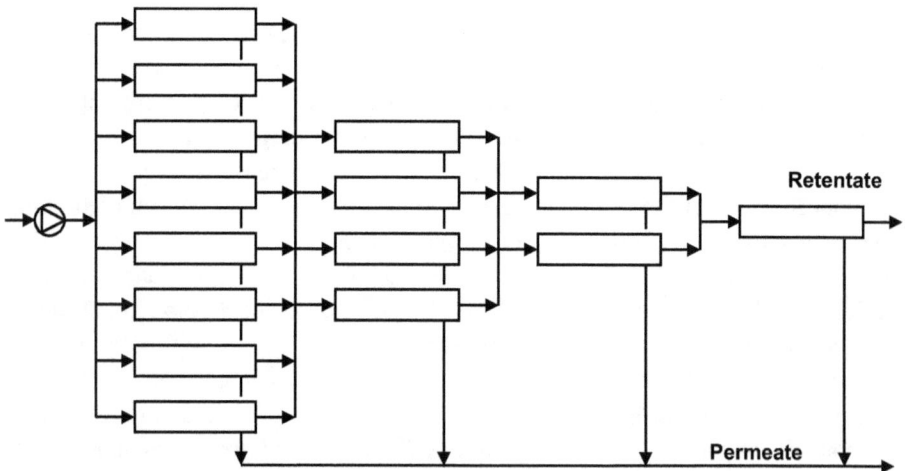

Fig. 12.11: Tapered design multistage single-pass system.

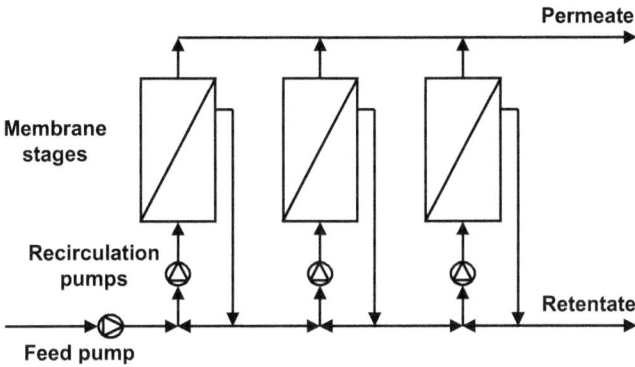

Fig. 12.12: Three-stage recirculation system.

The second system is the recirculation system or feed recycle system where the feed is pressurized by a pump and allowed to pass several times through one stage, consisting of several modules. As shown in Fig. 12.12, each stage is fitted with a recirculation pump which maximizes the hydrodynamic conditions. The feed recycle system is much more flexible than the single-pass system and is to be preferred in cases where severe fouling and concentration polarization occur. Often the single-stage design does not result in the desired product quality. The extent to which a feed mixture can be separated in a single stage is limited and determined by the separation factor. To achieve a higher degree of separation a countercurrent cascade of stages such as used in distillation can be applied. In a cascade operation where the permeate of the first stage is the feed of the second stage and so on, it is possible to obtain a very high product purity. An example of a two-stage permeate enrichment operation is given in Fig. 12.13. The type of design depends on

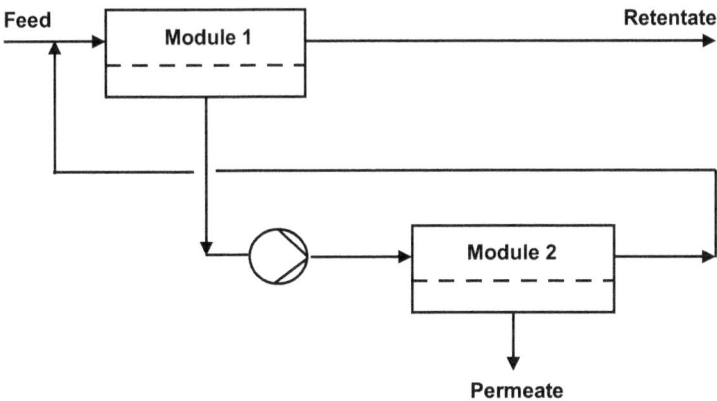

Fig. 12.13: Two-stage permeate enriching cascade.

whether the permeate or the retentate is the desired product. When more stages are required, the optimization of the process becomes more complex and difficult because of the large number of variable involved.

12.3 Membrane filtration processes

Various pressure-driven membrane filtration processes can be used to concentrate or purify a dilute solution. In these processes the solvent is the continuous phase and the concentration of the solute is relatively low. The necessary structure of the membranes employed is determined by the particle or molecular size. Particles with a diameter >0.1 µm are already retained by a rather open membrane structure in a process called microfiltration. The separation of macromolecules from an aqueous solution requires a more dense membrane structure and the process is called ultrafiltration. In reverse osmosis and nanofiltration low-molecular-weight components of approximately equal size are separated from each other. In this case a very dense (asymmetric) membrane is used. Going from microfiltration through ultrafiltration to reverse osmosis and nanofiltration, the hydrodynamic resistance of the membrane increases and consequently a higher pressure difference is needed. Typical values for applied pressures and fluxes have already been given in Table 12.1.

12.3.1 Microfiltration

Microfiltration is used in a wide variety of industrial applications where particles of a size >0.1 µm, such as colloids, dyes and microorganisms, have to be retained from a liquid or a gas. One of the main industrial applications is the sterilization and clarification of all kinds of beverages and pharmaceuticals. In fact, microfiltration was developed to separate microorganisms from water and is the oldest and largest application of membrane separation. New fields of application are biotechnology and biomedical technology. In biotechnology, microfiltration is especially suitable in cell harvesting.

The most important applications today are still based on dead-end filtration. In this process design the entire fluid flow is forced through the membrane that usually operates as a depth filter and collects the particles within the membrane. As particles accumulate in the membrane interior or on its surface, the pressure required to maintain the required flow increases. In cross-flow microfiltration systems screen filter membranes that collect retained particles on the surface are increasingly preferred. Cross-flow systems are more complex than dead-end filter systems because they require additional auxiliary equipment such as a recirculation pump, valves and so on. However, a screen membrane has a much longer lifetime than a depth membrane, and can be regenerated by back flushing. Cross-flow filtration is being adopted increasingly for microfiltration of high volume industrial streams containing significant particle levels.

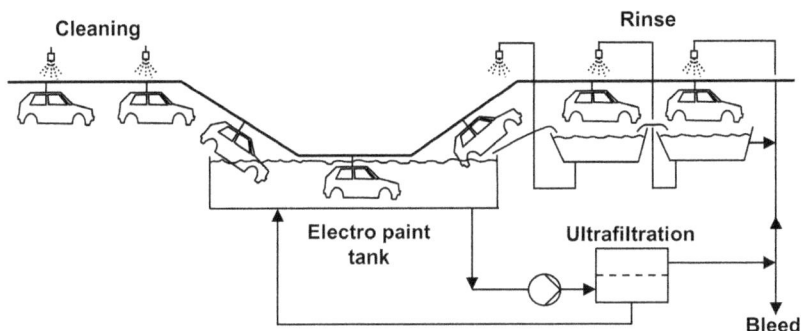

Fig. 12.14: Flow scheme of an electrocoat paint ultrafiltration system.

12.3.2 Ultrafiltration

Ultrafiltration is used over a wide field of applications involving situations where high-molecular-weight components have to be separated from low-molecular-weight components. The particle size being rejected from the permeate stream range from 0.1 μm to 1 nm. Early and still important applications were the recovery of electro coat paint from industrial coating operations shown in Fig. 12.14 and the clarification of emulsified oily wastewaters in the metalworking industry. More recent applications are in the food and dairy industry for the concentration of milk, recovery of whey proteins and clarification of fruit juices and alcoholic beverages. With the ultrafiltration process illustrated in Fig. 12.15, whey is separated into an aqueous solution of lactose and salts (permeate), retaining casein, butterfat, bacteria and proteins in the retentate. Other examples are the latex particle recovery from wastewater, separation of macromolecular solutions and the manufacturing of sterile vaccines by removing viruses from solution. In reverse osmosis and nanofiltration processes ultrafiltration as well as microfiltration are frequently employed as a pretreatment step.

An important difference with microfiltration is that ultrafiltration membranes have an asymmetric structure with a much denser top layer and consequently a much higher hydrodynamic resistance. The top-layer thickness is generally less than 1 μm. The membranes are characterized by their molecular weight cutoff, defined as the molecular weight of the globular protein molecule that is rejected 95% by the membrane. The rejection of linear polymer molecules of equivalent molecular weight is usually much less because they are able to sneak through the membrane pores, whereas more rigid globular molecules are retained. An important point which must be taken into account is that the process performance is not equal to the intrinsic membrane properties in actual separations due to concentration polarization and fouling.

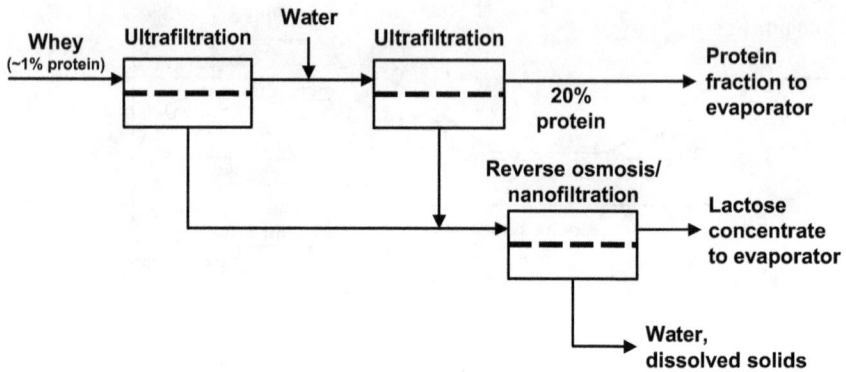

Fig. 12.15: Simplified flow scheme of an ultrafiltration/reverse osmosis process to extract valuable components from cheese whey.

12.3.3 Reverse osmosis and nanofiltration

Nanofiltration and reverse osmosis are used when low-molecular-weight solutes such as inorganic salts or small organic molecules such as glucose or sucrose have to be separated from a solvent. The difference with ultrafiltration lies in the size of the solute. Consequently denser membranes are required with a much higher hydrodynamic resistance. A much higher pressure must be applied to force the same amount of solvent through the membrane. Moreover, the osmotic pressure has to be overcome that arises when two solutions of different concentration are separated by a semipermeable membrane. This situation is illustrated schematically in Fig. 12.16. Because the solvent molecules in the dilute phase have a higher chemical potential than those in the concentrated phase, the solvent molecules want to flow from the dilute phase to the concentrated phase. This process continues until osmotic equilibrium has been reached. The resulting hydrodynamic pressure difference ΔP is called the osmotic pressure difference $\Delta \Pi$. In reverse osmosis this process is reversed by applying an external hydrostatic pressure to force water into the direction of the clean water product. A complication may be the increasing salt concentration in the remainder of the solution, causing the osmotic pressure to increase to such an extent that the required value of hydrostatic pressure gets too high for practical purposes. If it is assumed that no solute permeates through the membrane the effective water flow can be described by

$$N_{\text{water}} = A(\Delta P - \Delta \Pi) \tag{12.11}$$

where A is a proportionality constant, ΔP is the applied pressure difference and $\Delta \Pi$ is the osmotic pressure differential across the membrane. As this equation shows, water flows from the dilute to the concentrated salt-solution side by normal osmosis at low applied pressure $\Delta P < \Delta \Pi$. When the applied pressure is higher than the

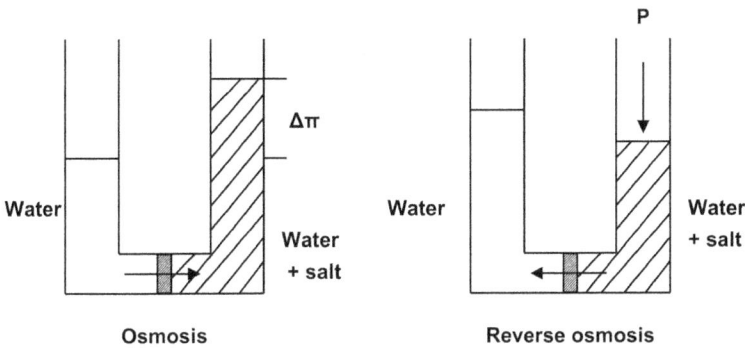

Fig. 12.16: Principle of osmosis and reverse osmosis.

osmotic pressure water flows from the concentrated to the dilute salt-solution side of the membrane.

The production of potable water was the first membrane-based separation process to be commercialized on a significant scale. The discovery of the asymmetric cellulose acetate membrane made desalination by reverse osmosis technically and economically feasible. The first cellulose di-acetate modules had a salt rejection of 97–98%. This was enough to produce potable water (<500 ppm salt) from brackish water (Fig. 12.17), but insufficient for the desalination of seawater. Seawater desalination became within reach by the development of interfacial composite membranes with salt rejections greater than 99.5%. The general trend of the desalination industry is toward spiral wound modules. Besides the application of reverse osmosis in drinking water, a second important market is the production of ultrapure water. Such water is used in steam boilers or in the electronics industry, where extremely pure water with a total salt concentration below 1 ppm are required to wash silicon wafers. Other applications where reverse osmosis may be used are the food industry (fruit juice, sugar, coffee), the galvanic industry (waste streams) and the dairy industry (milk).

While reverse osmosis and nanofiltration are closely related in terms of applications, their separation mechanisms are quite different. Where reverse osmosis is based on reversing the phenomena of osmosis, nanofiltration can be considered a filtration process through nanometers pores. The network structure of nanofiltration membranes is more open, implying that the retention for monovalent ions becomes much lower but the retention for bivalent ions remains very high. In addition the retention is high for microsolutes such as herbicides, insecticides, pesticides, dyes and sugars. Since the water permeability of nanofiltration membranes is much higher it will be preferred over reverse osmosis for somewhat larger microsolutes and when a high retention for monovalent salt is not required.

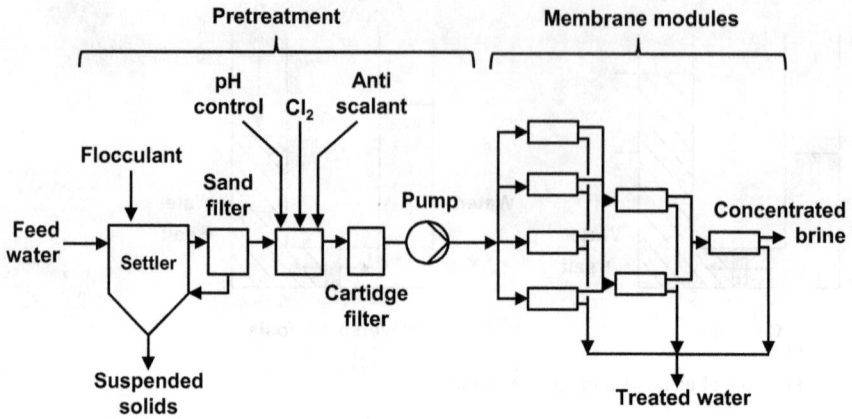

Fig. 12.17: Typical brackish water reverse osmosis desalination plant.

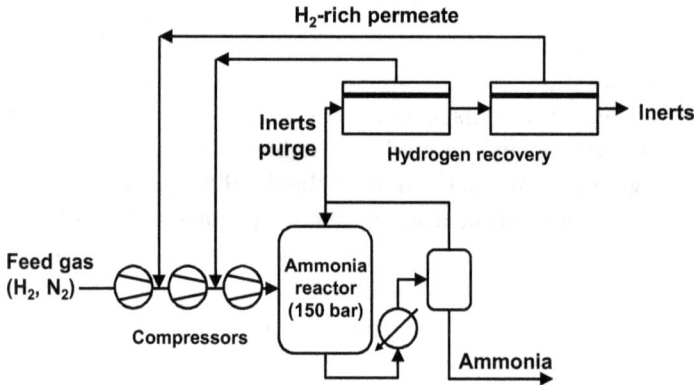

Fig. 12.18: Flow scheme of a membrane system to recover hydrogen from an ammonia reactor purge stream.

12.4 Solubility-driven processes

12.4.1 Gas and vapor permeation

In gas permeation, certain components of the feed permeate a permselective membrane at a much higher rate than the others. The driving force is the pressure difference between the pressurized feed gas and the lower pressure permeate. This pressure difference can be established by applying a high pressure on the feed side and/or maintaining a low pressure on the permeate side. Although porous membranes are also applicable, all current commercial gas separations are based on diffusion through dense polymer films. As was shown earlier the ability of a membrane to separate two gases, called the ideal membrane separation factor SF^*, is the ratio of their permeabilities:

$$SF^*_{AB} = \frac{P_{M,A}}{P_{M,B}} = \frac{K_A D_A}{K_B D_B} = \frac{H_A D_A}{H_B D_B} \tag{12.12}$$

The ratio of the diffusion coefficients represents the mobility selectivity, reflecting the different sizes of two molecules. In polymeric materials the diffusion coefficient decreases with increasing molecular size. Hence, the mobility selectivity always favors the passage of small molecules over large ones. The ratio of the sorption coefficients can be viewed as the solubility selectivity, reflecting the relative condensabilities of the two gases. If molecule 1 is larger than 2, the mobility selectivity will always be less than one while the sorption selectivity will normally be larger than one. The balance between the sorption selectivity and the mobility selectivity determines whether a membrane material is selective for large or small molecules in a gas mixture.

Both hollow-fiber and spiral wound modules are used in gas separation applications. Spiral wound modules are favored if the gas stream contains entrained droplets as in air or natural gas separations. The first large-scale commercial application was the recovery of hydrogen from ammonia purge gas streams. Hydrogen is a small, noncondensable gas, which is highly permeable compared to all other gases. A typical membrane system flow scheme is shown in Fig. 12.18. A two-step membrane design is used to reduce the cost of recompressing the hydrogen permeate stream to the very high pressures of ammonia reactors. Today the largest gas separation process in use is the production of nitrogen from air. The first membranes were based on polysulfone and ethylcellulose with oxygen–nitrogen selectivities of 4–5. With the second-generation materials selectivities in the range 7–8 and higher fluxes are obtained, making the economics of small-scale nitrogen production very favorable. A growing application is the removal of condensable organic vapors from air and other streams. These processes fall into three main categories and use rubbery membranes that are more permeable to the organic vapor. The first category concerns small systems to recover chlorinated, fluorinated and other high value hydrocarbons from process vent streams. The driving force for their installation is primarily economic. The second category is larger units to recover hydrocarbon vapors from petroleum transfer operations. Although these hydrocarbons have some economic value, the main driving force is environmental. The third category is the recovery and recycling of monomers in polymerization plants. Vent streams of these plants contain large amounts of monomer that can be removed by organic-vapor-permeable membranes and returned to the reactor. Figure 12.19 shows an example for the recovery of propylene from the nitrogen used for devolatilization in a polypropylene plant. Both the clean nitrogen and the propylene are recycled within the polypropylene plant.

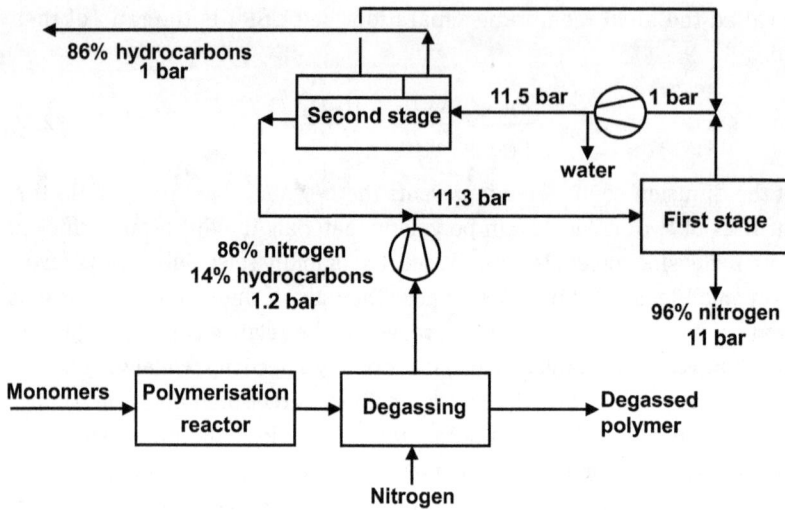

Fig. 12.19: Two-stage membrane unit for propylene and nitrogen recovery in a polypropylene plant.

12.4.2 Pervaporation

In this separation process, a multicomponent liquid stream is passed across a membrane that preferentially permeates one or more of the components. As the feed liquid flows across the membrane surface, the permeated components pass through the membrane as a vapor. Transport is induced by maintaining a vapor pressure on the permeate side of the membrane that is lower than the vapor pressure of the feed liquid. As illustrated in Fig. 12.20, the pressure difference is maintained by condensing the permeate vapor, which is removed as a concentrated permeate fraction.

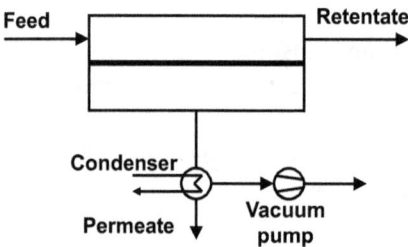

Fig. 12.20: Schematic of a pervaporation process with a downstream vacuum.

Pervaporation is a complex process in which both mass and heat transfer occurs. The separation principle in pervaporation is based on differences in solubility and diffusivity. The most convenient method to describe the separation principles of pervaporation is to divide the overall process into two steps. The first is an (imaginary) evaporation of the feed liquid to form a saturated vapor phase on the feed side of the membrane. The second is permeation of this vapor through the membrane to the low pressure permeate side of the membrane. The evaporation step produces a separation according to the partial vapor pressures of the components in equilibrium with the feed solution. The permeation of components through the membrane is largely analogous to conventional gas permeation. Separation is achieved by the combination of solubility and diffusivity in the dense membrane layer. The total separation factor is now proportional to the product of these two contributions:

$$SF_{pervap} = SF_{evaporation} \, SF_{permeation} \tag{12.13}$$

To achieve good separations both terms should be large. Although one might conclude that pervaporation is most suited to the removal of volatile components from relatively involatile components, membranes can be made sufficiently selective to make nonvolatile components permeate the membrane preferentially.

The three major applications of pervaporation are solvent dehydration, recovery organics from water, and the separation of mixed organic solvents. Most developed is the separation of water from concentrated alcohol solutions such as ethanol and isopropanol. A flow scheme for an integrated distillation-pervaporation plant is shown in Fig. 12.21. The distillation column produces a concentrated ethanol stream, which is fed to the pervaporation system to break the azeotrope and obtain nearly pure ethanol. A second interesting application of dehydration membranes is the shift of chemical equilibrium in esterification reactions:

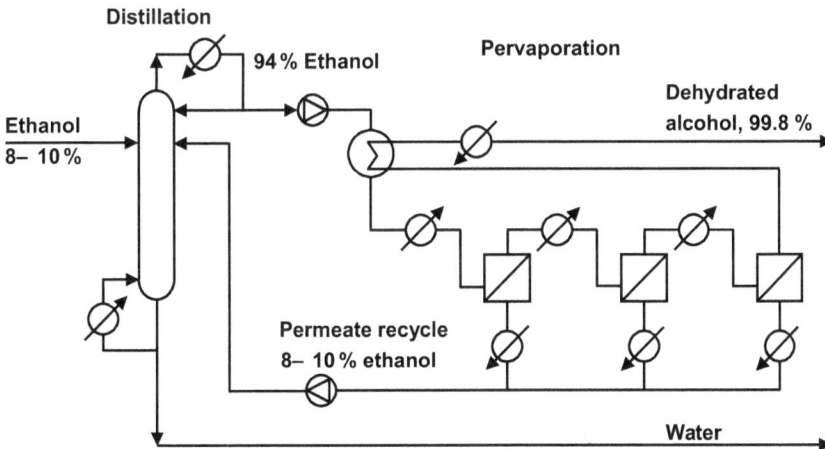

$$Acid + Alcohol \leftrightarrow Ester + Water \tag{12.14}$$

Fig. 12.21: Schematic of integrated distillation/pervaporation plant for ethanol dehydration.

In these reactions, the degree of conversion is limited by water buildup in the reactor. Continuous removal of water forces the equilibrium to the right and almost complete conversion can be achieved. The second application is the separation of small amounts of organic solvents from contaminated water. This separation is relatively easy because organic compounds and water exhibit distinct membrane permeation properties. Pervaporation for the separation of organic/organic mixtures is currently under development and could become a major application in the near future.

12.5 Electrodialysis

The principle of electrodialysis is depicted in Fig. 12.22. In this process an electric potential gradient is used to remove ions from an aqueous solution with charged, ion-selective membranes. A large number of anion- and cation-exchange membranes are placed in an alternating pattern between a cathode and an anode in a plate-and-frame stack. The positively charged sodium ions migrate to the cathode but are unable to pass the positively charged membranes. In the same manner the negatively charged sulfate ions migrate to the anode but cannot pass the negatively charged membranes. The resulting overall effect is that every second cell becomes depleted of salt, while the adjacent cells become concentrated in salt. The degree of concentration is determined by the flow rate of the feed solution through the stack.

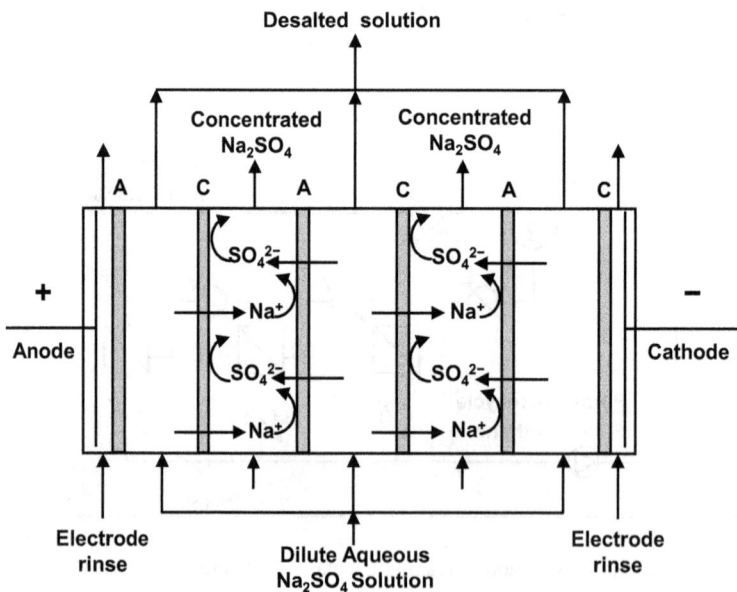

Fig. 12.22: Plate-and-frame electrodialysis stack.

Electrodialysis is used widely to desalinate brackish water. In desalination one of the most attractive features of electrodialysis is its energy efficiency compared to evaporation or reverse osmosis. Because the electric current needed to desalinate a solution is directly proportional to the quantity of ions transported through the membrane, electrodialysis is almost always the lowest cost process in the 500–2,000 ppm range. A very special reverse application of desalination is the production of a salt concentrate from seawater in Japan. The process is also used in the food industry to deionize cheese whey, deacidification of wine and fruit juices. A number of other applications exist in wastewater treatment, particularly regeneration of waste acids used in metal pickling operations and removal of heavy metals from electroplating rinse waters. A very recent industrial application is the removal of organic acids from a fermentation broth.

Bipolar membranes consist of an anionic and cationic membrane laminated together. At the interface between the two membranes ions are generated by the water splitting reaction. The liberated hydrogen ions migrate to the cathode while the hydroxyl ions migrate to the anode. When used in an electrodialysis process, bipolar membranes can divide a neutral salt into the conjugate acid and base. Unfortunately the process is limited to the generation of relatively dilute acid and base solutions.

Nomenclature

A	Proportionality constant	
C	Concentration	$[\text{mol/m}^3]$
D	Coefficient of diffusion	$[\text{m}^2/\text{s}]$
F	Feed flow	$[\text{mol/s}]$
H_i	Henry's coefficient of component i	$[\text{mol/bar}]$
J	Component flux	$[\text{mol/s m}^2]$
k	Proportionality constant	$[\text{kg/s m}^2 \text{ bar}^1, \text{mol/s m}^2 \text{ V}]$
K_i	Distribution coefficient of component i	$[-]$
N	Permeation rate	$[\text{kg/s m}^2, \text{mol/s m}^2]$
P	Pressure	$[\text{Pa, bar}]$
P	Permeate flow	$[\text{mol/s}]$
P_m	Membrane permeability	
$P_{M,i}$	Permeability of component i	
R	Retentate flow	$[\text{mol/s}]$
R_i	Retention of i	$[-]$
R_{tot}	Membrane resistance	
S	Solubility	$[\text{mol/m}^3]$
SF	Separation factor	$[-]$
x, y	Mole fractions	$[-]$
δ_M	Membrane thickness	$[\text{m}]$
ΔX	Driving force	$[\text{Pa, bar, V}]$
η	Viscosity	$[\text{Pa.s}]$
Π	Osmotic pressure	$[\text{Pa, bar}]$

13 Crystallization and precipitation

13.1 Introduction

Crystallization is one of the oldest unit operations in the portfolio of industrial separations. Large quantities of crystalline products such as sodium chloride, sucrose and fertilizer chemicals (ammonium nitrate, ammonium phosphates and urea) are manufactured commercially. In the production of organic chemicals, crystallization is used to recover product, to refine intermediate chemicals and to remove undesired salts. Crystalline products coming from the pharmaceutical, fine chemical and dye industries are produced in relatively small quantities but represent a high-value and important industrial sector.

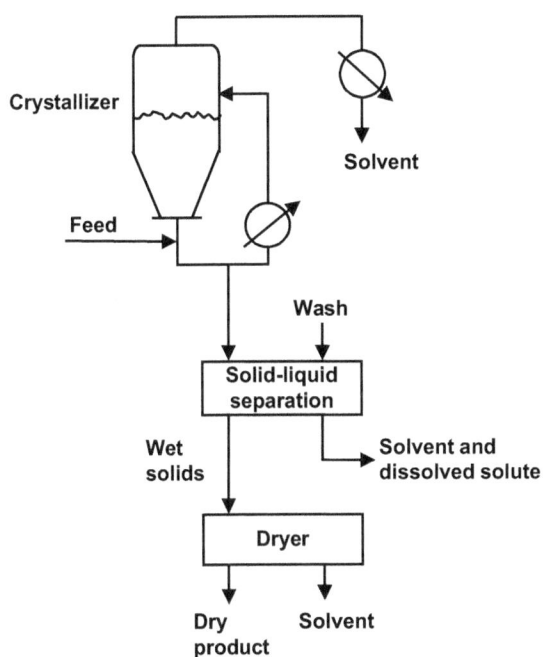

Fig. 13.1: Schematic of a crystallization-initiated solids processing sequence.

Crystallization distinguishes itself from other separation technologies in that a solid phase is generated from a minority component, referred to as the solute, dissolved in a majority component, referred to as the solvent. In the process, the feed consists of a homogeneous mixture of solute and solvent that is brought to supersaturated condition by cooling and/or evaporation of the solvent, to obtain the desired solute in a solid form. The process of solid phase formation is termed crystallization, and the operation occurs in a vessel called a crystallizer. Functions that can be achieved

https://doi.org/10.1515/9783110712445-013

by crystallization include separation, purification, concentration, solidification and analysis. Crystallization usually occurs on added seed crystals or on very small crystals, nuclei, broken off other crystals. The crystal particles grow and become larger until they are removed (harvested) from the crystallizer. Although the crystals are often of high purity, crystallizer performance is also characterized by the inherent crystal shape and size distribution. These solid phase properties are especially important, because crystallization is, often, the initial step in a solid processing sequence, similar to that shown in Fig. 13.1. After crystallization, the solids are normally separated from the crystallizer liquid, washed and discharged to downstream drying equipment. Since product size and suspended solids concentration are controlled to a large extent in the crystallizer, predictable and reliable crystallizer performance is essential for smooth operation of the downstream system.

The main advantages of crystallization are that nearly pure solid product can be recovered in the desired shape in one separation stage. With care in design, product purity in excess of 99.0% can be attained in a single crystallization, separation and washing sequence. This large separation factor is one of the reasons that make crystallization a desirable separation method. A second reason why crystallization is often the separation method of choice is that it can produce uniform crystals of the desired shape. During crystallization, the conditions are controlled, so that the crystals have the desired physical form for direct packaging and sale. This is important when a solid product is desired. The major disadvantages of crystallization are that purification of more than one component and/or full solute recovery are not attainable in one stage. Thus, additional equipment is required to remove the solute completely from the remaining crystallizer solution. Since crystallization involves processing and handling of a solid phase, the operation is normally applied when no alternative separation technique can be identified. Some important considerations that may favor the choice of crystallization as the preferred option over other separation techniques, such as distillation, include:

- Solute is heat sensitive and/or a high temperature boiler that decomposes at the high temperatures is required to conduct distillation,
- Low relative volatility between the solute and contaminants and/or the existence of azeotropes between solute and contaminants.
- Solid solute product: For example, after purification via distillation, a solute must be solidified by flaking or prilling, and crystallization may be more convenient to employ.
- Comparative economics favor crystallization: If distillation requires high temperatures and energy usage, crystallization may offer economic incentives.

Precipitation is an important separation method in production of many fine chemicals and in mineral processing. It is a related process, since solutes dissolved in a solvent precipitate out. However, the precipitate is usually amorphous and will have a poorly defined shape and size. Precipitates are often aggregates of several species and may include salts or occluded solvent. Thus precipitation serves as a "rough cut" to either remove impurities or to concentrate and partially purify the product.

13.2 Fundamentals

The design of a crystallization process requires information on the solute solubility, metastable limits, and crystal nucleation and growth characteristics. Solubility and phase relationships influence the choice of crystallizer and method of operation. Although equilibrium and equilibrium calculations are very important, they, alone, are not sufficient for a complete crystallizer design as for other equilibrium staged separations. Equilibrium calculations can predict the total mass of crystals produced after a time sufficiently long to reach equilibrium, but cannot tell the size and the number of crystals produced, especially if the process is executed in a finite amount of time. To include the crystal size distribution in the design calculations, crystal nucleation and growth rates must be incorporated together with population balances. Nucleation leads to the formation of small initial crystals that are large enough to not spontaneously melt again. Growth is the enlargement of these initial crystals caused by deposition of solid material on an existing crystal surface. The relative rates at which nucleation and growth occur determine the crystal size distribution. Acceptable operating conditions for the minimization of uncontrolled nucleation and encrustation of heat-exchange surfaces are defined by metastable limits.

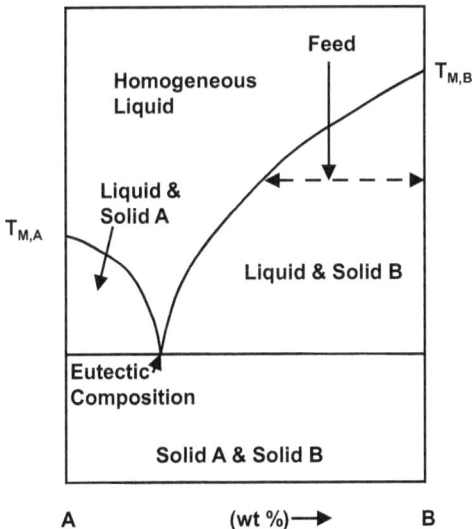

Fig. 13.2: Eutectic solid–liquid phase diagram. Horizontal is composition and vertical the temperature.

13.2.1 Solid–Liquid equilibria

Accurate solid–liquid equilibrium data are essential to evaluate the process design options for crystallization processes. A saturated solution is a solution that is in thermodynamic equilibrium with the solid phase of its solute at a specified temperature (the pressured dependence is typically very weak). The saturation solubility of a solute is given by the fundamental thermodynamic relationships for equilibrium between a solid and a liquid phase. When the effect of differences between the heat capacities of the liquid and solid is neglected, the saturation solubility is given by

$$\ln(y_s x_s) = \frac{\Delta H_M}{R}\left(\frac{1}{T_M} - \frac{1}{T}\right) \tag{13.1}$$

where x_S is the solubility (liquid mole fraction) of the solute in the solvent, y_S is the liquid phase activity coefficient of the solute, ΔH_M is the enthalpy of melting of the solute, also called heat of fusion, T_M (K) is the solid melting temperature and T (K) is the system temperature. For ideal solutions, eq. (13.1) reduces to the van't Hoff relationship:

$$\ln(x_s) = \frac{\Delta H_M}{R}\left(\frac{1}{T_M} - \frac{1}{T}\right) \tag{13.2}$$

With this equation, the solute solubility for an ideal solution can be calculated from the heat of fusion and the pure solid melting temperature of the solute. Thus, the

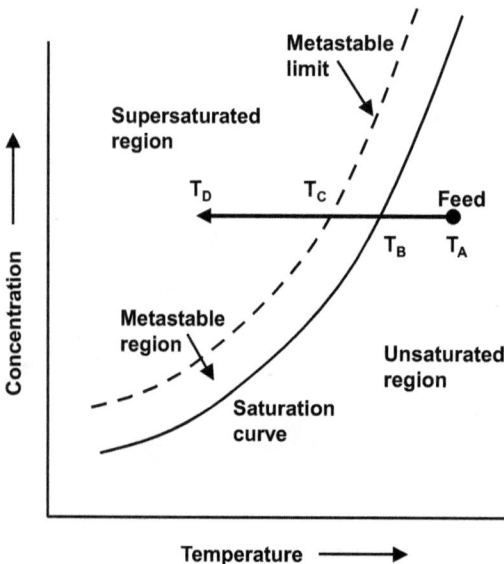

Fig. 13.3: Schematic depiction of supersaturation. In the metastable region, no spontaneous nucleation occurs over the time scale of the crystallization process.

solubility depends only on the properties of the solute and is independent of the nature of the solvent. A common type of solid–liquid phase diagram is the binary eutectic, shown in Fig. 13.2, in which a pure solid component is formed by cooling an unsaturated solution until solids appear. Continued cooling will increase the yield of the pure solid component. At the eutectic point, both components solidify, and additional purification is not possible.

13.2.2 Supersaturation and metastability

The kinetic phenomena that influence crystal size distributions are nucleation and growth. The driving force for both these phenomena is supersaturation, defined as the deviation from thermodynamic equilibrium. If a solution or melt is slowly cooled, the temperature will gradually drop, until the saturation point is reached. At this point, however, no spontaneous solidification will take place because the rate of nucleation is negligible. Change of phase only occurs when a certain degree of supercooling provides the required driving force to initiate nucleation. This region of supercooling is known as the metastable region, where the solution contains more dissolved solute than that given by the thermodynamic equilibrium saturation value. This is illustrated in Fig. 13.3 that shows the solubility curve, the supersaturation line of a solution and the metastable region in which the nucleation rate is so low that, effectively, no nucleation is observed over the time scale of the crystallization process. The degree of supersaturation can be expressed in many different ways. In crystallization, it is common to use the difference between the solute concentration and the concentration at equilibrium (at the current temperature):

$$\Delta c = c - c^* \tag{13.3}$$

where c is the actual solution concentration, and c* the temperature-dependent equilibrium saturation value. Another common expression is the relative supersaturation, s, given by the ratio of the difference between the solute concentration and the equilibrium concentration to the equilibrium concentration:

$$s = \frac{c - c^*}{c^*} \tag{13.4}$$

Solution concentration may be expressed in a variety of units. For general mass balance calculations, kilograms anhydrate per kilogram of solvent or kilograms hydrate per kilograms of free solvent are most convenient. The former avoids complications if different phases can crystallize over the temperature range considered. The importance of supersaturation in setting nucleation and growth rates is clarified in Fig. 13.4, which schematically shows the influence of supersaturation on nucleation and growth. The key aspects in this figure are the qualitative relationships of the two forms of nucleation to growth and to each other. Growth rate and secondary nucleation kinetics are

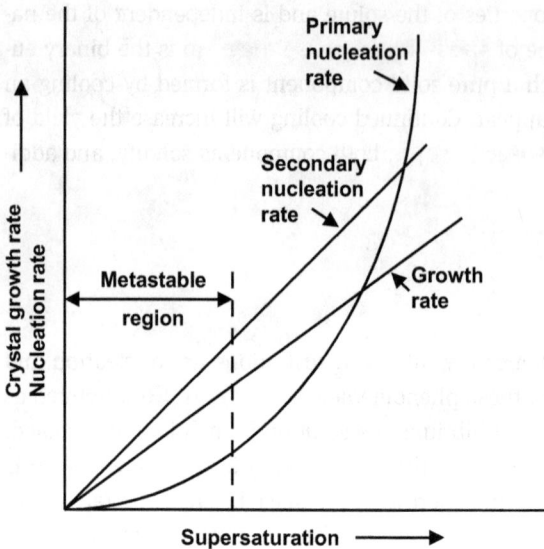

Fig. 13.4: Influence of supersaturation on nucleation and crystal growth rates.

low-order (shown as linear) functions of supersaturation, while primary nucleation follows a high-order dependence on supersaturation. Design of a crystallizer to produce a desired crystal size distribution requires the quantification of nucleation and growth rates to externally controlled variables and to supersaturation.

13.2.3 Nucleation

In crystallization, nucleation is the formation of initial small crystals that are large enough to not spontaneously melt again. The nucleation process differs from growth in that a new crystal is formed. It sets the character of the crystallization process and is, therefore, the most critical component in relating crystallizer design and operation to crystal size distribution. In primary nucleation, existing crystals are not involved in the nucleation process. Homogeneous primary nucleation occurs in the bulk liquid phase, in the absence of crystals or any foreign particles. Heterogeneous primary nucleation occurs on a foreign solid surface such as a dust particle or the vessel wall. Secondary nucleation is heterogeneous nucleation induced by existing crystals.

Homogeneous primary nucleation is only observed when the vessel is meticulously cleaned, polished and closed to avoid the presence of atmospheric dust that will act as heterogeneous nucleation sites. In classical homogeneous nucleation theory, a large number of solute units (atoms, molecules or ions) can aggregate together in a highly supersaturated solution to form a cluster. Such clusters continually form in the solution because of spontaneous thermal fluctuations. When the clusters are still very small, they are unstable (because of dominance of surface energy) and will spontaneously redissolve. Depending on the amount of supersaturation, still, a few clusters will survive and reach a certain critical size, for which the free energy change upon further growth is negative. Then, the cluster is stable and serves as a nucleus for further growth. The simplest expression for the homogeneous primary nucleation rate (assuming a spherical cluster and small relative supersaturation, $s < 0.1$) is

$$B^o = A \exp\left(-\frac{16\pi y^3 v^2}{3k^3 T^3 s^2}\right) \tag{13.5}$$

where k is the Boltzmann constant, T is absolute temperature, y is the surface energy per unit area, v is molar volume, s is the relative supersaturation and A is a constant. The high-order dependence on supersaturation is especially important, as a small variation in supersaturation may produce an enormous change in nucleation rate and can lead to production of excessive fines, when primary nucleation mechanisms are important.

In heterogeneous primary nucleation, the crystal forms around a foreign object such as a dust particle or a crack in the vessel wall. The foreign object lowers the surface energy because the solvent wets the solid. Although eq. (13.5) is applicable, the effectively lower surface energy leads to an increase in nucleation rate by 4 to 5 orders of magnitude. This greatly increases the rate of nucleation at much lower supersaturations, which is of specific importance to industrial systems that are never completely free of suspended solids, less perfectly cleaned and contain incompletely polished interior surfaces.

Secondary nucleation is the formation of new crystals as a result of the presence of solute crystals through several mechanisms, including initial breeding, contact nucleation and shear breeding. In most industrial crystallizers, secondary nucleation is far more important than primary nucleation. This is because continuous and seeded batch crystallizers always contain crystals that can participate in secondary nucleation mechanisms and are operated at low supersaturations to obtain regular and pure product. The presence of crystals and low supersaturation can only support secondary nucleation and not primary nucleation. Initial breeding results from adding seed crystals to a supersaturated solution to induce nucleation. The presence of growing crystals can also contribute to secondary nucleation by several different mechanisms. Contact nucleation results from nuclei formed by collisions of crystals against each other, against crystallizer internals or with an impellor, agitator or circulation pump. Shear breeding results when a supersaturated solution flows by a crystal surface and carries along crystal precursors formed in the region of the growing crystal surface.

The ease with which nuclei can be produced by contact nucleation is a clear indication that this mechanism is dominant in many industrial crystallization operations. For design purposes, the metastable limit can provide a useful semi-empirical approach to correlate the effective or apparent heterogeneous and secondary nucleation rate:

$$B^o = k_n \, (c - c_m)^n \quad c^* < c_m \tag{13.6}$$

where c is the solute concentration, c_m is the metastable limit, i.e., the solute concentration at which spontaneous nucleation starts to occur at the given temperature (see Fig. 13.3), and c^* is the equilibrium solute concentration at saturation. The metastable limit must be determined through experimentation. Fortunately c_m is very close to c^* for many inorganic systems, and satisfactory correlation can be obtained from

$$B^o = k_n(c - c^*)^n \tag{13.7}$$

where the parameters k_n and n must be evaluated from experimental data. The order for heterogeneous primary nucleation can range from 2 to 9. For secondary nucleation, n, the order is significantly lower and in the range of 0–3. Many commercial crystallizers operate in the secondary nucleation range.

13.2.4 Crystal growth

Crystal growth is a complex subject, which, like nucleation, is not completely understood and may be expressed in a variety of ways:
- linear advance rate of an individual crystal face
- change in a characteristic dimension of a crystal
- rate of change in mass of a crystal.

Adsorption layer or kinetic theories have proven to be quite fruitful in explaining crystal growth. They postulate that there is an "adsorbed" layer of units, which is loosely held to the crystal face. These adsorbed units are free to move on the two-dimensional surface, but they have essentially lost one degree of freedom. Thus, the unit must crystallize through two-dimensional nucleation. At low supersaturations, the energy required for two-dimensional nucleation is considerably less than for normal nucleation. Thus, existing crystals can grow under conditions where three-dimensional nucleation will not occur. Crystal growth will usually occur at supersaturation levels lower than can be explained by the two-dimensional nucleation theory. This happens because growth is much faster at any imperfection, where the surface is not a perfect plane. Since crystals are usually not perfect, growth usually proceeds by a "filling-in" process. Small crystals are more likely to be perfect than large crystals and, therefore, more likely to require two-dimensional nucleation. Large crystals are less likely to be perfect, and are more likely to be damaged by the impeller or baffles. Thus, the large crystals grow by healing kinks, pits and dislocations. These imperfections are also the reason why even two crystals of the same size may have different growth rates, although all conditions appear to be the same. This is called growth rate dispersion. If one of the crystals happens to be more perfect than the other, it will have a lower growth rate.

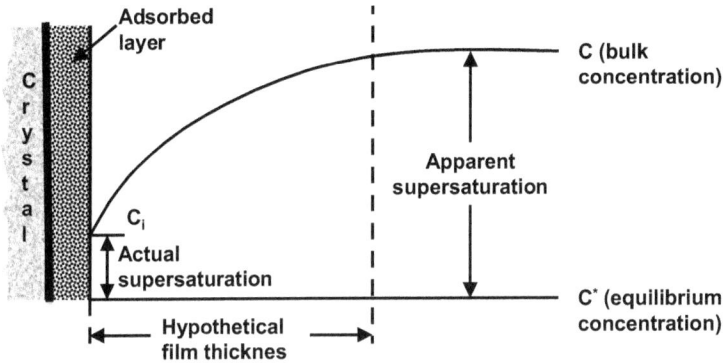

Fig. 13.5: Schematic of crystal growth mass transfer process.

The adsorption theory, which explains the crystallization once a unit is at the surface, is often incorporated into mass transfer theories that describe the movement of the unit to the surface. In crystallization, the picture of the mass transfer process is somewhat different from mass transfer theories for other processes, and different empirical expressions are used. Figure 13.5 shows a schematic of the mass transfer process that can be broken down in the following seven steps:

1. Bulk diffusion of solvated units through a film of reduced concentration
2. Diffusion of solvated units in the adsorbed layer
3. Partial or total desolvation
4. Surface diffusion of units to growth site
5. Incorporation into the crystal lattice
6. Counter diffusion of solvent through the adsorbed layer
7. Counter diffusion of solvent through the film

Since it is difficult to include all these steps in a theory, usually, a simplified model is used, which includes step 1, combines steps 2 to 5, and ignores the last two steps. This way, a linear growth rate G of the crystal is obtained, which is equal in all three dimensions:

$$G = \frac{dL}{dt} = k_G \left(c - c^* \right) \tag{13.8}$$

Due to the assumptions involved, it should not be surprising that many systems do not satisfy this law. In these cases, a common approach is to introduce an empirical "overall growth-rate order," n, yielding for the growth rate a power-law function, in which the constants k_G and n are determined by fitting experimental crystal growth data:

$$G = \frac{dL}{dt} = k_G (c - c^*)^n \tag{13.9}$$

Such an approach will only be valid over small ranges of supersaturation. In addition to supersaturation, temperature will also affect the crystal growth rate through film and surface diffusion and the crystal rearrangement process. If either mass transfer or surface reaction controls the temperature dependence of growth kinetics, the pre-factor, k_G in eq. (13.9) can often be expressed in terms of an Arrhenius expression:

$$k_G = k_G^o \exp\left(-\frac{\Delta E_G}{RT}\right) \tag{13.10}$$

where ΔE_G is an effective activation energy for growth. When both mechanisms are important, this equation is not valid, and Arrhenius plots of $\ln(k_G)$ versus $1/T$ tend to give curved instead of straight lines.

13.2.5 Effects of impurities

The presence of impurities can alter the growth rates of crystalline materials significantly. The most common is a decrease in growth rate, which is considered to be caused by adsorption of the impurity onto the crystal surface. Once located on the surface, the impurity forms a barrier to solute transfer from the solution to the crystal. Another important effect associated with the presence of impurities is that they may change the crystal *shape* (habit). Habit alteration is considered to result form unequal changes in the growth rates of different crystal faces.

13.3 Crystal characteristics

In crystallization processes, product requirements are a key issue in determining the ultimate success in fulfilling the function of the operation. Key parameters of product quality are the size distribution (including mean and spread), the morphology (including habit or shape and form) and purity. Crystal size distribution (CSD) determines several important processing and product properties. It is often important to control the CSD. The most favored is a monodisperse CSD, where all crystals are of the same size and dissolve at a known and reproducible rate. Critical phenomena that influence the CSD, besides nucleation and growth rates, are breakage and agglomeration. Breakage of crystals is almost always undesirable because it is detrimental to crystal appearance, and it can lead to excessive fines and have a deleterious effect on crystal purity. Agglomeration is the formation of a larger solid particle through sticking together of two or more smaller particles.

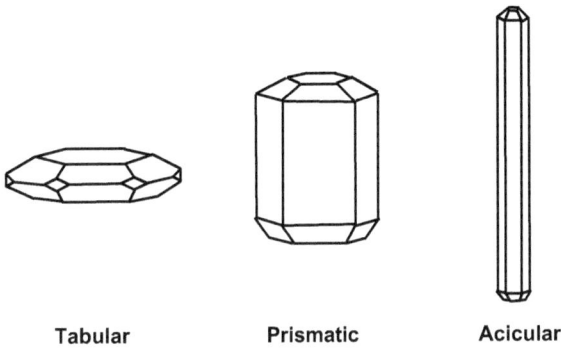

Tabular   Prismatic   Acicular

Fig. 13.6: Various shapes of a hexagonal crystal.

13.3.1 Morphology

Every chemical compound has a unique crystal shape that depends enormously on the conditions in the crystallizer. Some substances illustrate polymorphism, meaning that the substance can crystallize into two or more unique forms. A good example is carbon, which can crystallize into graphite and diamond. The general shape of a crystal is referred to as habit. A characteristic crystal shape results from the ordered internal structure of the solid with crystal surfaces forming parallel to planes formed by the constituent units. The unique aspect of a crystal is that the angles between adjacent faces are constant. The surfaces (faces) of a crystal may exhibit varying degrees of development, but not the angles. With constant angles but different sizes of the faces, the shape of a crystal can vary enormously. This is illustrated in Fig. 13.6 for three hexagonal crystals. The final crystal shape is determined by the relative growth rates of the different crystal faces. Faster growing faces become smaller than slower growing faces and may disappear from the crystal altogether, in the extreme. The relative growth process of faces and, thereby, the final crystal shape, are affected by many variables such as rate of crystallization, impurities, agitation, solvent used, degree of supersaturation, and so on. The appearance of the crystalline product, purity and its processing characteristics (such as washing and filtration) are affected by crystal habit. The effect on purity becomes even more important when crystallization is employed as a purification technique. Mechanisms by which impurities can be incorporated into crystalline products include adsorption on the crystal surface, solvent entrapment in cracks, crevices and agglomerates, and inclusion of pockets of liquid. An impurity having a structure sufficiently similar to the material being crystallized can also be incorporated into the crystal lattice by substitution or entrapment.

13.3.2 Crystal size distribution

Particles produced by crystallization have a distribution of sizes that varies in a definite way over a specific size range (Fig. 13.7). A crystal size distribution (CSD) is most commonly expressed as a population (number) distribution relating the number of crystals at each size to the size, or as a mass (weight) distribution expressing how mass is distributed over the size range. The two distributions are related and affect many aspects of crystal processing and properties. An average crystal size can be used to characterize a CSD. However, the average can be determined on any of several bases such as number, volume, weight or length. The dominant crystal size, L_D, is most often used as a representation of the product size. The coefficient of variation (cv) of a distribution is a measure of the spread of the distribution around the characteristic size. It is often used in conjunction with dominant size to characterize crystal populations through the equation:

$$cv = \frac{\sigma}{L_D} \tag{13.11}$$

where σ is the standard deviation of the distribution.

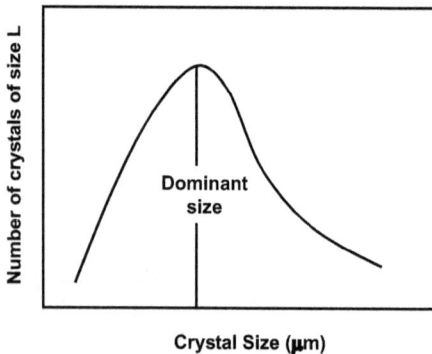

Fig. 13.7: Example of a crystal size distribution; in this case, number distribution as a function of crystal length.

13.3.3 Size control

CSDs produced in a perfectly mixed continuous crystallizer are highly constrained. The form of the CSD in such systems is entirely determined by the residence time distribution of a perfectly mixed crystallizer. Greater flexibility can be obtained through introduction of selective removal devices that alter the residence time distribution of materials flowing through the crystallizer. Clear-liquor advance is the removal of mother liquor from the crystallizer, without simultaneous removal of crystals. The primary objective of classified fines removal is preferential withdrawal of crystals whose size is below some specified value. A simple method for implementation of classified fines removal is to remove slurry from a settling zone in the crystallizer. Constructing a baffle that separates the zone from the well-mixed region of the vessel can create the settling zone. The separation of crystals in the settling zone is based on the dependence of settling velocity on crystal size. Such crystals may be redissolved and the resulting solution returned to the crystallizer. Classified product removal is carried out to remove, preferentially, those crystals whose sizes are larger than some specified value.

13.4 Types of solution crystallizers

13.4.1 Basic operations

The basic functions that a system for the crystallization from a solution should provide are:

1. A means of generating supersaturation, which may include provision for addition or removal of heat
2. A vessel to provide sufficient residence time for the process streams to approach equilibrium and the crystals to grow to a desired size
3. Mixing regime to provide an environment for uniform crystal growth
4. Possibly, the capability of selectively removing fines or coarse product to control the CSD

Solution crystallizers are generally classified into one of five categories, according to the method by which supersaturation is achieved. The technique employed to generate supersaturation in a solution is referred to as the mode of operation. The mode chosen by the designer is strongly influenced by the phase-equilibrium characteristics of the system, and it dictates the material and energy balance requirements of the system. The common techniques for producing solids from a solution include:

- Lowering the temperature of the feed solution by direct or indirect cooling. If solute solubility is strongly temperature dependent, this is the preferred approach.
- Adding heat to the system to remove (evaporate) solvent and. Thus. "salt out" the solute. This technique is effective if solubility is insensitive to temperature.
- Vacuum cooling the feed solution. without external heating. If solubility is strongly dependent on temperature. this method is attractive.
- Combining techniques: Vacuum cooling supplemented by external heating for systems whose solubility has an intermediate dependence on temperature is especially common.
- Adding a nonsolvent: This is a common technique for precipitating solute from solution and is useful as both a laboratory technique and as an industrial process for product recovery.

These methods can be employed in single or multistage crystallization or in batch operations. Generally, production rates over 50 ton/day justify continuous operation. The continuous operations tend to have higher yield and require less energy than batch, but batch is more versatile. Multistage operation is employed, where evaporative requirements exceed the capabilities of a single vessel and/or energy costs dictate staging of the operation. Another reason for staging may sometimes be the production of more uniform and/or larger crystals. Operation of crystallizers in series generates CSDs having narrower size spread than the same volume of crystallizers in parallel. Batch crystallizers produce a narrower CSD than continuous well-mixed units. Although numerous design methods and kinetic theories exist to analyze specific crystallizer configurations, no clear-cut guidance is available for the choice between crystallizer types and/or modes of operation.

13.4.2 Cooling crystallizers

Cooling crystallizers obtain supersaturation by cooling the saturated solution. They are very desirable when they work because of their high yield and low energy consumption. Cooling crystallization is applicable to systems where temperature has a large influence on solubility (e.g., ammonium aluminum sulfate and soda). The method is not useful for systems like NaCl, where temperature has little effect on solubility. When cooling is the selected mode by which supersaturation is generated, heat can be transferred through an external cooling surface or through coils or a jacket internal to the crystallizer body. The forced circulation crystallizer is a simple, highly effective unit designed to provide high heat-transfer coefficients in either an evaporative or cooling mode. A schematic diagram in which slurry is withdrawn from the crystallizer body and pumped through an external heat exchanger is shown in Fig. 13.8. Most systems are operated with a high recirculation rate to provide good mixing inside the crystallizer and high rates of heat transfer to minimize encrustation. The tendency of the solute to form encrustations on the cooling surface often limits their operation, by restricting the temperature of the cooling liquid and the temperature decrease of the slurry flowing through the heat exchanger. High heat transfer rates reduce the formation of encrustations considerably, by minimizing the temperature difference over the heat transfer surface. It is not uncommon to limit the decrease in magma (mixture of crystals and solution) temperature to about 3–5 °C. The feed is commonly introduced into the circulation loop to provide rapid mixing with the magma and minimize the occurrence of regions of high supersaturation, which can lead to excessive nucleation.

The use of a conventional heat exchanger can be avoided by employing direct-contact cooling (DCC) in which the product liquor is allowed to come into direct contact with a cold heat-transfer medium (coolant). Other potential advantages include better heat transfer and smaller cooling load. However, problems include product contamination from the coolant and the cost of extra processing required

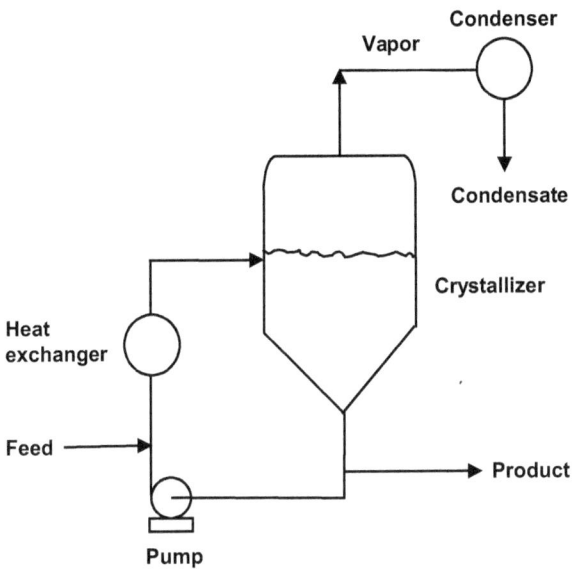

Fig. 13.8: Schematic of forced circulation crystallizer with external cooling.

for recovering the coolant for further use. Crystallization processes employing DCC have been used successfully for dewaxing of lubricating oils, desalination of water and production of inorganic salts from aqueous solutions.

13.4.3 Evaporating and vacuum crystallizers

Evaporative crystallizers supersaturate the solution by removing solvent through evaporation. These crystallizers are used when temperature has little effect on solubility (e.g., NaCl) or with inverted solubilities (e.g., calcium acetate). The crystallizer may be as simple as a shallow open pan heated by an open fire. Steam-heated evaporators are widely used to produce common salt from brine and in sugar refining. These applications often use an evaporative crystallizer containing calandria (steam chest). The magma circulates by dropping through the central downcomer and then rises, as it is heated in the calandria. At the top, some of the solution evaporates, increasing the supersaturation causing crystal growth. On a large scale too, many types of forced circulation evaporating crystallizers are used. These systems are similar to the one shown in Fig. 13.8 and, obviously, very similar to evaporators without crystallization. Operational problems with evaporative crystallizers can be caused by scale formation on the heat exchanger surfaces or at the vapor–liquid interface in the crystallizer. Such problems can be overcome by not allowing vaporization or excessive temperatures within the exchanger and by proper introduction of the circulating magma into the crystallizer. The latter may be accomplished by introducing the magma below the surface of the magma in the crystallizer.

The evaporators are often connected together to make a multiple-effect cascade, as shown in Fig. 13.9. Here, the vapor from the first unit is used as the steam to heat the second unit. This reduces the steam requirement per kg of product. The pressure must be varied as shown in the figure, so that the condensing vapor will be at a temperature greater than the boiling temperature in the second stage. A variety of ways that connect the different stages have been developed.

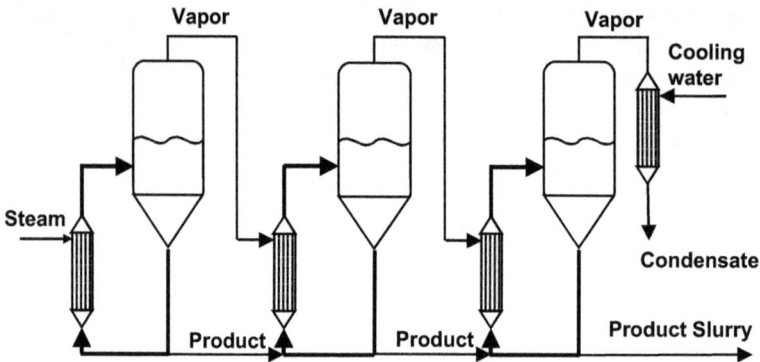

Fig. 13.9: Multiple-effect evaporating crystallizer cascade for ammonium sulfate.

Vacuum crystallizers utilize evaporation to both concentrate the solution and cool the mixture. They combine the operating principles of evaporative and cooling crystallizers. The liquid feed enters the crystallizer at a higher pressure and temperature, where it flashes, and part of the feed evaporates. This flashing causes adiabatic cooling of the liquid. Crystals and mother liquor exit the bottom of the crystallizer. Due to the vacuum equipment required to operate at pressures of 5 to 20 mbar, vacuum crystallizers are considerably more complex than other types of crystallizers and most common in large systems.

13.4.4 Continuous crystallizers

Many different continuously operated crystallizers are available. The majority can be divided into three basic types: forced circulation, draft-tube agitated and fluidized bed units. A forced circulation crystallizer has already been discussed and shown in Fig. 13.8. Figure 13.10 shows a *draft-tube-baffle* (DTB) crystallizer designed to provide preferential removal of both fines and classified product. A relatively low speed propeller agitator is located in a draft tube, which extends to a few inches below the liquor level in the crystallizer. The steady movement of magma up to the surface produces a gentle, uniform boiling action over the whole cross-sectional area of the crystallizer. Between the baffle and the outside wall of the crystallizer, agitation effects are absent. This provides a settling zone that permits regulation of the magma density and control of the removal of excess nuclei. Flow through the annular zone can be adjusted to only remove crystals below a certain size and dissolve them in the fines dissolution exchanger. Feed is introduced to the fines circulation line to dissolve nuclei resulting from feed introduction.

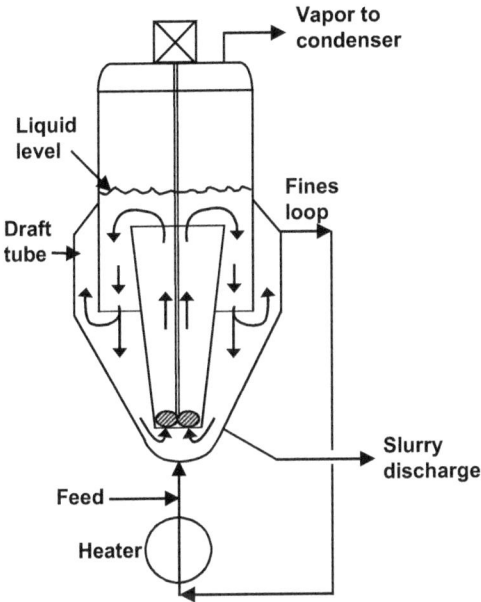

Fig. 13.10: Schematic of a draft-tube-baffle (DTB) crystallizer.

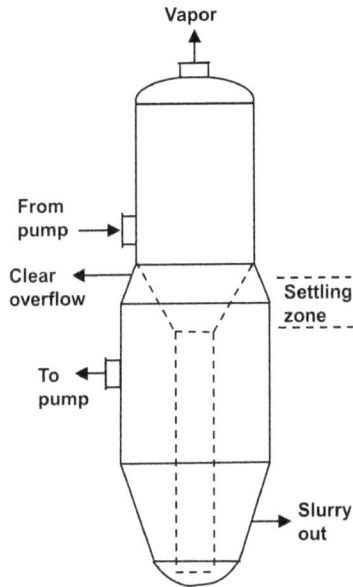

Fig. 13.11: Oslo fluidized bed crystallizer.

Another type of continuous crystallizer is the Oslo fluidized bed crystallizer shown in Fig. 13.11. In units of this type, a bed of crystals is suspended in the vessel by the upward flow of supersaturated liquor in the annular region surrounding a central downcomer. The objective is to form a supersaturated solution in the upper chamber and then relieve the supersaturation through growth in the lower chamber. The use of a downflow pipe in the crystallizer provides good mixing in the growth chamber.

13.5 Crystallizer modeling and design

The final design of a crystallizer is the culmination of the design strategy depicted in Fig. 13.12. In the conceptual design stage, equilibrium data and the operating mode (method of supersaturation generation) are surveyed. Solvent choice and processing conditions are determined in this step. After establishing the type of crystallizer, finally, the crystallizer functional design is made. The use of numerical models for the design of industrial crystallizers has lagged behind, compared to other unit operations, because of the complexities associated with rationally describing and interrelating growth and nucleation kinetics with the process configuration and mechanical features of the crystallizer. The design and analysis of crystallization processes for the continuous well-mixed suspension type have developed into formal design algorithms, which can now be applied in situations of industrial importance. Examples of specific process configurations that can be modeled rigorously include fines destruction, clear-liquor advance, classified product removal, vessel staging and seeding. The basic requirement for these rigorous process configuration models is a validated population balance CSD algorithm.

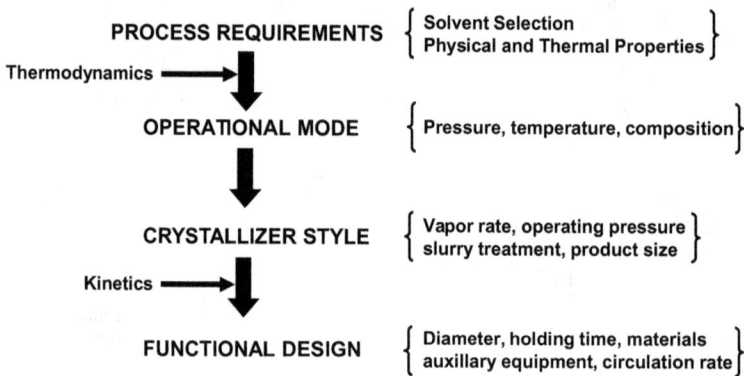

PROCESS REQUIREMENTS $\left\{ \begin{array}{l} \text{Solvent Selection} \\ \text{Physical and Thermal Properties} \end{array} \right\}$

Thermodynamics ➡

OPERATIONAL MODE $\left\{ \text{Pressure, temperature, composition} \right\}$

⬇

CRYSTALLIZER STYLE $\left\{ \begin{array}{l} \text{Vapor rate, operating pressure} \\ \text{slurry treatment, product size} \end{array} \right\}$

Kinetics ➡

FUNCTIONAL DESIGN $\left\{ \begin{array}{l} \text{Diameter, holding time, materials} \\ \text{auxillary equipment, circulation rate} \end{array} \right\}$

Fig. 13.12: Crystallization system design strategy.

13.5.1 Basic yield calculations

For crystallization, a high recovery of crystallized solute is generally the desired design objective. The theoretical maximum product recovery or yield from the crystallizer is defined by

$$\text{Yield (\%)} = 100 \times \frac{F_S X_F - S_S X_C}{F_S X_F} \tag{13.12}$$

where the meaning of the quantities F_S, X_F, S_S and X_C is illustrated in Fig. 13.13 and defined as

X_F is the mass of dissolved solute in the feed per unit mass of solvent
X_C is the mass of dissolved solute in liquid discharged from the crystallizer per unit mass of discharged solvent
F_S is the mass feed rate of solvent into crystallizer
S_S is the mass discharge rate of solvent leaving crystallizer

The solvent feed rate F_S and the ratio of dissolved solute to solvent in the feed X_F are usually fixed by upstream requirements. The composition of dissolved solute in the liquid leaving the crystallizer X_C is determined by solubility, which is set by the temperature of the operation. The other adjustable variable is the solvent discharge rate S_S. In evaporative systems, solvent is evaporated by adding heat to the system. Thus, S_S decreases and recovery increases as more solvent is evaporated and removed as a separate product stream. The conclusion from the preceding discussion is that, for a given feed concentration and solubility relationship, the mode of crystallizer operation governs the maximum recovery of solute. For a cooling crystallization system, where no solvent is removed, the only control that affects maximum solute recovery is crystallizer temperature. For evaporative systems, recovery is influenced by both the system temperature and by the quantity of solvent vaporized and removed from the system.

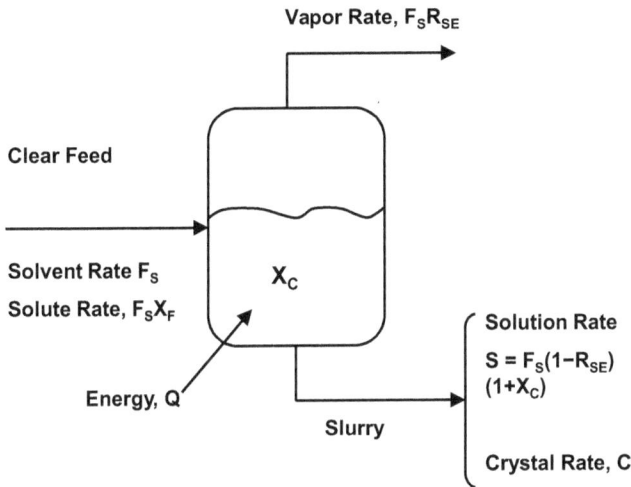

Fig. 13.13: Mass balance for a binary crystallization system.

For a system wherein a pure solid component is crystallized by cooling or evaporation, the overall mass balance provides the maximum solute yield and total suspended solids concentration. R_{SE} is the ratio of the mass of solvent evaporated per unit mass of solvent fed, whereas S and C are the solution and crystal removal rate. From the overall material balance, it follows that:

$$F_S + F_S X_F = F_S R_{SE} + S + C \tag{13.13}$$

The solution (solvent plus still dissolved solute) removal rate becomes

$$S = F_S (1 - R_{SE}) (1 + X_C) \tag{13.14}$$

Substitution of (13.14) into (13.13) gives the crystal production rate C

$$C = F_S[X_F - X_C(1 - R_{SE})] \tag{13.15}$$

The maximum yield is now defined by

$$Y_{MAX} = \frac{C}{F_S X_F} = 1 - \frac{X_C}{X_F}(1 - R_{SE}) \tag{13.16}$$

which can be simplified for specific operating modes as follows:

Cooling crystallization only: $\qquad R_{SE} = 0 \rightarrow Y_{MAX} = \dfrac{X_F - X_C}{X_F}$ (13.17)

Evaporation only (at constant T): $\quad X_C = X_F \rightarrow Y_{MAX} = R_{SE}$ (13.18)

These expressions can be solved, provided the composition X_C of the exit stream is known. In many instances, it is acceptable to assume that the exit stream composition corresponds to saturation conditions. Systems in which this occurs are said to exhibit fast growth or Class II behavior. Should growth kinetics be too slow to use all the supersaturation, the system is said to exhibit slow-growth behavior and is classified as a Class I system.

13.5.2 Population balances

A more precise analysis of a crystallization system requires the quantification of nucleation and growth kinetics, mass and energy balances and the crystal size distribution. Population balances are a major theoretical tool for predicting and analyzing crystal size distributions. A balance on the number of crystals in any size range, say L_1 to L_2, accounts for crystals that enter and leave that size range by convective flow into and out of the control volume, and those that enter and leave the size range by growth. Crystal breakage and agglomeration are often ignored, and it is assumed that crystals that are formed by nucleation have a size close to zero. The number of crystals in the size range L_1 to L_2 is given by

$$\Delta N = \int_{L_1}^{L_2} n \, dL \tag{13.19}$$

The population balance differs from a mass balance in that only the crystals within a given size range are included. It is assumed that the population distribution is a continuous function and that crystal size, surface area and volume can all be described by a characteristic dimension L. Population balances for crystallizers are usually described by a population density, n that gives the number of crystals, dN in the size range L to $L + dL$:

$$n = \frac{dN}{dL} = \lim_{\Delta L \to 0} \frac{\Delta N}{\Delta L} \qquad (13.20)$$

where ΔN is the number of crystals in size range ΔL per unit volume; see Fig. 13.14. It is important to note that either solvent or slurry volume can be taken as a basis for the definition of n.

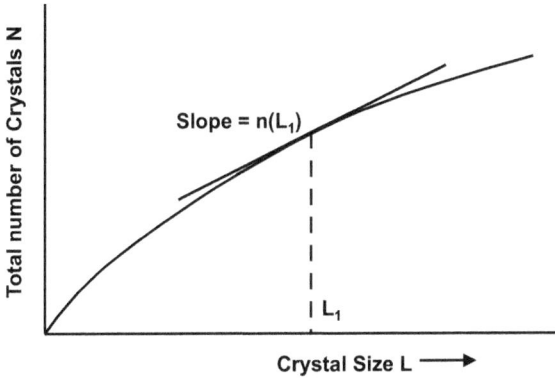

Fig. 13.14: The population density n is given by the derivative of N, which is the number of crystals (per unit volume) with a size equal to or smaller than L.

The population balance equation for a control volume over the slurry in a crystallizer is as follows:

$$\frac{dn(L,t)}{dt} = B(L,t) - D(L,t) + \sum_k \frac{Q_k n_k(L,t)}{V_T} \qquad (13.21)$$

The left-hand side indicates the change in population density with time, while the right-hand side indicates different causes for this change. B is the crystal birth rate, which is caused by nucleation (only for the smallest, critical, size L_{cr}) and by growth *into* the current size range, L to $L + dL$, and possibly also by agglomeration of two smaller crystals, or breakage of a larger crystal, into the current size range. Oppositely, D is the crystal death rate, caused by growth out of the size range L to $L + dL$, and possibly also by breakage or agglomeration of a crystal from the current size range. Usually, birth and death rates are expressed as integrals over the crystal population, n using so-called kernels. A lot of research is going into improving and tuning the dependence of these kernels on process conditions such as temperature, supersaturation and shear rate. The last term is a sum over different inlets and outlets of the crystallizer, each with a volumetric flow rate Q_k (negative for an outlet) and inlet/outlet crystal size distribution, n_k, where V_T is the total volume of the crystallizer. This gives the possibility of modeling the effect of feeding the crystallizer with a pre-existing crystal distribution (initial breeding through seeding). For more details, we refer to the literature; here, we limit ourselves to the simplest example of an MSMPR crystallizer.

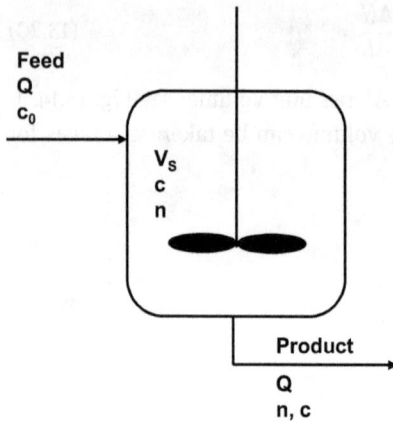

Fig. 13.15: Schematic diagram of a simple, perfectly mixed crystallizer.

13.5.3 The well-mixed MSMPR crystallizer

Application of the population balance is most easily described for the idealized case of a continuous, steady-state MSMPR (mixed-suspension mixed-product removal) crystallizer shown in Fig. 13.15. The assumptions made are that the system is in steady state ($dn/dt = 0$), that the crystal growth rate is independent of crystal size, that no crystals are present in the feed stream, and that the crystals leaving the crystallizer have the same size distribution as in the volume of the crystallizer. For a well-mixed and constant volume V_T in a crystallizer, the population balance (13.21) then becomes

$$0 = -G\frac{dn}{dL} - \frac{Q_0}{V_T}n \tag{13.22}$$

where Q_0 is the volumetric feed and discharge rate, G is the crystal growth rate (dL/dt) and n is the population density for this size range. Replacing Q_0/V_T by $1/\tau$, with τ the mean residence time, and supplying the appropriate boundary condition for crystal nucleation, this can be solved to predict an exponentially decaying size distribution:

$$n(L) = n_0 \exp\left(-\frac{L}{G\tau}\right) \tag{13.23}$$

where n_0 is determined by the ratio of nucleation and growth rate, B^0/G. Although this is an idealized model, the distribution of the larger crystals produced by some industrial crystallizers is in qualitative agreement with eq. (13.23).

13.6 Precipitation

Precipitation is an important industrial process, which is closely related to crystallization from solution. In precipitation, a solute is made supersaturated so that dissolved solutes precipitate out. Precipitation differs from crystallization in that the precipitate is usually amorphous, has a poorly defined size and shape, is often an aggregate and is usually not pure. Precipitation can be carried out in one step to remove a large number of compounds in a single precipitate. This is the easiest application and is most common. Fractional precipitation requires a series of steps, with each step optimized to precipitate the desired component. It is not possible to do extremely sharp purifications by fractional precipitation. Precipitation and crystallization are often complementary processes, since precipitation is used for a "rough cut," while crystallization is used for final purification. Precipitation is commonly used to manufacture pigments, pharmaceuticals and photographic chemicals. In the production of ultra-fine crystalline powders, precipitation is often considered an attractive alternative to milling, particularly for heat-sensitive substances.

Similar to crystallization processes, precipitation consists of three basic steps — the creation of supersaturation, followed by the generation of nuclei and the subsequent growth of these nuclei to visible size. The equipment used is often very similar to that used for crystallization. Precipitation operations usually have a mixing operation where various reagents are added to make the solution supersaturated. This is followed by a holding period, which allows for an induction period before nuclei form and a latent period before the supersaturation starts to decrease. Once the precipitate starts to grow, desupersaturation often occurs at a constant rate. Towards the end of the desupersaturation period, the rate decreases, as the concentration approaches equilibrium. During this period, aging often occurs. During aging, small crystals redissolve and the solute is redeposited onto the larger precipitates. The result is a uniform crop of fairly large precipitates, which is relatively easy to separate from the solution by centrifugation.

Precipitation processes are conveniently classified by the method used to produce the supersaturation: cooling, solvent evaporation, reaction, salting out, antisolvents and pH. Supersaturation caused by cooling and or evaporation of solvent is very similar to solution crystallization. Many products are made by mixing reagents that react to form an insoluble precipitate. A representative example is the formation of gypsum by addition of sulfuric acid to a calcium chloride solution:

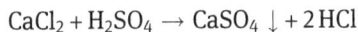

$$CaCl_2 + H_2SO_4 \rightarrow CaSO_4 \downarrow + 2\,HCl$$

In aqueous solution, high inorganic salt concentrations also often cause the precipitation of solute. Antisolvents or nonsolvents can be added to achieve the same effect. The solubility of many compounds is affected by pH.

13.7 Melt crystallization

In some crystallization systems, the use of a solvent can be avoided. Melt crystallization is the process where crystalline material is separated from a melt of the same (crystallizing) species by direct or indirect cooling, until crystals are formed in the liquid phase. Two basic techniques of melt crystallization are gradual deposition of a crystalline layer on a chilled surface and fast crystallization of a discrete crystal suspension in an agitated vessel. Zone melting relies on the distribution of solute between the liquid and solid phases to effect a separation. Normal freezing is the slow solidification of a melt. The impurity is rejected into the liquid phase by the advancing solid interface. The basic requirement of melt crystallization is that the composition of the crystallized solid differs from that of the liquid mixture from which it is deposited. The process is usually operated near the pure-component freezing temperature. High or ultrahigh product purity is desired in many of the melt purification processes. Single-stage crystallization is often not sufficient to achieve the required purity of the final product. Further separation can be achieved by repeating the crystallization step or by countercurrent contacting of the crystals with a relatively pure liquid stream.

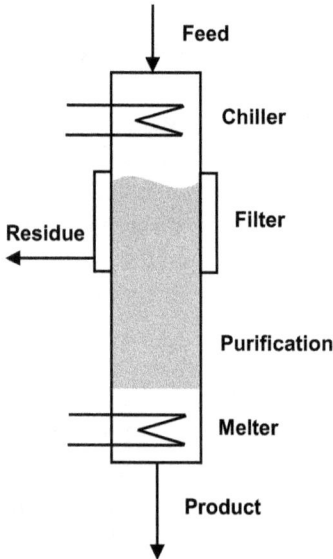

Fig. 13.16: Vertical column melt crystallizer.

Since there is no solvent, melt crystallization has the advantage that no solvent removal and recovery is required, and contamination by the solvent is impossible. However, there is also no way to influence the melt properties (viscosity, diffusivity), and the chemicals being purified must be stable at the melting point. For salts with very high melting points, solution crystallization is less expensive. At the lower temperatures possible in solution crystallization, distribution coefficients can be more favorable by orders of magnitude. These comparisons show why crystallization from solution is much more common for bulk separations. Melt crystallization becomes advantageous when the presence of a solvent would be detrimental or when highly pure products are desired. Currently, the interest in wider application of melt crystallization is stimulated by the energy-saving potential in large scale processing, because solvent evaporation is absent and the heat of fusion is several times less than the heat of vaporization.

Multiple countercurrent contacting stages are possible by conducting melt crystallization inside a column, in a manner somewhat analogous to distillation. The main incentive is to attain higher purity product than can be achieved in a single stage of conventional crystallization. The concept of a column crystallizer is to form a crystal suspension and to force the solids to flow countercurrently against a stream of enriched reflux liquid by gravity or rotating blades. At the end of the crystallizer, the crystals are melted. A portion of the melt is removed as product, and the remainder is returned to the system as enriched reflux to wash the product crystals. One of the early column crystallizers, shown schematically in Fig. 13.16, was developed for the separation of xylene isomers. In this unit, p-xylene crystals are formed in a scraped surface chiller above the column and fed to the column. The crystals move countercurrently downward to impure liquid in the upper portion of the column and melted p-xylene in the lower part of the column. Impure liquor is withdrawn from an appropriate point near the top of the column of crystals, while pure product, p-xylene, is removed from the bottom of the column. An inherent limitation of column crystallization is the difficulty of controlling the solid-phase movement, because the similar solid- and liquid-phase densities make gravitational separation difficult.

The Sulzer MWB system, schematically depicted in Fig. 13.17, is an example of a commercial melt crystallization process that uses the gradual deposition of solids on a chilled surface. Crystal growth is on the inside of a battery of tubes through which the melt is flowing. During crystallization, the front of the crystals advances into the direction of the mother liquor. This buildup of a solids layer requires sequential operation. Steps include partial freezing of a falling film of melt inside vertical tubes, followed by slight heating, a "sweating" operation, and complete melting and recovery of the refined product. The recovered product melt can be put through the cycle again to increase purity, or fresh feed can be introduced to the cycle. The process has been used on a large scale in the purification of a wide range of organic substances.

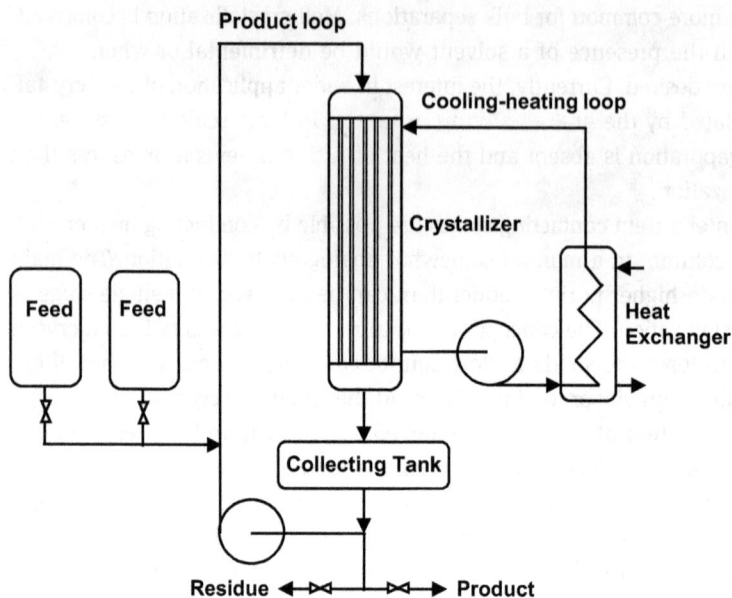

Fig. 13.17: Sulzer MWB melt crystallizer system.

Nomenclature

B^0	Homogeneous primary nucleation rate	[nuclei/unit time · unit volume]
c	Actual solution concentration	[mol/m^3]
c^*	Equilibrium saturation concentration	[mol/m^3]
C	Crystal production rate	[kg/unit time]
cv	Coefficient of variation	[–]
ΔE_G	Activation energy	[J/mol K]
F_S	Mass feed rate of solvent, solute free	[kg/unit time]
G	Crystal growth rate	[m/unit time]
ΔH_M	Enthalpy of melting	[J/mol]
k	Rate constant	
L	Crystal size	[m]
L_D	(Dominant) crystal size	[m]
n	Order of nucleation or growth rate	[–]
n	Population density	[number of crystals/m m^3]
N	Number of crystals	[number of crystals/unit volume]
Q	Volumetric discharge rate	[m^3/unit time]
R	Gas constant	[8.314 J/mol K]
R_{SE}	Unit solvent evaporated per unit solvent fed	[–]
s	Relative super saturation	[–]
S	Solution removal rate	[kg/unit time]
S_S	Mass discharge rate of solvent, solute free	[kg/unit time]
t	Time	[s]
$T\ (T_M)$	Temperature (at melting point)	[K]

V	Molar volume of solute	$[m^3/mol]$
V_T	Volume of magma (suspension or clear liquor)	$[m^3]$
x_S	Solubility of solids, mole fraction	$[-]$
X	Mass of dissolved solute/mass solvent	$[kg/kg_{Solvent}]$
Y	Yield	$[-]$
γ_S	Activity coefficient of solvent	$[-]$
γ	Surface energy (for critical nucleus)	$[J/m^2]$
σ	Standard deviation of distribution	$[m]$

14 Solids finishing technologies

14.1 Overview

Solids technology is concerned with the production of disperse solid products. Characteristically, the properties of solid products depend not only on chemical composition but also on the state of dispersion. The production processes are composed of individual unit operations, which include all necessary activities from preparation of the solid to packaging the product for sale.

The production of disperse solid products is characterized by the generation and repeated modification of dispersed states. A typical example of a solids processing chain, illustrated in Fig. 14.1, includes all required unit operations to convert a homogeneous solution into a packaged product with the desired particle morphology. Every unit operation in the processing chain modifies the physical and dispersed state of the product. The state changes from solution to a suspension that contains individual particles or loose flakes in crystallization. These particles are considerably compacted to form wet, loose agglomerates during solid/liquid separation. Drying results in particles that can be formed into larger secondary agglomerates or pulverized at the size adjustment stage.

Users of solid chemicals demand that the end products have clearly defined properties, which depend on the state of dispersion of the material. These application properties, on which the performance of a dye, medicine, sweetener or fertilizer may depend include dispersibility, dissolution rate, floating capacity, abrasion resistance and absence of dust. Modifying the state of dispersion throughout the solids processing chain leads to a product with the desired application properties. However,

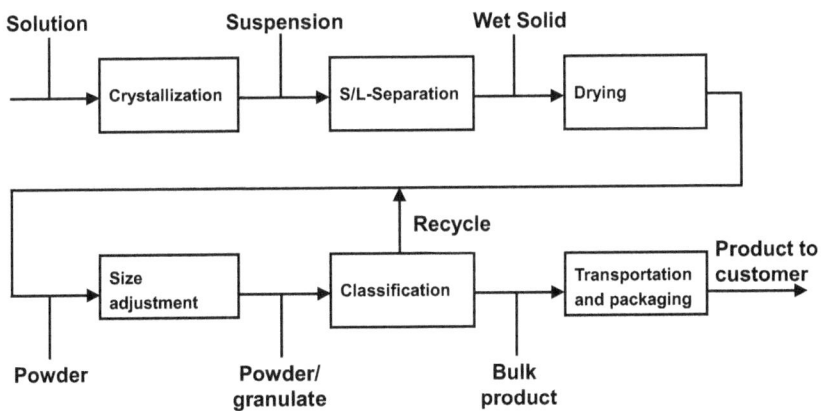

Fig. 14.1: Typical solids technology processing chain.

https://doi.org/10.1515/9783110712445-014

the processing chain affects the application properties not only in predetermined ways, but, often, also accidentally.

14.2 Drying

Drying is an operation in which volatile compounds (usually liquids) are removed from solids, slurries and solutions to yield solid products. The separation may be carried out in a mechanical manner, without phase change or by evaporation through the supply of heat. Thermal drying consists of two steps. First, heat is supplied to the material to evaporate the moisture out of the product. Secondly, the vapor is separated from the product phase and, if necessary, condensed outside of the dryer. This thermal drying is discussed in the following sections. If possible, mechanical predrying is installed upstream of thermal drying, because solids handling is made easier, and liquid separation without evaporation requires less energy and is therefore less costly (Fig. 14.2).

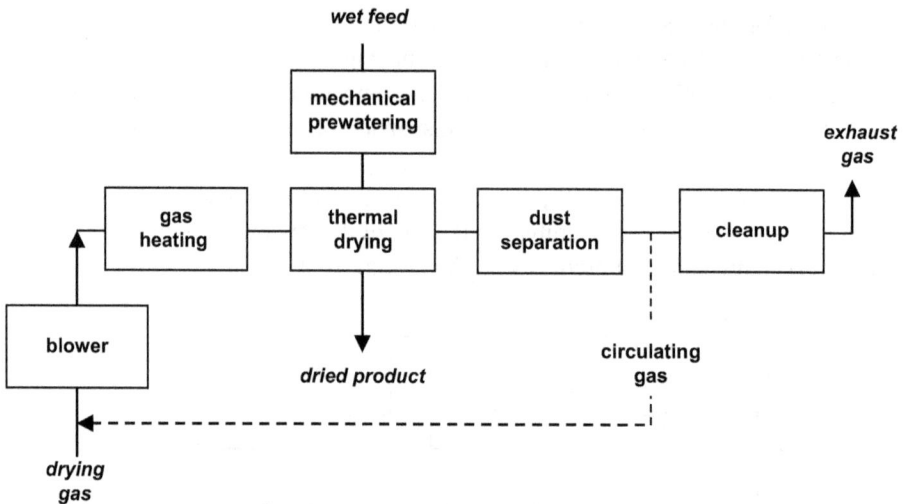

Fig. 14.2: Open and partial air recirculation (- - -) convective drying system.

The goal of most drying operations is not only to separate a volatile liquid, but also to produce a dry solid of a desirable size, shape, porosity, density, texture, color or flavor. Drying is usually the final processing step before packaging and makes many materials such as soap powders and dyestuffs, more suitable for handling. Drying or dehydration of biological materials, especially foods and biopharmaceuticals, is used as a preservation technique. The product that is to be dried is denoted as the moist solid. In most cases, water is removed as the liquid, but it could also be a solvent such as alcohol or hexane. The substance that carries the necessary heat is called the drying agent. This substance can be air, an inert gas such as nitrogen, or superheated steam. Radiation, hot surfaces or microwaves can also be used to supply the required heat. Since all drying operations involve processing of solids, equipment material handling capability is of primary importance. In fact, most industrial dryers are derived from material handling equipment designed to accommodate specific forms of solids. Environmental factors, such as emission control and energy efficiency, increasingly influence equipment choices.

14.2.1 Classification of drying operations

Drying methods and processes can be classified in several ways. One classification is as batch, where the material is inserted into the drying equipment and drying proceeds for a given period of time; another is continuous drying, where the material is continuously added to the dryer and dried material continuously removed. A batch dryer is best suited for small quantities and for use in multiple-product plants. This dryer is one into which a charge is placed, the dryer runs through its cycle, and the charge is removed. In contrast, continuous dryers operate best under steady-state conditions, drying continuous feed and product streams. Optimum operation of most continuous dryers is at design rate and steady state. Continuous dryers are unsuitable for short operation runs in multiproduct plants.

Drying processes can also be classified according to the physical conditions used to add heat and remove water vapor. The most frequently applied heat transfer mechanisms are convection drying and contact drying. In convection drying, the sensitive heat of a hot gas is supplied to the material surface by convection. The drying agent flowing past or through the body also removes the evaporated water and transports it from the dryer (Fig. 14.2). To save energy, partial recirculation of the drying medium is also used. Drying operations involving toxic, noxious or flammable vapors employ gas-tight equipment, combined with recirculating inert gas systems having integral dust collectors, vapor condensers and gas reheaters (Fig. 14.3). In contact drying, the heat is supplied to the wet material by conduction from heated surfaces such as bands, plates, cylinders or the dryer wall. The amount of heat transferred depends not only on the thermal conductivity of the heating surface but also on the heat transfer coefficient from the heating medium to the surface. Common heating mediums include steam, organic liquids and molten metals. Since all the heat for moisture evaporation passes through the material layer, the thermal efficiency of contact drying is higher than convective drying, where most of the heat is flowing over the material and wasted into the outlet air.

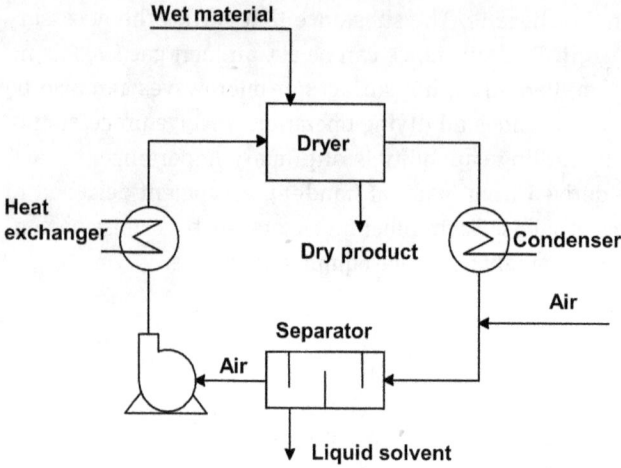

Fig. 14.3: Convective drying in a closed system.

14.2.2 Drying mechanisms

In the design of drying operations, an understanding of liquid and vapor mass transfer mechanisms is essential for quality control. Since no two materials behave alike, this understanding is usually obtained by measuring drying behavior under controlled conditions in a prototypic, pilot plant dryer. From these experiments, the drying rate is usually determined as the change of moisture content, with time. For most moist solids, especially those having capillary porosity, the drying rate depends upon the moisture content in a manner similar to that shown in Figs. 14.4 and 14.5. In the initial

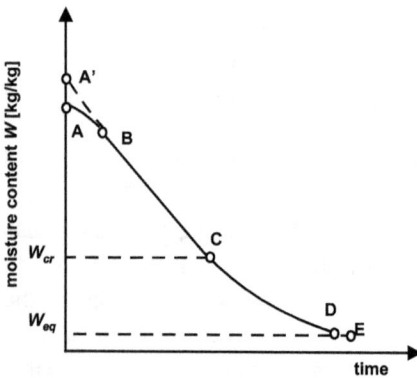

Fig. 14.4: Drying curve for a typical porous material.

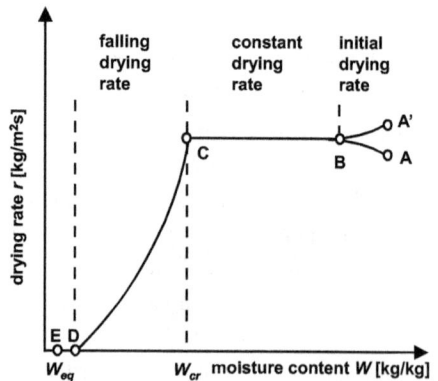

Fig. 14.5: Drying rate curve for the same material.

drying period, the solid and the liquid have a lower temperature and need to be brought to their ultimate drying temperature by the hot gas stream. This initial heating period is usually quite short and can often be ignored.

After reaching the equilibrium temperature, in most applications, the drying rate remains practically constant for a period of time. During this period, the surface of the solid is very wet and a continuous film of liquid exists on the drying surface. This liquid is entirely unbound and acts as if the solid underneath the liquid film were not present. Under these circumstances, the rate of evaporation is independent of the solid and is essentially the same as the rate of evaporation from a free liquid surface. More precisely, the evaporation rate is controlled by the heat transfer rate from the hot gas to the wet surface. The mass-transfer rate adjusts to the heat transfer rate, and the wet surface reaches a steady-state temperature. This temperature is called the wet bulb temperature and is controlled by the maximum amount of vapor that can be carried by the drying gas. As a result, the drying rate remains constant as long as the external conditions are constant. Under these conditions, all principles relating to simultaneous heat and mass transfer between gases and liquids apply. The steady state drying rate, dw/dt (kg/s), can be obtained from the condition that the heat flux from the drying gas to the solid is equal to the product of the drying rate and the vaporization enthalpy ΔH_v (J/kg):

$$-\Delta H_v \frac{dw}{dt} = k_g A(p^* - p)\Delta H_v = hA(T_{gas} - T_{wet}) \tag{14.1}$$

Here, k_g is the mass transfer coefficient for the gas phase (kg/(s m^2 Pa)), where the mass transfer across a surface area A (m^2) is driven by the difference between the moisture partial vapor pressure p (Pa) in the drying gas and the moisture saturation pressure, p^* (Pa) at the temperature of the wet film T_{wet}. Since p^* must be evaluated at the wet film temperature, the wet bulb temperature can only be determined self-consistently, by balancing both sides of the last equal sign in eq. (14.1). If the solid is sufficiently porous, most liquid evaporated in the constant drying rate period is supplied from the interior of the solid. The constant rate period continues as long as the liquid can be supplied to the surface as fast as it is evaporated. During constant rate drying, the material temperature is controlled more easily in a direct heat dryer than in an indirect heat dryer, because in the former drying process, the material temperature does not exceed the wet bulb temperature, as long as all surfaces are wet.

When the moisture content is reduced below a critical value, (internal) mass transfer limitations within the drying solid begin to dominate over the (external) mass transfer rate to the gas phase. At that point, the surface of the solid dries out and further evaporation starts taking place in the interior of the porous solid. This is called the falling drying rate period. First, the wetted surface area decreases continuously until the surface is completely dry, and the plane of evaporation slowly recedes from the surface. Heat for evaporation is transferred through the solid to the zone of evaporation. The drying rate is controlled by internal material moisture transport and decreases with decreasing moisture content. The amount of moisture removed in the falling rate period may be relatively small but the required time is usually long.

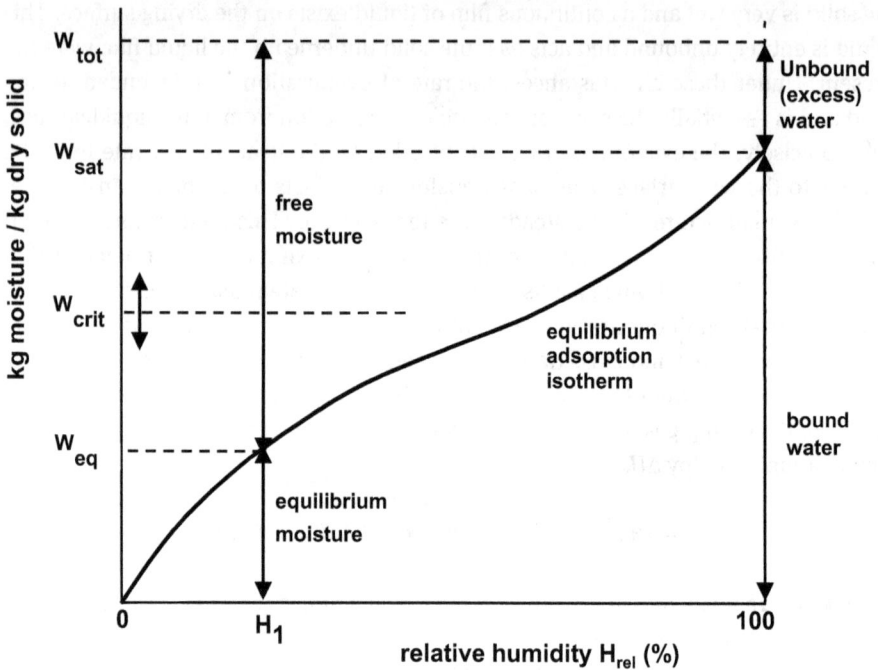

Fig. 14.6: Graphical illustration of the various kinds of moisture.

At the end of the falling rate period, the remaining moisture in the solid is bound by sorption. The drying rate decreases rapidly with decreasing moisture content and tends to zero, as the hygroscopic equilibrium moisture content is approached. This equilibrium moisture content is the steady-state equilibrium reached by the gain or loss of moisture, when material is exposed to an environment of specific temperature and humidity for a sufficient time. The equilibrium state (Fig. 14.6) is independent of the drying method or rate. It is a material property that relates to the moisture partial vapor pressure in the drying gas, p, by Henry's law:

$$p = Kx \tag{14.2}$$

where K is Henry's constant and x is the equilibrium dry basis moisture content. From this relation, it is clear that low residual moisture contents are obtained at low partial vapor pressures in the drying gas. This is usually achieved by removing the moisture from the drying gas at a low temperature and heating the gas before drying. Instead of using the moisture partial vapor pressure p, the equilibrium moisture content can also be related to the relative humidity of the drying air, which is defined as

$$H_{rel} = 100 \frac{p}{p^*} \qquad (14.3)$$

where p^* is the moisture saturation pressure at the operating temperature of the drying gas. Note that p^* in eq. (14.1) was evaluated at the wet film temperature, but at this point in the drying process, the temperature of the solid will essentially be equal to the temperature of the drying gas. Introduction in eq. (14.2) results in an expression where an alternative Henry's constant, K^* is used that is almost independent of temperature:

$$H_{rel} = K^* x \qquad (14.4)$$

In drying operations, it is useful to know the maximum weight of vapor that can be carried by a unit weight of dry gas. This maximum weight is reached at saturation, when the partial pressure of vapor in the gas equals the vapor pressure of the liquid at the specific temperature:

$$H_S = \frac{p^*}{(P-p^*)} \frac{M_V}{M_G} \qquad (14.5)$$

where H_s is the saturation humidity, i.e., the weight ratio of moisture/kg dry gas, P is the total system pressure and M_V/M_G the molecular weight ratio of vapor to dry gas. At any condition less than saturation, the humidity H is expressed similarly:

$$H = \frac{p}{(P-p)} \frac{M_V}{M_G} \qquad (14.6)$$

14.2.3 Direct heat dryers

In direct heat dryers, steam heated, extended surface coils are used to heat the drying gas up to temperatures as high as 200 °C. Electric and hot oil heaters are used for higher temperatures. Diluted combustion products are suitable for all temperatures. In most direct heat dryers, more gas is needed to transport heat than to purge vapor. Some of the most commonly used direct heat dryers are listed below.

14.2.3.1 Batch compartment dryers
Direct heat batch compartment dryers are often called tray dryers because of frequent use for drying materials loaded in trays on trucks or shelves. Figure 14.7 illustrates a two-truck dryer. The compartment enclosure comprises insulated panels

designed to limit exterior surface temperature to less than 50 °C. Slurries, filter cakes and particulate solids are placed in stacks of trays. Large objects are placed on shelves or stacked in piles. Unless the material is dusty, gas is recirculated through an internal heater as shown. Just enough purge is exchanged, so as to maintain needed internal humidity. For inert gas operation, purge gas is sent through an external dehumidifier and returned. These dryers are economical only for single-product rates less than 500 t/year, multiple product operation and batch processing. Variable speed fans are employed to provide higher gas velocity over the material, during early drying stages. To minimize dusting, the fans reduce velocity after constant drying when gas-side heat transfer at the material surface is no longer the limiting drying mechanism. Shallow tray loading yields faster drying, but care is needed to ensure depth uniformity and labor is increased.

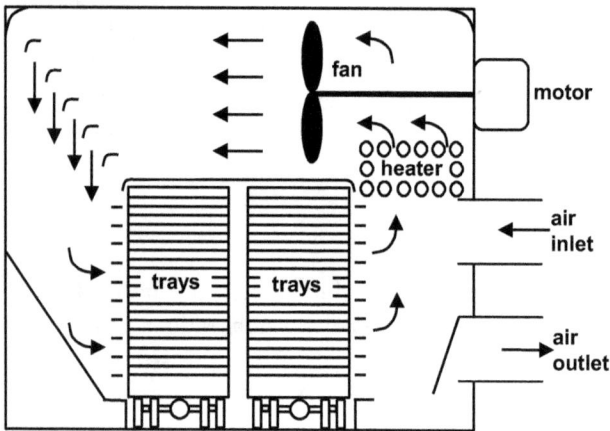

Fig. 14.7: Two-truck tray dryer (adapted from [7]).

14.2.3.2 Belt dryers

In belt dryers, a loading device that is especially designed for the product is used to place the moist solid on the surface of a circulating belt, which passes through a drying chamber that resembles a tunnel (Fig. 14.8). The solid remains undisturbed while it dries. At the end of this chamber, the material falls from the belt into a chute for further processing. In some installations, the material falls onto another belt that moves in the opposite direction to the first one. Centrifugal or axial flow blowers are used to aerate the moist materials. The air stream can enter the solid from below or above. Belt dryers are ideal for friable, molded, granular or crystalline products that require long, undisturbed drying time. They are used in all branches of industry.

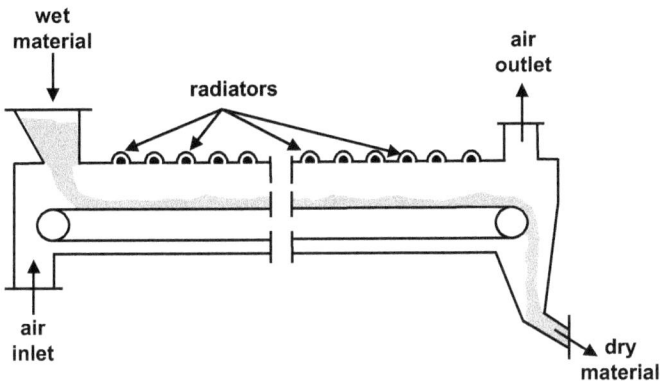

Fig. 14.8: Radiation belt dryer.

14.2.3.3 Rotary dryers

A direct heat rotary dryer is a horizontal rotating cylinder through which gas is blown to dry material that is showered by lifting flights attached to the inside of the cylinder (Fig. 14.9). Shell diameters are 0.5–6 m. Batch rotary dryers are usually one or two diameters long. Continuous dryers are, at least, four and, sometimes, ten diameters long. At each end, a stationary hood is joined to the cylinder by a rotating seal. These hoods carry the inlet and exit gas connections and the feed and product conveyors. For continuous drying, the cylinder may be slightly inclined to the horizontal to control material flow. Dry product may be recycled for feed conditioning, if material is too fluid or sticky, initially, for adequate showering. Slurries may also be sprayed into the shell in a manner that the feed strikes and mixes with a moving bed of dry particles. Material fillage in a continuous dryer is 10–18% of cylinder volume. Greater fillage cannot be showered properly and tends to flush towards the discharge end.

Direct heat rotary dryers are the workhorses of bulk solids industries. Most particulate materials can somehow be processed through them. These dryers provide reasonably good gas contacting, positive material conveying without serious backmixing, good thermal efficiency, and good flexibility for control of gas velocity and material residence time. Gas flow in these rotary dryers may be cocurrent or countercurrent. Cocurrent operation is preferred for heat-sensitive materials, because gas and product leave at the same temperature. Countercurrent allows a product temperature higher than the exit gas temperature, which increases the dryer efficiency. To prevent dust and vapor from escaping at the cylinder seals, most rotary dryers operate at a slightly negative internal pressure.

Wet product in

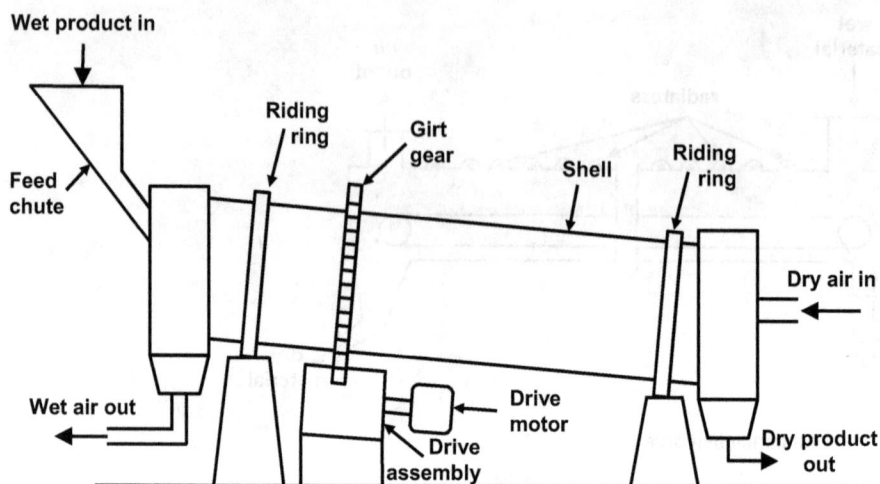

Fig. 14.9: Direct heat rotary dryer.

14.2.3.4 Flash dryers

In flash dryers, materials are simultaneously transported and dried. The simplest form, illustrated in Fig. 14.10, consists of a vertical tube in which granular or pulverized materials are dried, while suspended in a gas or air stream. The available drying time is only a few seconds. Only fine materials with high rates of heat and mass transfer or coarser products with only surface moisture to be removed can be used in such dryers. Solids that contain internal moisture can only be dried to a limited extent, by this method. Flash dryers are well suited for drying thermally sensitive materials and are widely used for drying organic and inorganic salts, plastic powders, granules, foodstuffs, etc.

14.2.3.5 Fluidized bed dryers

These units are known for their high drying efficiency. A two-stage model is illustrated in Fig. 14.11. The solid moves horizontally in a chute and the drying agent flows vertically through a perforated floor to fluidize the solid. These machines can operate continuously, because the solid is transported while suspended in the drying agent. Materials that can be suspended in the drying agent are usually powders, crystals and granular or short-fibred products that remain finely divided. Pastes and slurries are mixed with previously dried material and are then easily fluidized.

14.2.3.6 Spray dryers

Spray drying is used to convert suspensions, slurries or pastes into a powder that can be entrained by a gas stream. A spray dryer is a large, usually vertical chamber through

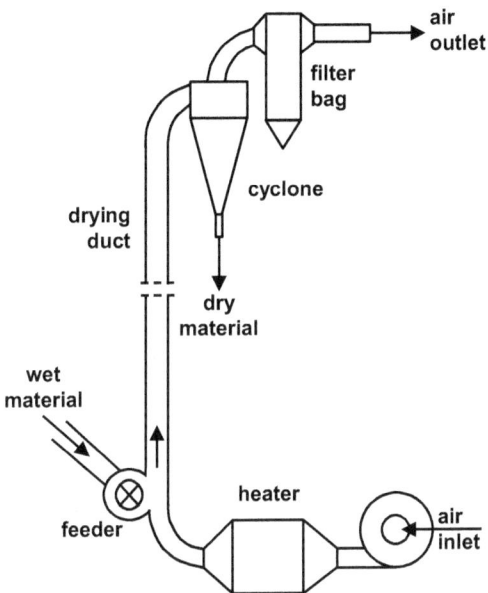

Fig. 14.10: Flash dryer.

which hot gas is blown and into which a suspension, slurry or pumpable paste is sprayed by a suitable atomizer. The largest spray-dried particle is about 1 mm and the smallest, about 5 μm. Since all drops must reach a nonsticky state before hitting a chamber wall, the largest drop produced determines the size of the drying chamber. The chamber shape is determined by nozzle or disk spray pattern. Nozzle chambers are tall towers, usually having height/diameter ratios of 4–5. Disk chambers are of large diameter and are short. A spray dryer may be cocurrent, countercurrent, or mixed flow. Cocurrent dryers are used for heat-sensitive materials. Countercurrent spray dryers yield higher bulk density products and minimize hollow particle production. Figure 14.12 shows an open-cycle, cocurrent, disk atomizer chamber with a pneumatic conveyor following for product cooling. Spray dryers are often followed by fluidized beds for second-stage drying or fines agglomeration. Spray dryers are particularly suitable for drying solids that are temperature sensitive. Applications include coffee and milk powders, detergents, instant foods, pigments and dyes.

14.2.4 Contact dryers

In contact dryers, most of the heat is transferred by conduction. Common heating media are steam and hot water, depending on the actual operating temperature. Based on dryer costs alone, contact dryers are more expensive to build and install than direct heat dryers, but they avoid the need for expensive air treatment units. For environmental concerns, contact dryers are more attractive because they are

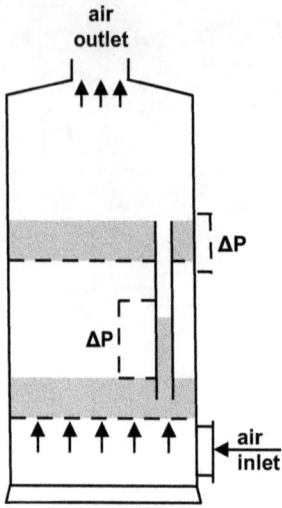

Fig. 14.11: Two-stage fluidized bed dryer.

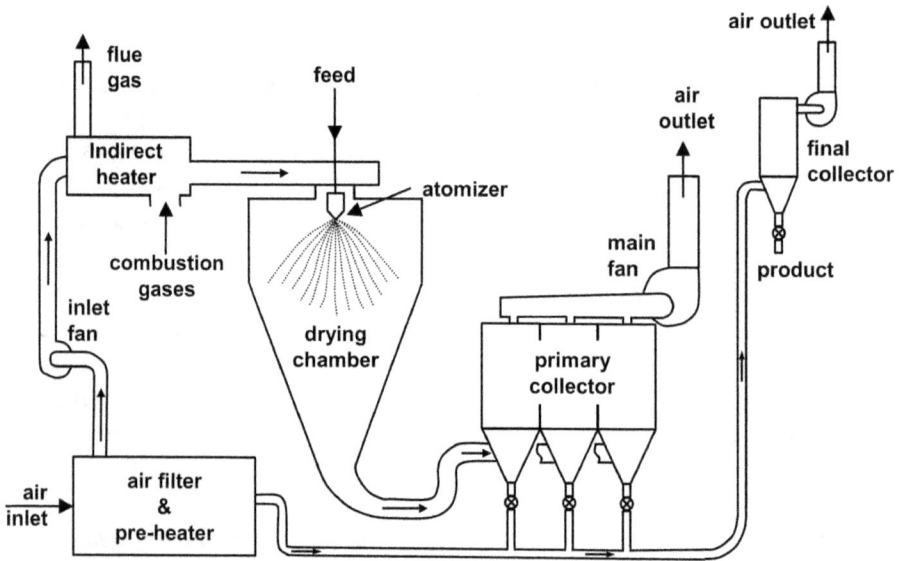

Fig. 14.12: Spray dryer system (adapted from [7]).

more energy efficient and use only small amounts of purge gas. Dust and vapor recovery systems for contact dryers are smaller and less costly. Here, some of the commonly encountered contact dryers are described.

14.2.4.1 Rotary and agitator dryers

The heat necessary for drying is transferred through the peripheral walls of these dryers in contact drying. Only a small amount of air is necessary to carry away the moisture that is taken from the solid. Accordingly, the air velocity in these units is quite low. This is advantageous when drying materials that dust easily or form dust during drying. The rotary steam tube dryer is a horizontal rotating cylinder in which one or more circumferential rows of steam-heated tubes are installed. In agitator dryers (Fig. 14.13), the vessel is stationary, and the solids product is slowly agitated. Material holdup may vary from a few minutes to several hours. Agitator speeds rarely exceed 10 min^{-1}, because, at higher speeds, the mechanical stress and power demand become too large.

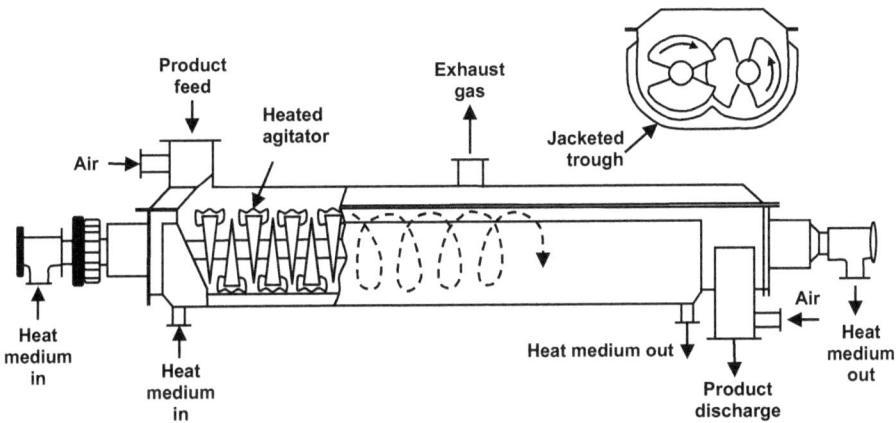

Fig. 14.13: Indirect heat paddle-type agitator dryer.

14.2.4.2 Vacuum dryers

Technically, the principal differences between vacuum dryers and other types of contact dryers are their seals and means to produce a vacuum. Drying under reduced pressure is advantageous for materials that are temperature sensitive or easily decompose because the evaporation temperature is reduced. Vacuum dryers are most often used to dry pharmaceutical products and foodstuffs. The simplest form of a vacuum dryer for batch drying is the vacuum shelf dryer, where the moist solid lies on a heated plate. Improved heat transfer with higher efficiency is obtained in the acuum tumble dryer (Fig. 14.14), in which the moist solid is constantly agitated and mixed. These tumble dryers are also used for the final drying step during the production of various polyamides.

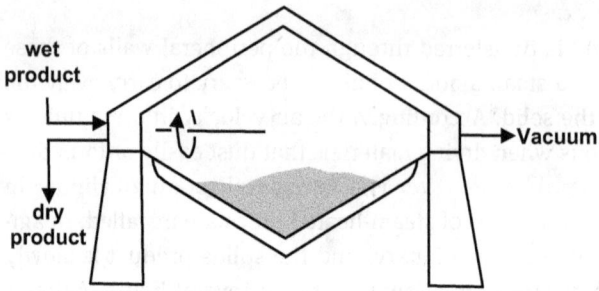

Fig. 14.14: Tumble vacuum dryer.

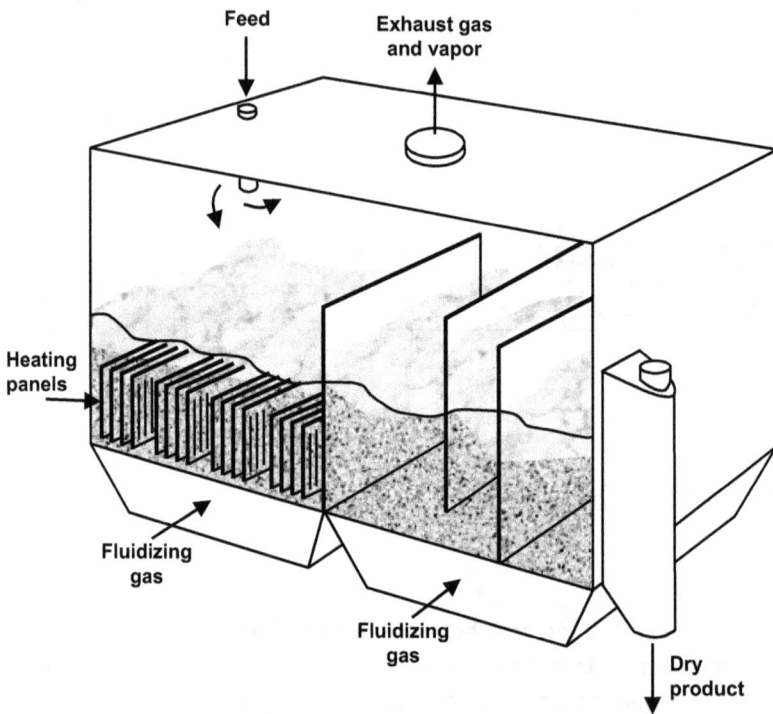

Fig. 14.15: Indirect heat fluid bed dryer.

14.2.4.3 Fluidized bed dryers

Indirect heat fluidized bed dryers are usually rectangular vessels in which vertical pipe or plate coils are installed. Figure 14.15 shows a diagram of a two-stage, indirect heat fluid bed incorporating pipe or plate coils heaters. The general design is used for several particulate polymers. Excellent heat transfer is obtained in an environment of intense particle agitation and mixing. Due to its favorable heat and mass transfer capabilities, flexibility for staging and lack of rotary seals, an indirect heat fluidized bed dryer is an ideal vessel for vapor recovery and drying in special atmospheres.

14.2.5 Other drying methods

Some other, less frequently used, drying methods are radiation heating, dielectric drying, electrohydrodynamic drying and freeze drying. In radiation heating, the energy is supplied from an electromagnetic (typically, infrared) radiation source that is remote from the surface of the solid. The electromagnetic radiation energy is absorbed in the form of increased molecular vibrations in the wet product, which dissipates to other types of molecular fluctuations and leads to an increase in temperature and evaporation of moisture. Radiation drying is expensive because of the high cost of electrical power or combustion gas to heat the radiators. Since the penetration depth of infrared waves is relatively small, it is mainly used for short drying times of thin product layers such as films, coatings and paints.

With high-frequency or dielectric drying, the wet product forms a dielectric between the electrodes of a plate capacitor exposed to a high frequency electric field. The generated intermolecular friction causes the generation of heat inside the product. This is used to heat up the product and to evaporate the moisture. High-frequency drying allows gentle thermal drying of the product. Substantial deformation and shrinkage cracks are avoided. Therefore, dielectric drying is employed for gentle drying of products such as fine wood, ceramic products, foods, pharmaceuticals and luxury goods.

Electrohydrodynamic drying is a new technology to dry materials convectively, without using additional heat. Electrohydrodynamic airflow is generated by a high-voltage difference between an emitter electrode, such as a thin wire and a grounded collector electrode, such as a plate. Due to the high voltage, the air around the emitter is ionized locally, and the ions are driven towards the collector. Collisions between ions and air molecules induce a wind of the order of 0.1–10 m/s, enhancing moisture removal from a wet material placed between the electrodes. Since it is nonthermal, electrohydrodynamic drying is particularly suitable to dehydrate heat-sensitive biomaterials.

Freeze-drying is a vacuum sublimation drying process. At temperatures below 0 °C and under vacuum, moisture sublimes from a frozen wet product directly from the solid to the gaseous state. Firstly the wet product has to be frozen from −15 °C to approximately −50 °C. The freezing rate and final temperature essentially determine the drying time and final quality of the product such as structure, consistency, color and flavor. The frozen product is usually granulated, sieved and then charged to the dryer. Under vacuum, it is then dried, either discontinuously on heated plates or continuously while being mixed and moved over a heated surface. Sublimed moisture vapor solidifies to ice on a cooling agent operated condenser. Drying rates are low, because the low allowable rate of heat flow controls the process. Freeze-drying or sublimation drying is carried out in a chamber freezer, vacuum disc dryer, vibrating film dryer, cascade dryer or spray freezing dryer. The high investment and operating costs of freeze-drying are only worthwhile for high-grade, thermally sensitive products. Certain important properties of the products such as flavor, taste, color and protein conformation are retained so that biologically active proteins maintain their biological activity (e.g., vaccines).

14.3 Size reduction

Size reduction or comminution of solids is an important industrial unit operation that is used to reduce the size of individual pieces. The goal of size reduction depends on the final application but is typically covered by preparing raw materials for subsequent processing. Some examples are:
– ore preparation to allow concentration of the valuable fraction,
– preparation of raw material for subsequent reactions,
– production of a defined particle size distribution necessary for a final application (fillers for plastics, rubber and paint, food, pharma),
– preparation of waste materials for recycling (shredding of old tires and waste plastic granulation),

The size of operation ranges from a few kilograms per hour for specialty products to hundreds of tons for metallurgical extraction purposes. In many operations, lumps up to a meter in size must be reduced to a fine powder of less than 100 µm. It is clear that this size reduction cannot be efficiently achieved in a single machine. A sequence of different types are used, each designed for efficient operation on a particular feed size. The science of size reduction still relies heavily on experience. It is important that requirements are discussed intensively with equipment manufacturers and large-scale test work carried out before decisions are made on the most suitable methods for a given requirement.

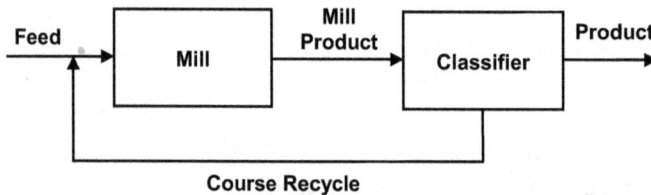

Fig. 14.16: Normal closed mill circuit.

In industrial practice, mills are frequently operated in closed circuit. Figure 14.16 illustrates that the milled product is passed through a size classifier that separates the milled product in the final product and a coarser stream that is returned to the mill feed. The general reason for closed circuit operation is to remove particles that are already fine enough, to prevent energy being wasted on grinding them even finer. The return of fine material back to the mill feed decreases efficiency as a result of overgrinding. The function of efficient classification is to reduce the proportion of fine material by avoiding overgrinding in-size material or fines. The concept of indirect inefficiency is that, although a mill may be operating efficiently in transferring input energy to breakage, it can be inefficient if that energy is used to break material that already meets specifications.

Size reduction equipment cannot be operated without auxiliary devices, such as bunkers, feeders, feed control and conveying equipment. This auxiliary equipment is often responsible for weaknesses and disturbances in size-reduction as well as drying plants.

14.3.1 Particle breakage

In size reduction processes, lumps of solids must be fractured by subjecting the material to contact forces or stresses. The stresses are applied by transmitting mechanical force to the solid. In addition, most grinding machines have some degree of impact stress that propagates stress waves through the solid, activating flaws to fracture in the process. These forces cause deformation, which generates internal stress in the particles followed by particle breakage when this stress reaches a certain level. It is important to differentiate between elastic and plastic deformations within materials. With brittle materials, the response to imposed stress is mainly elastic deformation, until the fracture point is reached, at which breakage occurs. An elastic material can be stressed, producing deformation, and the material returns to its original shape when the stress is removed. However, if the solid is strained too far, catastrophic failure occurs, and the solid fractures at a stress termed the tensile strength. Plastic or inelastic deformations are encountered in ductile materials that undergo a partially irreversible stretching before failure occurs. Figure 14.17 compares the stress–strain characteristics of brittle and ductile materials.

The total area under the stress–strain curves is the strain energy stored in a body. This energy is not uniformly distributed throughout the material but concentrated around the tips of existing cracks or flaws. This local stress concentration initiates crack propagation and gives rise to particle failure. A crack propagates when the overall stress around the crack reaches a critical value. As crack propagation progresses, the strain energy released exceeds the energy associated with new surface formation. The excess energy concentrates around other cracks in the material, causing multiple fractures. This is the typical behavior of elastic materials such as rocks, ores and coals that undergo brittle fracture through preexisting Griffith flaws. The strength or grindability of these materials correlates only weakly with the chemical bond strength because the number, size and orientations of the flaws basically determine particle strength.

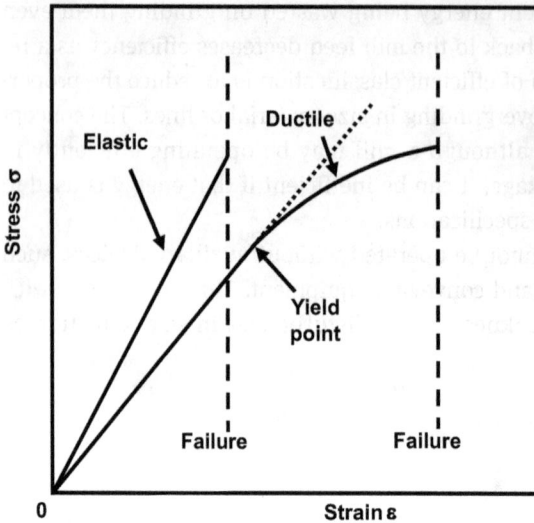

Fig. 14.17: Illustration off stress–strain curves for elastic and ductile materials.

A comparison between the failure of brittle and ductile materials shows the following major features:

1. Purely brittle failure is almost independent of temperature. Ductile materials exhibit a decrease of strength with increase of temperature, owing to greater mobility of dislocations.

2. For failure from Griffith cracks, a smaller particle has a smaller probability of containing a large flaw and will be relatively stronger. In other words, as brittle materials break, the remaining fragments are stronger because the larger flaws have broken out. This explains why it is more difficult to break small particles than large particles. In practice, this is seen where a limit of grindability is reached with many materials and with subjection to further grinding, no decrease in particle size can be observed.

3. The rate of stress application is more important with ductile materials than with purely brittle materials, because a high rate of stress application may give brittle failure, whereas the same stress reached by slow steps would give time for ductile behavior.

4. Ductile materials demonstrate work hardening, that is, initial deformation produces movement and pileup of dislocations, making further deformation more difficult. They also demonstrate stress fatigue, again owing to the gradual accumulation of dislocations on repeated cycles of stress.

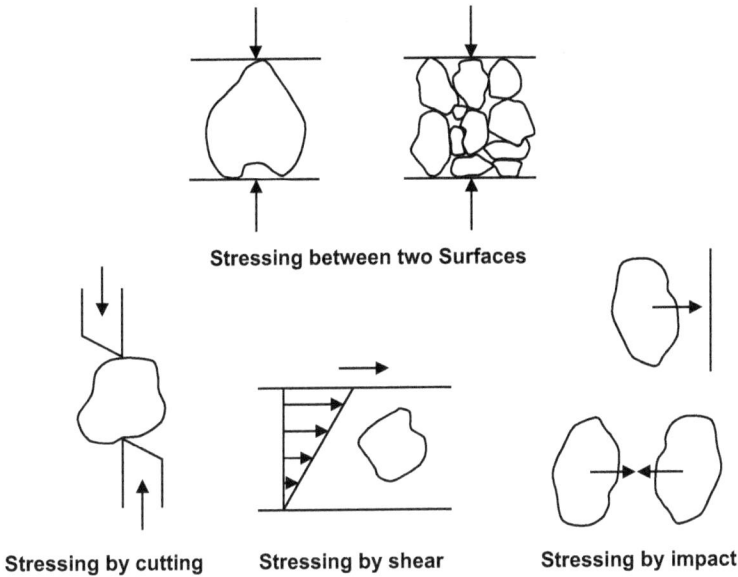

Fig. 14.18: Various methods of stress application for size reduction.

14.3.2 Methods and selection criteria for size reduction

Equipment for size reduction can be categorized by the method in which the necessary stress is applied to the particles. The various methods are illustrated in Fig. 14.18:

- Stressing between two solid surfaces (crushing): Either single particles or a bed of particles are crushed between two solid surfaces. The force applied to the solid surfaces determines the amount of stress that can be applied.
- Stressing by impact: Size reduction is achieved by the impact of a particle against a solid surface or other particles. The particle can be accelerated to impact against a surface, or the surface can be accelerated to impact the particle, as in an impact mill. The momentum transferred is limited by the mass of the particle and the achievable impact velocity.
- Stressing by cutting: This method is useful for materials that exhibit plastic behavior. An example is tough, rubbery materials, where the best stress application for size reduction is the scissors type of action.
- Stressing by the surrounding medium: Size reduction is affected by shear forces or pressure waves. The amount of energy that can be transferred is very limited, and this method is mainly used to break agglomerates.

Selection of the most suitable machine for a given requirement is an extremely complex process. The principle factors which must be addressed are:

- Toughness/brittleness: In ductile materials, the excess strain energy causes plastic deformation, whereas, in brittle materials, new cracks are propagated. Brittle materials can be reduced relatively easily, whereas ductile materials present challenges. It is sometimes possible to cool a tough material to a temperature low enough to display brittle behavior, as is sometimes used for grinding polymer granules and active pharmaceutical ingredients. The action of the cooling is to reduce the flexibility of the bonds joining the groups making up the polymer chains. It is often necessary to cool to very low temperatures, e.g., using liquid nitrogen (boiling point of 77 K or −196 °C).
- Hardness: There are several hardness (qv) scales. In selecting size reduction equipment, hardness is generally expressed according to the Mohs' scale (Tab. 14.1), where a body in one range scratches a body in the immediately previous range. High-speed machines, such as impact mills, begin to suffer high wear rates when processing materials above Mohs' hardness of 3, unless very special wear-resisting measures can be taken.

Tab. 14.1: Mohs' scale of hardness.

Mohs' hardness	Material
1	Talcum
2	Gypsum
3	Calcite
4	Fluorite
5	Apatite
6	Feldspar
7	Quartz
8	Topaz
9	Corundum
10	Diamond

- Abrasiveness: Abrasion is a special type of fracture responsible for the tearing out of small pieces of material from the surfaces of the components used to apply stress to the material being fractured. This property is closely related to hardness in homogeneous materials, but is also affected by particle shape, such as the presence of sharp corners. Although grinding components are designed strong enough to stress the material being comminuted without bulk fracture to themselves, this is no guarantee that their surface will be abrasion resistant. The fracture mechanics of abrasion involves high local surface stresses, owing to asperities in the particles and in the grinding surface. High rates of surface stressing in shear caused by high relative speed between stressing and stressed agents, undoubtedly, assist abrasive fracture.
- Feed size: The acceptable feed size for a given machine is governed by the type of feed device and physical characteristics of the machine.

- Cohesion: Many materials stick together and adhere to machine parts, depending on their composition, condition, particle size and temperature.
- Particle shape and structure: Some materials exhibit particular properties, owing to their shape or form. It is often desired to maintain particle shape. In such cases, an impact type mill is usually preferred over a ball mill, as the latter tends to alter the original particle shape more strongly.
- Heat sensitivity: Only maximally 1–2% of the applied energy is effectively used for size reduction. The remainder is mainly converted to heat, which is absorbed by the grinding air, product and equipment. As a result, materials can become sticky (e.g., fat containing products), can degrade chemically (active pharmaceutical ingredients, foods) or lose flavor.

14.3.3 Size reduction equipment

A broad selection of different types of equipment is available as a result of historical development and the wide range of applications based on particle size, throughput and material properties. No uniform criteria are available for classifying size reduction equipment. The classification scheme used in this section is based on the method of stress application, particle size and similarities between process types and machine construction.

14.3.3.1 Crushers

Crushers can process particles up to 1.5 m in edge length. Stress is applied by either crushing single particles or a bed of particles between two solid surfaces. Crushers are used in mines, for ore preparation and for the production of road construction materials. Most crushers are used for coarse and medium-size reduction to produce particles ready for transport, or to perform a primary reduction step prior to subsequent comminution. The principal requirement for crushers is a mechanical one — the machines must be very robust because of the high stresses required to crush a large lump. Crushers can be classified into four types: jaw crushers, gyratory crushers, impact crushers and roll crushers.

Figures 14.19 and 14.20, respectively, show examples of jaw and gyratory crushers that both compress the feed between a stationary and a movable surface. The reciprocating action of the movable jaw in a jaw crusher strains lumps of feed to the point of fracture, as does the nonsymmetric movement of the rotating mantle in a cone or gyratory crusher. The size reduction ratio is of the order of 10 and is varied by the adjustable gap setting. The basic action is that the entering brittle material is crushed, the broken products fall due to gravity into a narrower space, and bigger fragments are crushed as the metal-lump-metal space closes again, with material moving down, until all of it falls through the gap. The crushing surfaces of jaw crushers are often ribbed

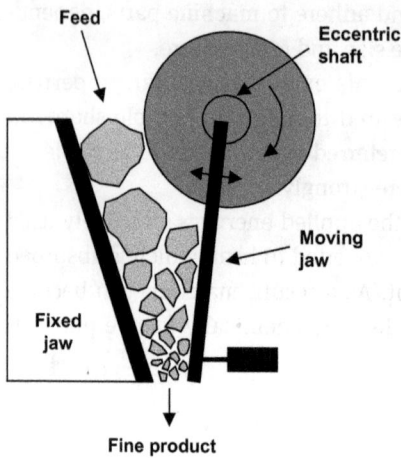

Fig. 14.19: Jaw crusher.

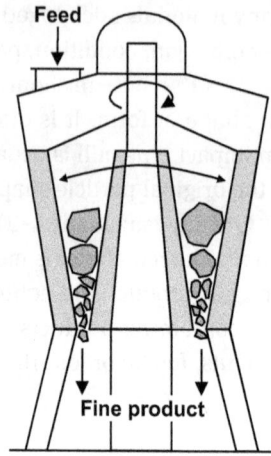

Fig. 14.20: Gyratory crusher.

or toothed to help prevent slippage of the lumps/particles as it is compressed and to give higher local stress at the surface of the material.

Figure 14.21 illustrates a hammer crusher. Material is broken by direct impact of the hammers, by being thrown against the case or breaker bars, and by compression and shear when nipped between the hammers and the case. The hammers are mounted on a heavy rotor, and/or the shaft is attached to a heavy flywheel, to give high inertia of the rotating mass. In most cases, the crushing zone is surrounded by grate bars, so that fragments that are larger than the openings of the grating are retained in the crushing zone. This type of crusher is best suited for soft and moderately hard nonabrasive but brittle materials. However, these crushers are simpler and, therefore, cheaper than jaw and cone crushers. Consequently, impact crushers have been developed for the processing of fairly abrasive rock that tend to wear the machine.

Double roll crushers consist of two rollers that rotate towards each other and are separated by an adjustable gap (Fig. 14.22). In a single roll crusher, a shaped crushing roller acts against a crushing plate. The material is fed in from above. The rollers, which often have a corrugated surface, draw in the feed and apply stress to it. The roller profile improves drawing in of the feed and, therefore, helps increase the capacity. Due to abrasive wear, the rolls have to be resurfaced at frequent intervals, if the crushed material is strong and abrasive. With soft to moderately hard brittle materials, throughputs of up to 3,500 t/h may be obtained.

14.3.3.2 Grinding media mills

These mills consist of a vessel containing a moving grinding medium such as balls, rods, short bars or coarse particles of the feed material itself. Tumbling ball mills, illustrated in Fig. 14.23, are widely used for dry and wet grinding to relatively fine sizes.

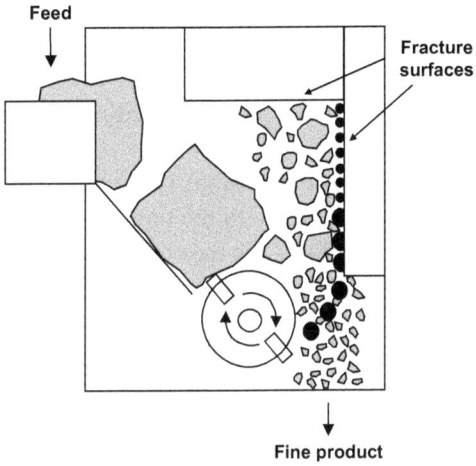

Fig. 14.21: Hammer crusher.

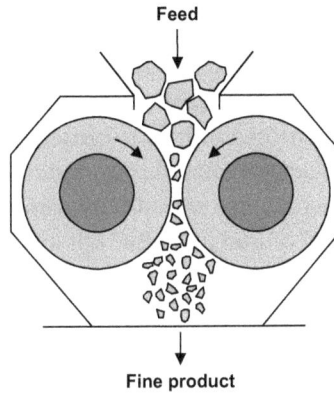

Fig. 14.22: Double roll crusher.

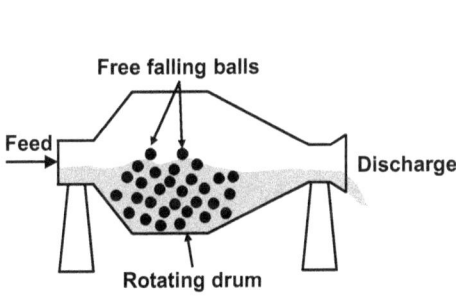

Fig. 14.23: Tumbling ball mill.

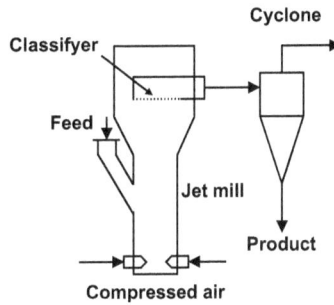

Fig. 14.24: Opposed jets fluid jet mill.

They have been used since the late 1800s, and the construction and principles remain essentially very simple. The machine consists of a cylindrical or conical tube into which loose grinding balls are filled up to a certain level. Size reduction is achieved by rotating the tube, so that the balls roll against each other or, if the speed is sufficient, they are lifted and fall. The grinding action ensures that a very high local stress can be applied to the particles so that a high portion of ultrafine product is produced.

14.3.3.3 Roller and rod mills

In roller mills, stress is applied to the feed in the gap between two counter-rotating rollers. The design is equivalent to that of roll crushers. The rollers are typically 0.5–1.5 m in diameter and, unlike crushers, are smooth or grooved. The feed must have a narrow particle-size distribution to ensure that stress is not applied to the coarsest particles in the mill gap only. As a result of the mode of operation, fine particle agglomerates are formed, usually. They are employed in a wide variety of applications for medium-coarse, down to fine size reduction, including in the abrasives industry and in the milling of cereals.

The rod mill acts, in principle, like a multiple set of rolls, as the cylinder rotates. The bed of rods is carried up until it lies at an angle to the horizontal. It is then unstable, and rods start to roll down the bed surface. The rods rolling over one another act like sets of rolls, stressing particles in a similar manner. The power to the mill is used to lift the rods against gravity.

14.3.3.4 Jet and impact mills

In this equipment group, stress is applied by transferring kinetic energy by either particle-particle contact (jet mills) or machine–particle contact (impact mills). In impact mills, the particles fly against an impact plate, or slowly moving particles are impacted by a rapidly moving grinding tool.. Stress can also result from interparticle collisions. Stress application differs from that in previously described mills, where stress is primarily applied by means of pressure. The main differences between impacting and compression are that stress is applied to individual particles and that the velocity at which stress is applied is considerably higher. Very fine particles can be generated under dry conditions with impact and jet mills.

In jet mills, the feed particles are accelerated by means of propellant jets (air, gas or steam) with high velocities. Comminution occurs by interparticle collisions. These mills are always operated in closed circuit with an air classifier. Figure 14.24 shows a schematic of an opposed jet mill in which the particles are accelerated through two or more pipes and directed against each other, in the central grinding chamber. Since the working principle of jet mills is based on particle-particle collisions, jet mills are very suited for milling abrasive materials.

The range of products processed in impact mills is extraordinarily large. Impact mills are found in all branches of industry that deal with bulk materials or powders. Impact mill designs can be divided into beater mills, hammer mills, and mills with or without peripheral grinding faces. Hammer mills are basically similar to hammer crushers (Fig. 14.25). The hammers are usually simple iron plates. Screen openings surrounding the rotor range between 2 and 10 mm. The impact devices used in pin mills (Fig. 14.26) are large numbers of round pins or elongated squares. The devices are arranged in concentric rings of varying radius. The degree of fineness in a pin mill is determined primarily by the peripheral speed of the rotor. Since the number of collisions is relatively low, pin mills are well suited for temperature-sensitive or caking materials.

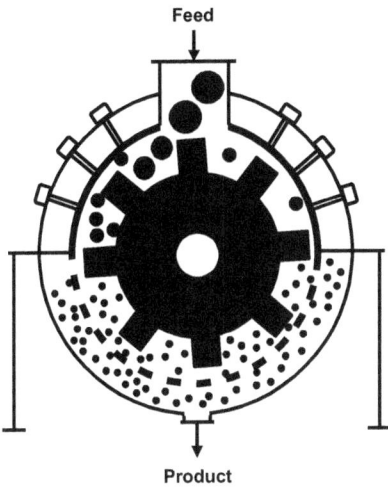

Feed

Product

Fig. 14.25: Hammer mill.

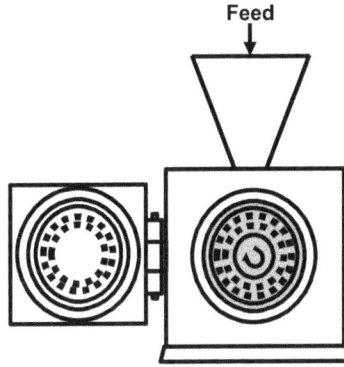

Feed

Fig. 14.26: Pin mill.

14.3.3.5 Cutting mills

Tough nonabrasive materials such as polymers, rubber, paper and wood waste cannot be comminuted with any of the mills described so far. A whole class of mills is designed specifically for these applications. These rely on the cutting action between rotating and static sharp edges with narrow clearance. The efficiency of this type of mill is highly dependent on maintaining sharp cutting edges. Examples are shredders, chippers, knife choppers and pulp machines. The common design of these machines is to have a rotor equipped with several knife blades that cut the product against stationary knife bars.

14.4 Size enlargement

Size enlargement includes all processes that bring together fine powders into larger agglomerates in order to improve the powder properties. Usually, agglomerates are relatively permanent entities in which the original particles can still be identified. The resulting collection of particles is called an agglomerate or a granule. Size enlargement is used in many industries such as fertilizer, light-weight aggregate and pharmaceutical production. Benefits gained from size enlargement are extremely diverse and strongly dependent on the application for which the operation is used. An overview of the advantages of agglomerated products is listed in Tab. 14.2.

The desired product properties determine which enlargement process is used. Size enlargement processes are classified by the principal mechanisms that are

Tab. 14.2: Advantages of agglomerated products.

Benefit	Applications
Reduced dust content and thereby increased handling safety and fewer product loss	Pharmaceuticals, dyes, pigments
Freely flowing	Fertilizer granulation
Improved storage and handling characteristics	Fertilizer granulation, carbon black
Improved metering and dosing capabilities	Food additives
No segregation of coagglomerated materials	Pharmaceuticals
Increased bulk density and lower bulk volume Defined shape	Carbon black Catalyst particles
Defined weight of each agglomerate	Drugs in pharmaceutical tablets
Control of porosity or density and herewith dispersibility, solubility, reactivity, heat conductivity, etc.	Food additives, pigments, dyes
Improved product appeal	Fuel briquettes
Increased sales value	

used to agglomerate the particles. The selection of a specific process is possible only if the user clearly defines the required product properties.

14.4.1 Agglomeration principles

During production and processing of solid materials, adhesion phenomena become more and more important, when particles with sizes below approximately 100 μm are handled. Adhesion of finely divided materials takes place during all operations of mechanical process engineering and can either be desired or undesired. On the other hand, adhesion during size reduction is always undesired, because it reduces the grinding effect. However, when it is desired to increase the particle size of fine powders by agglomeration, adhesion is systematically promoted. In these desired particle agglomerates, binding forces must act to keep the individual particles together. The mechanisms by which particles bond together and grow into agglomerates are affected by the size enlargement method of choice. Nevertheless, there are certain aspects of the bonding process that are essentially independent of the equipment and method. As shown in Fig. 14.27, the binding mechanisms can be divided into five major groups.
- Solid bridges may develop at elevated temperatures by diffusion of molecules from one particle to another at the points of contact ("sintering"). Solid bridges

can also be formed by chemical reaction, crystallization of dissolved binder substances, hardening binders and solidification of melted components.

–   Interfacial forces and capillary pressure in liquid bridges can create strong bonds that disappear if the liquid evaporates and no other binding mechanism takes over.
–   Adhesion and cohesion forces in highly viscous, not freely moving binders can form bonds very similar to those of solid bridges. Thin adsorption layers are immobile.
–   Attraction forces between solid particles such as van der Waals, electrostatic or magnetic can cause solid particles to stick together, if they approach each other closely enough. Decreasing particle size favors this mechanism.
–   Form-closed bonds by geometric interlocking or folding of fibers, platelets or bulky particles.

In most processes, more than one bonding mechanism is likely to act at the same time. Although geometric interlocking of particles influences agglomerate strength, its contribution is generally considered small in comparison to other mechanisms.

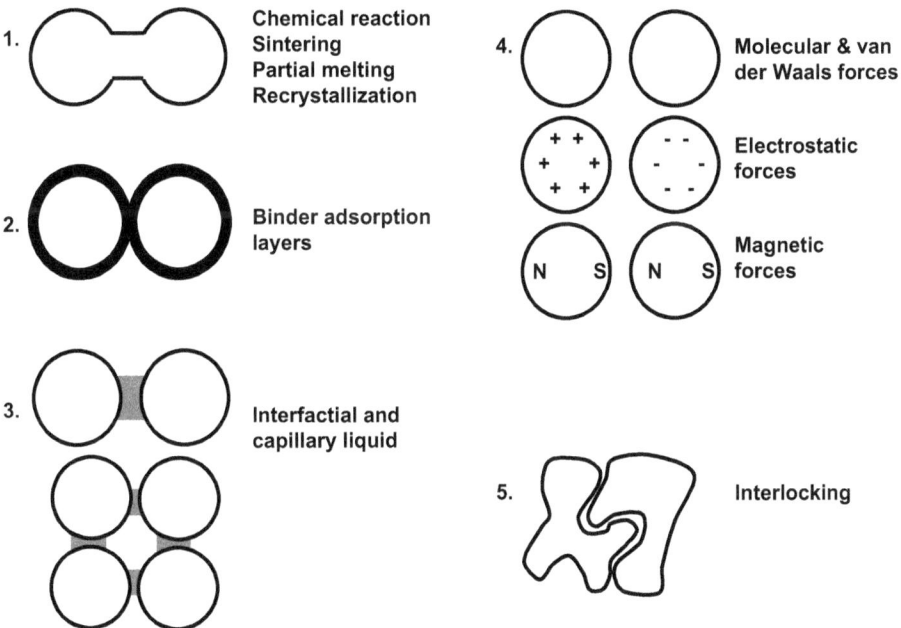

Fig. 14.27: Classification of binding mechanisms of agglomeration.

14.4.2 Methods of size enlargement

The most commonly used size enlargement methods can be classified in two types of processes:
– Growth/tumble agglomeration (no external forces)
– Pressure agglomeration (external forces)

In these processes, four principal mechanisms are used to bring fine particles together into larger agglomerates:
– Tumbling and other agitation methods
– Pressure compaction and extrusion methods
– Heat, reaction, fusion and drying methods
– Agglomeration from liquid suspensions

When required, the agglomerate strength can be increased by the addition of binders.

14.4.3 Growth/tumble agglomeration

Growth/tumble agglomeration is the "most natural" of all size enlargement processes and uses mechanisms similar to natural agglomeration. When fine particles are brought into intimate contact through agitation, binding forces come into action to hold the particles together. In these agglomerates, capillary binding forces caused by wetting with water or aqueous solutions is the most common binding mechanism. Because the particles to be agglomerated are larger, binders are often added to enhance the particle-to-particle adhesion and maintain strength after drying.

Figure 14.28 shows that the mechanism of growth agglomeration comprises several phases. In the nucleation phase, microagglomerates are produced by the coalescence of fine particles. Subsequently, these nuclei grow into larger aggregates by further coalescence of nuclei and/or layering with other particles. To survive and grow in an agitated system, the formed agglomerates must withstand the destructive forces generated by the agitated particles. These forces become larger with increasing size of the particles to be agglomerated, until size enlargement by tumbling is no longer possible. This limitation to relatively small particles, and the fact that only temporarily bonded conglomerates are formed, are the most important drawbacks of all tumble agglomeration methods. A curing step must follow to obtain permanent bonding by solid bridges, e.g., by sintering, chemical reaction, partial melting and solidification or crystallization of dissolved substances via drying. This curing step is normally the expensive part of a tumble agglomeration process. However, for very large amounts of solids and fine particles, tumble agglomeration is a preferred technology. Another reason for the application of tumble agglomeration may be the high porosity of the agglomerates, resulting in a large surface area and easy solubility.

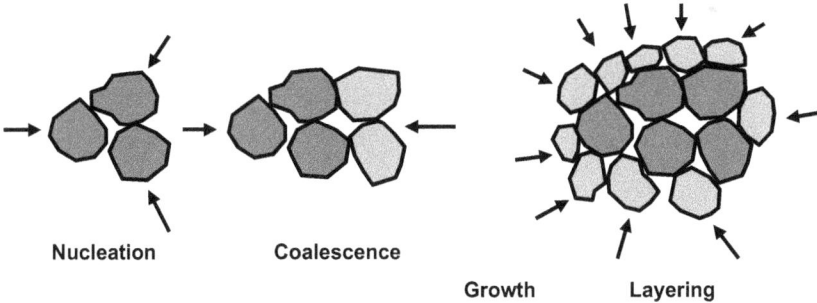

Nucleation   Coalescence

Growth   Layering

Fig. 14.28: Main mechanism of growth/tumble agglomeration.

The most common types of agitators are rotating drums, inclined discs, powder mixers and fluidized beds. Rotary drums and inclined disks or pans are the most important equipment. They are mainly used to convert finely dispersed material to a coarser product. Rotary drum agglomerators (Fig. 14.29) consist of an inclined rotary cylinder powered by fixed- or variable-speed drive. Feed material containing the correct amount of liquid agglomerates under the rolling, tumbling action of the rotating drum. The inclination of the axis assists material transport through the cylindrical drum. Liquid may be introduced either before or immediately after the solids enter the cylinder. Agglomerates of the desired size must be screened out of the exit stream. The oversized material must be reduced in size and fed back into the drum along with the fines to form a closed loop. The principal users of agglomeration drums are the iron and fertilizer industries.

An inclined disk or pan agglomerator (Fig. 14.30) consists of a tilted rotating plate equipped with a rim to contain the agglomerating charge. Solids are fed continuously from above and product agglomerates discharge over the rim. Moisture or other binding agents can be sprayed on at various locations on the plate surface. In the disk agglomerator, nuclei and small agglomerates move towards the base of the disk. The particles are transported higher due to their largely irregular shape. The larger, more rounded agglomerates roll easily over the smaller ones. In this way, the pan classifies the agglomerates. The overflow product is so uniform in size that subsequent classification is usually unnecessary. This inherent classification action offers an advantage in applications that require accurate agglomerate sizing. Other advantages of the inclined disk include less space requirements and lower costs. Drum agglomerators have the advantage of greater capacity, longer residence times, less sensitivity to upsets and more easy handling of dusty materials.

In theory, all solid and solid–liquid mixers are suitable for agglomerate production. In contrast to disk and drum agglomeration, a mechanical impeller is used

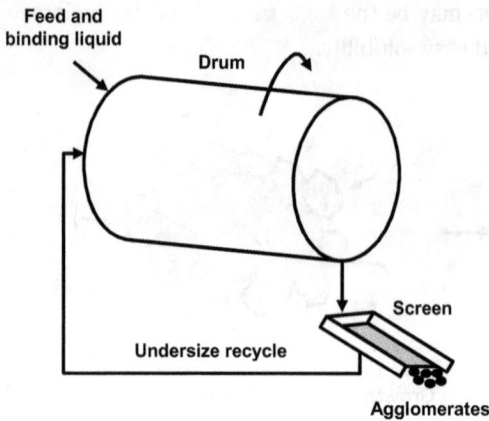

Fig. 14.29: Rotary drum agglomerator.

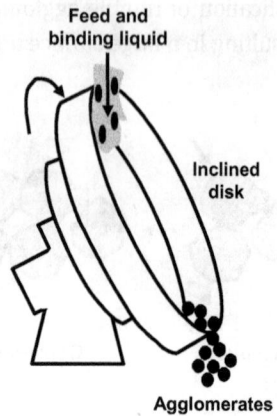

Fig. 14.30: Inclined disk or pan agglomerator.

to agitate the powder. When a suitable amount of agglomeration liquid is sprayed in, most of the particles form agglomerates due to the stresses generated by stirring. Horizontal pan mixers, pin mixers (Fig. 14.31) and other types of intensive agitation devices are used. Shaft mixers operating at very high rotational speeds are also used to granulate extremely fines.

In fluidized bed agglomeration (Fig. 14.32), pumpable suspensions, pastes, or melts are converted into agglomerates, mostly in combination with drying. Externally heated gas (air) is forced or drawn upward through a porous distributor plate and then through the particle bed. The solid particles are intensively pneumatically mixed, and the agglomeration fluid is sprayed with nozzles that are located above the bed surface. The motion of the bed serves to distribute the fluid uniformly among the particles. A portion of the agglomerates is drawn through an opening beneath the bed surface. The product is separated from oversized particles and fines that are recycled to the fluidized bed.

14.4.4 Pressure agglomeration

Pressure or press agglomeration represents a large share in commercial applications of size enlargement by agglomeration. As far as applicability is concerned, high-pressure agglomeration is largely independent of feed particle size and the forces acting upon the particulate feed may be very high with certain equipment. Moreover, a curing step is usually not necessary. Therefore, it constitutes the most versatile group of size enlargement processes by agglomeration. Advantages of pressure agglomeration are that essentially dry solids are processed and that the

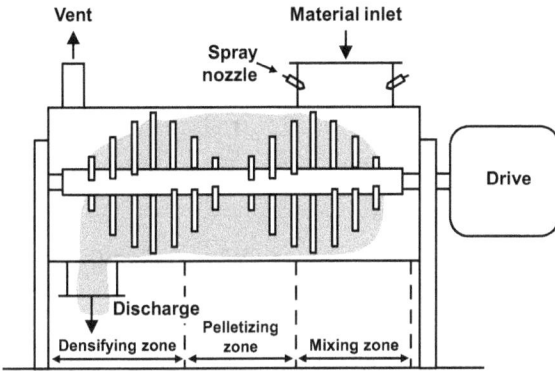

Fig. 14.31: Pin mixer used in carbon black pelleting.

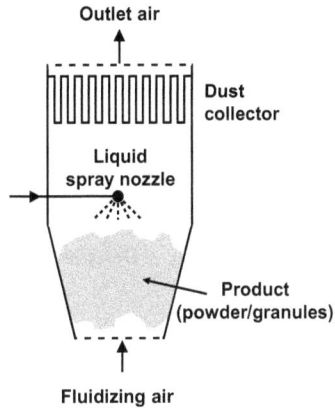

Fig. 14.32: Batch fluid bed agglomerator.

amount of material in the system is relatively small. Therefore, pressure agglomeration methods are particularly suitable for batch and shift table production operations as well as multiproduct applications. Due to the relative complexity of the equipment and its comparatively small capacity per unit, these techniques are most commonly used in low to medium capacity applications.

The compression techniques for size enlargement produce agglomeration by application of suitable forces to particulates held in a confined space. The compaction process of void reduction may be considered to occur by two essentially independent mechanisms (Fig. 14.33). When a particulate solid is placed into a die and pressure is applied, a reduction in volume will occur due to rearrangement of particles at low pressure followed by elastic and plastic deformation of the particles at higher pressures. This high-pressure compaction continues until the compact density approaches the true density of the material. The various methods in use differ in both the means of pressure application and the method used to confine the powder. Tablet production, pressing, molding and extrusion operations are commonly used to produce agglomerates of well-defined shape, dimensions and uniformity.

Tablet machines have already been in use for a very long time. They may be either single-punch or rotary presses, but all compact the powder to a single tablet by high mechanical force. In compacting presses such as roll presses, particulate material is compacted by squeezing as it is carried into the gap between two rolls rotating at equal speed. This is probably the most versatile method of size enlargement because most materials can be agglomerated by this technique with the aid of binders, heat and/or very high pressures.

Pellet mills (Fig. 14.34) differ from tablet machines in that the particulates are compressed and formed into agglomerates by extrusion through a die. The fresh products usually need a curing step to arrive at the final product. Several types of

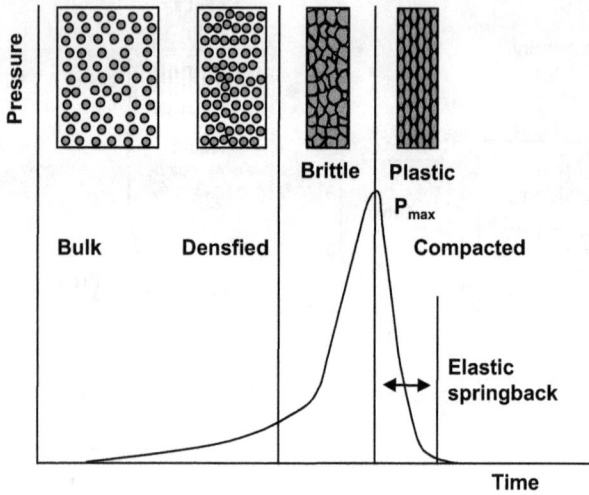

Fig. 14.33: Mechanism of compaction in pressure agglomeration.

equipment that use the extrusion principle are available. The die may be a horizontal perforated plate with rollers acting on its upper surface to press material through the plate. This pressing of particles through a die is also used in extruders (Fig. 14.35). The rotation of the screw in the extruder barrel develops sufficient pressure to agglomerate the particles and force them through the die.

14.4.5 Other agglomeration techniques

The remaining size-enlargement techniques depend on heat transfer to accomplish particle bonding. Sintering is a curing technology that uses high temperatures to accomplish agglomeration by ceramic bond formation and grain growth by diffusion. Drying and solidification on surfaces can produce granular products directly from fluid pastes or melts. On the surface of drum dryers, solutions, slurries or pastes are dried in a thin film. A blade scrapes the product off in flake, chip or granular form. In flakers, a thin film of molten feed is applied to a polished cooled surface. Virtually any molten material that solidifies rapidly with cooling can be treated with this forming method. The cooled solid is scraped from the surface as a flaked or granular product. In suspended particle techniques, granular solids are produced directly from a liquid phase by dispersion in a gas to allow solidification through evaporation of the liquid. Equipment used includes spray dryers, prilling towers, spouted and fluidized beds. Agglomerate formation occurs by hardening of the feed droplets into solid particles.

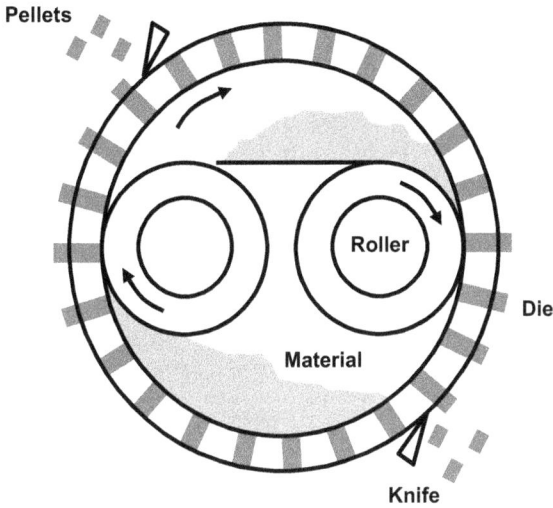

Fig. 14.34: Operating principle of a pellet mill.

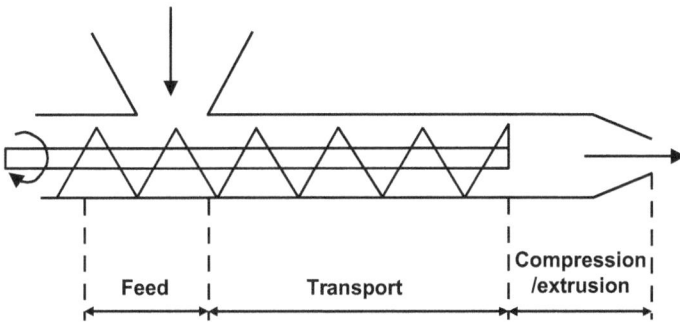

Fig. 14.35: Operating of a simple axial screw extruder (single).

14.5 Conveying

14.5.1 Transportation systems

Conveying is a term used for the transport of bulk solids. The movement of bulk solids from one location in an industrial process operation to another has often been a source of considerable worry and expense. While gases and liquids can flow rather easily from one location to another, granular solids are much harder to transport. Conveyors are machines that transport material fed in a gravimetric or volumetric fashion at a controlled rate. By addition of appropriate weight sensing and control modules, conveyors can also be used as gravimetric feeders. The selection of the type of conveyor for a specific application is dependent on the:

1. required capacity
2. conveying path
3. handling characteristics of the material

The effective operation of most modern processing plants depends on the ability to move raw materials, semi-processed and fully processed products, both within the manufacturing environment and, ultimately, to the end user.

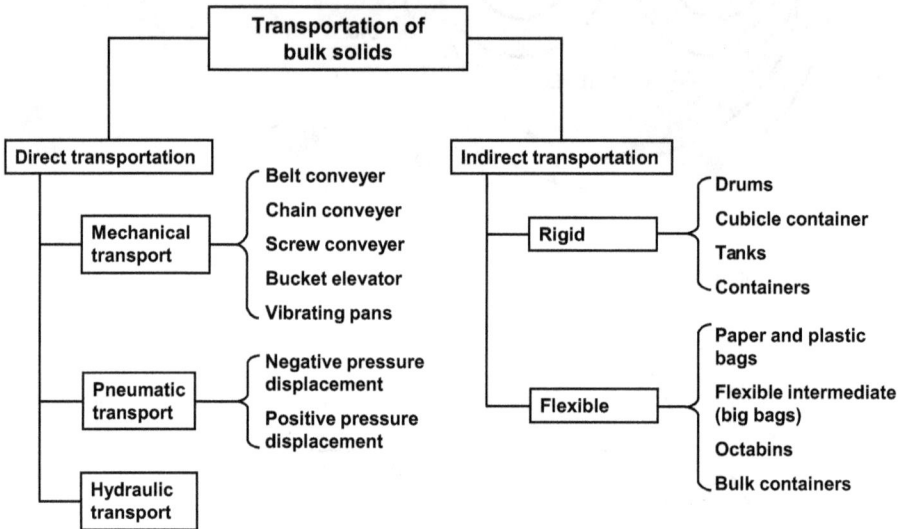

Fig. 14.36: Overview of solids transportation systems.

Essentially, two categories of bulk materials handling systems exist: direct and indirect transportation. The type of equipment can sometimes be suitable for both types of operation. The categorization of the transportation systems is illustrated in Fig. 14.36. Three of the most salient features for selecting a suitable handling system are knowledge of the distance, the required throughput and need for intermediate storage. When specifying distance, details of the route profile are important: elevation, horizontal and vertical conditions to be traversed, head room and route flexibility are critical issues for effective system selection. The handling system will often incorporate storage facilities, which also have to be carefully designed to meet a number of requirements.

14.5.2 Mechanical conveyors

Mechanical conveying techniques are the most widely used form of materials handling in the process industry. Although a large number of devices are available, most systems conform to the basic elements in which either belts, chains or moving flights are used to move material. Mechanical conveyors have distinct advantages, in terms of the ability to exercise accurate control in the monitoring of material from one process to another. Further, it is possible to design systems that meet a wide variety of operating conditions, including corrosive environments and extremes of temperature.

14.5.2.1 Belt conveyors

A belt conveyor is made up of an endless fabric or elastomer covered belt that traverses between two or more pulleys, and is supported at intermediate points by idler rollers. These conveyors can handle a wide range of materials, from fine powders to large, lumpy stone and coal. Large amounts (5,000 t/h) of material can be transported over very long distances at lower cost than most other forms of conveying systems. Versatility, reliability, and range of capacities have made belt conveyors the most widely used mechanical conveying systems in bulk solids industry. A typical belt conveyor arrangement is shown in Fig. 14.37. The belt conveyor system consists of several main components:

- Belt: The belt consists of a carcass and a cover. Numerous forms of construction are used.
- Idlers: Troughing idlers enable the belt load to be distributed and the belt volume and mass to be optimized.
- Drive unit: The belt is driven by one drum at the turnaround side of the conveyor.
- Loading and discharging: The belt is usually loaded by a gravity feeder such as a hopper, a preceding conveyor or a vibratory feeder. Discharge of the bulk material is usually by gravity dumping.
- Belt cleaners: This accessory to the belt conveyor is, probably, the most important. There are several types of belt cleaners designed to minimize the amount of carryover, increasing the life of the conveying system.

Belt conveyors can be arranged horizontally and with inclined or declined sections combined with convex and concave curves. The desired path of travel is limited only by the strength of the belt and the permissible angle of incline or decline for the particular situation.

14.5.2.2 Chain conveyors

A chain conveyor consists of an endless chain or cable, pulling a series of spaced skeleton or solid plug flights through an enclosed casing or housing. Material is introduced through an opening in the casing, where it is captured by the flights and drawn through the casing, until it reaches an opening in the housing, where it is discharged by gravity. Unique advantages of chain conveyors are: compactness in

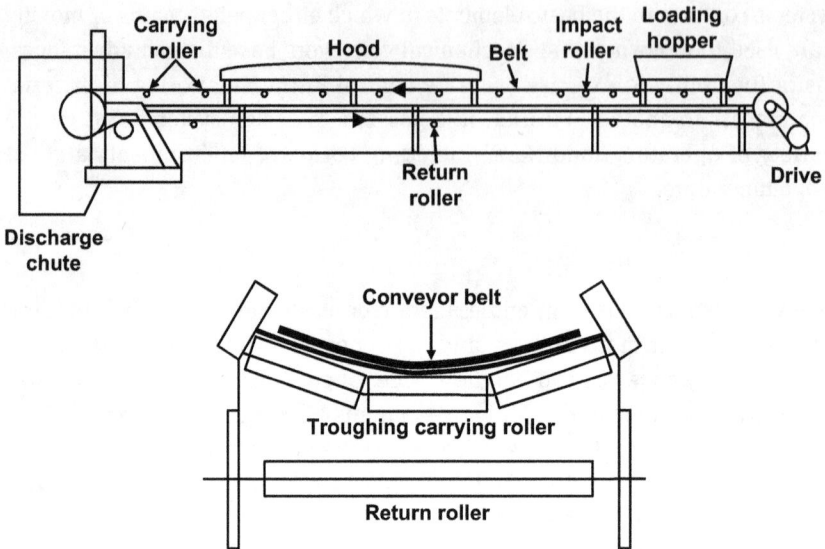

Fig. 14.37: Belt conveyor and support assembly.

cross section; total enclosing of the bulk material; combining feeding, conveying and elevating in one machine; and multiple inlet and discharge openings.

14.5.2.3 Screw conveyors

A screw conveyor consists primarily of a rotating helicord placed in a stationary tubular or U-shaped trough (Fig. 14.38). The helicord flight conveyor screw is made from a helix formed from a flat bar. This helix is then mounted on a pipe or round bar. As the screw rotates, material heaps up in front of the advancing flight and is pushed through the trough. Screw conveyors are of simple, relative low cost construction and can handle a wide variety of solid particles, ranging from lumps to powders within a completely enclosed housing. Conveying distances are limited by the torque capacity of available drive shafts. Power requirements are relatively high, and conveying efficiency is considerably reduced when screws are inclined or mounted vertically. There are a diversified number of uses for screw conveyors, such as controlled heating or cooling, mixing and blending.

14.5.2.4 Vibrating conveyors

Figure 14.39 shows that a vibrating conveyor consists of a trough supported by tuned springs and/or hinged links, having a drive system. The oscillating movement of the vibratory conveyor accelerates the particles in both the horizontal and vertical directions, causing solid particles to be moved along the trough. These

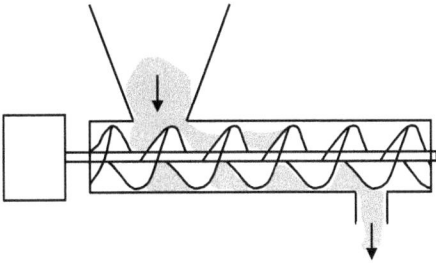

Fig. 14.38: Arrangement of a screw conveyor.   Fig. 14.39: Schematic of a vibrating conveyor.

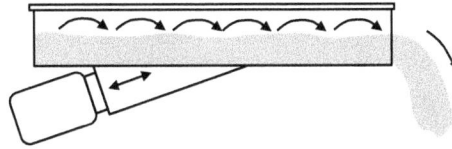

conveyors are ideally suited to the handling of granular, not free-flowing materials and other materials which are awkward to handle. One of the biggest advantages of vibratory conveyors is that they can convey products such as sugar and milk powder under conditions that almost eliminate degradation. They are uniquely suited for handling abrasive, hot and dusty materials and can be designed to withstand heavy impact loads from materials such as rocks, iron and steel castings. Equipment operates at frequencies ranging from 5 to 15 Hz; stroke range from 5 to 50 mm; and lengths up to 50 m. There are two types of oscillating conveyors: reciprocating and vibrating.

On a reciprocating conveyor, material is carried forward in a horizontal direction by frictional contact with the trough. These conveyors are useful for handling granular free-flowing materials with a minimum of attrition. On a vibrating conveyor, material is moved along the trough in a series of hops. The particles are accelerated from the trough in an upward and forward trajectory as the trough moves forward. Because of its flexibility and ability to handle a wide range of materials, the oscillating vibrating conveyor is the most commonly used type. Material must be fed onto a vibrating conveyor at a controlled rate. They are not designed for operation under a head load of solids in a storage silo or hopper. In addition to horizontal conveying, these conveyors can be used to perform other functions such as elevating, heating, drying, cooling, fluidization, agglomeration, screening and dewatering.

14.5.2.5 Bucket elevators

Bucket elevators are the most commonly used device in the continuous unloading of ships, tankers and other large storage depots. They consist of a series of buckets attached to an endless belt or chain that are filled with material and lifted vertically to a head pulley or sprocket, where the material is dumped. The buckets are then returned back down to a tail pulley or sprocket at the bottom. Bucket elevators are not self-feeding. They must be fed at a controlled rate to avoid overfilling the buckets and damaging the machinery. In the usual arrangement, the chain or belt path is vertical or steeply inclined in a single plane. There are four broad classifications of bucket elevators: centrifugal, continuous, positive and internal discharge. Centrifugal and continuous discharge elevators are, by far, the most commonly used. The centrifugal type bucket elevator is used for free-flowing or granular products. The various elevator types are shown in Fig. 14.40. Bucket elevators have a distinct advantage over other high load conveyors in that they are completely maneuverable, simple, efficient and reliable.

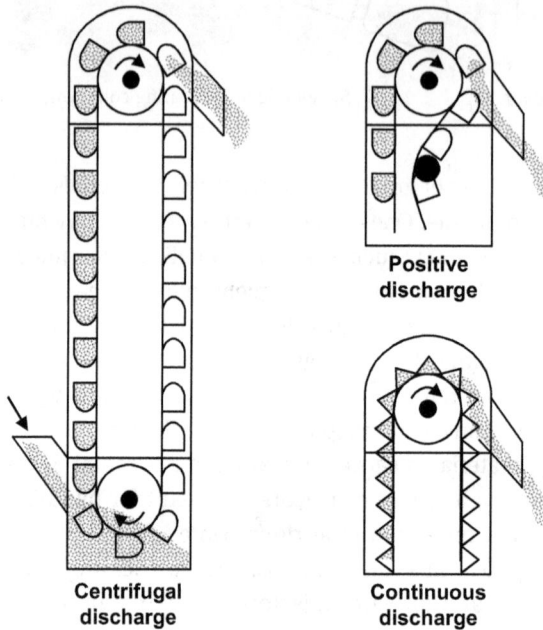

Centrifugal discharge

Positive discharge

Continuous discharge

Fig. 14.40: Arrangement of bucket elevators: (a) centrifugal,(b) continuous and (c) positive.

14.5.3 Pneumatic conveying

Man has always recognized the ability of the wind to move solid matter. Nowadays, pneumatic transport of solids represents the most widely used material handling technique in the process industry. Of all the solids moving equipment alternatives, pneumatic conveying is probably the most suitable method for the continuous transport of small-sized solids. A major advantage of pneumatic conveying is its extreme flexibility with regard to space design. As long as the conveying pipe is properly designed, the pneumatic conveyor can move the solids over, under and around buildings, large equipment and other obstructions. In comparison to other solids moving methods, pneumatic transport has a very decided advantage with regard to safety. Mechanical transport systems offer much potential for accidents and elaborate precautions must be taken for their safe operation.

In the past, the major disadvantage of pneumatic conveying has been the relatively high cost of power consumption compared to other bulk solids moving systems. However, nowadays these power costs have been reduced considerably and are counterbalanced by the steeply rising labor costs encountered with the other more manually operated solids moving methods. To date, pneumatic conveying is applicable for relatively short distances, usually less than 4,000 m. The type of material to be conveyed has a very great bearing on whether a gas-blown system can be utilized. Although pneumatic conveying can be used for many solids conveying applications, these solids must be relatively dry and somewhat free flowing. In general, solid matter that is fragile and easily crumbled is not amenable to pneumatic conveying, when breakage cannot be tolerated in the final product. Hygroscopic or agglomerating materials are not easily air transported.

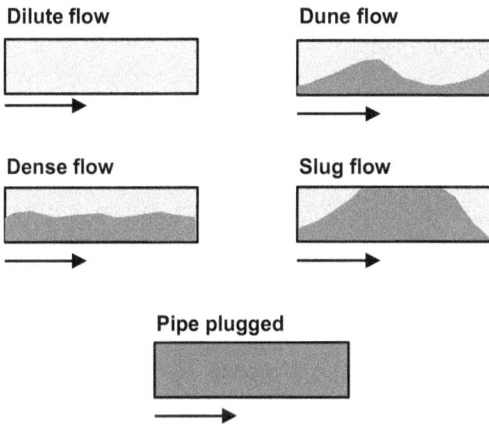

Fig. 14.41: Solid-gas flow patterns.

14.5.3.1 Classifications

The particle transport in pneumatic conveyors is commonly classified as being either dilute phase or dense phase. As illustrated in Fig. 14.41, dilute phase flow is characterized by fully suspended flow in which all particles are moved along the pipe supported by the gas flow. When the solid–gas ratio is increased, the individual particles tend to settle at the bottom of the pipe and slide over other particles. This transition between dilute and dense phase flow is known as the saltation point. Dense phase flow is thus characterized by the solids moving as a bed along the bottom of the conveying pipe. When the particle segregation reaches a certain limit, the solids move from one dune to the next. Additional increases in the solid loading can result in a slug flow, which is characterized by intermittent flow of gas and solids in alternating slugs. An even higher solid-gas ratio may cause the solids to fill up a considerable portion of the pipe, ultimately leading to complete blockage. These flow profiles are generally a function of the solids being conveyed.

Dense phase conveying is often called high-pressure, or low-velocity conveying. The term low-velocity conveying is quite descriptive because typically velocities around 10 m/s are used. Dense phase systems have been used to transport up to 50 t/h and range from about 150–300 m in length. In general, dense phase flow, which employs far less conveying air, is more energy efficient than dilute phase flow. Additional advantages over dilute phase systems are gentler handling of materials and longer conveying distances. The added pressure requirements, however, dictate the need for more expensive high-pressure equipment. As a result, dense phase systems are at a cost disadvantage when multiple feed points are required, or when incremental compressed air for conveying is not available from existing sources.

Although the current trend is towards dense phase conveying, the most popular pneumatic conveying systems still operate in the dilute phase mode. Dilute conveying systems operate as positive pressure systems at pressures up to 1 bar, or as negative pressure systems at pressures up to −0.5 bar, relative to atmospheric pressure. Typical flow air velocities are in the range of 25 m/s.

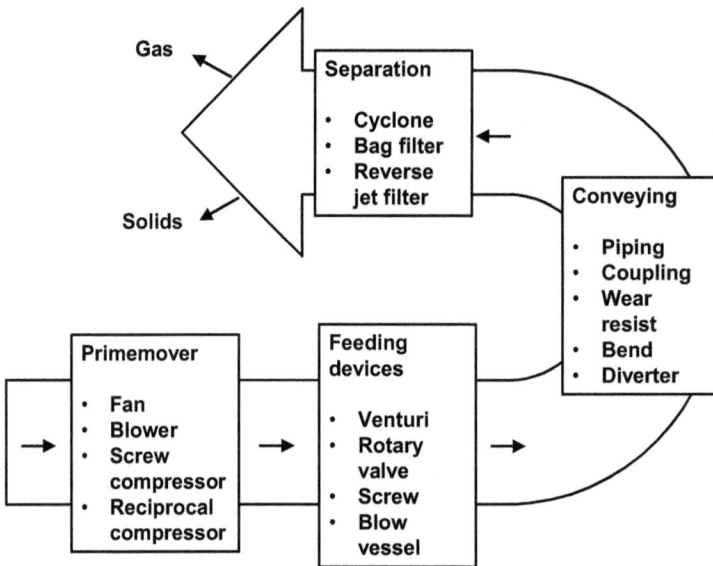

Fig. 14.42: Principle components of a pneumatic conveying system.

14.5.3.2 Pneumatic conveying systems

Certain basic components are required for a pneumatic conveying system. The solid product at atmospheric pressure must be fed into a flowing gas stream, conveyed over a prescribed route and, upon reaching its destination, separated from the conveying gas. As illustrated in Fig. 14.42, a basic pneumatic conveying system consists of a prime mover such as blowers or compressors, the solids feeder, the conveying conduit system and the solids collecting equipment. The solids feeding system is the most important item in any pneumatic conveying system. It must uniformly and continuously inject the solid particles into a positive or negative pressure gas stream. The conveyor lines must be designed such that the passage of solid–gas mixtures is relatively free of any flow-restriction regions, where the solids might settle and stop moving, causing partial or complete flow plugs and blockages. The collection device must separate the solids from the gas medium, as completely and efficiently as possible. Gas–solid separations systems vary from cyclone separators to highly sophisticated reverse jet bag filters.

There are two major classifications of pneumatic conveyors: vacuum and pressurized. The designation depends on whether the pressure within the conveyor is more or less than the atmospheric pressure. In general, the gas mover in a vacuum system is located behind the final solids collector, while in a pressurized conveyor, the air blower is located in front of the equipment feeding the solids into the gas stream. A positive pressure system facilitates the feeding of the product into a positive pressure air stream. These systems are normally used to move solids from a fixed feeding location to multiple delivery points, using a single pipeline and diverter valves to direct the flow. Positive pressure conveyor equipment is usually arranged as shown in Fig. 14.43, with the feeder discharging the solids into a pressurized moving-gas stream. The major

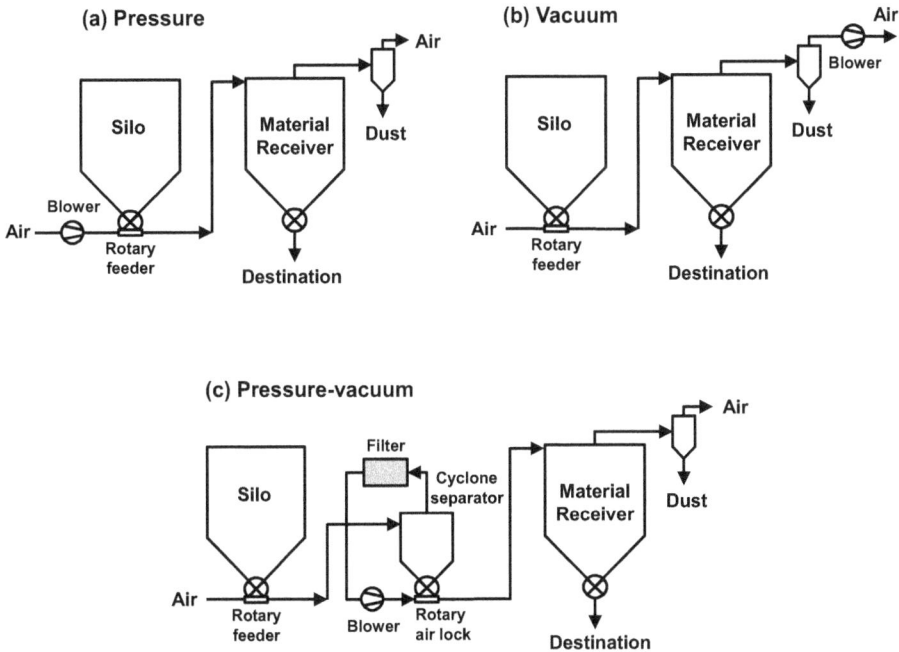

Fig. 14.43: Typical arrangements for pneumatic conveying systems.

advantage of the pressurized system is the minimum amount of equipment required at the endpoint receiver.

Negative pressure (vacuum) systems are well suited to applications where the product has to be picked up from several feed points and delivered to a single receiving point. This conveyor system, illustrated in Fig. 14.43, is composed of a gas intake and a mechanical solids feeder, a solids transport line, a gas–solids separation device, followed by the gas mover. Vacuum systems are extensively used for the movement of toxic substances, where any hole in the pipeline would cause pollution problems. However, the lines should still be as leak free as possible, since undesired air inlet flow requires a larger air blower capacity and can cause undesirable contamination of the solid product.

Nomenclature

A	Surface area for drying	$[m^2]$
H	Humidity	$[kg/(kg\ dry)]$
H_{rel}	Relative humidity	$[\%]$
ΔH_v	Heat of evaporation	$[J/kg]$
h	Heat transfer coefficient	$[W/m^2\ K]$
K	Henry's constant	$[bar/kg\ m^3]$
k_g	Mass transfer coefficient in gas phase	$[kg/s\ m^2\ Pa]$
M	Molar weight	$[kg/mol]$
p	(Partial) pressure	$[Pa]$
t	Time	$[s]$
T	Temperature	$[K]$
w	Weight	$[kg]$
x	Equilibrium dry basis moisture content	$[kg/m^3]$

15 Product technology

15.1 Cheese-coating technology

15.1.1 Cheese production

The raw material for cheese manufacture is fresh raw milk. Most production techniques are intended to selectively increase the casein, fat and dry matter contents in cheese by expelling whey. Moreover, they create appropriate conditions that permit lactose to lactate fermentation and ripen to develop in a way that is specific to each type of cheese. The starter cultures used for lactate fermentation consist of lactic acid bacteria and are introduced in amounts ranging from 0.05 to 5 wt.%. Cheese yield depends primarily on the fat and protein content of the milk. The cheese-making process begins with the coagulation of milk in a vessel. The salt and protein concentrations as well as the acidity level are very important for coagulation. Rennet enzymes, acid or combinations of enzymes and acid coagulate the casein, and the resulting casein particles form a three-dimensional network, enclosing the other milk components. As soon as the coagulum has attained the desired consistency, it is cut into cubes. In this curd-making process, the solid curds are separated from the liquid whey containing the dissolved lactose, whey protein and salts. Whey expulsion is subsequently regulated according to the final water content desired in the ripe cheese. When the curd is firm enough, it is separated from the whey by transferring to a perforated metal or plastic mold. The next stage is the further expulsion of whey by pressing and turning the cheese in the mold. After salting, the cheese is submitted to further curing and ripening procedures. The characteristics of the end product are determined by the lactic acid breakdown, proteolysis and lipolysis. The principal cheese quality criteria are odor, taste, body texture, shape appearance and, in some cases, eye formation.

Ripening takes place in special rooms where temperature, relative air humidity, air motion and treatment of the cheese are essential factors. One of the most important points during ripening and in storage of cheese is the prevention of mold and bacteria growth on the cheese. Molds attack the cheese crust, which, in turn, becomes sensitive to attack by other parasites, such as cheese mite, and the product becomes unattractive to the customer's eye. Furthermore, molds excrete components that have a negative effect on the taste of the cheese. Some of these components may even be toxic. Although many measures can be taken to minimize the growth of molds, it is very difficult, in practice, to keep the cheese completely free of mold growth. This has resulted in the use of so-called plastic coatings that are applied to the cheese when it is drying. After drying the plastic coating, the layer forms a strong barrier that prevents molds from taking nutrients from the cheese crust. Moreover, many of these coatings contain additives that prevent mold growth.

https://doi.org/10.1515/9783110712445-015

15.1.2 Coatings

Coating agents are compounds that are used to cover or coat foods that are added to food surfaces for their protection. Coatings are not packaging materials and are not food additives either because coatings are not intended for consumption. Coating agents are used to protect foods or their surfaces against undesirable changes, such as drying out and loss of aroma. In some foods, coating agents may prevent oxidation or the growth of undesirable microorganisms on the surface. Other coating agents give an attractive look to the surface of a food that would otherwise not look very appealing. Coatings are applied as liquid, paste or powder products, which are applied to surfaces in layers of given thickness, which form adherent films on the surface of the substrate. These coatings generally have a pronounced yellow, ochre, red, brown or black color, permitting the standardization of visual display of marketed products.

For most uncooked pressed cheeses, treatments aim to limit or prevent the development of wild flora spontaneously settling on the surface of the cheese. An important technology to prevent the growth of wild flora is applying, at the beginning of ripening, films of a varied nature (wax, paraffin wax, plastic film or varnish), by immersion or spraying. Polyvinyl acetate (PVA) coatings are useful for protecting high-quality hard and semihard naturally ripened cheeses against mechanical damage and mold growth. PVA is an emulsion of copolymers in water that enhances the natural ripening qualities of cheese and leads to optimum standardization in appearance and characteristics. The function of the coating on ripening and storage is to protect the cheese from mechanical, physical and microbial damage or spoilage, and to ultimately improve the final presentation and consumer appeal of the cheese. When cheese coatings are applied, it is important to achieve a uniform distribution of coating on the cheese, and the correct amount and distribution of the antimicrobial agent (natamycin) to protect against molds and yeast growth, both in the coating and on the cheese, as well as to allow the coating to dry properly before turning and subsequent treatments. As long as the film is intimately in contact with the cheese, without discontinuities caused by the presence of air or residual whey, protection is efficient.

15.1.3 Application techniques

Many techniques have been developed for the industrial application of coatings. The analysis of the criteria for choosing a coating method is a complex task. Economic factors, in relation to the performance targets, generally have top priority for choosing an application method on a commercial basis. Coating systems and processes are usually preferred that best satisfy the demands and requirements for thin coatings, high degree of material utilization, low energy costs and good automation.

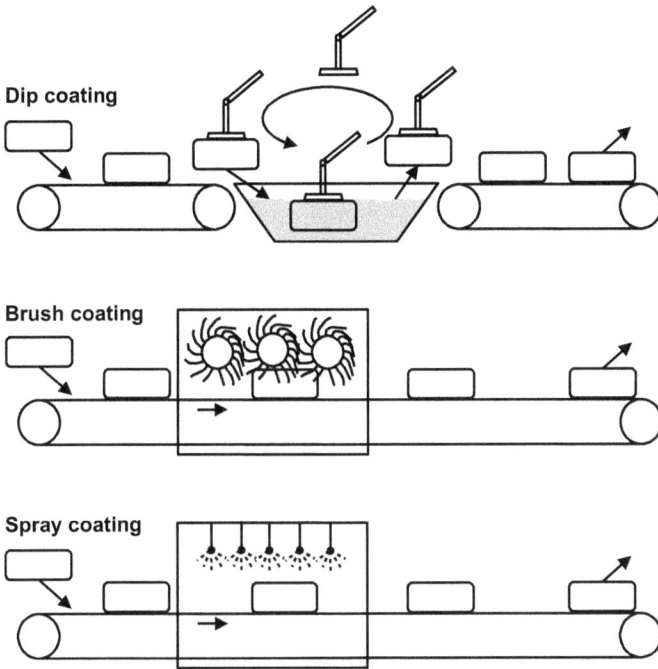

Fig. 15.1: Schematic of an automated dip (top), brush (middle) and spray (bottom) coating machine.

15.1.3.1 Dip and brush coating

Dip and brush coatings are among the oldest coating methods. In conventional dip coating (Fig. 15.1), the item is fully immersed in the coating fluid and then withdrawn. The liquid paint adheres to the surface and is then dried or heat cured. Care should be taken to ensure that the items do not float during dipping and that air bubbles are not trapped. The speed at which the item is removed from the bath must be selected so that the excess coating fluid adhering to the surface can run off. Draining and evaporation times must be sufficiently long to ensure satisfactory evaporation of the solvents (if necessary, a hot air zone should be included for waterborne paints). A benefit of all dipping processes is that the items are completely coated with only low coating losses, and the processes are very easily automated. However, at the same time, variations in coating thickness and tearing cannot always be avoided, so requirements with respect to surface quality should not be too tight.

Application of a coating with a brush is commonly used for small parts and for retouching sections of surfaces. Advantages of brushing (Fig. 15.1) are the good wetting and covering of surface defects, low coating losses and flexibility towards shape. Disadvantages and limitations include the labor-intensiveness, criticality to flow and nonuniformity of the applied coating layer. The material and shape of the brush used should be selected according to the coating material and the object. For example, water-borne coatings and solvent-based coatings, and even high- and low-viscosity materials require different brushes.

15.1.3.2 Spraying (atomization)

In spraying of either liquids or powders, two fundamentally different methods are used: conventional spraying and electrostatic spraying. In conventional spraying (Fig. 15.1), compressed air flows through an annular gap in the head of a spray gun. The gap is formed between a bore in the air cap and the concentric paint nozzle. Further, air jets from the air-cap bores regulate the jet shape and assist atomization. The expanding compressed air leaves the paint nozzle at a high velocity. A low-pressure area is formed in the nozzle aperture that exerts a suction effect and assists in the outflow of the coating fluid. The difference between the velocities of the compressed air and the exiting paint atomizes the paint into particles that are conveyed as spherical droplets in the free jet. The size of the droplets produced depends on the pressure applied in the process. At relatively high pressures, in the range of 2–7 bar, the droplets can be as small as 10 μm in diameter, whereas a lower pressure results in droplet sizes of 20–300 μm.

15.1.3.3 Drying

Although the vast majority of coatings are applied in the fluid state, coatings are only functional when dry. Thus, the drying process is as important as the coating process, as it can either enhance or deteriorate the properties of the coating. The purpose of the drying process is to produce a uniform dry coating from the applied wet coating, by evaporating the inert solvent from the functional coating ingredients (polymer, binder, pigments, dyes, hardener, coating aids and so forth), so that only the desired coating remains on the substrate. Solvents used in the coating process can range from water to volatile organic materials. Usually, thermal drying is used, in which, heat supplied to the fluid coating evaporates the solvent. If the coating is applied uniformly, the dryer must immobilize it and maintain the coating's uniformity throughout the drying process. However, some coatings are applied with nonuniformities and level out during the drying process, as in brush and spray coating. Solvent removal must be done without adversely affecting the coating formulation or interfering with the desired uniformity of the coating.

15.1.3.4 Film formation

Even though most emulsion paints contain pigments, fillers and additives in addition to the polymer dispersion, only the role of the polymer dispersion in describing the film-forming process, schematically illustrated in Fig. 15.2, will be considered. When the water evaporates from the emulsion paint after application, the polymer particles move closer together. As soon as they touch one another, the water starts to

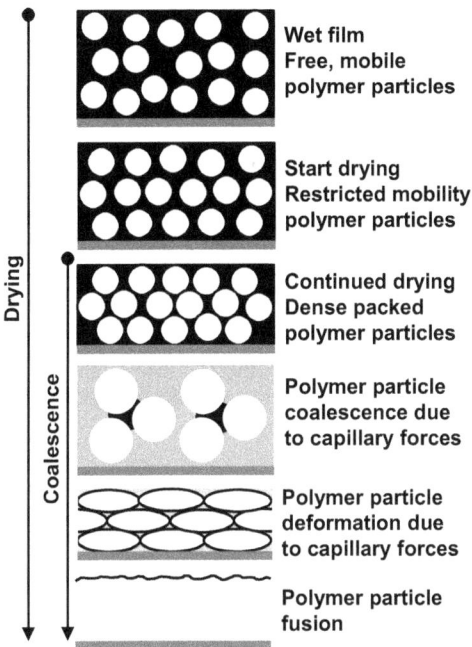

Fig. 15.2: Model representation of film forming in a polymer dispersion.

evaporate from the space between them and capillary forces start to develop. The polymer particles are deformed by these capillary forces and congregate even more closely together. Upon further drying, increasingly large forces thus develop in the capillaries, which likewise become narrower. Under the resulting pressure, the liquid is expelled and the polymer particles ultimately fuse.

15.2 Enzyme formulation

15.2.1 Introduction

Industrial applications of enzymes form an important branch of biotechnology. Enzymes are biologically active catalysts, which can induce very specific reactions under gentle conditions. They are used in the detergent industry and important industrial processes, such as the conversion of starch to glucose syrups (α-amylase and amyloglucosidase) and the production of fructose syrups from glucose (glucose isomerase). When an enzyme is recovered from a fermentation broth, it is usually present in an aqueous solution or processed to a dried state. Both types of products have to be formulated to comply with requirements appropriate to their final application. Requirements related to the storage of enzymes until the time of application include

enzyme, microbial and physical stability. Some applications demand special requirements, such as having no precipitate, off-odor/taste and off-color.

Any formulation is a compromise between the mentioned requirements. For example, the pH necessary for good microbial or physical stability may differ from the
pH that gives the optimum enzyme stability. Product formulation is an important subject in the production of an enzyme containing biotechnological products. It is the
final and probably most essential step between production and application. There is a
variety of formulation objectives, such as dissolving, suspending, reuse, stabilization, conservation, dust prevention and protection against an aggressive environment.
As is the case for the encapsulation of detergent enzymes, the formulation of an enzyme is normally considered a way to store and transport the enzyme until its application. One common exception is enzyme immobilization, where the formulation is an
active part of the application.

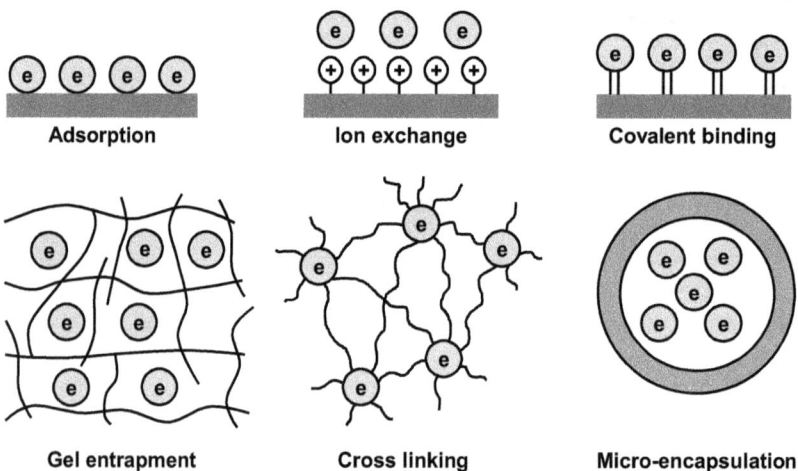

Fig. 15.3: Schematic representation of several techniques for immobilizing enzymes.

15.2.2 Glucose isomerase immobilization

Enzymes have been of great value to the starch industry since the 1960s, when the enzyme amyloglucosidase made it possible to completely breakdown starch into glucose. A few years later, most glucose production switched to the enzymatic process,
which gave better yields, higher purity and easier crystallization. Nowadays, glucose
isomerase is the most successful immobilized product used to convert glucose to high
fructose corn syrup (HFCS) by the isomerization of glucose to approximately 42%
fructose, which is fractionated chromatographically to 55% fructose and 45% glucose.
The breakthrough for immobilized glucose isomerase for the production of HFCS was
initiated by a number of factors. The intracellular glucose isomerase is expensive and

cannot be applied economically without re-use. A solution to this is the immobilization of the enzyme on an insoluble carrier that can be considered as a particular form of formulation. It allows continuous processing in packed bed reactors, enabling control of substrate-to-product conversion by adjusting the flow rate. Moreover, particles with physical properties good enough to survive in large enzyme columns were needed to handle the huge volumes generated by the starch industry, and no enzyme is left in the solution, preventing the contamination and toxicity problems. The enzyme particles are packed in large columns with packed bed heights above 4 m and column diameters of up to 2 m. The substrate solution contains approximately 50% w/w carbohydrates and is isomerized at around 60 °C to prevent infection of the syrup.

Numerous enzyme immobilization techniques have been described in literature. The choice of a suitable immobilization method for a given enzyme and application is based on a number of considerations, including previous experience, new experiments, enzyme cost and productivity, process demands, chemical and physical stability of the support, approval and safety issues regarding support and chemicals used. The main immobilization principles are shown in Fig. 15.3. Enzyme characteristics that greatly influence the approach include size, surface properties, lysine content and polarity. In immobilization by adsorption or via electrostatic interaction, the enzyme is adsorbed and interacts electrostatically with a large surface area of the carrier material, such as alumina, clay, carbon, ion exchange resins, cellulose or glass. In immobilization by covalent attachment, the functional groups of a number of amino acids are attached to the surface of a chemically modified support material. Those functional groups of the enzymes are used that can interact with other molecules, such as amino and carboxyl groups. Synthetic polymers, cellulose, agarose and porous glass are frequently used as support materials and chemical agents are often used to link the enzyme to the support. In immobilization by occlusion or entrapment, the enzyme molecule is captured in a polymer matrix. Polyacrylamide, starch, gelatin, carrageenan and alginate have been used for the polymer matrix. To stabilize the gels, cross-linking with bi functional or polyfunctional agents may be necessary. Immobilization by direct cross-linking is an alternative. In immobilization by micro encapsulation, the enzyme molecules are enclosed in polymeric capsules of 10–250 µm in diameter.

The process depicted in Fig. 15.4 uses a whole cell immobilization technique. This is possible because the substrate to be converted (glucose) can easily diffuse into the destroyed cells. The immobilization principle is based on gel inclusion, followed by cross-linking. The enzyme-containing cells are mixed with gelatin at a temperature of about 40 °C. The final gelatin concentration is about 8% (w/w). The mixture is then prilled in a column containing a cold, water immiscible organic solvent, like butyl acetate. While falling through the column, the enzyme-containing droplets will solidify. The spherical particles are collected at the bottom of the column, dehydrated, cross-linked with glutaraldehyde and washed. Finally, the product is sieved and preserved in propylene glycol. The same process can be used for immobilizing living cells. The used gel is permeable for substrate and formed products.

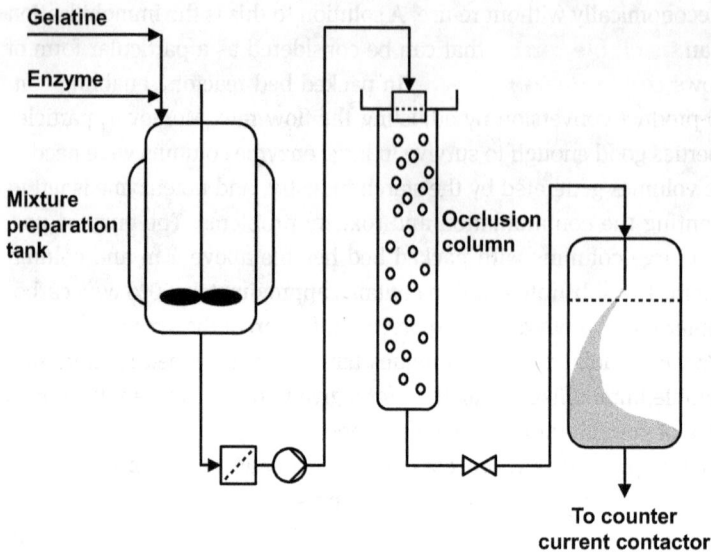

Fig. 15.4: Flow sheet of a production unit of immobilized enzymes.

15.2.3 Detergent enzymes

Nowadays, the detergent industry is the largest user of industrial enzymes. Penetration of enzymes in laundry detergent products is very high. Since the introduction of compact powder detergents, the use of enzymes as detergent ingredients has grown considerably. The trend toward more compact products has changed enzymes from being minor additives to being key ingredients, in line with surfactants and bleach systems. Several changes in washing habits and detergent formulations underlie this development. Lower washing temperatures, required by modern textiles made from synthetic fibers and for reason of energy conservation, have caused a demand for more effective detergent ingredients.

Stains with good water solubility are easily removed during the washing process. All other stains are partially removed by the surfactant/builder system of a detergent, although the result is often unsatisfactory. In most cases, a suitable detergent enzyme may help. The catalytic nature of enzymes makes them highly efficient as detergent components. Contrary to the purely physical action of the surfactant system, enzymes work on degrading the dirt, which are often fats or proteins, into smaller and more soluble fragments. Typically, only 0.5–2% of the granulated enzyme product is required in a compact powder detergent formulation. The actual enzyme content is in fact much lower because enzyme granulates generally contain less than 10% enzyme protein; the rest being inert encapsulation material. The resulting concentration of enzyme protein in the washing liquor will be in the range of, at most, a few milligrams per liter. Currently, four different classes of enzymes are used in detergents:

- Proteases are enzymes that are able to hydrolyze peptide bonds in proteins and, therefore, remove proteinaceous stains. Most detergent proteases are stable during the wash cycle in the presence of such active-bleach systems.
- Amylases catalyze the degradation of starch stains and improve cleaning by hydrolyzing the starch glue that binds other dirt and stains to the fabric. They are fully compatible with detergent proteases and work together in the wash process. During storage in powder detergents, the amylases are very stable in the presence of proteases.
- Lipases improve the removal of fats/oils of animal and vegetable origin.
- Cellulases remove the cellulose fibrils from the surface of cotton yarn to improve the removal of particulate soil.

Not all enzymes with a potential for stain degradation and/or removal are suitable for inclusion in detergent products. A detergent enzyme must have good activity at the pH of detergent solutions (between 7 and 11) and at the relevant wash temperatures (20–60 °C), and must be compatible with detergent components during storage as well as during the wash process. In particular, such an enzyme must be resistant toward protease degradation under these conditions. With enzymes like proteases and lipases, a broad substrate specificity is demanded.

In the early days, the enzyme was added to the detergent as a dry powder. However, various allergenic and skin reactions seemed to appear with people who worked in the production of the proteolytic enzyme. This problem was overcome by a proper formulation, which would encapsulate the proteolytic enzyme in order to prevent the release of enzyme dust. Since then, the washing process has changed considerably, in particular, the washing temperature. To an increasing extent, the washing process occurs at lower temperatures. In this situation, however, the bleach components are less effective. This has led to the introduction of increasingly more bleachers and bleach activators (such as TAED, tetraacetylethylenediamine) to facilitate bleaching at low temperature. As a result, the formulation had to offer protection against this type of aggressive components in the washing powder. At the same time, other market demands were also made with respect to the properties of the particles. These were, dissolve quickly, be of uniform particle size with no release of odor and a color contrast with the detergent. Moreover, the formulation had to be adapted in such a way that the product would have a long shelf life in a high relative humidity environment. Furthermore, a form of controlled release is desired to prevent the release of the total enzyme activity at once. Additionally, the different enzyme activities should not be released at the same time, because they are not compatible with each other. Therefore, multilayer formulations have been developed that bring the different enzymes in solution.

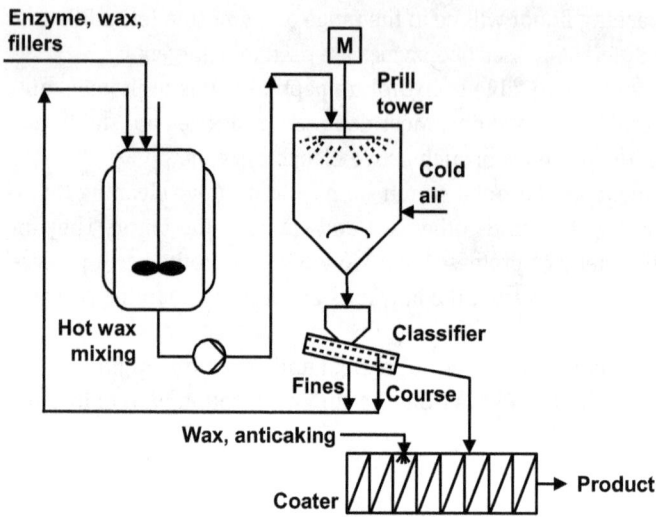

Fig. 15.5: Process for formulating dust-free detergent enzymes.

Today, microorganisms grown in fermentors with volumes ranging up to several hundreds of cubic meters produce most industrial enzymes. After fermentation, the enzyme is recovered from the fermentation broth. Often, the first step is the removal of whole cells and other particulate matter by centrifugation or filtration. Most microorganisms that are used to produce industrial enzymes release the enzymes from the cell into the growth medium, which facilitates the process. Ultrafiltration and diafiltration are increasingly used to remove water, salts and other low molecular weight products.

After fermentation and recovery, the purified enzyme must be formulated into a dry product before it can be introduced into powder detergents. The challenge faced by enzyme producers is thus to provide a mechanically robust and rapidly soluble dry product. A common way is to precipitate the enzyme from the cell-free fermentation broth and collect the precipitated enzyme by filtration. The wet particles are then dried, sieved and the pure enzyme stored as powder. As presented in Fig. 15.5, the first step in the enzyme formulation process is mixing the enzyme powder with several additives in a nonionic melt (ethoxylated-C_{18}-fatty alcohol). The mixture is fed to a prill tower where the droplets quickly solidify to give roughly spherical particles. The product obtained consists of matrix particles, i.e., the enzyme is uniformly distributed throughout the inert material. The coarse and fine fractions are re-melted, while the proper fraction (250–600 µm) is coated to prevent dust formation during handling.

In layered granulates, inert carrier particles are used as a starting material. On the surface of these cores a layer of enzyme is applied, and on top of that, additional layers can be applied such as a coating layer to improve the dust properties. Today, various kinds of fluidized beds are used in the production. On the basis of this fluid

bed coating technology, Genencor International has developed a completely different technique. The starting point is a carrier material such as sugar or salt crystals that
are sprayed with the enzyme solution in the fluid state, while the solvent (water) is
evaporated. When sufficient enzyme activity has been applied, a coating is sprayed.
The advantage of this technique is that various layers with different enzymes can be
applied. A disadvantage is that high demands are posed on the particle size of the carrier material and the correct dosing of the enzyme activity on the particles. A typical
particle obtained from the fluid bed process is depicted in Fig. 15.6.

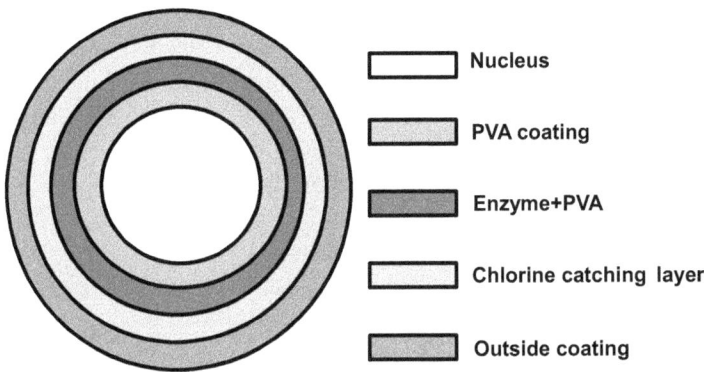

Nucleus

PVA coating

Enzyme+PVA

Chlorine catching layer

Outside coating

Fig. 15.6: Enzyme containing granule with various coatings.

The different layers are clearly indicated. The nucleus consists of the sugar agglomerate. A layer of polyvinyl alcohol (PVA) is attached to strengthen the particle. On
this PVA layer, the enzyme is sprayed. The enzyme solution also contains PVA to
improve the adhesion on the particle. Subsequently, an ammonium sulfate layer
has been applied to catch the chlorine present in the tap water during the washing
process, because chlorine would directly deactivate the enzyme at the beginning of
the washing process. The last layer consists of titanium dioxide or a pigment to give
the particle its final color. Again, PVA is added to this layer to minimize dust formation during handling.

15.2.4 Application research

In addition to formulation, application research is of utmost importance in the development and marketing of bioactive preparations. In application research, it must
be verified whether the product meets the requirements in specific cases of application, and the formulation will have to be adjusted in the light of the demands of specific customers.

After the immobilization of glucose isomerase, a lot of additional questions have to be answered before application of the immobilized enzyme. It is important to know the activity and lifetime of the immobilized enzyme as a function of the process conditions and its properties. This would include average particle size (distribution), pH, temperature and substrate concentration. Because of its application in packed bed columns, the product must have sufficient mechanical strength to withstand the force induced at high linear flow velocities. Furthermore, the immobilized enzyme must maintain its physical strength at temperatures of 60–65 °C. The rate of the equilibrium reaction between glucose and fructose can be described by Michaelis Menten kinetics. Because the rate constant is temperature-dependent, the reaction will go faster at higher temperatures. However, simultaneously, the deactivation rate of the enzyme will increase at higher temperatures. When both processes are combined, the total productivity plot versus time in Fig. 15.7 is obtained. It can be seen that although the initial production rate is higher at higher temperatures, the total productivity is higher at lower temperatures because of less rapid enzyme deactivation.

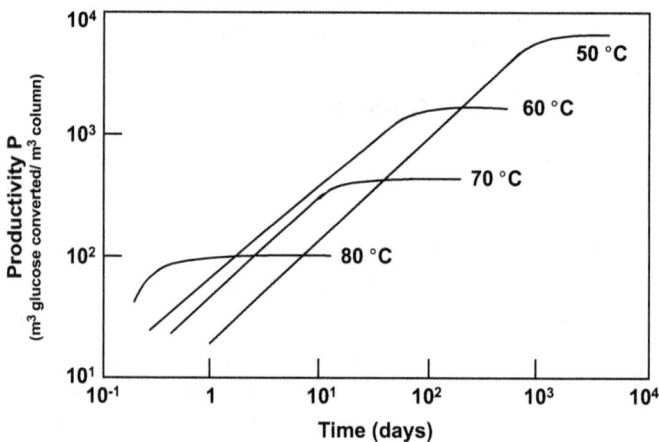

Fig. 15.7: Productivity of a commercial glucose isomerase as a function of operating time for various temperatures.

For detergents, it should be self-evident that the main part of application research with detergent enzymes is aimed at the contribution of the enzyme to the washing activity. Any laundering process is an interplay between the equipment used, the materials entering the washing process and the procedure followed. Thus, elaborate washing experiments are necessary. A given enzyme may be assayed by its action on soluble substrates under chemical and physical conditions different from those encountered in a real-life wash. Such experiments indicate the enzyme's performance with respect to pH and temperature variations, or in conjunction with other soluble substances, etc. Wash trials are carried out by the use of solid test pieces. Laboratory wash trials are usually conducted in small-scale models of washing machines. The degree of stain removal is determined by the percentage of remission of light that is reflected from a dirty patch. Typically, the intensity of the light remitted at 460 nm when illuminating the test pieces with a standardized daylight source is expressed as a percentage, R, of the intensity of the incident light at the same wavelength. The ΔR value is then a measure of the enzyme effect. It is defined as the difference in R between the fabric washed with and without the enzyme. The remission value is known to correlate well with the visual impression of the whiteness of the fabric. Figure 15.8 records the results of a protease performance test as a function of enzyme dosage. Detergent enzyme performance is often reported in the form of such dose–response curve. The performance increases dramatically with enzyme concentration at the beginning, but reaches a maximum level at higher enzyme concentrations. Also, many washing tests are carried out with "real" laundry to examine the effect of the enzyme.

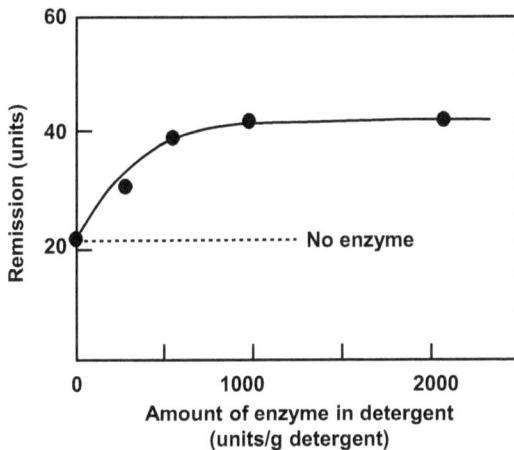

Fig. 15.8: Washing test for proteases.

15.3 Compounding

15.3.1 Introduction

Before a polymer can be used to make a product, it is usually mixed with various ingredients, which serve a variety of purposes. The mixing processes also provide an opportunity to alter the physical form of the polymer so that it is readily handled at the final conversion stage of its processing. There are two important reasons for compounding. The first is that additives are sometimes needed to alter the properties of the material, e.g., by making it harder or more flexible or cheaper. The second is that it is often important to prevent the degradation of the polymer in service or during processing by means of appropriate additives.

Most of the polymers made by the larger polymer producers leave their plants as grains or powders, which can be further processed by other manufacturers. From the original polymer, a variety of products with different properties can be designed by adding several additives. For example, dyes can be added or ultraviolet stabilizers. Because of the nature of polymers, special methods are needed to achieve a good dispersion of the additives in the polymer matrix. This chapter deals with several of these methods. For a better understanding of the processes, thermal properties of the polymers are needed. Most important is the temperature dependence of the modulus of elasticity for which a typical dependence is given in Fig. 15.9. In this figure, one can identify the glass temperature (t_g), melt temperature (t_m) or trajectory and the rubber plateau in between. The glass temperature represents the transition from a crisp state to the rubber stage, where the polymer is more flexible. The polymer remains flexible over a range of temperatures, represented by the rubber plateau. After further increasing the temperature, the polymer will melt, as can be seen from the figure. Polymers can only be processed in the rubber state or when they are molten.

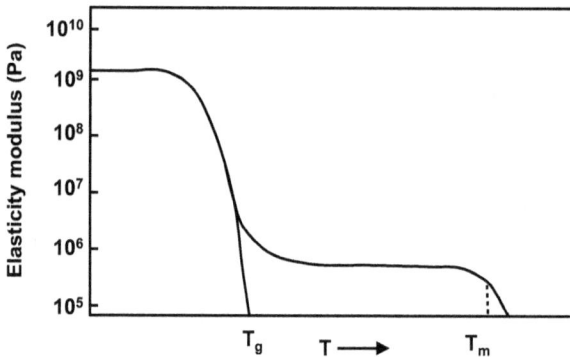

Fig. 15.9: Modulus of elasticity of a polymer as a function of temperature.

15.3.2 Compound formulation

Since the desired properties for materials vary over a broad range, it is likely that a pure polymer does not meet these demands. The polymer may not be strong enough, have the wrong color or might be too expensive to be used as a pure substance. Fortunately, the properties of polymers can be adjusted to almost every need by adding other polymers, plasticizers, dyes, fillers, antioxidants and so on. These additives must be incorporated in the polymer before final shaping takes place; this step in the production is called compounding. The typical equipment used in compounding is discussed in the next section. This section deals with the formulation of the mixture supplied to the compounding process.

Probably, the best known example of formulation is the addition of soot to rubber. This is done to reduce the wear of the polymer. Without the use of soot as a filler, a car tire has to be replaced after 10,000 km, while, with the addition of soot, this can be up to ten times as long. This is mainly due to the very high surface area of the soot, which efficiently levels out the tension on the polymer. In general, the addition of particles, like soot, to a resin improves the stiffness and not the tensile strength, as is the case with soot in rubber. The price of the raw material can also be a reason to formulate a cheaper mixture of polymers that approaches the properties of the (more expensive) single polymer. It is also possible to improve the properties of a cheap resin by adding a polymer with better properties. If the polymers are miscible on a molecular scale, the properties of the mixture can be gradually varied between the properties of both polymers. In this way, a broad range of materials can be designed. Molecular miscibility of polymers is more an exception than a rule; dispersions are more common. Dispersions consist of a continuous phase with a second phase (finely) divided in the first phase. Ideally, the second phase consists of fine droplets in the continuous phase but often the droplets are not completely spherical. Properties of dispersions also vary between the properties of the pure polymers, but the relation is not as straightforward as is the case with miscible polymers. For example, the modulus of elasticity exhibits different relations for real mixtures and dispersions (Fig. 15.10A and B). As shown in Fig. 15.10C and D, it is also possible to achieve different relations if the polymers are partially miscible.

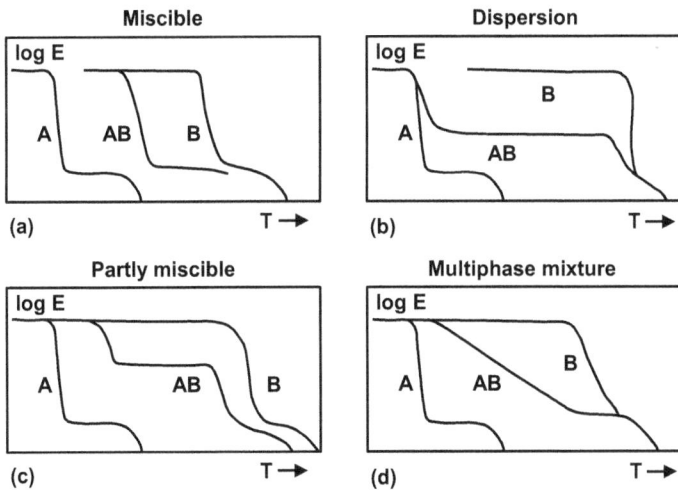

Fig. 15.10: Modulus of elasticity for various polymer mixtures.

Adding fibers to the polymer which increase the tensile strength can also influence the modulus of elasticity, in most cases. The fibers added can be either short or long and oriented or disoriented. The fiber length should be long enough to be able to take the pressure from the polymer matrix. Disoriented fibers result in a toughening of the polymer in all directions. When the fibers are oriented in one direction, this is the only direction in which reasonable toughening takes place.

15.3.3 Additives

Different additives can be used to improve the properties of polymers. Three main groups can be distinguished. Additives that:
1. protect plastics against degradation and aging, like antioxidants, UV stabilizers and heat stabilizers;
2. facilitate or control processing, such as lubricants, mold release and blowing agents;
3. impart new, desirable qualities to plastics, like pigments, dyes and flame-retardants.

In addition to the chemical and physical properties, economics also plays a large role in selecting the optimal additive for a specific polymer. The most important additives are discussed in this section.

15.3.3.1 Protective additives

Although a very large number of additives are used in this category, the most important groups are antioxidants, heat stabilizers and UV stabilizers. Antioxidants are used to protect the polymer structure against atmospheric oxidation and minimize the associated damage (e.g., discoloration, reduction in gloss, cracking and embrittlement) during processing and during the service life of the final product. Many polymers have sites on their molecular chains that are susceptible to attack by oxygen, often catalyzed by catalyst residues and contaminants. Exposing the polymer to high temperatures and severe shear during production and processing accelerates the oxidation reactions. It is the function of the antioxidant additives to combat oxidative attack by interfering chemically with the series of reactions that lead to chain scission. Most antioxidants are oxidized themselves and consumed in performing their function, so the oxidation behavior of the additive in a given polymer is crucial for its effectiveness. Important antioxidants include sterically hindered phenols, secondary aromatic amines, sterically hindered amines, phosphites, and metal salts.

Heat stabilizers perform a similar function in preventing degradation at high processing temperatures. Their function is to stop the other types of degradation reactions, for example the tendency of some polymers to depolymerize. Combinations of stabilizers and costabilizers are normally used to obtain optimal effects. These additives are particularly important in the processing of polyvinyl chloride (PVC), which readily degrades and darkens when heated under the evolution of hydrogen chloride. Heat stabilizers for PVC are divided into metal-containing and metal-free costabilizers such as phosphates, epoxy compounds and polyols. Typical stabilizers consist of heavy metal compounds, containing lead, organotin, cadmium or zinc. The choice of a stabilizer system depends on the applied processing methods, the intended use and its compatibility with other additives.

UV stabilizers are mainly employed to extend the service life of plastics for outdoor service, such as garden furniture, stadium seating, window frames, floor coverings, films for protective coverage and agriculture, light fixtures and illuminated signs and automotive parts such as bumpers and body paints. UV stabilizers often work in conjunction with antioxidants because the attack at the reactive site on the polymer by UV radiation in sunlight is one of the initiating reactions that starts oxidative degradation. Many plastics, therefore, suffer from yellowing, surface cracking, embrittlement, reduction of gloss or chalking after a short time in outdoors service and, ultimately, disintegrate. UV stabilizers can be incorporated in the bulk plastic or applied as a coating. They work by absorbing UV. preferentially, and re-emitting the energy harmlessly at a lower wavelength. Ideally, a UV stabilizer should not be consumed but operate in a closed cycle so that it remains in its active form even after a long period of weathering or use. Typical concentrations used range from 0.05% to 0.5%, reaching 3% in some critical applications.

15.3.3.2 Processing aids

Plastic surfaces can become electrostatically charged by friction during processing or service. Such charges can make processing difficult and their discharge can be unpleasant. Discharges can even create a potentially dangerous spark in some cases. Surface-active antistatic agents are used to prevent the build-up of undesirable static charges by creating a conductive surface layer that allows the charge to dissipate quickly and prevent it from building up. In some cases, other additives must be used to impart volume conductivity throughout the bulk of the plastic. Surface-active antistatic agents are applied by solution spraying or incorporated into the plastic mass. They are typically used at levels of 0.1–2.5%.

Processing lubricants and mold-release agents are widely used to assist the passage of a material through the processing machinery. They are often oils and waxes, classified into internal and external lubricants. Internal lubricants lubricate the polymer granules during processing and allow for easier and cooler melting. These materials are often at least partially miscible with the polymer melt. External lubricants are essentially immiscible. They lubricate the mix against friction with the processing machinery, allowing the correct degree of friction for the process to work while preventing too much friction that will cause local high temperatures and degradation. Lubricants also improve the dispersion of pigments and fillers in plastics resulting in more uniform colors and improved material properties.

15.3.3.3 Property modifiers

Property modifiers, as their name suggests, alter the physical properties of the polymer. There are many types and several examples, including reinforcing and nonreinforcing fillers, chemical additives, other polymers, flame retardants and colorants. Nearly all thermosetting and thermoplastic resins are filled or reinforced for special applications. Filled (or reinforced) plastics contain large amounts (20–50%) of fillers, divided into extenders, which are added to cheapen the mix, and reinforcements, which improve several of the polymer properties. Calcium carbonate and china clay are quantitatively the most important extenders. Other fillers include ground dolomite, calcium sulfate (gypsum), barium sulfate, wood flour, cork flour, starch, metal oxides and metal powders. Coupling agents improve the adhesion between filler and polymer, preferably via chemical bonds. Their use confers reinforcing properties on inexpensive extenders, improves the performance of reinforcements and allows the filler content to be increased. Important examples of reinforcing fillers are carbon black added to rubbers and fine particle silica. Carbon black is obtained by the incomplete combustion of gaseous and liquid hydrocarbons. Silica is used as a natural crystalline product (sand, quartz powder) or amorphous synthetic product. The reinforcing effect of fillers depends on their chemistries, shapes (fibers, flakes, spheres) and sizes (fiber length, particle size). A wide variety of inorganic and organic fibrous fillers and reinforcements are used. Important examples are short (3–6 mm) carbon and glass fibers. Glass fibers are milled to short fibers (e.g., 0.2–0.6 mm) when the reinforcement is blended with the polymer in the extruder.

A second important example of property modification is the addition of small amounts of another polymer as impact modifier. Because several plastics such as PVC, polyolefins or polystyrene are brittle, impact modifiers are used to improve the impact strength, especially at low temperatures. Impact modifiers are elastomeric copolymers with low glass transition temperatures. They are dispersed as discrete soft phases in the thermoplastic polymer. Transfer of impact energy to the elastomer phases requires not only a good distribution but also adhesion at the interface between the two phases by chemical bonding or physical cross-linking. Important examples are methyl methacrylate–butadiene–styrene copolymers, graft copolymers of acrylonitrile and styrene on polybutadiene (ABS), ethylene–propylene–diene copolymers and styrene–butadiene rubber.

The application of combustible plastics in buildings, equipment and vehicles requires the incorporation of flame retardants to reduce the fire risk. In plastics, flame retardants reduce ignitability, combustion rate and heat release. Although flame retardants can be especially effective in the early stages of a fire and can reduce flammability, plastics that are modified with these additives are still combustible. Because flame retardants generally lower the quality and increase the price of a plastic, they are only used when mandated. Selection criteria are the effectiveness and price of the flame retardant, processability of the plastic and influence on the polymer properties. Flame retardants may be inorganic substances, halogenated organic compounds, organophosphorus compounds or other organic substances.

Colorants (dye, pigment) are added for the decorative character of the polymer. They are chosen from a wide range of organic and inorganic compounds. Pigments are particulate organic and inorganic solids that are virtually insoluble in the polymer and must, therefore, be dispersed in the matrix. Many color formulations are derived from a combination of inorganic and organic pigments whose properties complement one another. Dyes are organic compounds that are usually dissolved in the substrate. Polymer-soluble dyes are usually suitable for transparent plastics but not for polyolefins. Incompatibility is manifested by "crocking"–the migration of colorant to the free surface of the plastic from which it can be rubbed off. Colorants are introduced into plastics by several methods. The dry colorant may be blended with the polymer before processing (dry blending) or through the addition of concentrated predispersed colorant pellets (masterbatches). Masterbatches offer numerous advantages over powdered colorants, particularly improved dispersion and dissolution of the colorants and more hygienic working conditions.

15.3.4 Polymer mixing mechanisms

Most polymers produced undergo processing that involves mixing in one way or other, for example for the addition of stabilizers or when another resin is added to achieve the desired properties. The best (or most constant) properties are usually achieved when the dispersion is as homogeneous as possible. As mixing is such an important feature in polymer processing, a general idea of the mixing process is essential. In general, the purpose of mixing polymers will be to achieve a sufficiently stable dispersion of two nonmixable resins. As is always the case with nonmixable liquids, dispersion will consist of a continuous phase and a phase where droplets are dispersed in the continuous matrix. Mixing, in this is case, is essentially the deformation and breaking up of the droplets. The two main driving forces are shear stress on the dispersed phase and the interfacial tension of the dispersed phase with the continuous phase. Depending on the ratio of these two forces, two different mixing processes can be observed: distributive and dispersive mixing. The first process mainly involves deformation of the droplets and the second mainly involves size reduction of the dispersed phase. The size of the droplets in, for example, an extruder is reduced from several millimeters at the feed to the micrometer range at the die.

15.3.4.1 Distributive mixing

If the shear stress on the droplets of the disperse phase is very large when compared to the surface tension of the droplets, deformation of the droplets takes place (almost) exclusively. During this stage, droplets are deformed due to the shear stress applied. Since the shear stress in an extruder varies with the distance from the wall, the droplets are deformed to an ellipsoid. Ultimately, the droplets will be deformed into a long strand of the polymer (which is actually a very long ellipsoid); of course, of the same total volume as the original droplet. If the "droplet" is perpendicular to the direction of deformation, the effect of deformation will be as large as possible. During deformation, the shape of the droplet will change to a strand, which is oriented in the direction of the shear stress. Therefore, in most applications, the droplets (and matrix) are folded and reoriented after the first stretching step. Apart from improving the effectiveness of the shear stress applied, it also improves the speed of deformation, as can be seen from Fig. 15.11. Instead of a linear increase in length, the increase becomes exponential, speeding up the process.

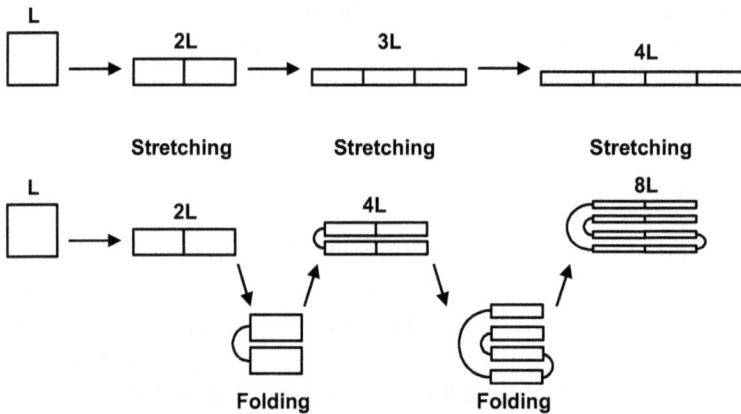

Fig. 15.11: Deformation with and without folding.

In practical situations, the equipment used is designed to perform this stretching and folding in one way or other. The Sulzer SMX is an often used mixer because of its good mixing properties and low pressure drop.

15.3.4.2 Dispersive mixing

The diameter of the strands formed during distributive mixing will decrease with time as long as the shear stress is applied. In this process, time and the level of shear stress are interchangeable. A long time with a low shear stress will have the same affect as a high shear stress applied for a short time. Ultimately, the strand will reach a diameter at which it will break up into several droplets. The formation of these droplets is not completely understood yet, but several mechanisms were found empirically. Of course, the diameter of the strand will not be completely homogeneous; some parts will be larger, some smaller. It has been shown that if the wavelength at which these inhomogeneities appear is larger that the circumference of the original strand, the amplitude will grow. If the wavelength is smaller, the amplitude will eventually approach zero. If the amplitude reaches the value of the original diameter of the strand, the strand will break and the droplets are formed. In this description, the strand is assumed to be at rest and that there is no flow around it. If there is a flow around or along the strand, the process remains almost the same. The only difference is that the growth of the amplitude will be suppressed by the flow. Because of this, more time is needed before the strand breaks up and the resulting droplets will be smaller since the strand has been stretched more.

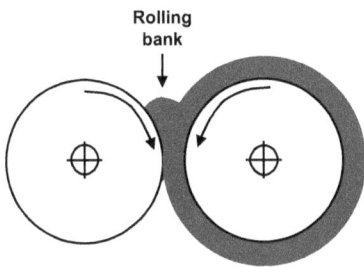

Fig. 15.12: Two-roll mixer.

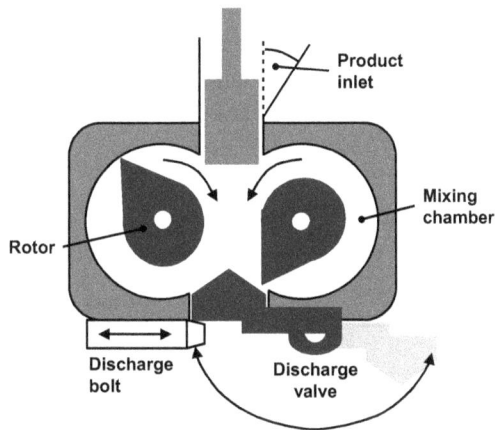

Fig. 15.13: Banbury mixer.

15.3.5 Compounding equipment

Since almost all products require additives in the polymer, it can be understood that compounding is a vital step in polymer processing. The main goal of compounding is mixing all the ingredients necessary to achieve the desired properties and to result in a mixture, as homogeneous as needed, for the final application. Frequently used methods for compounding are roll-mill mixers, banbury-mixers and extruders.

The simplest machine for intensive mixing is the two-roll mill (Fig. 15.12). Roll-mill mixers consist of two steam- or oil-heated counter-rotating rolls, with a small gap between them. It mixes the polymer at a relatively high viscosity, corresponding with the rubber plateau in Fig. 15.9. The two rolls have different rotation speeds and the polymer bypasses the fastest roll and again enters the gap between the rolls. When a two-roll mill is used for mixing, the objective is to pass the appropriate loading of the matrix material through the nip a few times until it warms up, softens and forms a smooth band around one of the rolls. The nip is adjusted once the band is formed to give a small "bank" of polymer rolling along the top to the nip. As soon as this condition is achieved, the additives are introduced and the mill starts to incorporate them into the material. This mixing process is assisted manually by cutting the band with a knife, so that a flap is formed which can be folded to the other side. To achieve a relatively homogeneous mixture, cutting and folding has to take place up to 20 times. The resultant mix is an intimate one, below the resolution of the eye. Two-roll mills are good at dispersive mixing in the radial direction but poor at distributive mixing along the length of the rolls. The Banbury mixer is the best-known mixer. Nowadays, it is mainly used in the production of polyolefins, whereas it was originally designed for rubber compounding. It is highly useful when highly viscous polymers have to be blended. This type has two counter-rotating rotors that very tightly fit to the vessel walls. A compression ram is used to confine the polymer in the mixing compartment (see Fig. 15.13). Because of the very high friction, temperature can rise very quickly in the mixer, so cooling is needed. After a predefined time, the lumps of polymer are discharged and another batch of polymer can be mixed.

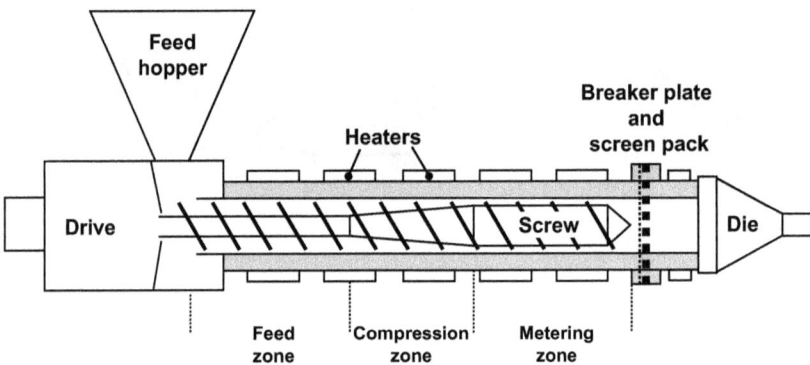

Fig. 15.14: Mean features of a single-screw extruder.

Since polymers have to be melted before they can be extruded, extruders are often used to combine compounding with the actual extrusion process. The most widely used machine is the single-screw extruder shown in Fig. 15.14. The single-screw extruder is primarily a drag pump, suitable for working with highly viscous fluids and capable of operation at high pressures and temperatures. The design is mechanically simple and essentially consists of a screw fitting closely in a cylindrical barrel. The screw and barrel can be modified to meet the specifications needed for specific polymers with different viscosities and volatiles-content. The usual screw configuration has 20 or more turns with a pitch similar to diameter, giving the appearance of a long slender machine. Solid polymer is fed at one end and the molten extrudate emerges from the other. Inside, the polymer melts and homogenizes. In the feed zone, the screw depth is constant and the solid polymer is preheated while being conveyed at a steady rate to the subsequent zones. Since it is a mixture of solids (polymer) and gas (usually air), the feed has a low bulk density and a relatively high volume. This is the reason for the high channel volume in the first section. The second zone has a decreasing channel depth. This compression zone melts the solid polymer particles by the shearing action of the screw, expels the air trapped between the granules, improves the heat transfer from the heated barrel wall and accommodates the density change during melting. A consequence of the decreasing channel depth is an increasing pressure along the extruder (Fig. 15.9) that finally forces the molten polymer through the die. In the metering zone, we again find a constant screw depth. Its function is to homogenize the melt and supply it to the die zone. The drag flow or volumetric conveying capability of the extruder for the plastic melt depends only on the screw speed, N, and the geometry, A, of the screw:

$$Q_D = A \times N \tag{15.1}$$

At the discharge end of the metering section, sufficient pressure must be generated to overcome the resistance of the transfer piping and the shaping die. This pressure causes the plastic melt to flow back down the channel and, possibly, also over the flight tips. This pressure flow is dependent on the screw geometry, B, pressure driving force ΔP, filled length of the screw, L, and melt viscosity, η:

$$Q_P = \frac{B \Delta P}{\eta L} \tag{15.2}$$

The net flow of an extruder is simply the difference between the drag and the pressure flow:

$$Q = Q_D - Q_P \tag{15.3}$$

Several additional screw sections can be incorporated into the metering zone, such as a stage for venting volatiles or pins to improve the mixing action. In addition to distributive mixing, high shear stresses over the flight tip give a degree of dispersive mixing for breaking up solid agglomerates such as pigments. The drag mechanism also causes internal shear of the viscous material being pumped, leading to a temperature rise in the polymer. This inefficiency is utilized beneficially in melting the polymer by internal rather than by external heating. The single screw extruder is highly suitable for continuously processing a wide range of synthetic thermoplastic polymers into finished or semifinished products. Due to the reasonably good mixing in the extruder, they are even used only for compounding if the final product has to be produced by another method. This is mainly due to the ease of operation and low capital costs.

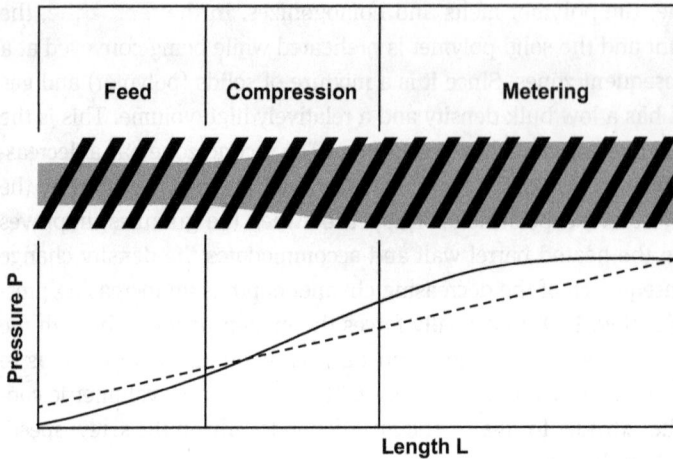

Fig. 15.15: Zones in a single-screw extruder.

Several companies produce extruders in which adjustments have been made to improve the mixing abilities, compared to a "normal" extruder. Two designs frequently applied in compounding processes are the Buss Ko-Kneader and the twin-screw extruder. As illustrated in Fig. 15.16, the Buss Ko-Kneader is an extruder with stationary pins attached to the barrel and interrupted screw flights. In this way, high shear is generated in the narrow gap between the pins and the screw flights. A recirculating flow is generated, which provides very good dispersive mixing and with that an excellent control over the melt temperature. The Ko-Kneader is used for a wide range of polymers, but is particularly successful with temperature-sensitive materials, such as PVC and thermosetting plastics.

Twin-screw extruders are used where superior mixing or conveying is important. Instead of a cylindrical barrel, the twin-screw extruder has an eight-shaped barrel to accommodate the two screws. They are characterized by their two main properties: the rotation of the two screws and the position of the screws. The screws can be either co-rotating or counter-rotating and they can be positioned either intermeshing or non-intermeshing. As the names indicate, the difference is in whether the two screws rotate in the same or in opposite directions. The nonintermeshing types consist of essentially two single screws side by side and work in a similar manner to the single-screw machines. The majority of twin-screw extruders have the two screws intermeshing to a greater or lesser extent (Fig. 15.17).

The twin-screw extruder acts as a positive displacement pump, whose output is roughly proportional to the speed and is less affected by the back pressure and friction. A general advantage of using twin-screw extruders is the fact that they have a larger heat transfer area available and, thereby, provide a better temperature control. For compounding, the corotating extruder with low conjugation is the best option, since it

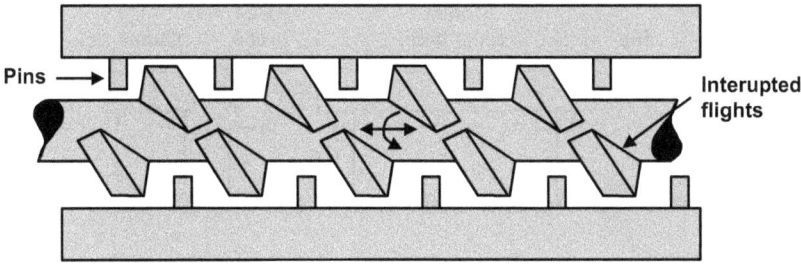

Fig. 15.16: Buss Ko-Kneader.

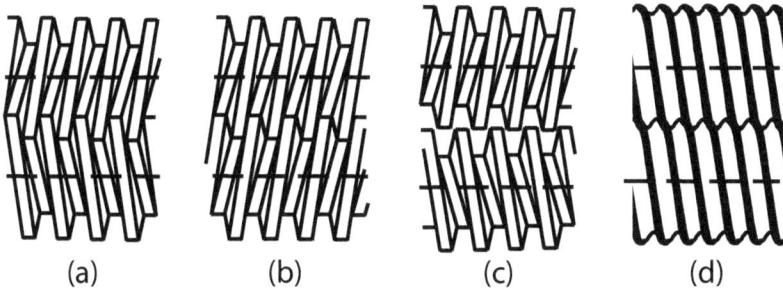

(a) (b) (c) (d)

Fig. 15.17: Types of twin-screw extruders: (a) intermeshing counter-rotating, (b) intermeshing corotating, (c) nonintermeshing counter-rotating and (d) sinusoidal self-wiping screws.

has the best mixing properties. The maximum rotation speed of the screws can be used as an indication of the mixing abilities of the extruder. If the rotation speed is high (about 300 rpm), the positive conveying in the extruder is poor and this is generally an indication of better mixing properties. If a self-wiping extruder is used (Fig. 15.17d) with a relatively high degree of openness, these effects are larger and the rotation speed can be up to 600 rpm. The twin-screw extruder is thus good for compounding, especially of heat- or shear-sensitive polymers. However, its high initial costs and relatively low output are obstacles to its use.

15.4 Polymer processing

Depending on the shape desired and the dimensions of the product, a large variety of techniques is available for shaping the polymer end product. By outlining a selection of the techniques used, we hope to convey some idea of the broad scope of polymer processing. However, because of size limitations, only the processes that use extruders as the initial step will be discussed.

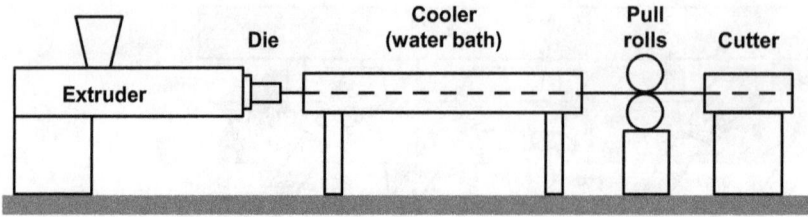

Fig. 15.18: Extrusion setup.

15.4.1 Extrusion processes

Profile extrusion is the direct manufacture of shaped products, such as rods, strips, profiles and sections from the extruder die. Hollow sections include circular tube and pipe, square tube for racking and light furniture and complex hollow sections such as window framing. In continuous operation, extrusion would produce a very long object with a shape determined by the die used in the process. Therefore, additional operations are required to deliver a fully satisfactory product. With the use of a cutter at the end of the cooling section, objects of well-defined lengths can be produced. Pull rolls are used to assist transporting the shape through the cooling section to the cutter and can also be used to stretch the shape or fiber before it solidifies. For example, grains are produced by extruding a fiber with a relatively high thickness and cutting it into very small pieces. A complete setup for (industrial) extrusion is given in Fig. 15.18.

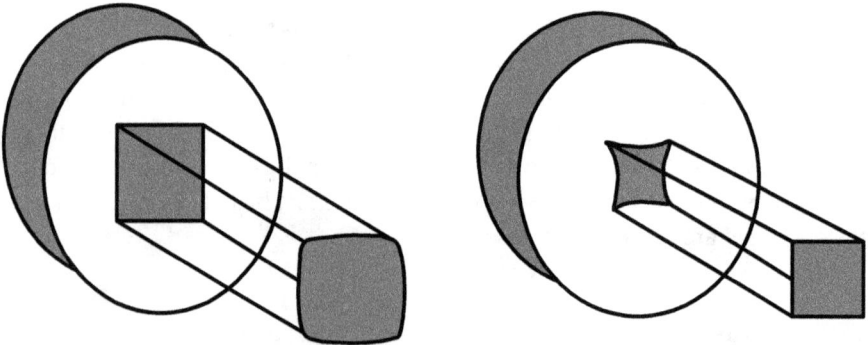

Fig. 15.19: Different die shapes with corresponding product shapes.

In profile extrusion, the shape and dimensions of the die determine the shape of the final product; a circular opening results in a solid fiber. However, due to the effect of die swell, the design of dies for extrusion processes is never an easy task. Die swell is the effect in which the polymer expands as soon as it leaves the die. The result is an extrudate, which differs in its dimensions from those of the die orifice. Thus, an extruded rod would be of a larger diameter and a pipe would have thicker walls. Die swell is the combined result of the elastic component in the overall response of a polymer melt to stress and the pressure difference between the extruder and the surroundings. It results from a recovery of the elastic deformation as the extrudate leaves the constraint of the die channel and before it freezes. Therefore, the size of the die is never equal to the size of the resulting fiber and the shape of the die can also differ from the shape of the desired product. Figure 15.19 shows a square die with the resulting product shape and a die with the resulting square product. For similar reasons, the die dimensions determine the pipe dimensions, only approximately. In pipe extrusion, it is common to apply some "drawdown" to pull the extrudate away from the die exit and counteract the swell. If more exact pipe dimensions are required, a sizing mandrel is used. If the exact inner diameter is required, the extruded pipe is passed over a mandrel of an appropriate size. When the outer diameter is specified, an external sizing device is required. Two basic versions exist. In one, the pipe is pressurized against the external mandrel by air injection, whilst in the other, a vacuum outside the mandrel allows the normal atmospheric pressure inside the pipe to hold it against the mandrel.

A derivation of the conventional profile extruder is that in which the melt turns through 90° before emerging from its die. This is used in electrical cable manufacture where polymer insulation, often polyethylene (PE), is applied to electrical conductors by this means. Figure 15.20 shows this process schematically. The wire is pulled through the coating head of the extruder by the haul-off gear of the plant. The molten polymer enters at right angles and is forced around the cable by the die pressure. The wire takes along the polymer coating, whose thickness depends on the balance between the polymer flow rate and the wire speed. A typical single wire may consist of a copper conductor of 0.45 mm diameter, with an insulation covering of low-density PE (LDPE) 0.22 mm thick, produced with extrusion rates in excess of 1,500 m/min.

15.4.2 Injection and blow molding

Two different types of molding are used in commercial processes: injection molding and blow molding. In injection molding, a batch of molten polymer is injected under pressure into a steel mold. After the plastic solidifies, the mold is opened and the product is removed from the mold. This can be done either by gravity or by ejector pins. An injection-molding machine consists of an injection unit and a clamp unit, which houses the mold. This basic layout of an injection-molding machine is shown schematically in Fig. 15.21. In the first section, the process is virtually the same as an extrusion process. The main difference is that the screw is displaced rearward as it pumps the melt into the reservoir formed by the injection cylinder. An additional complication is that the screw can run intermittently and reciprocate, piston like, within the barrel to act as the ram for injection. The sequence of operations of the

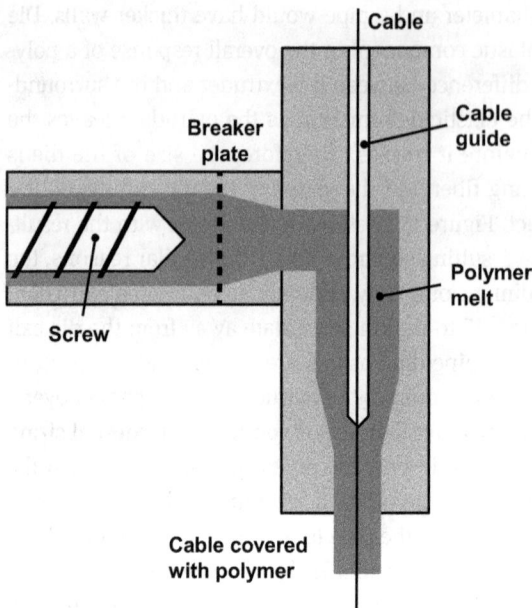

Fig. 15.20: Cable covering by cross-head extrusion.

injection molding process starts with an empty closed mold and a shot of melt ready in the injection unit. Injection occurs by opening the valve and the screw forces the melt through the nozzle into the mold. Pressure is maintained during the early stages of cooling to counteract contraction. Once cooling progresses, the pressure is released by closing the valve and the screw rotation is restarted. Pressure develops against the closed-off nozzle and the screw moves backward to accumulate a fresh shot of melt in the front. Meanwhile, the molding in the mold has to be cooled, the press and the mold opened and the molding removed. Finally, the mold closes again and the cycle repeats. A large proportion of the total cycle time is taken by cooling, including the hold-on time. As a consequence, cooling rates are an important concern in the economics of injection molding.

The mold contains the cavity in which the molded product forms, the channels along which the melt flows as it is injected, the cooling channels through which the cooling water is pumped and the ejector pins which remove the molding from the cavity. Molds are custom made and polished to a very high gloss, because every detail of the mold will be visible in the finished product. Usually, a full shot, as it is called, is not completely used. Often, a sprue, runner and gate are present, as can be seen from Fig. 15.22. These are separated from the desired part and fed back to the injection unit where they are melted and used in the process again. The mold is mechanically fastened in the clamp unit, but is interchangeable to allow different products to be molded. The clamp unit is essentially a press, closed by a hydraulic or a mechanical toggle system. The clamping force available to it must be great enough to resist the force generated by the melt as it is injected. The pressure in this melt can

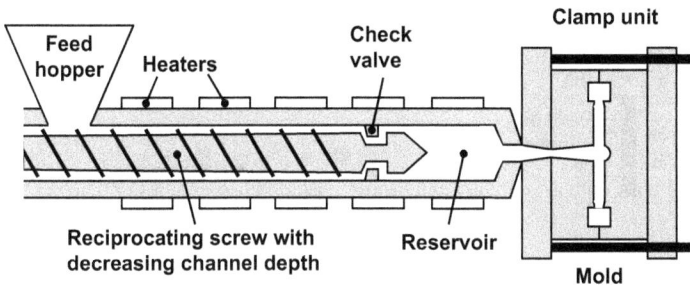

Fig. 15.21: Schematic of an injection-molding machine.

be in excess of 1,000 bar, so that for moldings with a large projected area, the force required can be as high as several thousand tons.

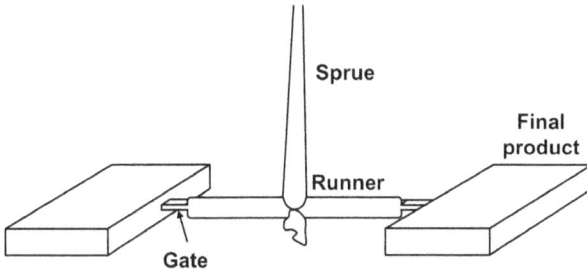

Fig. 15.22: Injection molded product.

Blow molding is the established technique for producing bottles and other hollow objects. There are two major sub-divisions: extrusion blow molding and injection blow molding. In extrusion blow molding, the resin is used directly from the extruder and the process takes place in four steps, as can be seen from Fig. 15.23. In the first step, an extruder runs continuously to produce parisons at a single cross-head die. The mold moves to the die and closes on a length of the parison, sealing one end. A parison cutter separates it from the die and the mold moves away from the die to allow extrusion of the next parison. The mold arrives at the blowing/cooling station, where a blow-pin is inserted into the open end of the parison and seals it. Compressed air is fed through the blow-pin into the parison, inflating it against the inside of the cooled mold surfaces. The bottom of the bottle is formed by the "pinch-off" of the mold and a characteristic of blow molded plastic bottles is the scar caused by this mold closure weld. Once the molding has sufficiently solidified, the product is recovered by opening the mold and the molding cycle begins again. Although the extrusion process may be intermittent, the continuous arrangement, where the parison is cut off and moved away in the mold, is the most used because it allows for higher production rates.

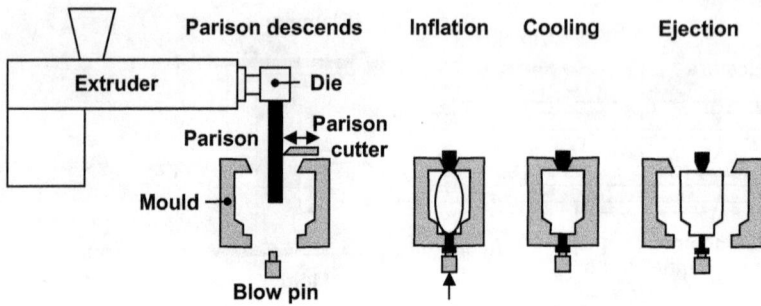

Fig. 15.23: Extrusion blow molding.

In recent years, injection blow molding has emerged as a major method for the production of PET bottles for carbonated drinks. It differs from extrusion blow molding, using an injection molded preform instead of a directly extruded parison. The tube is first extruded and quenched rapidly to prevent crystallization of the PET. Next, the tube is heated to just above its glass transition temperature, usually 90–100 °C, and stretch-blown. As can be seen from Fig. 15.24, the stretch blowing is accomplished by pushing down the blow-pin to stretch the preform downwards and simultaneous blowing to give a radial expansion. The actual blowing introduces a circumferential orientation that results in a rapid crystallization and solidification of the PET.

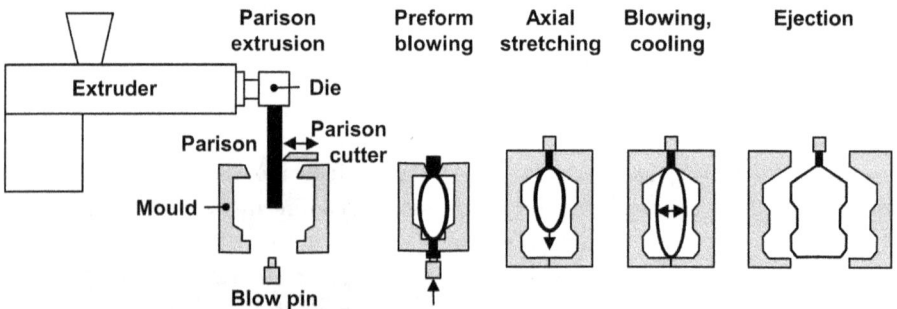

Fig. 15.24: Stages in injection blow molding.

15.4.3 Melt and gel spinning

Many synthetic fibers are produced by extrusion through fine holes, known as melt spinning. Although melt spinning is the most economical spinning process, it can be applied only to polymers that are stable at temperatures sufficiently above their melting point, such as nylon, polyester, PE and polypropylene (PP). Most melt-spinning plants melt the polymer with the aid of screw extruders, which can be fed directly with the solid polymer grains and forward a molten polymer of high melt viscosity (Fig. 15.25). The polymer melt is extruded under pressure through a spinneret, which is a flat plate made of stainless steel containing orifices. The number of holes may range from a few to several thousand, with individual hole diameters in the range of 175–750 μm. The feed rate to individual spinning units is controlled by an accurately machined metering gear pump, which delivers molten polymer at a constant rate into a filter assembly. As molten polymer passes through the spinneret, the emerging bundle of fibers is quenched in water or in a countercurrent flow of air, under a high draw-down rate. Usually, the product obtained from the spinning process is not directly suitable for commercial use. The fibers are then further cold-drawn as they pass to the windup, to orientate the molecules and develop the necessary linear strength. The dimensions of the produced filament or yarn are expressed in a unit called tex. This measure of fineness or linear density is defined as 1 g/1,000 m. The filament linear density is related to its diameter by

$$d(\mu m) = 2 \times 10^3 \sqrt{\frac{1}{\pi} \frac{\text{linear density (tex)}}{\text{bulk density (kg/m}^3)}} \tag{15.4}$$

The linear density is controlled by the throughput of the polymer per hole, W (kg/s), and the windup speed V (m/s). The melt-spun filament linear density (tex) is then:

$$\text{linear density (tex)} = \frac{10^3 W}{V} \tag{15.5}$$

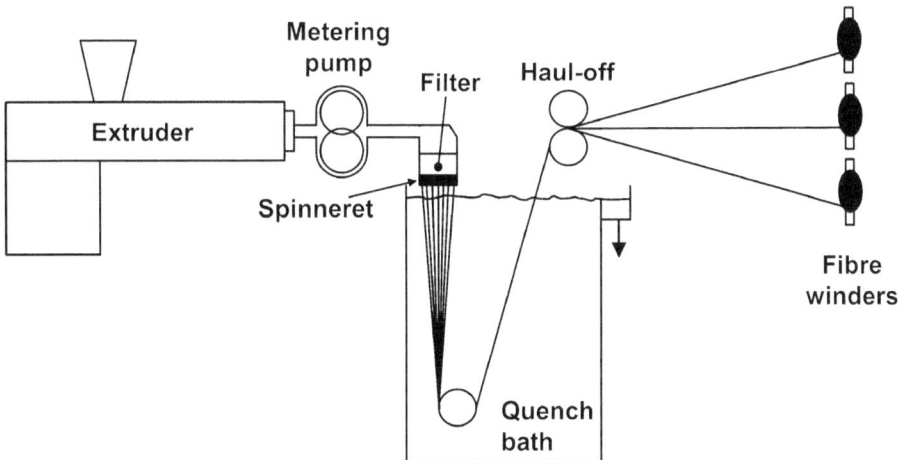

Fig. 15.25: Melt spinning of fibers.

The production of DSM's high-strength fiber, Dyneema®, required a new method. Its strength is due to the orientation of the polymer molecules in the fiber; they are all parallel to the length of the fiber. Melt-spinning cannot be used in this case because of the very high melt viscosity of the ultra-high-molecular-weight (UHMW) PE. In order to be able to use the desired polymer, a new process called gel spinning was developed. An additional advantage of the use of a solvent is that the degree of entanglement in the polymer decreases. This is especially true if dilute solutions are used. With this reduced degree of entanglement, it is a lot easier to achieve a highly oriented polymer in the final fiber and a higher degree of orientation is achieved.

Gel spinning is essentially a two-step process. First, a fiber is spun from a solution of the polymer and after that, the fiber is stretched and dried. For the production of Dyneema®, a dilute solution of 1–2% of PE in an organic solvent is used for spinning the fibers at 150 °C. A typical solvent is decalin, a higher aromatic compound, yielding fibers with good mechanical properties. The PE used has an UHMW of around $1-4 \times 10^6$, one of the reasons why a high strength fiber results. The spinneret used in this process has openings with diameters of 0.5–2 mm, depending on the desired fiber diameter. The fibers are quenched to room temperature in a water bath and a disoriented gel fiber remains. The solvent content of the fiber has to be at least 50% and can be as high as 98%, which has to be completely removed to produce a strong homogeneous fiber. This is done by increasing the temperature to 50 °C, prior to stretching. At this stage, the molecules are still disoriented and the fiber does not have a high strength. Next, the fiber is stretched further and the temperature is gradually increased from 120 to 160 °C. Now, the orientation of the polymer takes place and the fiber attains its high strength. The remainder of the solvent also evaporates in this stage. It is carried by a hot gas and can be recycled after cooling and/or cleaning. A schematic overview of the gel-spinning process is given in Fig. 15.26.

15.4.4 Film production techniques

Polymeric films can be produced by calendering, extrusion, casting and blowing. Calenders are used for making rubber sheeting and similar products from plasticized PVC and ABS, with varying thicknesses of less than 1 mm. The process requires the polymer to be in the rubbery state. A sheet of accurate thickness is produced by passing the compound between rotating rolls. Often, more than one nip is required and to produce a sheet of a specific accuracy, multiroll machines are used. The rolls are often placed in an inverted L-configuration, which can be recognized in the typical layout of a calendering plant shown in Fig. 15.27. The rolls tend to bend because of the pressure and this has to be taken into account in the design. They are also equipped with cooling channels to control the temperature during processing. Each roll is driven separately by a variable-speed drive. The first sheet is formed at the first nip directly from the feed, which may be supplied from a two-roll mill or extruder. The second gap

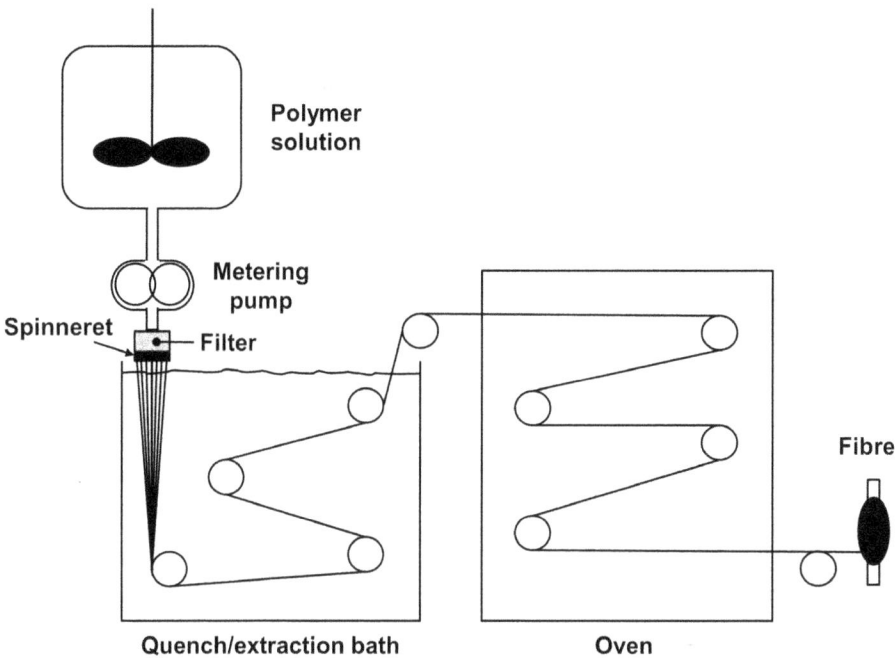

Fig. 15.26: Gel spinning process.

usually allows the sheet through, without nipping. The sheet is then remade at the third gap where a thin "pencil" bank is kept rolling. Calenders are mainly used for resins with high melt-viscosity, which results in high pressures on the rolls. Under good conditions, the thickness of the sheet can be controlled to ±0.02 mm.

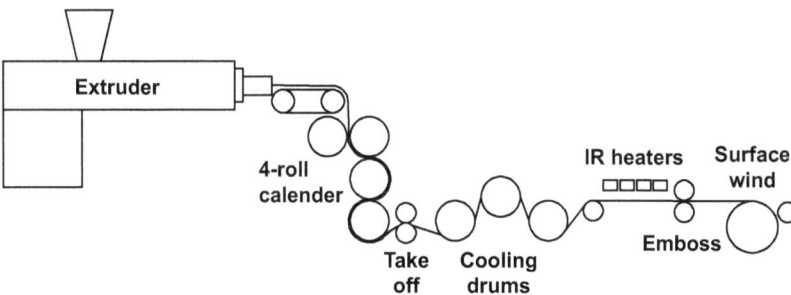

Fig. 15.27: Calendering plant containing an inverted L four-roll configuration.

Film and sheet can also be produced by direct extrusion in a manner similar to profiles (Fig. 15.28). Thicknesses below 0.5 mm (approximately) are usually called film and thicknesses of more than 0.5 mm, sheet. As widths between 1 and 2 m are frequently required, the design of dies for the production of sheet or film presents certain challenges. The requirement is to deform the melt from its essentially cylindrical shape to a wide, thin form and obtain a uniform distribution of the polymer melt across the die.

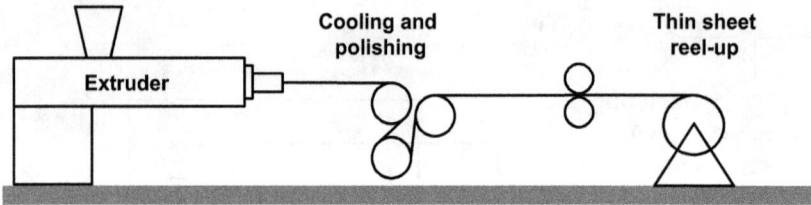

Fig. 15.28: Sheet extrusion.

In casting, the film is extruded directly onto a roll through a long straight slit, with an adjustable gap in the order of 0.4 mm (Fig. 15.29). The die is positioned very carefully with respect to the casting roll, which is used for controlled cooling and to impart some degree of drawdown to enhance the film properties by orientation and reduce the film thickness. The cooling roll is highly polished to achieve a smooth and flawless surface of the film produced. Usually, the die for cast film is deflected downward to effect the best approach to the casting rolls. Since the web of resin narrows as it is drawn from the die, the die is wider than the desired film. The edges of the film are removed because of thickening during the process. The edge trim can be recycled, if desired. The process of extrusion coating is similar. The film emerging from the die contacts the substrate to be coated at a nip and the resulting laminate proceeds round a cooling train to the windup. Cardboard coated in this way is extensively used for foodstuffs.

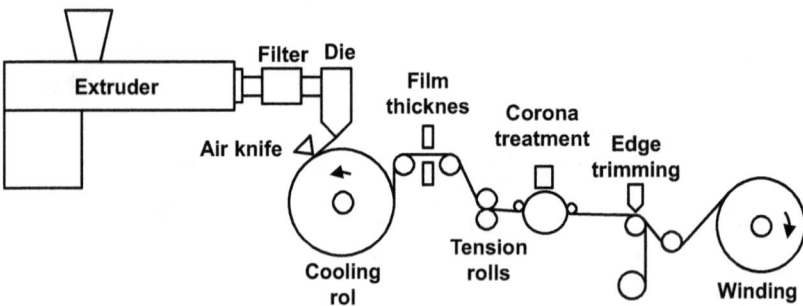

Fig. 15.29: Film extrusion.

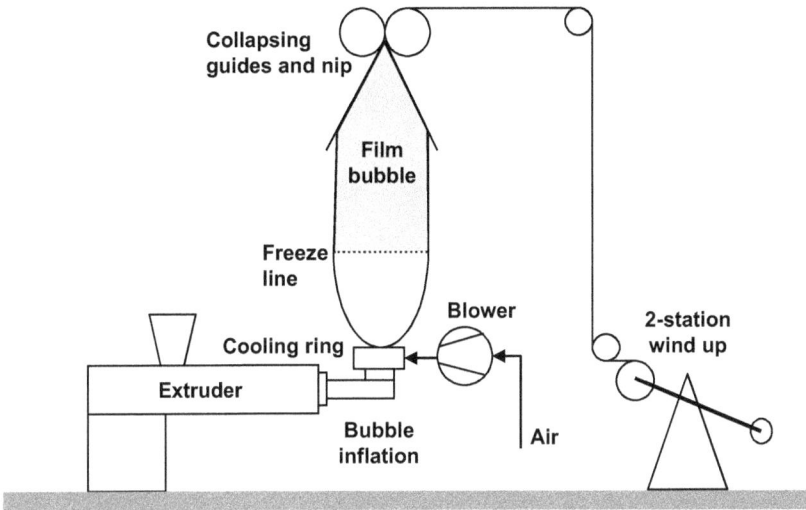

Fig. 15.30: Diagram of the tubular blown film extrusion process.

Widely used for the productions of thin films from LDPE, HDPE (high-density PE), PP and PET is the tubular blown film process. Blown film is made by extruding the polymer melt through an annular die, producing a tube that is inflated with air to yield a bubble of the desired diameter and at the same time drawn upward. The result is that the produced film becomes biaxially orientated and this orientation is made permanent by the crystallization that freezes the orientation. The blown tube is cooled and solidified by blowing the air upward along the outside of the expanded tube and guided by converging boards or sets of roller onto the nip rollers. (Fig. 15.30). It may be wound directly as a flat tube or slit at both sides and wound into two flat reels. Extrusion is usually vertically upward to avoid buoyancy effects on the blown tube, and permitting the heavy extruder and windups to be floor mounted.

The minimum thickness of a film which can be produced on a continuous basis without a lot of rejected films depends on the viscosity of the polymer used. Figure 15.31 shows a schematic representation of the minimum thickness dependency of PE films on the melt index. The melt index can be regarded as the reciprocal value of the melt-viscosity at a constant shear stress. As can be seen, the minimum thickness generally decreases with an increasing melt index. Exceptions are resin numbers 1 and 3, both representing another important factor in film casting: the molecular weight distribution. Resin 3 has a broad distribution, which results in better flow characteristics, whereas number 1 has a narrow distribution, resulting in worse flow characteristics.

For polymers that run well in this process, such as PEs (HDPE, LDPE), the produced bubble is stabilized by the tension stiffening of the polymer melt and also a favorable tension-induced crystallization rate. An interesting contrast exists in the

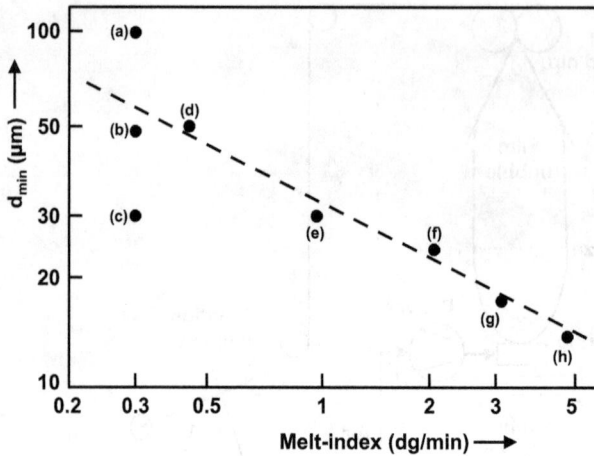

Fig. 15.31: Minimum thickness of PE films.

production of PP film. PP is tension thinning in the melt and its crystallization rate during cooling is also rather slow. A different technique is adopted in which the extrudate from the melt is quenched with ice water to give a rubbery amorphous tube. This tube is then reheated to the temperature at which crystallization is at its maximum and blown. Blowing the reheated tube avoids the problems associated with a tension-thinning melt and a slow crystallization rate, which would result in an unstable bubble. This process for PP further differs from PE in that it is run vertically downward.

15.4.5 Thermoforming

Thermoforming refers to a group of processes that involve clamping a sheet of thermoplastic above an open mold, heating it so that it becomes soft and rubbery and then forcing the sheet to take up the contours of the mold, where it cools and solidifies. Deformation of the sheet is achieved either by applying a mechanical force or compressed air to the top of the sheet (pressure forming), or by drawing a vacuum between the sheet and the mold (vacuum forming). This processing method uses relatively low pressures and temperatures, because high temperatures would result in a polymer that is not self-supporting. In principle, a product can be made by the two different molds shown in Fig. 15.32. The female mold uses a cavity for the main formation, whereas the male mold uses a projection.

As can be seen from lower part of the figure, this type of thermoforming produces a product with nonuniform thickness. This is due to contact chilling of the polymer, which results in excessive thinning of the parts of the sheet that contact the mold last. A relatively easy adjustment to the process, especially in vacuum forming, can

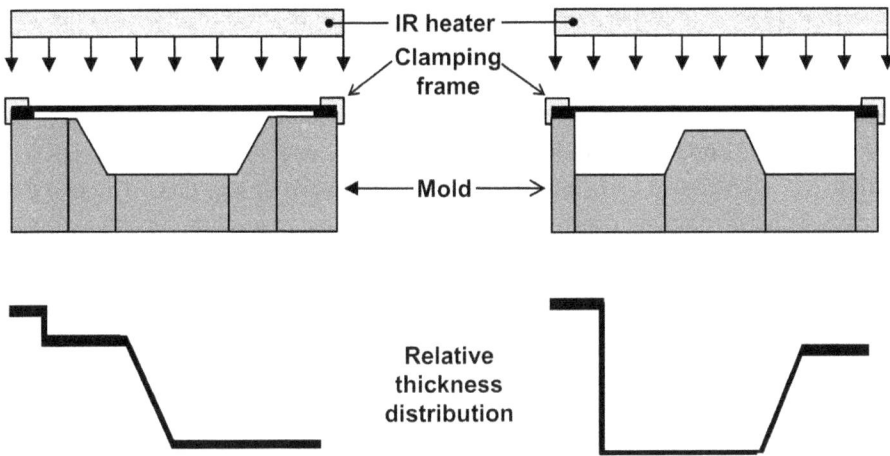

Fig. 15.32: Female (left) and male (right) thermoforming molds.

be made to achieve a product with a more uniform thickness. Simply blowing a bubble of the sheet before applying a vacuum does this. In this way, the sheet is pre-stretched to the desired dimensions and in the actual forming stage, the thickness of the sheet remains intact.

Thermoformed products are mainly of a relatively simple geometry, most being thin-walled panels, trays, cups and boxes of various shapes and sizes. Advantages of this technique are its low capital cost, low molding pressure and reasonably short cycle times. For small series or for research purposes, it is even possible to make the mold out of wood. Balanced against these advantages are a number of disadvantages, among which are the limited product geometry and the difficulty in obtaining a narrow thickness distribution. Plastics used for thermoforming are usually resins that have rubber-like properties over a wide temperature range, mostly the amorphous polymers. Examples include polystyrene, ABS, PVC and polycarbonate (all amorphous) and significant quantities of semicrystalline HDPE and PP.

15.4.6 Foam extrusion

Foamed sheet is an important product for thermal insulation, packaging and cushioning. It is produced by extrusion under carefully controlled conditions. Cylindrical shapes are extruded for pipe insulation. The foams are formed by the action of gas-producing blowing agents that are mixed into the polymer melt. These agents may consist of low boiling solvents, azo compounds that release nitrogen or isocyanates that release carbon dioxide. Owing to the high pressure, the blowing agent remains dissolved until the pressure is reduced at the die exit and instantaneous foaming occurs. Foaming requires a fairly constant output for control of the product density and close control of the temperature to achieve the correct degree of decomposition of the foaming agent. Critical to the success of most foaming extrusion operations are a good mixing within the extruder and a cooling of the melt just prior to the entry of the die to prevent premature release of the blowing agent and to control the density and surface finish of the sheet. Cooling is most effectively accomplished with a tandem arrangement of two extruders, as shown in Fig. 15.33. The first extruder assures a complete dissolution of the blowing agent and the second extruder is operated at a slow speed for optimum cooling.

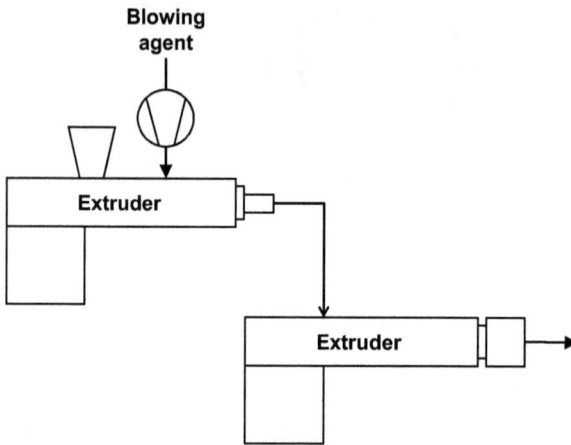

Fig. 15.33: Foam extrusion line.

Nomenclature

A,B	Screw geometry	[m^2]
d	Filament diameter	[m]
L	Filled length	[m]
N	Screw speed	[m/s]
P	Pressure	[Pa]
Q	Flow	[m^2/s]
R	Remission value	[%]
T	Temperature	[K]
V	Windup speed	[m/s]
W	Throughput per hole	[kg/s]
η	Melt viscosity	[Pa s]

16 Development, scale-up and engineering

16.1 Introduction

Process development can be defined as the study of development of a process from a laboratory to a commercial scale. It is not an independent field in chemical engineering, but contains and integrates all fields in the process technology, such as the basic fields of physical transport phenomena, kinetics and thermodynamics and the applied fields of chemical reaction engineering, physical/chemical separations and equipment design. The starting point generally consists of laboratory results that concern a chemical reaction whose translation into a commercial plant appears viable. To go directly from a laboratory scale to the industrial scale is rarely feasible. As a rule, one or more additional experiments are necessary to reproduce the laboratory results on a larger scale. The main goal of process development is to find out which steps in the process may be expected to present difficulties, and which additional research needs to be performed to solve these problems at minimum cost and as quickly as possible. It is here that the methodology of process development, and hence of scale-up becomes decisive for the success of the operation. As most operations are scale dependent, scale modification is an essential part of the work of the process engineer. By studying processes, the so-called scaling rules can be defined. Scaling rules allow us to apply knowledge and experience from small laboratory equipment to a larger industrial unit.

16.1.1 General aspects of scaling up

The necessity for scaling up in the process industry lies in the growth of the consumer market and in the competition between producers. On expansion of capacity, it is financially more favorable to build one large apparatus, instead of several parallel units. In the latter case, the cost of investment I increases proportionally to the production capacity C, while in the first case, the so-called six-tenth rule applies for rough estimates on process installations:

$$\left(\frac{I}{I_0}\right) \approx \left(\frac{C}{C_0}\right)^m \qquad m \approx 0.6 \qquad (16.1)$$

The exponent m is called the degression coefficient. It varies from case to case. Larger installations are advantageous, as long as m is smaller than unity. From a process viewpoint, everything is allowed during the designing and construction of an apparatus and installations. However, there are constructional limitations. These often determine the maximum size of fitting such as castings, stopcocks, etc. If equipment becomes so large that transportation is not allowed, it has to be constructed at the

https://doi.org/10.1515/9783110712445-016

spot. Duplication of installations has the advantage of a more rapid design and a shorter delivery time. For the salary costs in the process industry, a similar relation as between I and C can be used:

$$\left(\frac{K}{K_0}\right) \approx \left(\frac{C}{C_0}\right)^n \qquad n \approx 0.2 - 0.25 \qquad (16.2)$$

Hence, scaling up reduces the salary costs per unit product but is disadvantageous for employment.

A different aspect of scaling up is the vulnerability of the production. If one large installation and an identical installation built from parallel units are compared, large installations will sooner give losses when not running at 100% capacity, since the fixed costs are related to the investments. In case of a breakdown or maintenance of one large installation, everything will stop. In a multiple unit installation, a part can still continue. The organization of the maintenance and construction of large installations should be tight, as delay and retardation can result in important financial losses.

Raw materials can often be managed more economically in large installations as opposed to multiple installations. Furthermore, energy and material losses tend to be proportional to the surface area, which decreases relative to the volume when the geometrical dimensions increase. The larger the volume of the streams in a plant, the more worthwhile recycling and recovery of energy and materials becomes. On the other hand, more security measures are necessary because of the higher risk with severe consequences of a calamity.

16.1.2 Ways of scaling up

There are three common ways of scaling up. These are used both for scaling up of separate pieces of equipment and for complete processes. One can enlarge an apparatus or installation by adding one or more identical units in parallel to the existing system. This is called a multiple train installation for which the total capacity, C_N is described by the following simple formula:

$$C_N = N\,C_0 \qquad (16.3)$$

where N is the number of identical installations and C_0 the capacity of a single unit. Examples are the furnaces in the naphtha crackers and the polymerization of polypropylene. Another option is to enlarge the equipment geometrically. Here, the shape remains the same but the absolute dimensions are increased. A complete factory can be enlarged by enlargement of the individual different parts. The capacity usually increases with the scale ratio to the third power:

$$C_N = \left(\frac{L_N}{L_0}\right)^3 C_0 \tag{16.4}$$

Examples where this has been applied are the oxidation of cyclohexane, ammonia production, high-pressure polyethylene polymerization, caprolactam process and the aspartame plant. The third way of scaling up is by simply increasing the throughput of an apparatus or installation. This is usually possible because there is a certain design margin in a plant, such as in many older polypropylene and caprolactam plants. The word debottlenecking is often used in this context.

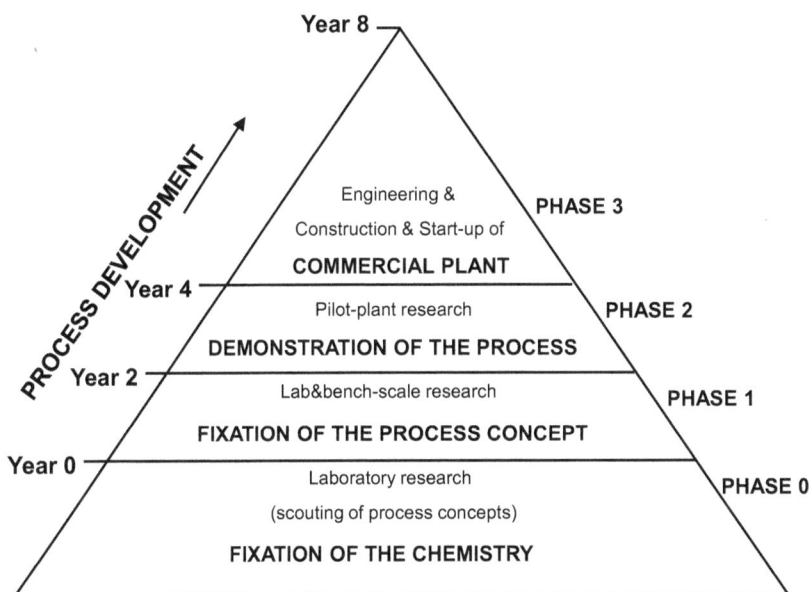

Fig. 16.1: Phases and time line involved in developing a process.

16.2 Development and scale-up in the bulk chemical industry

16.2.1 Basic course of process development

The development of chemical processes is a complex procedure. The first hurdle in establishing a new process is overcome when a promising synthetic route, usually with associated catalysts, is discovered.

To line up the capabilities and viability of a new process, one starts thinking about a full-scale plant. Based on an optimized laboratory synthesis, an initial process concept is developed. If the economical outlook is good, one can start a development project. By referring to the scale of experimentation, different phases can be distinguished in the development of a process. A rough outline, divided into three types and illustrated in Fig. 16.1 is as follows:

1. Laboratory and bench-scale experiments, in which certain aspects of the process are investigated by handling relatively small amounts of raw materials, in order to reduce the material and resource constraints to the minimum.
2. Pilot-plant research, at a scale that varies within wide proportions, but in which all the industrial constraints are taken into account. It is during pilot-plant experiments that scale-up problems must be dealt with.
3. Design and construction of a commercial plant or a demonstration unit. In general, demonstration units are very expensive, and all the efforts made in the process development attempt to avoid this very expensive intermediate step.

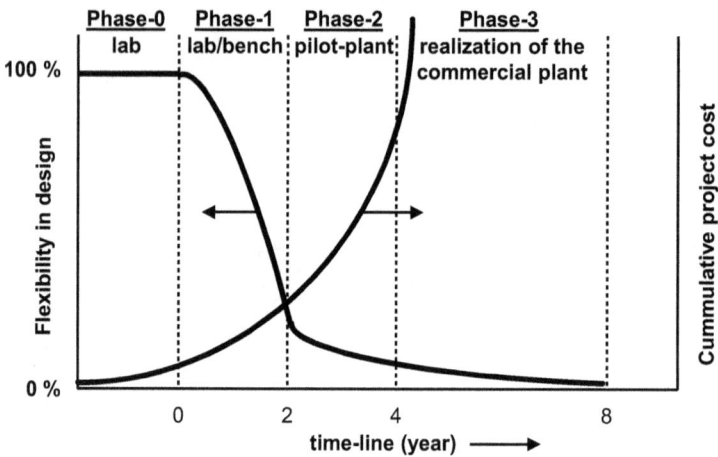

Fig. 16.2: Flexibility in design and cumulative project costs during a process development project.

Process development does not, however, take place in a one-way street. Assumptions are made for the individual development stages that are only confirmed or refuted when the next stage is being worked on. It may be necessary, therefore, to go through the individual stages several times with modified assumptions, resulting in a cyclic pattern. The most important task in each stage is to find the weak points and subject them to particularly close scrutiny. The entire process will then be examined again with the improved data obtained in this way, and so on. The fact that many decisions must be taken with incomplete knowledge is a fundamental but inevitable difficulty. To delay development until all uncertainties have been eliminated would be just as wrong as starting industrial development on the basis of laboratory discoveries alone. A start should be made on choosing between various processes or process variants as early as possible, so that consideration of a larger number of possibilities is restricted to the laboratory. Prolonged investigation of two variants on a trial-plant scale should be avoided. Most mistakes are made at the beginning of the activity, but it is still relatively easy and cheap to eliminate them at the mini-plant stage. However, Fig. 16.2 illustrates that the further process development advances, the more expensive it becomes to eliminate mistakes. In the final production plant, corrections can only be made with an enormous expenditure of time and money. Each development stage is followed by an evaluation to decide whether development should be continued, stopped or started again, at an earlier development level. A more detailed scheme of this chemical process development process is given in Fig. 16.3.

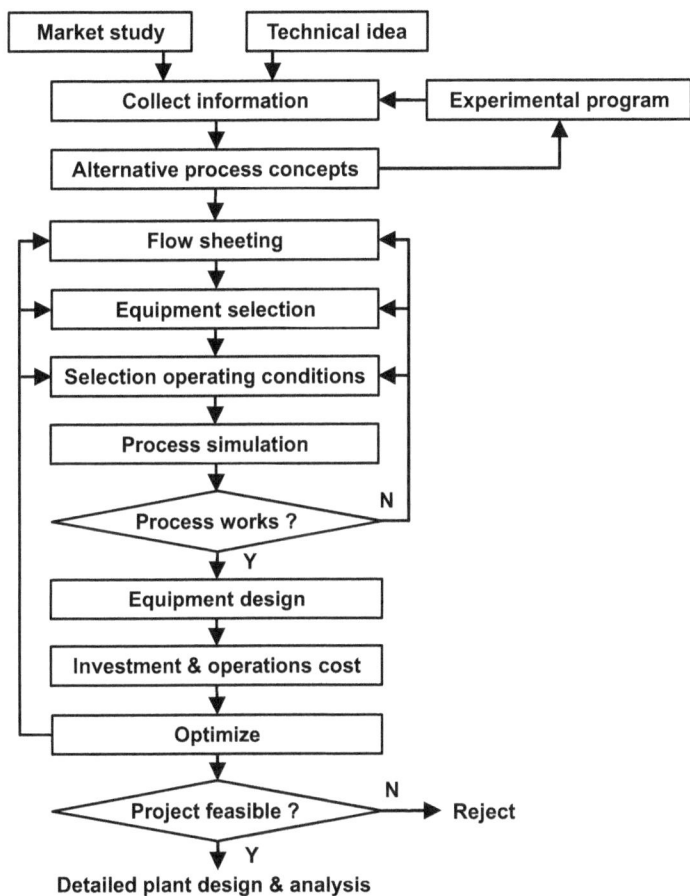

Fig. 16.3: Steps in the development of a process.

16.2.2 Laboratory and bench scale research

When the laboratory phase has been completed and before the actual process development is started, further information must be obtained, since the latter stage is normally associated with high costs. The core of a chemical plant is the reactor. Its input and output decide the structure of the entire plant built up around it. Therefore, detailed knowledge on thermodynamic equilibria, kinetics, selectivity, conversion and heats of reaction of the chemical reaction must be available at the earliest possible stage. Secondly, a database of reliable physicochemical data, such as densities, viscosities, vapor pressures and phase equilibria, has to be created for use in computer simulations and safety studies. The first step is to collect all available literature data on a compound. Many companies now maintain their own compound data banks. If no property values can be found in the literature, which is often the case for binary and ternary data, they should initially be estimated by using empirical formulae or realistic values that have proven themselves for many substances. However, experimental determination or confirmation of the most important values on which the plant design is based is unavoidable.

It is essential to determine the demand for the end product as well as its specification and achievable price. The product specification is of particular significance for process development. In the simplest case, it consists of a minimum purity, for example, that the end product should contain at least 99% of a particular chemical compound. However, in most cases, the end product cannot be specified so simply, since its subsequent use may be affected in different ways by individual impurities.

Once all the information has been assembled, the creative part of the development process can be started to draw up the initial versions of alternative process concepts for analysis, evaluation and selection. When considering possible alternatives, the researcher will be constrained by many factors, which will narrow down the number of possible designs. Some of these constraints will be fixed, such as those that arise from physical laws, economics, safety and governmental regulations. Such constraints that are outside the designer's influence are called external constraints, setting the outer boundary of possible designs, as shown in Fig. 16.4. Within this boundary, there will be a number of potential designs bounded by the internal constraints over which the designer has some control. For instance, the time available for completion of a design will usually limit the number of alternative designs that can be considered. Finally, the project team should discuss the remaining initial versions of the process. This discussion should take account of all the available information and, after all the advantages and disadvantages have been reviewed, only one version of the process should remain. On the basis of this initial version, an industrial plant is designed. The individual unit operations (reactor, absorber, distillation, etc.) are designed on the basis of the existing information (approximate size and diameter of columns, etc.). Initially, these operations can be examined in the laboratory individually and independently of one another to determine whether they are feasible in principle. Laboratory

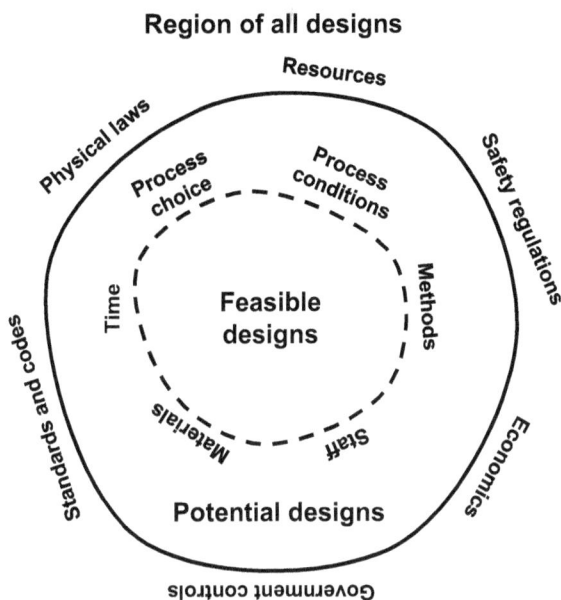

Fig. 16.4: Design constraints (adapted from [2]).

experiments are generally carried out with pure, well-defined materials. This differs, of course, from the subsequent situation, where the solvent must be recycled for cost reasons and, inevitably, becomes enriched with by-products, some of which are unknown, however, at this point in time. Even at this stage, new requirements may be imposed on the reaction stage. For example, difficulty in removing a by-product may make it necessary to alter the reaction conditions or modify the catalyst. As long as the chemical reaction is subject to extensive modification, there is little point in paying a great deal of attention to further processing. It is only when the reaction mixture is being produced in a representative manner that an initial separation and purification procedure can be devised. Once all the individual steps have been successfully tested, it is possible to draw up a reliable flow sheet of the entire process. At the same time, computer simulations are used to draw up an initial complete mass and energy balance sheet for the industrial plant. However, if any subsidiary step is found to be infeasible at this stage, the process concept must be changed. This preliminary flow sheet, which has not as yet been fully tested as a whole, can be used to draw up an initial cost estimate.

When the individual steps are put together, the created recycle streams raise new process engineering problems. Material recycled to the reactor may drastically affect the activity and service life of the catalyst, or the solvent circuit may become enriched with by-products and this may lower product purity. Testing of such process integration effects usually requires the experimental testing in a mini-plant, because many of the quantities that are required for a computer simulation are unknown. The mini-plant concept has been developed to predict the performance of large-scale equipment from data generated by low-cost experiments in small equipment, rather than relying solely on fundamental data. It consists of a scaled-down version of the hypothetical large-scale plant where most of the characteristics and still unknown parameters of the process can be measured. For plate-type distillation columns, the accepted mini-plant procedure has been to use vacuum-jacketed and silvered glass columns to measure the plate efficiencies and predict the performance of large-diameter plate columns. Since the result must allow prediction of the performance on a larger scale, the mini-plant approach is limited to those steps in the process that can be extrapolated directly to the industrial-scale plant, without a large risk. For many operations (extraction, crystallization) however, scale-up factors of ca. 10^4 are still not feasible. As a result very few processes can be constructed without additional testing of certain critical process sections on an intermediate pilot stage.

An alternative to in-house mini-plant research is testing at equipment manufacturers. Equipment manufacturers can be excellent sources of scale-up information, since they have years of experience with the particular pieces of equipment they sell. Many of them have facilities with their equipment miniaturized to the smallest unit size where reliable design data can still be obtained and provide performance guarantees. Important examples are: filtration, drying, solids conveying, crystallization, agitation, flaking, gas-solid separations and liquid-solid separations. Since equipment manufacturers expect to sell a large piece of equipment, the test charge is usually low or, sometimes, the test is free. One of the greatest uncertainties in selecting equipment from manufacturer test data is the possibility that the material used for the tests is not truly representative of what will be made in the commercial plant. Quite often, the test period by the manufacturer is short, and the risk of what may happen after extended periods of operation is not completely understood. Similarly, recycle streams cannot generally be incorporated in a manufacturer test run. Overcoming these problems requires leasing one of the manufacturer test units and installing it in the pilot plant or as a side stream in a production plant. Here, the equipment can be tested at the actual operating conditions, and the data that is generated will be more meaningful for scale-up.

16.2.3 Pilot plant research and demonstration plants

Once the process concept has been confirmed on bench scale, the question arises whether the process should be tested on pilot plant scale before designing the commercial plant. Generally, the answer is "yes" for new products, due to the high risks if the pilot plant stage is skipped in the process development stage. As illustrated by Tab. 16.1, the scale of this so-called pilot plant is between that of the mini-plant and that of the commercial plant. Typical quantities produced are a few

Tab. 16.1: Comparison different scales and operation capacities.

Type of scale	Capacity (order of magnitude, in $m^3.day^{-1}$)
laboratory	< 0.001
bench scale	< 0.001
mini-plant	0.01
pilot plant	0.1
test-plant	1
commercial plant	1000 ($tons.day^{-1}$)

kilograms per hour or tons per annum to enable application tests to be carried out on the product or large-scale deliveries to be made to customers. Operation of this pilot plant makes it possible to complete and verify data and document information obtained at an earlier stage of process development. Before a pilot plant program is undertaken, the objectives and scope of the pilot plant must be clearly established. As indicated below, there may be many reasons to construct a pilot plant for testing a process:

- Generate market development quantities of a new product.
- Evaluate the feasibility of a new process.
- Generate design and scale-up data for a commercial plant.
- Demonstrate a new process.
- Investigate construction materials.
- Investigate accumulation characteristics, fouling and congestion.
- Investigate dynamic behavior.

Only the commercial reason of market evaluation can be enough to build a pilot plant in order to obtain the desired product and to test the market with the new product. If an objective of the pilot plant is to demonstrate a new process while producing significant quantities of product, then a totally integrated pilot plant may be the most appropriate. The key is to design, construct and operate a pilot plant containing all the recycle streams and ecology control systems. Recycle streams always hide the danger of impurities that are not discovered on a laboratory scale but may have a tremendous effect on the chemical reactions and the physical separations. Additionally, materials of construction, maintenance, operating labor, instrumentation, process control, raw material handling and waste disposal can be tested.

Operation of the pilot plant should clarify all the issues that have not been fully dealt with in the mini-plant. Prediction of the process performance in pilot plants and production plants has been a part of the art and science of chemical engineering for many decades. Each segment of a unit operation in a pilot plant may have a different degree of predictability for a design based on first principles. The goal is to model and predict what will happen, while conducting the minimum amount of experiments. A research program should be available to determine the relevant data against a minimal time to decrease the costs. Generally, a development team working closely with the design team conducts the pilot plant tests. The pilot plant should be designed as a scaled-down version of the industrial-scale plant and not as a larger copy of the existing mini-plant. This is an expensive, time-consuming step that needs to be anticipated and planned for by the development team, as early as possible. Construction and operation may cost millions of euros. Equipment costs vary, but, generally, the costs of the instrumentation can go up to 50% of the pilot plant costs. The operating costs are also high because technicians and operators are needed. Once process development on a pilot scale has been successfully concluded, the pilot plant must be kept on standby until the industrial-scale plant is running satisfactorily. Normally, when the larger plant is started up, the pilot plant is operated simultaneously, so that any problems that occur in the former can be dealt with rapidly.

When the feasibility of a process cannot be proven on a pilot plant scale, a demonstration plant may be needed to gain experience with the new process and/or new product. A demonstration plant differs from a pilot plant in that it can approach the size of a full-scale production plant. As a result of the integrated design concept and relatively large scale of operation, the demonstration plant is relatively expensive to construct and operate. For this reason, demonstration plants are built only when the unknowns in process scale-up or demands in the market place preclude selection of other pilot plant alternatives. This is often the case with solids handling processes consisting of filters, grinders, conveyors and solid dryers. The minimum size of a particular process unit operation or piece of equipment then determines the minimum size of a demonstration unit. Where pilot plants concentrate on process fundamentals such as the chemistry, kinetics and thermodynamics of the process, a demonstration plant is built to a scale that is practical for studying the engineering and mechanical aspects of the process. In planning the construction of a demonstration plant, the same design considerations as those used in the construction of a full-scale plant apply. In most cases, data for scale-up to the demonstration plant size will have been obtained from one of the smaller units discussed previously. The critical consideration in the design of the demonstration plant is placing emphasis on those portions of the plant that experienced process uncertainties. That is, one must carefully design those parts of the process that precluded scale-up to the full-sized commercial plant, to begin with.

16.2.4 Feasibility evaluation

Increasingly complex markets require critical examination of the project after each development stage for feasibility, economic relevance and environmental impact. As a rule, this is done with the aid of a feasibility study that includes documentation of preliminary design work and evaluation of the process status. At the start of a project, the bases used for planning and evaluation of the project are still very inexact. Development costs often constitute a considerable proportion of the total costs of a project. If the process is completely new, these costs are often around 50% of the investment costs for the industrial plant, and perhaps more, if a pilot plant must be built. Therefore, no major costs should be incurred, until it is known whether the new process will be feasible or not. A first indication is to verify that the target product(s) at least provide more value than the costs incurred by the minimal consumption of raw materials. Such economic potential evaluation requires nothing more than first indications of possible sales prices, raw material prices and targeted production volume, which are typically obtained from market data.

After initial studies have shown that investment is desirable and the market analysis led to a tentative capacity figure, a first estimate of investment costs and subsequent manufacturing costs is performed. The approximate determination of investment costs is subdivided into simple global methods and detailed methods. At this stage of development, generally, global order of magnitude methods are used, which permit investment costs to be estimated relatively easily and with an accuracy between ± 35% (study estimate) and ± 50% (order of magnitude) (see also Fig. 16.10). The manufacturing cost generally consists of three main categories:

1. Direct costs: annual cost of each raw material, consumables, utilities, operating labor, maintenance, operating supplies, laboratory charges, royalties
2. Indirect costs: plant overhead (general, safety, medical, payroll, logistics, restaurant, facilities), depreciation, property taxes, insurance, interest, rent
3. General costs: administration, marketing, sales, research and development

All these costs are usually integrated and included into a cost sheet in which the raw materials and utilities comprise the so-called Bill Of Materials (BOM), indicating exactly what is needed as input to produce one kilogram or metric ton of product and is a main means of communication between the technical and financial people within a company.

At this stage of development, with first estimations on investment, manufacturing costs and sales revenues established, a more detailed economic feasibility assessment can be made. Economic indicators commonly used involve payback time, return on investment, net present value and internal rate of return. Next to economic feasibility, the HSE (health, safety, environmental) and sustainability performance of the project are also commonly systematically assessed against current and future design constraints (Fig. 16.4) as well as company policy. Sustainability evaluation typically involves Life Cycle Assessment (LCA), which can be performed at various levels of detail, depending on the state of development. The most simple is a carbon footprint estimation providing a first indication whether the envisaged process has a lower footprint than existing processes.

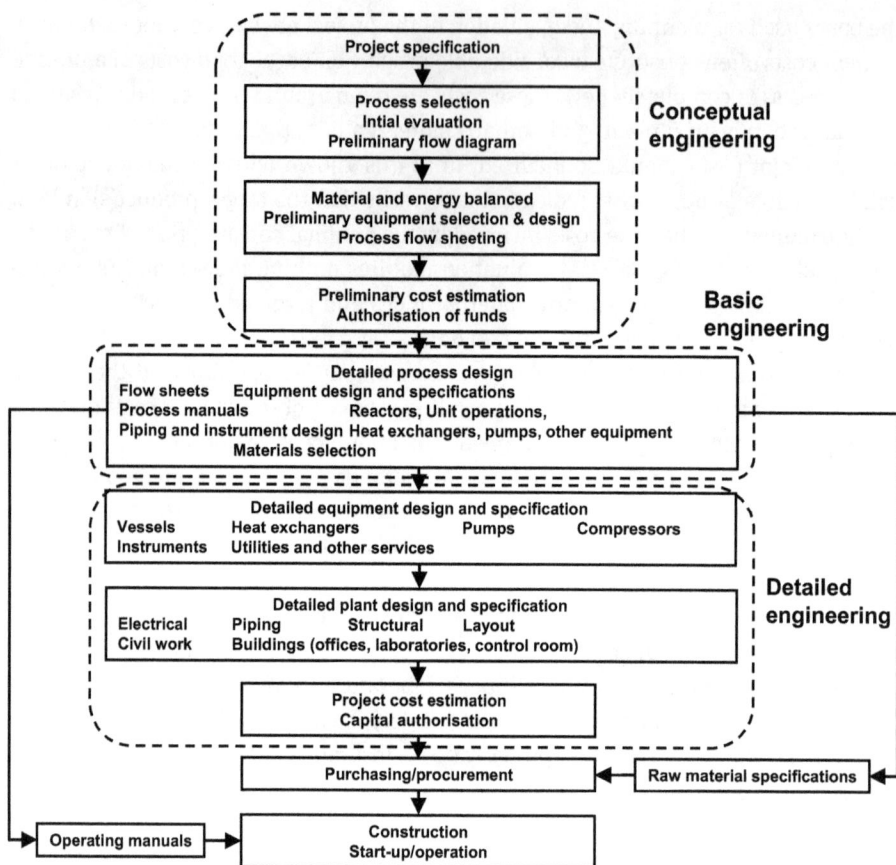

Fig. 16.5: The structure of a chemical engineering project (adapted from [2]).

16.3 Engineering and construction

16.3.1 Introduction

Once the decision has been made to go ahead with the design and construction of a new plant, the preliminary design resulting from the feasibility study needs to be transformed into detailed plans and specifications for all the components of a chemical plant. The sequence of steps in the design, construction and start-up of a typical chemical process plant is shown diagrammatically in Fig. 16.5. Three major project phases can be distinguished:

1. Conceptual Engineering covers the process design, which covers the steps from the initial selection of the process to be used, through to the issuing of the process flow-sheets, and includes the selection, specification and chemical

engineering design of equipment. In a typical organization, this phase is the responsibility of the process design group.

2. Basic Engineering concerns the detailed mechanical design of equipment, the structural, civil and electrical design, and the specification and design of the auxiliary services. These activities will be the responsibility of the specialist design groups. Other specialist groups will be responsible for cost estimation and the purchase and procurement of equipment and materials.

3. Detailed Engineering, Procurement and Construction (EPC) consists of the engineering of the chemical plant, procurement of plant equipment and material, construction and commissioning. Purchasing agents procure equipment from specialist manufacturers. Construction and installation firms are put under contract to build the plant.

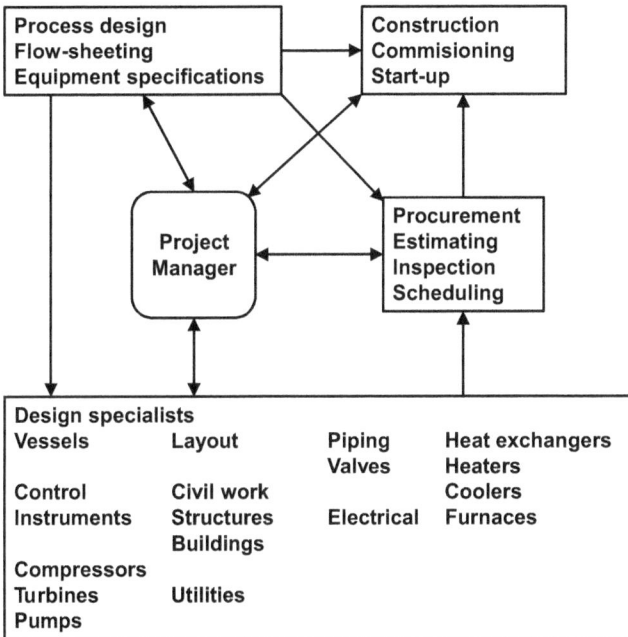

Fig. 16.6: Project organization (adapted from [2]).

Chemical plant design and construction projects are complex operations comprising many interrelated activities and personnel from a wide range of disciplines working together in an interdisciplinary setting.

Completion of the project within a predefined time and budget calls for careful organization of the people working on the project, a clear definition of their responsibilities and competences, and an appropriate management concept. A project manager, often a chemical engineer by training, is usually responsible for the coordination of the project as shown in Fig. 16.6. For the customer, the project manager is the first representative of the engineering contractor for the project. In larger projects, project engineers are put under an experienced project manager and take responsibility for the execution of portions of the project. Care should be taken that the final construction and operation of chemical plants are governed by many environmental protection and safety regulations. These requirements have major consequences for plant design and construction. After starting the preliminary talks with regulatory authorities, it may take two or more years before the application for a construction permit is approved. As this is a substantial fraction of the 2–4 years total project duration, the engineering and authority engineering activities must be well coordinated to avoid future delays in the work at the construction site.

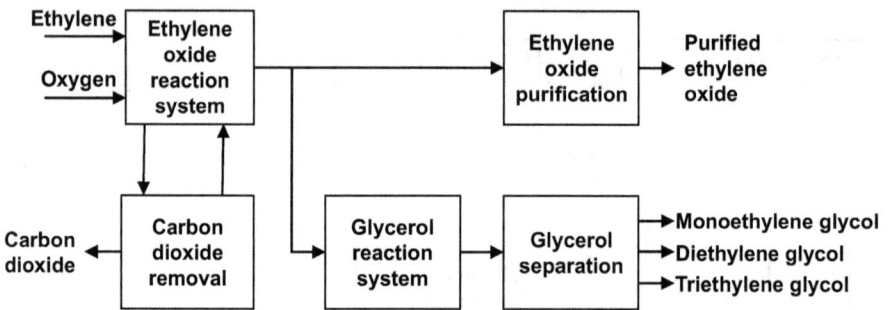

Fig. 16.7: Block diagram of an ethylene oxide and ethylene glycols plant.

16.3.2 Conceptual engineering

Based on the outcome of the feasibility study, one can decide to proceed to the conceptual design phase. The feasibility study defines the process objective, which provides the basis for process selection. The main task in conceptual design is to obtain a more exact calculation that takes into consideration all costs, up until commissioning. The first step toward this objective is to work out the engineering details. Process selection can be done most simply in the form of a block flow diagram (Fig. 16.7), in which each block represents a unit operation or, in complex plants, a plant section containing several unit operations. Lines representing the principal material and energy streams connect the blocks. The first step in process design is to establish the operating parameters for the major stages in the process. The next step is to prepare a process flow diagram from the block flow diagram (Fig. 16.8). Standard symbols are used to represent reactors and other apparatus. The next step is to compile the specifications for all feed stocks, auxiliaries, catalysts, utilities and products. On the basis of the process flow diagrams, the preliminary process parameters and the specifications are used to prepare material and energy balances for the process steps and, finally, for the entire process. This objective is achieved through an iterative procedure to modify the process flow diagram and/or the process parameters, so that a closed material and energy balance is attained. Process design calculations for multistage, interconnected processes with material and energy recycle, soon become very complex. Computer tools such as flow sheeting programs (Aspen Plus, Process) enable designers to model unit operations and allow them to be interconnected. A plant can thus be represented as a network of unit operations with material and energy streams in the simulations. The last step in preliminary design is the preparation of a report containing the updated feasibility study, schedules for financing and personnel requirements, time schedule for the phases of project execution and project manual containing all the studies and results from preliminary design.

16.3.3 Basic engineering

Basic engineering is based on the process design, which was developed during the conceptual design phase. During basic engineering, the geometric dimensions of individual equipment items, the design capacities, temperatures and pressures, construction materials and the layout of the entire plant are established. These design data are entered in process engineering data sheets that contain all relevant specifications for the specialist engineer. Material selection should be solved by close collaboration between the materials specialist, the designer and the process engineer. A process and instrumentation diagram (P&ID) (Fig. 16.9), based on the process flow diagram is needed for more accurate calculations at the preliminary design stage. For the basic engineering package, it should contain dimensions and construction materials for all equipment, machinery and piping as well as the field instruments and control devices required for the operation and control of the plant. The plant layout should not only include approximate data on the position and sizes of the main plant items, but also include pipe bridges, roads, control room, compressor buildings, etc. The accessibility of plant equipment for repair, maintenance, construction, safety and inspection must be considered from the very beginning. If the existing information is adequate, the layout should be drawn to scale.

Once the engineering documents are available, detailed plant costs can be determined. From Fig. 16.10, it can be seen that due to a much more detailed knowledge, significant improvement in the accuracy of the estimated investments costs has been achieved, compared to when the development phase was completed. This improved accuracy is achieved at the expense of investing considerable time and money during basic engineering. The investment calculation is subdivided into equipment, bulk materials and indirect costs. Equipment includes all itemized plant equipment such as towers, reactors, heat exchangers and pumps. The best way to get accurate equipment prices is to submit enquiries to the manufacturers. Such inquiries are

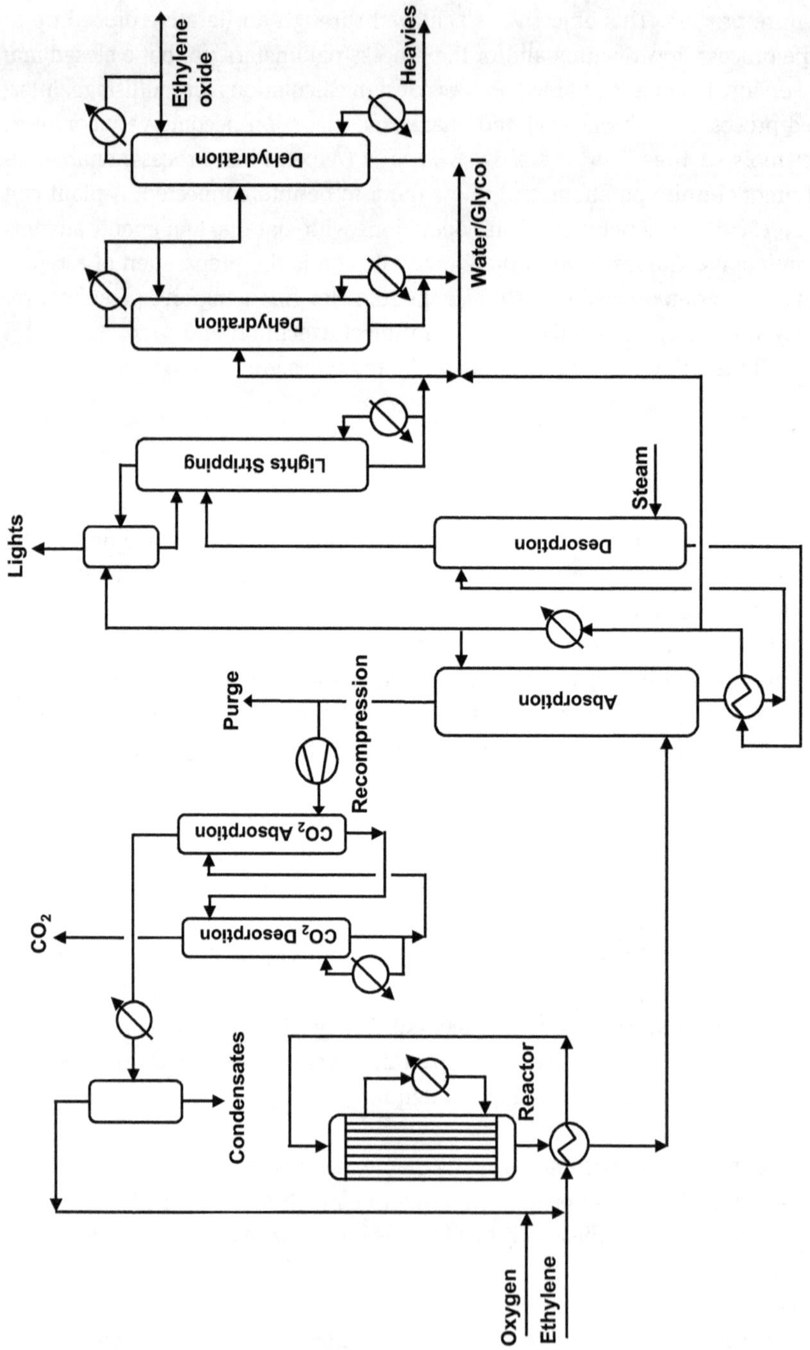

Fig. 16.8: Example of process flow diagram for an ethylene oxide plant.

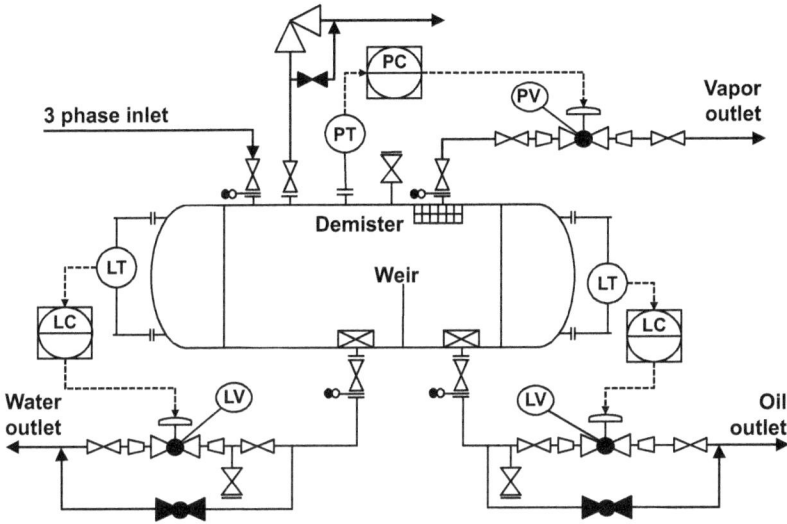

Fig. 16.9: Example of a P&ID diagram.

time-consuming and, hence, unwelcome. Therefore, inquiries are only made with complicated equipment. Simpler items are estimated with in-house documents, based on the costs of previously purchased equipment. Bulk materials cover items such as pipes, control systems, electrical equipment, insulation and paint. Provided the engineering documentation is exact enough, bulk material costs can be calculated fairly accurately, by applying unit prices. Other costs include engineering, procurement, construction supervision, commissioning, insurance, price escalation and contingencies. As opposed to bulk material costs, the other costs are calculated for the project as a whole, by employing multiplication factors.

The final basic engineering documentation includes the process flow diagrams, piping and instrumentation diagrams, plant layout, equipment list, utilities distribution scheme, process engineering data sheets, noise protection concept, electrical equipment list, summary of electrical consumers, data sheets for control equipment and instrumentation, functional process control and instrumentation plan, description of the process control and soil report. The content and level of detail in basic engineering documentation is such that detail engineering can be done without significant difficulties.

16.3.4 Detailed engineering, procurement and construction

In detailed engineering, the engineering and procurement teams prepare detailed plans, drawings, specifications, calculations and descriptions needed for bid invitations, selection of manufacturers, vendors and subcontractors, shipping of plant

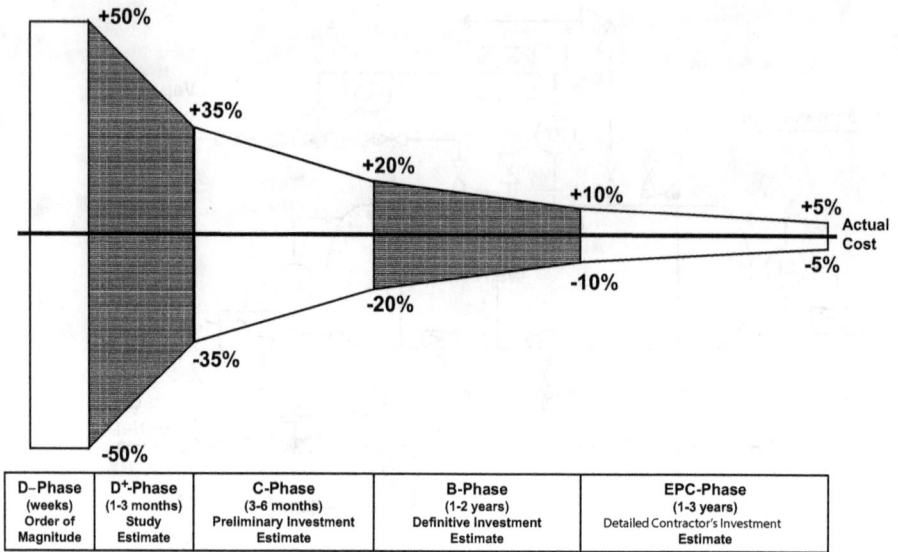

D–Phase (weeks) Order of Magnitude	D⁺-Phase (1-3 months) Study Estimate	C-Phase (3-6 months) Preliminary Investment Estimate	B-Phase (1-2 years) Definitive Investment Estimate	EPC-Phase (1-3 years) Detailed Contractor's Investment Estimate

Fig. 16.10: Effect of project phase on accuracy of estimated investment costs.

equipment, execution of civil work, erection of plant equipment and commissioning. Most of the process engineering of a project would have been performed at the basic engineering stage. In addition, the process engineers devise the plant control concepts, complete piping and instrumentation diagrams, prepare the final process description and write detailed startup instructions needed for commissioning. The plant layout (plot plan) and piping and instrumentation diagram created during basic engineering are continually updated in the course of detailed engineering. The preliminary P&I diagram from the basic engineering is developed and elaborated in detailed engineering. The final P&I diagram describes the whole plant in detail. The P&I diagram contains all essential information developed by the individual disciplines and also includes data provided by the manufacturers of equipment and machinery. Since this information becomes available over a prolonged span of time, the P&I diagram is revised several times in the engineering stage.

All important process engineering data for equipment and machinery have been specified during basic engineering. In detailed engineering, the equipment and machinery engineers complete this information. The equipment engineers prepare so-called guide drawings (Fig. 16.11), which are scaled drawings indicating all dimensions dictated by process engineering. The guide drawings and supplementary information form the technical portion of the bid invitation, which is sent to selected manufacturers. The design office of the manufacturer prepares detailed workshop drawings. Once these drawings have been approved, production can begin. The specification and procurement of machinery are similar to that for apparatus. During piping design, all drawings and specifications needed for procurement and installation of the piping components are prepared. The engineering of piping systems is closely linked with the engineering of all other disciplines. Piping accounts for a relatively high proportion of chemical plant costs, and piping engineering may represent as much as 20–40% of total engineering. Refineries and petrochemical plants lie at the upper end of this range. Reliable and economic supply of electric power to all consumers is covered during electrical design. Design begins where high-voltage power enters the plant and covers the planning of power generation and distribution, planning of electric utilities and installation planning. Electrical installation accounts for 6–10% of the total chemical plant costs, so detailed planning is a prerequisite for economic execution.

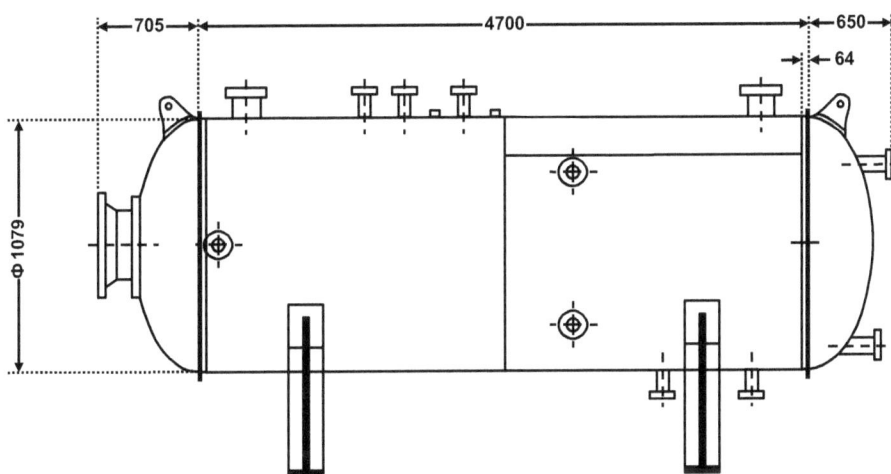

Fig. 16.11: Guide drawing for a vessel.

Procurement activities include preparing bid invitations, comparing bids, purchasing of plant components, expediting during the fabrication of plant components and shipping of plant components to the construction site. The procurement and engineering activities are closely linked together. The main execution phases of a project up to mechanical completion (engineering, procurement, civil work and erection) overlap one another, in time. The sequence of engineering work should guarantee that plant equipment with long delivery times is ordered as early as possible and civil work is begun early, so that equipment erection is not delayed. Civil work and erection of a facility are generally subcontracted to specialist firms by the engineering contractor or by the investor. The planning of civil work and erection is part of detail engineering and done by engineers in the civil engineering department assigned to the project team. Often, detailed civil design is assigned to engineering firms in the country where the plant will be built. These firms are familiar with local conditions, know the local regulations and have short lines of communication to the construction site and the firm performing the civil work. As soon as the engineering work is 25–30% complete, civil work should be started.

Planning the installation of plant equipment starts at a relatively early stage in the engineering process. In large plants, separate schedules are worked out for each plant section. The engineering firm generally directs the construction work performed by specialist subcontractors with the construction manager and his team supervising, coordinating and directing construction. An overall construction schedule is created during detail engineering. Work at the construction site begins with preparation of the terrain, followed by excavation and foundation work. The first step is the erection of heavy equipment and steel structures. Piping installation at a chemical plant is often the most labor intensive and longest phase of installation. Insulation work starts at vessels, towers and reactors. Pipes should not be insulated until a given plant section has a sufficiently large number of pipe runs that have been approved for insulation. The installation of electronic devices and control systems takes place after a section of piping has been completed. Devices in control rooms and substations can be installed independent of other work as soon as the buildings have been completed. Functional tests of the installed equipment mark the end of the erection work. These tests are done with the plant in cold condition and with no product.

After mechanical completion of the plant, most of the installation personnel leave. Some specialist engineers remain on site to solve problems that arise during commissioning. Commissioning of a plant comprises all the work done after mechanical completion up to certification of the guarantees embodied in the contract. Commissioning must be considered even during basic and detail engineering. Faulty process design can have serious effects on the time required for commissioning and the amount of corrective work needed. The start of production may be significantly delayed, and the owner may suffer a substantial loss of production and revenue. Responsibility for commissioning generally lies with the party granting the process license. Experienced startup engineers occupy the key positions in the commissioning team. Commissioning takes place according to the steps specified in the operating manual. The first units to be started are utilities and off sites. During start up of the plant, initial disorders are almost always encountered. An attempt should be made, however, to get the plant running first and start up all systems, provided the safety of personnel and equipment is not endangered. After operation has stabilized, conditions are optimized. When the planned values of product quantity and quality have been attained, the guarantee test is carried out. When the guarantee tests defined in the contract have been passed, the plant is handed over to the owner.

16.4 Development and scale-up of fine chemical processes

16.4.1 Differences between bulk and fine chemical industry

Table 16.2 summarizes a few important differences between the bulk chemical industry and the fine chemicals industry. The design of batch processes for fine/specialty chemicals production differs significantly from the design of large-scale continuous processes. Economic constraints result into a rather narrow window in fine chemical process development. Product innovation absorbs considerable resources in the fine chemicals industry, in part, because of the shorter life cycles of fine chemicals as compared to commodities. Process R&D is focused on existing installations, which are multi process-, multipurpose- and multi product-oriented. The reaction media are usually organic in reagents, solvents and products, and the chemistry is generally based on a flavored mix of mostly classical organic synthesis combined with emerging bioorganic synthesis and catalysis. Processing is based on batchwise operations in batch reactors and auxiliary equipment, while the productivity is limited due to relatively small-scale reactors. Batch processes are chosen because of their flexibility, but also because continuous production of small amounts is difficult to perform and often not economic. The expensive controlling equipment combined with the fact that each continuous process requires its own separation and purification equipment, makes small continuous processes usually more expensive than batch processes. Another advantage of batch processes is the higher degree of conversion, which can often be achieved in comparison with continuous processes. Due to these typical characteristics of the fine chemicals industry, special attention is paid to the process development and scale modification in this area.

Tab. 16.2: Differences between the bulk industry and the fine chemicals industry.

Bulk industry	Fine chemicals industry
Relative cheap raw materials	Expensive raw materials
Low added value	High added value
Large reactor volume	Small reactor volume
1–3 reaction steps	3–15 (or more!) reaction steps
Continuous processes	Batch or fed-batch processes
Equipment specific for one process	Multipurpose equipment

16.4.2 Fine chemical process development

The main task of R&D in fine chemicals is scaling up laboratory processes, so that the processes can be transferred to pilot plants and subsequently to industrial-scale production. Thus, the R&D department of a fine chemicals manufacturer,

```
┌─────────────────────┐
│   Route             │
│   selection         │
└─────────────────────┘
          │
          ▼
┌─────────────────────┐
│ Process development │
│   (chemistry &      │
│    technology)      │
└─────────────────────┘
          │
          ▼
┌─────────────────────┐
│ Production decisions│
│   (make/buy/toll/   │
│  custom manufacture)│
└─────────────────────┘
          │
          ▼
┌─────────────────────┐
│  Detailed design by │
│ operator, contractor or│
│     toll/custom     │
│    manufacturer     │
└─────────────────────┘
          │
          ▼
┌─────────────────────┐
│  Construction &     │
│  production         │
└─────────────────────┘
```

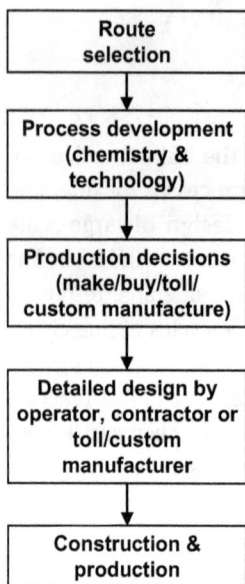

Fig. 16.12: Stages of Batch Process Development.

typically, is divided into a laboratory or process research section and a development section. Figure 16.12 shows the typical stages of design for manufacturing an organic chemical by a batch process. A central feature of batch process design for organic chemicals production is the importance of chemistry. During process development, it is not unusual to solve problems by a change in chemistry to avoid engineering difficulties. This is particularly the case during the early stages of development, the 'route selection' stage. The chemical route is the set of chemical reactions, which turn the feed stocks of a process into the products. In fine/specialty chemicals production, route selection is a major activity in process development. The number of feasible chemical routes depends most strongly on the type of product. Key features of the route selection decisions are that they are taken early in a project based on limited information. The decisions are complicated by the poor quality of available information. The challenges of route selection are complex due to multiple decision criteria such as process economics, potential safety and environmental compatibility.

Once a route has been selected, it will still require significant development to maximize yields and other performance measures. This work will usually be carried out in the development section that serves as an intermediary between laboratory and industrial scale, and operates the pilot plant. Both equipment and plant layout of the pilot plant mirror those of an industrial multipurpose plant, except for the size (typically 0.1 to 2.5 m^3) of reaction vessels and the degree of process automation. The resulting process description or recipe is, in essence, the equivalent of a mass balance and a flow sheet for a continuous process. It provides the necessary data for the determination of preliminary product specifications, the manufacture of semi-commercial quantities in the pilot plant, the assessment of the ecological impact, an estimation of the manufacturing cost in an industrial-scale plant, the validation of the process and determination of raw material specifications. Once a laboratory process has been adapted to the constraints of a pilot plant, has passed the risk analysis, and demonstration batches have been successfully and repeatedly run, the process is ready for transfer to the industrial-scale plant.

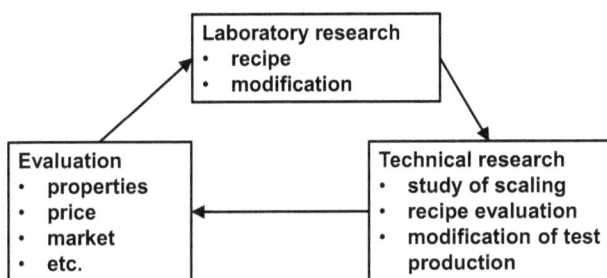

Fig. 16.13: Scale modification cycle in the fine chemicals industry.

16.4.3 Scale-Up challenges in multipurpose plants

Traditionally, the fine chemical process chemist is focused on route selection, optimization and critical process parameter identification, whereas the chemical engineer and the chemist are actively involved in the scale-up of fine chemical processes and are dealing with the principle that chemical rate constants are scale-independent, whereas physical parameters are not. Scale-up does not occur in a single step, but is typically performed by repeating the operation cycle displayed in Fig. 16.13, several times. A practical cycle would include testing of the laboratory recipe, stepwise increase of the scale until the desired production is achieved and continuous formulation and modification of the recipe. The final objective is to obtain a maximal yield of the pure product via a clean and safe process, at minimal costs. Physical changes, most commonly, cause a drop in process performance during scale-up, resulting in a poor chemical selectivity, reduced chemical yield or poor quality. Encountered scale-up problems are often related to the fixed design and absence of geometrical similarity between the small and the large-scale reactor in a multi-purpose installation. Adaptation of the bulk-chemical way of technology and process development will never be feasible, because the short-time-to-market approach often prevents a smooth realization of the fine-chemical process development trajectory. As an illustration, modeling techniques as such are hardly applicable, when time is the most important performance indicator in process development. In contrast, early recognition of scale-up problems in the lab phase and modification of the fine chemical batch reactor could boost the development speed to the required levels.

In 1995, a benchmark study was initiated to define the scale-up performance of the fine chemical industry. A knowledge base called ALICE was set up containing above 70 (non)-flyer examples and various scale-up factors. ALICE stands for *Analysis Like Interactive Concurrent Evaluation*. A (non)-flyer is a surprise in scale-up that needs a correction on an initially bad performance, resulting in chemical yield and product quality on large scale, which is, at least, identical to lab-scale process performance. Because fine chemical process R&D is highly focused on multi process implementation in an existing installation containing the same batch reactors and auxiliary equipment allowing only minor investments, it was expected that some kind of systematics in these scale-up surprises would appear. Evaluation of the (non)-flyers in Fig. 16.14 shows that in approximately 55% of the cases, the surprise in scale-up is based on longer processing times, heat and mass transfer and mixing, while on various typical fine chemical issues in the remaining cases. Filling a large-scale reactor simply takes more time than a small-scale reactor. What can be added in a few seconds on lab-scale may take hours in 10 m³ reactors. Serious delays may also occur if heat transfer rates are overestimated. A scale-up factor of 100 gives a 20-fold reduction in surface area to volume ratio. Then, the need to apply large temperature differences between wall and reaction mixture may cause fouling of the reactor wall due to coating and decomposition. Such fouling layers inside the reactor generally cause additional delay in heating and cooling rates. Large scale mixing takes time. What can be mixed in seconds on a lab-scale takes minutes in a 10 m³ reactor. Various factors, such as reactor shape, stirring rate and agitator shape, can be used to influence the mixing time. The power input is an important parameter in scale-up as well as the type of agitator. Mass transfer, chemical reaction or both may limit the rate of two-phase reactions. Possible mass transfer limitation may lead to difficulty in scale up. Also, isolation of solids is a point of concern during scale-up, as it is for maintaining the quality of the recycled materials.

16.4.4 Future developments

Various computer and robot-aided design methods as well as statistical methods are nowadays available to assist the chemist in finding the middle of the road in scale-up of fine chemical batch processes. Once the chemistry and chemical rate constants have been established on lab-scale in a master recipe, the resulting operating window is a good indication for the robustness of the selected process and a good starting point for the scale-up study. Improved scale-up characteristics can be expected from a multi-flexible batch reactor (Fig. 16.15), constructed in such a way that it is highly flexible in stirring rates, agitator types, dip-pipe feeding position and a multi-nozzle system for feeding chemicals, flexible baffles and loop facilities to increase mixing with a static mixer and also to increase heat exchange. Having such a batch reactor installed in a

Analysis Like Interactive Concurrent Evaluation

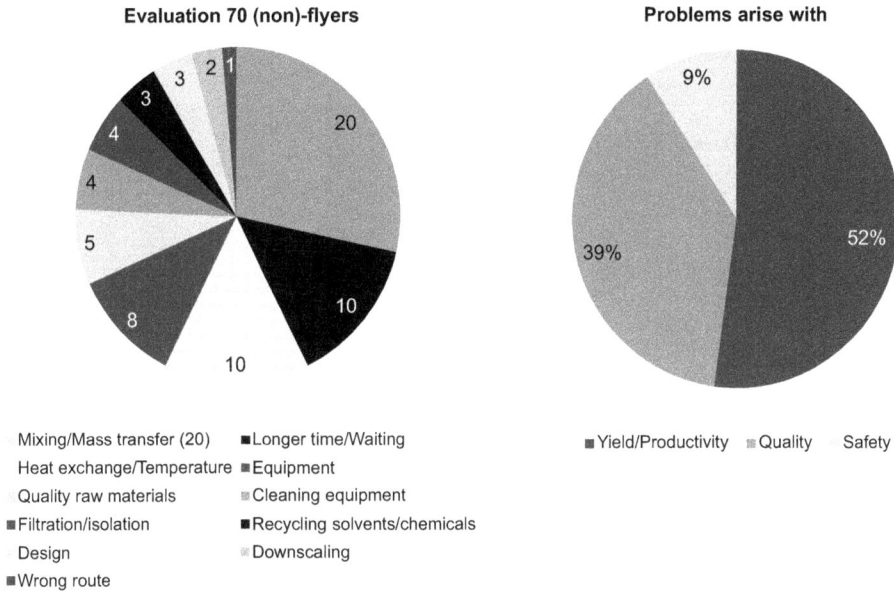

Evaluation 70 (non)-flyers

3 3 2 1
20
4
4
5
8
10
10

Mixing/Mass transfer (20)   ▪Longer time/Waiting
Heat exchange/Temperature ▪Equipment
Quality raw materials   ▫Cleaning equipment
▪Filtration/isolation   ▪Recycling solvents/chemicals
Design   ▫Downscaling
▪Wrong route

Problems arise with

9%
39%
52%

▪Yield/Productivity ▫Quality   Safety

Fig. 16.14: ALICE in process wonderland.

multi-purpose plant, the remaining surprises in scale-up can be reduced to a minimum, and immediate correction on large scale during the first campaign can be carried out, without lengthy downscaled experimentation on lab or pilot-plant scale.

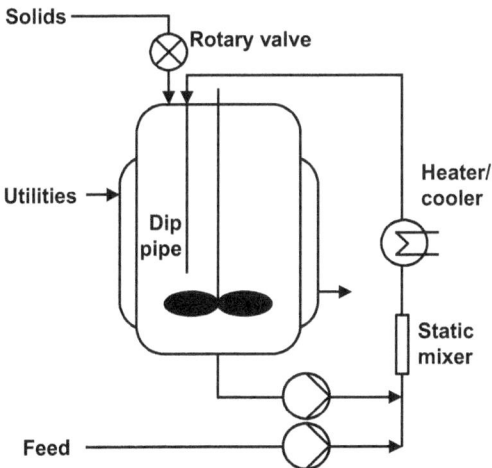

Solids —— Rotary valve

Utilities →

Heater/ cooler

Dip pipe

Static mixer

Feed ——

Fig. 16.15: Optimized design of a flexible, multipurpose batch reactor.

Nomenclature

C	capacity	[ton year^{-1}, kg s^{-1}]
I	investment capital	[\$, €]
L	equipment characteristic length/size	[m]
K	salary cost	[\$ year^{-1}, € year^{-1}]
m	degression coefficient investment capital	[–]
n	degression coefficient salary costs	[–]
N	number of parallel units	[–]

17 Hydrodynamic aspects of scale-up

17.1 Introduction

In this chapter, an introduction to scale-up in the process industry is given. Understanding the hydrodynamics of flows plays an essential role in scale-up. In the first part, some basic principles of single-phase and multiphase flows are given, with an overview of the hydrodynamic flow phenomena that occur during mixing, stirring, bubbling and fluidization operations. In the second part, the tools to perform scale modification are discussed, emphasizing the importance of dimensionless numbers. The last part consists of a short introduction to different types of computational fluid dynamic (CFD) simulations, which can aid in scale-up of process equipment.

Scale-up is a very important step in the design of a new plant and new process equipment. Because of the enormous costs involved, it is not possible to perform experiments for process design on full plant scale as the process would be, when finally operational. Therefore, experiments have to be carried out on a small scale, but these have to be representative of the full-scale process. To make sure the experiments represent the full-scale process, several aspects have to be considered. A simple example of the problems encountered in scale-up is the decrease of surface area relative to the volume of an apparatus, which is important in cooling, for example. If a sphere is considered, the surface area and volume can be calculated by $4\pi r^2$ and $(4/3)\pi r^3$, respectively, and thus their ratio (area/volume) is $3/r$. Cooling a larger vessel will, therefore, only be possible by introducing extra coils inside the vessel. Moreover, the hydrodynamics of a set-up i.e., the fluid flow can also differ significantly by increasing the equipment size, leading to different heat and mass transfer rates. This has been subjected to a lot of research, as it has influence on the product yield and quality. Examples of common hydrodynamic problems in process engineering are the occurrence of dead zones, channeling and bypasses in mixing tanks as illustrated in Fig. 17.1.

To scale up in a structured way, and design representative experiments, dimensional, regime and similarity analysis are very useful (and necessary) tools. Dimensional analysis is a theory in which several dimensionless numbers are used to describe a process. These can be geometrical ratios, ratios of relevant forces and so on. When scaling up, it is important to keep the most relevant dimensionless numbers, the same. Scale-up is often supported by CFD simulations, which make it possible to predict the fluid flow and heat and mass transfer in a scaled-up setup before it is actually built, facilitating its design and optimization.

https://doi.org/10.1515/9783110712445-017

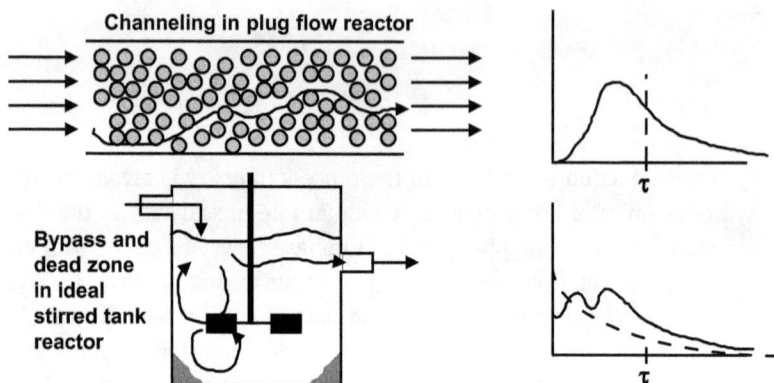

Fig. 17.1: Important examples of hydrodynamic problems in scale-up.

17.2 Hydrodynamic flows in process equipment

17.2.1 Single-phase flows

The simplest processes involve single-phase flows. Examples include gas-phase or liquid-phase reactors and mixers for miscible liquids. One of the most important numbers characterizing such a flow is the Reynolds number, named after Osborne Reynolds (1842–1916), who studied fluid flow in pipes, among other phenomena. The Reynolds number represents the ratio of inertial force and viscous force. In the case of fluid flow through a pipe, which can be either a gas or a liquid, the Reynolds number Re is given as follows:

$$\text{Re} = \frac{\rho v d}{\eta} \tag{17.1}$$

where ρ_F is the fluid density, v the (average) fluid velocity, d the diameter of the pipe and η the dynamic viscosity of the fluid. The Reynolds number is an indication of the flow regime present, in this case, in the pipe. Since the flow regime has a large influence on the rate at which, for example, heat and mass- transfer occur, it can be understood that it is very important to have a control over this dimensionless number. If Re < 2,000 (low velocity or narrow pipe), the flow is laminar (Fig. 17.2a) and if Re > 2,000 (high velocity or wide pipe), the flow is turbulent (Fig. 17.2b).

In the laminar flow regime, almost no transport of heat or mass takes place in the radial direction. Therefore, heat and mass-transfer is relatively poor in laminar regimes and heating or cooling a fluid in a pipe will almost always take place in a turbulent regime.

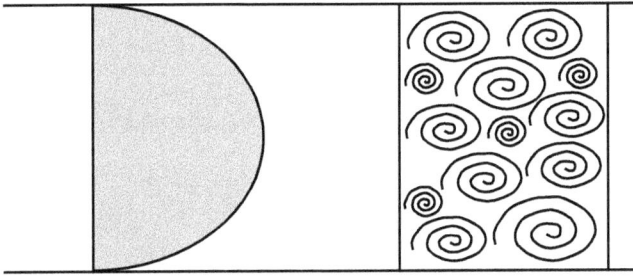

Fig. 17.2: Velocity profiles for different flow regimes: (a) laminar flow and (b) turbulent flow.

17.2.2 Multiphase flows

In most process equipment, more than one phase is present in the flow. Examples include stirred tank and fluidized bed reactors, which maximize mass transfer of reactants from a fluid phase to dispersed catalyst particles, bubble columns, which maximize mass transfer between a liquid and a gas, and extraction equipment, in which two immiscible liquids are vigorously stirred to enhance mass transfer from one liquid to another.

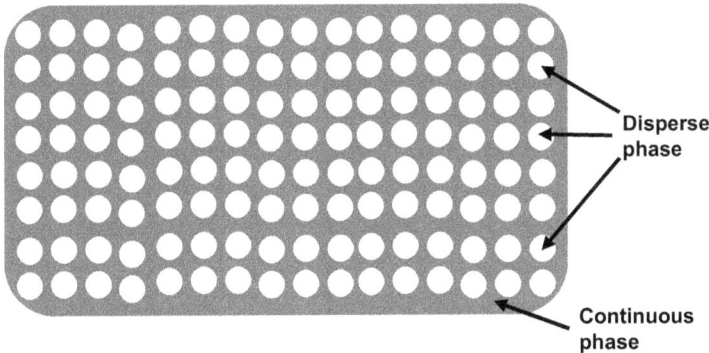

Fig. 17.3: Schematic representation of a multiphase mixture with a continuous and a dispersed phase.

Each phase can be classified as either continuous, if it occupies a connected region in space, or dispersed, if it occupies disconnected regions in space. See Fig. 17.3. Because we are considering flowing multiphase mixtures, at least one of the phases must be a continuous gas or a liquid. The dispersed phase can be a solid, liquid or gas, i.e., in the form of particles in a continuous gas or liquid phase, bubbles in a continuous liquid phase, or droplets in a continuous gas or another continuous liquid. For solid particles, it is obvious that they form a dispersed phase, but whether a gas or liquid phase

will be continuous or dispersed, and in the case of the latter, how finely it will be dispersed depends on
- the phase fraction of each phase
- physical properties (density, surface tension, viscosity) of each phase
- process design (sparger, atomizer, stirrer or pipe dimensions)
- operating conditions (pressure or flow rates through spargers, atomizers or pipes, stirring speed)

The phase fraction, α_p, of a phase, p, indicates the relative amount of volume occupied by that phase. Because the sum over all phase fractions must be unity, a two-phase flow can be characterized by the fraction of only one of its phases. Although there is no fixed definition, dispersions with a phase fraction up to 10^{-3} are considered dilute, between 10^{-3} and 0.1 as semidilute and larger than 0.1 as dense dispersed multiphase flows.

The physical properties of the phases and the operating conditions are very important in determining the type of multiphase flow. In particular, for systems with large density differences and operating conditions that lead to nondilute phase fractions, the hydrodynamics of each phase is significantly influenced by the presence of the other phase. Moreover, in dense dispersions, direct interactions between the particles, droplets or bubbles lead to significant resistance to flow and cluster formation. As a consequence, compared to single-phase flow, many more dimensionless numbers are needed to properly characterize a multiphase flow. Before discussing what these dimensionless numbers are, we will discuss some relevant hydrodynamic phenomena in a few selected types of process equipment.

17.2.3 Flows in mixers and stirrers

Mixing is an important unit operation carried out to homogenize materials, in terms of concentration of components, temperature and physical properties to intensify heat transfer and to create dispersions, suspensions or emulsions of mutually insoluble phases. Mixing can be achieved actively in mechanically agitated vessels or passively in static mixers.

17.2.3.1 Macromixing and micromixing
Mixing is considered to occur at the macro- and the microlevel. Macromixing is established by means of convective flow. When macro-scale variables are involved, every geometric design variable can affect the role of shear stresses. They can include items such as power, impeller speed, diameter, blade shape, width, height or thickness, number of blades, baffle location and number of impellers. Mixing time will influence the final result, i.e., reactant conversion and/or selectivity.

Microscale variables are involved when the particles, droplets, baffles or fluid clumps are in the order of 100 μm or less. In this case, the critical parameters are usually power per unit volume, distribution of power per unit volume between the impeller and the rest of the tank, velocity fluctuations, energy spectra, dissipation length, the smallest microscale eddy (swirl) size for the particular power level, and viscosity of the fluid. Micromixing occurs by turbulent diffusion and is primarily dependent on the energy dissipation per unit volume. The mixing energy is transferred from the largest eddies to the smallest ones, until it is eventually dissipated through friction in turbulent and viscous shear stresses and finally appears as heat in the system.

The rate of turbulent flow is greatest, close to the impeller. Here, there is a high shear rate due to the trailing vortices associated with all impellers. Furthermore, a high proportion of the energy introduced by the impeller is dissipated here. In general, the specific energy dissipation around the impeller is approximately 100 times higher than in the rest of the tank. This results in a velocity fluctuation ratio to the average velocity in the order of 10:1 between the impeller zone and the rest of the tank. Thus, the rate of homogenization of miscible liquids is greatest in this region, and gas and liquid–liquid dispersion occurs predominantly here. Therefore, processes that are particularly dependent on turbulent eddies and their associated forces are likely to be well correlated by the energy dissipation rate. Bubble formation and micromixing fall in this category. However, processes that are dependent on an-isotropic main flow and for which the nonhomogeneous nature of stirred tanks turbulence is significant, solid suspension and solid–liquid mass transfer, for example, are not well correlated that way. These processes primarily benefit from a longer mixing time.

17.2.3.2 Mechanically agitated vessels

Agitated vessels are frequently used in batch as well as in continuous service. Liquids are most often agitated in some kind of a tank or vessel, usually cylindrical in form and with a vertical axis. The proportions of the tank vary widely, depending on the nature of the agitation problem. A standardized design such as that shown in Fig. 17.4 is applicable in many situations. The tank bottom is rounded, instead of flat, to eliminate sharp corners or regions into which fluid currents would not penetrate. The liquid depth is approximately equal to the diameter of the tank. Mechanical agitation utilizes a rotating impeller immersed in the liquid to accomplish the mixing and dispersion. The impeller is mounted on a shaft supported from above, which is driven by a motor that is sometimes directly connected to the shaft but more often through, a speed-reducing gearbox. Accessories, such as inlet and outlet lines, coils, jackets and wells for thermometers or other temperature-measuring devices are usually included. The design of the mixer can be based on the discharge capacity of the propeller and on liquid entrainment, in relation to the tank volume and the desired blending time. The operating conditions are chosen to maximize the liquid throughput at a given power. This is achieved by a large stirrer diameter and a low stirrer speed.

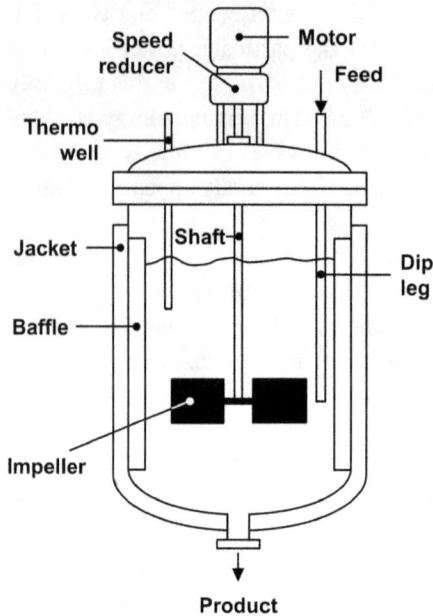

Fig. 17.4: Characteristic features of an agitated vessel.

The impeller causes the liquid to circulate through the vessel and eventually return to the impeller. If an axially positioned stirrer is operated in a vessel without inserts, the liquid is set into rotation and a vortex is formed. This is generally undesirable because gas can be entrained in the liquid, the degree of filling is reduced and bulk rotation of the liquid counteracts mixing. In large tanks with vertical agitators, the preferred method of reducing rotation and vortex formation is to install baffles, which impede rotational flow without interfering with the radial or longitudinal flow. Installing vertical strips on the wall of the tank attains a simple and effective baffling. The standard baffle configuration consists of four vertical plates having a width equal to 8–10% of the tank diameter. A small spacing between the baffles and the tank wall is allowed to minimize dead zones particularly in solid-liquid systems. Although the presence of wall baffles causes an increase in power consumption, they generally enhance the process result.

Some of the commonly used impeller agitators are shown in Fig. 17.5. They are divided into two classes. Those that generate fluid flow parallel to the axis of the impeller shaft are called axial-flow impellers and those that generate flow in a radial or tangential direction are called radial-flow impellers. The three main types of impellers for low-to-moderate viscosity liquids are propellers, turbines, paddles and

Liquid Viscosity (kg.m⁻¹s⁻¹)

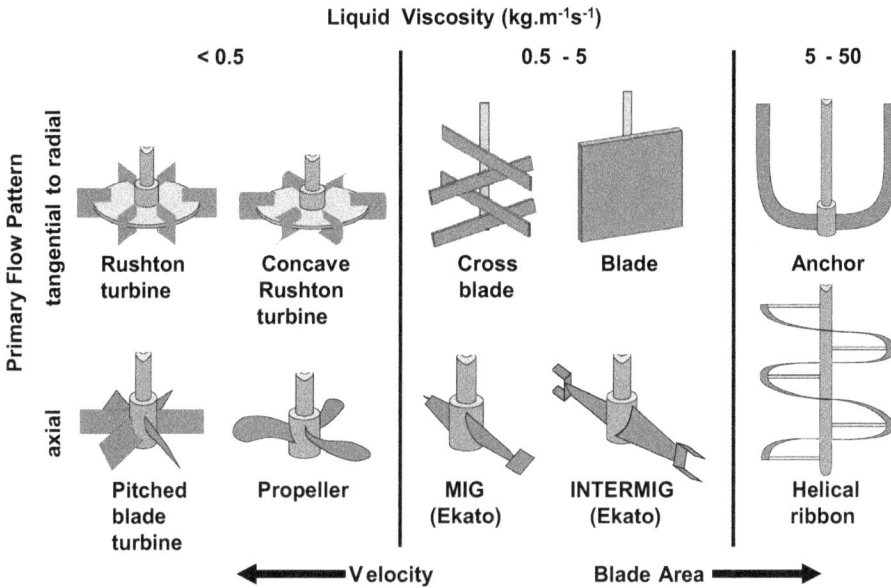

Fig. 17.5: Schematic of important industrial impeller types.

high-efficiency impellers. A propeller is an axial-flow, high-speed impeller that is suitable for bulk mixing of low viscosity liquids. Flat-bladed turbines are essentially radial-flow devices, suitable for processes controlled by turbulent mixing. Anchor and helical ribbon agitators, and other special shapes are used for more viscous fluids. The diameter of the helix is very close to the inside diameter of the tank. To provide good agitation near the floor of the tank, an anchor impeller may be used. Because it creates no vertical motion, it is a less effective mixer than a helical ribbon, but it promotes good heat transfer to or from the vessel wall.

As mentioned in the previous paragraph, the two main classes of turbine impellers are based on the flow patterns they generate: axial flow and radial flow. Axial flow impellers produce a flow pattern involving the full tank volume as a single stage (Fig. 17.6b and c). Radial flow impellers, however, produce two circulating loops, one below and one above the impeller (Fig. 17.6a and d). Mixing occurs between the two loops, but less intensely than within each loop. These differences in flow patterns cause variations in the shear rate distributions in the tank so that the mixing result is highly impacted by the impeller flow patterns. The flow patterns within a given impeller-type are altered by parameters, such as impeller diameter, liquid viscosity and the use of multiple impellers. For example, the flow pattern with a pitched blade turbine comes closer to radial as the impeller diameter is increased or as the liquid viscosity is increased. Multiple impellers are used when the liquid depth-to-tank diameter ratio is higher than 1. In that case, more circulation loops are formed, e.g., two loops with two pitched blade turbines (Fig. 17.6d).

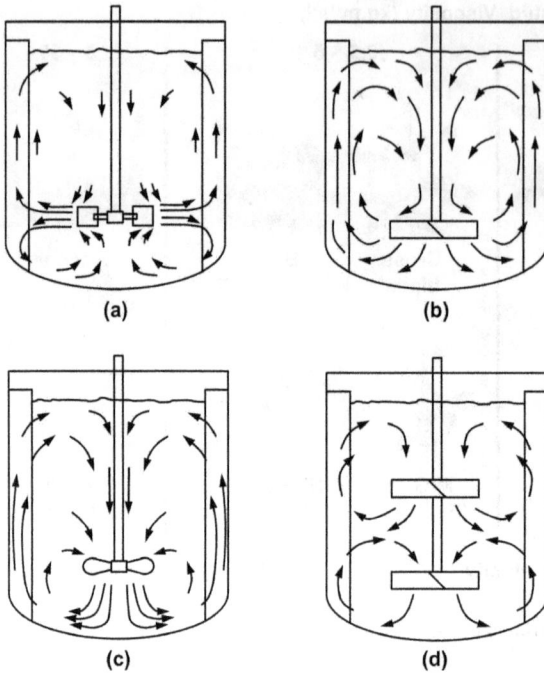

Fig. 17.6: Flow patterns with different impeller types, sizes and liquid viscosity:
(a) flat blade turbine, (b) pitched blade turbine, (c) propeller and (d) two pitched blade turbines.

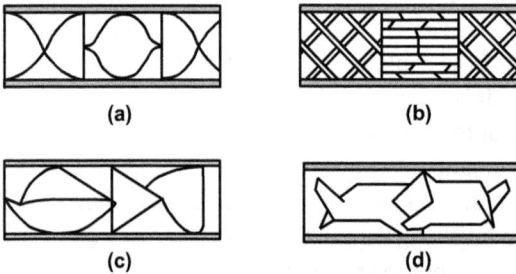

Fig. 17.7: Some proprietary static mixer designs: (a) Kenics, (b) Koch/Sulzer SMX,
(c) Komax and (d) Etoflo.

17.2.3.3 Static mixers

Inline motionless mixers derive the motion or energy dissipation needed for mixing from the flowing fluid itself. These mixers include orifice mixing columns, mixing valves and static mixers. Static mixers are used in the process industries for plastics and synthetic fibers, e.g., continuous polymerization, homogenization of melts and blending of additives in extruders, food manufacture, heat transfer, cosmetics, etc.

These motionless mixers provide complete transverse uniformity and minimize longitudinal mixing; therefore, their performance approaches perfect the plug-flow conditions. Static mixers are usually classified as operating either under laminar or turbulent flow conditions. They consist of repeated structures called mixing elements attached inside a pipe. There are many proprietary designs, of which a few are shown in Fig. 17.7. These mixing elements generate a process of division, rotation and reversal of fluid, that create shear and mixing. Homogeneity is accomplished by alternate division and recombining of fluids passing through (Fig. 17.8). The number of layers produced equals 2^n, where n is the number of elements. The mixing is, therefore, a progressive phenomenon, because the specific surface increase is proportional to the specific surface already created. This results in an exponential increase of the mixing process in time and is an example of the so-called distributive mixing (exchange of layers). As a result, they create a shearing action at the cost of pressure drop that causes mixing of single- and multiphase systems.

The power consumed by a static mixer is given by $P = Q \cdot \Delta P_{sm}$. A pump supplies this power that is used to create a flow rate Q of the fluid through the mixer. For homogenization of two or more liquids, static mixers reduce the variance of the local liquid fractions. The reduction of the variance is a product of shear rate and time and, therefore, equal to the ratio of the length over the diameter of the static mixer. The pressure drop over the static mixer, ΔP_{sm}, depends on the flow rate and the liquid viscosity. For a Kenics static mixer, ΔP_{sm} is about six times that of an empty pipe in laminar flow. For a Koch/Sulzer SMX mixer, it is 64 times higher.

After element number

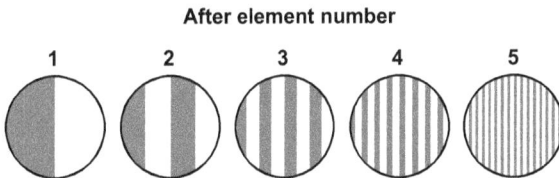

Fig. 17.8: Mechanism for laminar blending in Kenics static mixer. The gray zone represents liquid A and the white zone represents liquid B.

17.2.3.4 Mixing of single-phase systems

One of the most common mixing tasks is the mixing of two miscible liquids; this is also one of the most straightforward mixing problems. The main goal in this operation is to homogenize the mixture in order to improve heat and/or mass transfer. The criteria used in selecting an impeller for these mixing operations are the power consumption and the mixing time. Multiplication of the power consumption with the mixing time will result in the total amount of energy used in the mixing operation. It is desired to keep this amount as low as possible for a given system of liquids, while meeting the mixing requirements. In general, mixing energies required will be higher if the original liquids in the mixture do not have the same density, since these differences will tend to segregate the mixture. In a turbulent flow regime and at high power consumption, the standard impeller used is the Rushton turbine, while at low power, either a propeller or a hydrofoil is used. If mixing has to be performed in a transition range between the laminar and turbulent flow, either migs or intermigs are used. At high viscosities, most often, helical ribbons or anchors are used. This is done because other impellers will provide almost no fluid flow near the wall of the vessel. The helical ribbon is particularly favorable because it provides better top-to-bottom mass transfer and can also generate a helix near the center of the vessel, in which the liquid flows from the bottom of the vessel to the top.

17.2.3.5 Mixing of multiphase systems

For multiphase systems, the main goal in mixing tasks is to create a large area for mass transfer, which can lead to higher reaction rates or speed up the dissolution of solids. Three cases are discussed in this paragraph: mixing of immiscible liquids, suspending of solids and dispersion of gases in liquids.

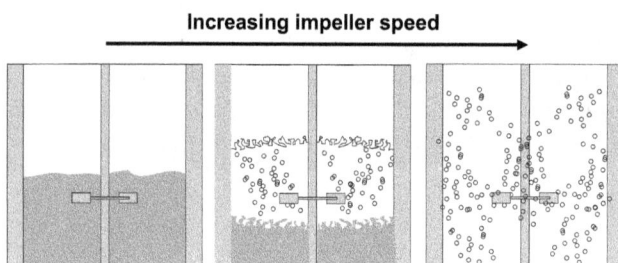

Fig. 17.9: Schematic of effect impeller speed on suspension of particles in a liquid.

Suspending of solids is needed in, for example, dissolution of solids and, in reactions, in the liquid phase with heterogeneous catalysts. For both these tasks, it is important to achieve a (reasonably) homogeneous suspension in order to speed up dissolution or to improve catalyst efficiency by reducing gradients in reactants. Most designs are based on the so-called Zwietering-criterion, in combination with the criterion of a homogeneous suspension. The Zwietering criterion states that the solids in the suspension will be at the bottom for no longer than one second to ensure good suspension. Research usually concentrates on determining the critical rotation speed needed for suspending the solids, as illustrated in Fig. 17.9. The default impeller used is the pitched blade turbine, relatively close to the bottom of the tank, in combination with a tank without a flat bottom to prevent the formation of dead zones where solids will hardly move.

The mixing of immiscible liquids can take place in extraction, for example. The main goal is to speed up mass transfer between the two phases by increasing the mass transfer area, thus achieving a larger recovery in the time available or shortening the mixing time needed to meet the criteria. The interfacial area is increased by the break-up of droplets of one of the fluids dispersed in the other. Depending on the viscosities of the two phases involved, break-up of the droplet will evolve in different ways. If the viscosities are almost identical, droplets will break to only a few smaller droplets (see Fig. 17.10) and the increase in interfacial area will be relatively slow. If the viscosity of the dispersed phase is low compared to the continuous phase, several smaller droplets will be formed and the area will increase significantly.

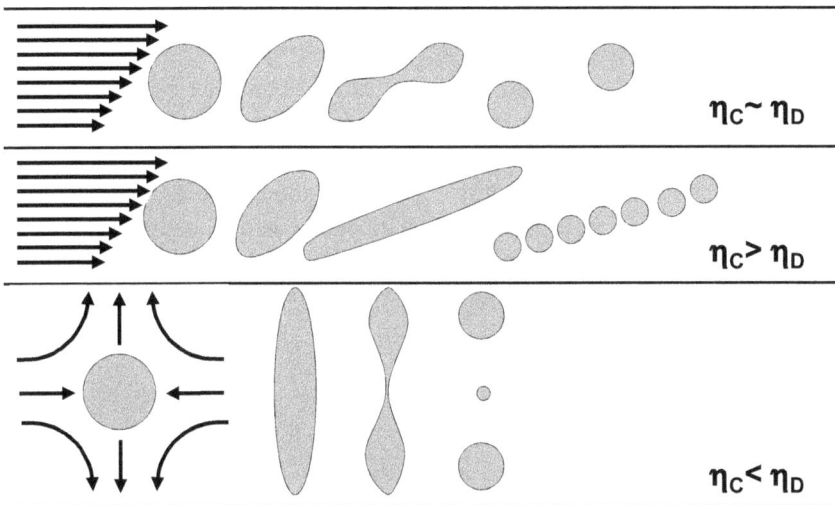

Fig. 17.10: Break-up mechanisms for different viscosities.

As in single-phase liquid mixing, the standard impeller used is the Rushton turbine. Scale-up, however, is more complicated because of several factors. The emulsion created, which consists of a continuous phase in which the second phase is finely dispersed as droplets, is very sensitive to impurities present. Some substances, like surfactants (soap-like molecules), have an enormous effect on the stability of the emulsion. It is also desired to know which of the two liquids will be in the continuous phase, because the process conditions depend on this to a large extent. As a rule of thumb, it can be presumed that the phase in which the impeller starts up will be the continuous phase. Concentration gradients, for example, will be smaller for the continuous phase; so if it is desired to keep a parameter constant (like the pH), it is desired to have this phase as the continuous phase. If a constant pH with a dispersed water phase is desired, a base is added to increase the pH. There will be droplets of base present in the mixture (with very high pH) and droplets of the original water phase (with a low pH). The pH of the droplets will only change if a droplet of the base and a droplet with low pH coalesces, which is a very slow process. Conversely, if water is the continuous phase, the base will be equally distributed within the mixture in a very short time and the pH will remain high. It is also important to know at which conditions a phase transition takes place. At different phase fractions, different phases will most likely be the continuous phase and a very fast change of these phases can occur. The transition can take place under the influence of changes in temperature, volume ratio and solute concentrations. Because of these reasons, the design is preferably based on the scale-up of pilot plant results.

In gas–liquid mixing, the main goal is the same as in liquid–liquid mixing: to create a large mass transfer area. Most applications involve reactions between a gas and a liquid, for example, in hydrogenation, oxidation and fermentation (where air is provided to the micro-organisms). In this mixing operation, the power input to achieve a certain degree of mixing is (roughly) independent of the impeller type. In practice, an impeller is operated at a high rotation speed to obtain a homogenous dispersion of the gas over the vessel (Fig. 17.11). Often, the Rushton turbine is used, because it is relatively cheap, but hydrofoils and concave Rushton turbines are also used. The position of the gas inlet, however, has a very large influence on the quality of mixing. If the inlet is positioned close to the surface of the liquid, almost no entrainment of gas takes place and the mixing is very poor. If the inlet is positioned very close to and below the impeller, an almost homogeneous dispersion of very fine gas bubbles is achieved.

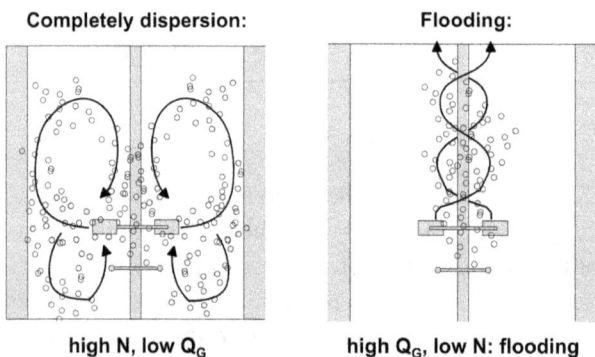

Completely dispersion: **Flooding:**

high N, low Q_G high Q_G, low N: flooding

Fig. 17.11: Effect of rotation speed and gas flow on gas dispersion in a liquid.

17.2.4 Flows in bubble columns

In a bubble column, a gas phase is bubbled through a column of liquid (Fig. 17.12), usually to promote heat and mass transfer between the gas and the liquid. During their rise through the column, the bubbles drive a convective flow in the liquid. The advantage of a bubble column is that intense gas-liquid contact is achieved without any moving parts. The main disadvantage is a possible occurrence of back-mixing, i.e., bubbles may be trapped in a downward recirculation flow, which leads to an increase in the spread of residence times of different bubbles, which is disadvantageous to reactive systems.

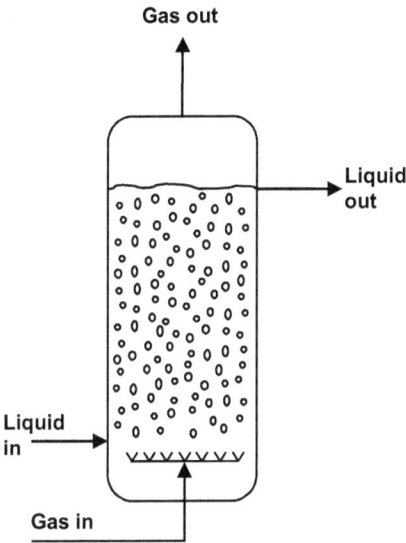

Fig. 17.12: Schematic of a bubble column.

Bubbles are created at the bottom of a bubble column, usually by perforated plates, perforated ring, or perforated spider spargers. The bubble diameter depends on a number of parameters, including liquid density and viscosity, gas-liquid surface tension, gas superficial velocity, the sparger opening diameter and the bubble column diameter. The bubble diameter is usually in the range of 1 to 10 mm.

At low superficial gas velocities and with narrow sparger openings, a homogeneous bubbly flow regime is observed, characterized by a narrow bubble size distribution and limited interactions between the bubbles. Most biochemical applications are performed in this regime because of the mild conditions. To increase productivity and to make sure that embedded small catalyst particles remain suspended throughout the reactor, most bubble column reactors in the chemical industry are operated at relatively high superficial gas velocities. For narrow bubble columns, this leads to a slug flow regime, while for bubble columns of sufficiently large diameter, which are typically used in the industry, this leads to a so-called churn-turbulent flow regime. The churn-turbulent flow regime is characterized by a broad bubble size distribution, significant bubble break-up and coalescence and turbulent flow in the liquid phase. In this regime, even when the gas is sparged evenly across the bottom, the gas fraction is typically higher near the center than near the side walls of the column, leading to strong fluid circulation flows and backmixing of gas bubbles along the sides.

Because bubble columns are intensely used in the process industry, significant research efforts are undertaken to understand the transitions between the different flow regimes and the consequences of the bubble residence time distribution.

17.2.5 Flows in fluidized beds

Similar to bubble columns, fluidized beds are used to maximize mass and heat transfer, but now between solid particles and a fluid and without the use of any moving parts. As already discussed extensively in Chapter 4, fluidized beds display different flow regimes, depending on the size, density and viscosity of the particle, and density and superficial velocity of the fluidizing medium. To achieve good mixing, fluidized beds often operate in the bubbling regime, using superficial fluid velocities between 1 and a few times the minimum fluidization velocity v_{mf}. In this regime, the excess fluid rises through the fluidized bed in the form of bubbles, which enhances the mixing by continually "stirring up" the bed.

In the design of fluidized beds, it should be taken into account that the fluid bubbles tend to coalesce, leading to an increase in the average bubble diameter with an increase in height. The increase in bubble diameter has been captured by numerous correlations, most of which are nonlinear with height. Moreover, wall friction and confinement effects have a relatively large impact on the motion and distribution of both solid and fluid in a fluidized bed. All these effects lead to sensitive and non-linear dependence of the fluid and solid flows on the dimensions of a fluidized bed, complicating their rational design. For instance, a pilot-scale fluidized bed of 10 cm diameter may experience slug flow with a relatively low rise velocity, while a full-scale bed of 1 m diameter with the same particles, same superficial velocity and the same bed height-to-diameter ratio, the flow would remain comfortably in the bubbling regime, with much larger bubble rise velocities.

17.3 Scale-up

The previous section has shown that a major problem in process equipment design is to scale up from a laboratory or pilot plant unit to a full-scale unit. For many problems, adequate mathematical models and/or scale-up correlations are not available. Various methods of scale-up have been proposed for these situations; many based on geometric similarity between the laboratory and the plant equipment. However, it is not always possible, and often not even desirable, to have the large and small equipment geometrically similar. It is often not desirable because, even if geometric similarity is obtained, dynamic and kinematic similarity is not obtained. Thus, the results of the scale-up are not always fully predictable. Important techniques to assist an engineer in the scale-up of process equipment include dimensional, similarity and regime analysis.

17.3.1 Dimensional, similarity and regime analysis

Dimensional analysis is a mathematical tool for the scale-up or scale-down of physical and chemical process equipment. It proceeds from the general principle that physical laws must be independent of the units of measurement used to express them. If one quantity is related to a group of other quantities, the quantities comprising the group must be related in such a manner that the net units or dimensions of the group are the same as those of the dependent quantity. Division can then form a dimensionless group immediately. The final result is a complete set of dimensionless quantities that describe the physical process and that indicates the conditions for which this process behaves "similarly" to the model and at the desired scale. For example, we already introduced the most important dimensionless number in process engineering, the Reynolds number, which is easily shown to be dimensionless:

$$\mathrm{Re} = \left(\frac{\rho v d}{\eta}\right) = \cdot \left[\frac{\left(\mathrm{kg}/\mathrm{m}^3\right)\left(\mathrm{m}/\mathrm{s}\right)\mathrm{m}}{\mathrm{Ns}/\mathrm{m}^2}\right] = \times \left[\frac{\left(\mathrm{kg}/\mathrm{m}^3\right)\left(\mathrm{m}/\mathrm{s}\right)\mathrm{m}}{\left(\mathrm{kgm}/\mathrm{s}^2\right)\left(\mathrm{s}/\mathrm{m}^2\right)}\right] = \left[\frac{\mathrm{kg}/\mathrm{ms}}{\mathrm{kg}/\mathrm{ms}}\right] = [-]$$

(17.2)

The most important strength of dimensional analysis lies in its ability to reduce the number of parameters required to describe complex problems and provide deeper insight into the physical nature of a process. It can even handle the lack of geometrical symmetry by using ratios of important dimensions as additional dimensionless numbers. This limits the work that needs to be done to handle situations in which, the governing equations are not known. However, the weakness of dimensional analysis lies in the need to know, in advance, which variables must be included and which can be ignored. This reflects the fact that dimensional analysis is not capable of describing the functional relationship among the dimensionless groups and the extrapolation of their empirical relationship outside the range covered by the original data. Achieving the simplest, most meaningful relationship and to judge how it might best be extrapolated, still relies heavily on previous experience and physical insight of the engineer.

After the dimensionless numbers are obtained, one has to make sure that these numbers do not change when scaling up. Two processes of different scale are considered similar if they take place under geometrically similar conditions and all dimensionless quantities describing the process have the same numerical value. This is the general idea behind similarity analysis and ensures that the full-scale process behaves similar to the experiments performed. However, in practice, it always appears impossible to keep all dimensionless numbers and geometric ratios constant during scale-up, because they depend differently on the variables or due to mechanical constraints. In such situations, a compromise needs to be found and another technique is needed to solve these problems: regime analysis. This is a formalized theory to predict which dimensionless numbers are (more) important, and thus have to remain constant and which are not (or less) important.

Regime-analysis deals with determining the rate-limiting step in the process. Because dimensionless numbers often represent ratios of transport mechanisms, process changes may occur when a dimensionless number has a value of around one. Regime analysis investigates which dimensionless numbers are most important for a certain regime (i.e., have a value between 0.1 and 10). The advantage of knowing the regime is that one only has to control the dimensionless groups belonging to that regime. In such a case, only partial similarity can be obtained, i.e., the important dimensionless numbers are kept constant, while the others are allowed to vary. A phenomenon that may occur in practice is that another regime is rate-limiting for a small-scale problem than for the large-scale situation. In such cases, it might be necessary to change the circumstances for the small-scale situation such that the same regime is limiting at both scales or release the requirement of geometrical similarity. An example is a tubular reactor with a mixed thermal and chemical regime. The problem is that the specific surface decreases during pure geometrical scale-up and the ratio of heat transfer and heat production changes. This can be compensated by using the so-called elements at small scale, as illustrated in Fig. 17.13. The small-scale model has the same height but smaller diameter as the larger scale prototype and the dimension of packing elements and catalyst particles is not decreased, to ensure equal rate-limiting steps in both setups.

17.3.2 Important dimensionless numbers for mixing vessels

Let us consider the example of a mixing tank (Fig. 17.14). The main objective of scale-up is to achieve the same quality of mixing in a commercial-size mixing tank as in a laboratory tank, or at least have an understanding of the differences expected in the commercial installation. Unfortunately, it is not possible to maintain the same combination of flow and shear distributions in commercial mixers as in small-scale tanks. The most obvious dimensionless numbers that are used in scale modification of mixing processes are geometrical ratios (height of the tank divided by the diameter, for example) in the case of geometrical scale-up. Most important values have been investigated thoroughly and for most situations, optimal ratios have been found, which can be applied in other cases as well. These are summarized in Tab. 17.1. Other important ratios are flow number, circulation time, power density and the power number. The flow number, ϕ, represents the ratio between the flow supplied and the flow from the impeller:

$$\phi = \frac{Q_m}{\rho N D^3} \tag{17.3}$$

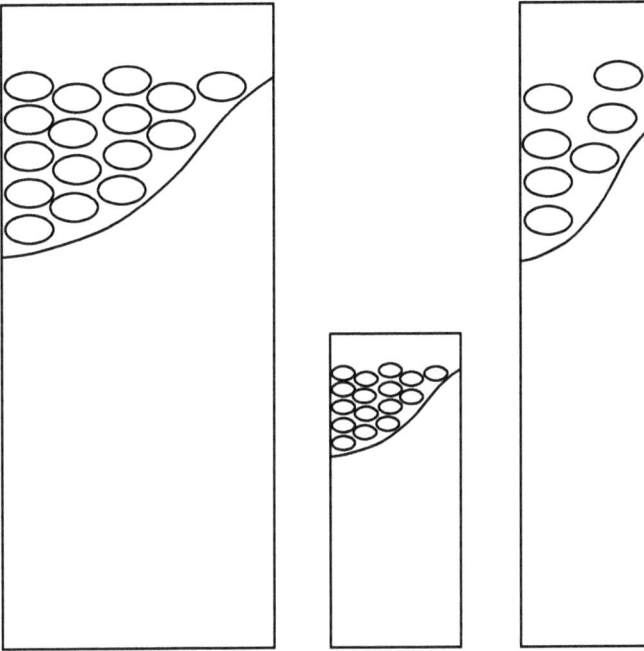

Fig. 17.13: Example of a non-geometrical similarity: The full-scale model is at the left, the geometrical similar small-scale model in the middle and the setup actually used at the right.

The circulation time is the ratio of the volume of the tank divided by the flow from the impeller. It indicates how long it takes for the impeller to move a volume of liquid equal to the volume of the tank and, thus, gives an indication of the mixing intensity. In a process that is controlled by micro-mixing, it is important to keep the power density ε constant. It represents the amount of energy applied for mixing per unit of mass of the mixture:

$$\varepsilon = \frac{P}{\rho_F V} \tag{17.4}$$

Finally, the power number represents the power required to drive the impeller and is often given as a correlation of dimensionless numbers to allow the determination of the required power when scaling up or down.

17.3.3 Important dimensionless numbers for bubble columns

Next, let us consider the example of a bubble column. To achieve similar heat and mass transfer rates in a scaled-up or scaled-down bubble column, it is important to achieve the same phase fractions, same bubble size and shape and remain in the

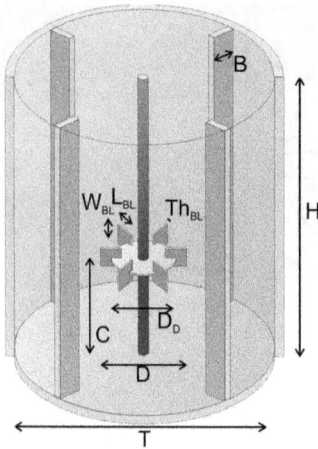

Fig. 17.14: Standard mixing vessel.

Tab. 17.1: Important geometrical ratios for a stirred tank and their value for a standard mixing vessel are shown in Fig. 17.13.

Ratio	Common value
Height:diameter tank (H/T)	1:1
Diameter impeller:diameter tank (D/T)	1:3
Baffle length:diameter tank (B/T)	1:12
Distance impeller bottom:diameter tank (C/T)	1:3
Length of a blade:impeller diameter ($L_{BL}:D$)	1:4
Blade thickness:impeller diameter ($Th_{BL}:D$)	1:50
Height of a blade:impeller diameter ($W_{BL}:D$)	1:5
Number of blades on impeller	6
Number of baffles	4

same regime of homogenous bubbling flow, slug flow or heterogeneous churn-turbulent flow.

The bubble size and shape of a single bubble rising through a liquid are interrelated through a balance between the buoyancy force (driving the bubble upward), surface tension force (driving the bubble shape towards a sphere), and the drag force (friction because of relative motion); the drag force itself can be subdivided into viscous and inertial forces. With these four forces, three independent dimensionless numbers can be defined, namely, the bubble Reynolds number (ratio of inertial to viscous forces), the Eötvös number (ratio of buoyancy to surface tension forces) and the Morton number (combination of inertial, buoyancy, surface tension and viscous forces), which need to remain the same:

$$\mathrm{Re_b} = \rho_L d_b v_b / \eta \qquad (17.5)$$

$$\mathrm{Eo} = g d_b^2 (\rho_L - \rho_G) / \sigma \qquad (17.6)$$

$$\mathrm{Mo} = g \eta^4 (\rho_L - \rho_G) / (\rho_L^2 \sigma^3) \qquad (17.7)$$

where d_b is the (sphere volume equivalent) bubble diameter, v_b the velocity of the bubble relative to the liquid, ρ_L and ρ_G the liquid and gas densities, η the liquid viscosity, σ the gas–liquid surface tension and g is the gravitational acceleration. Sometimes, the Weber number (ratio of inertial to surface tension forces) is used, which can be expressed in terms of the three dimensionless numbers mentioned earlier:

$$\mathrm{We} = \rho_L d_b v_b^2 / \sigma = \mathrm{Re_b}^2 \sqrt{\mathrm{Mo}/\mathrm{Eo}} \qquad (17.8)$$

At relatively high gas velocities, slug flow is observed in narrow columns, while churn-turbulent flow is observed in wider columns. The transition between these flow regimes can be explained as Rayleigh–Taylor instability for a column-wide slug. Such an instability is driven by a balance between the buoyancy forces and the surface tension forces at the scale of the diameter D of the column, leading to the following relevant dimensionless number, which may be interpreted as an Eötvös number for slugs of diameter D:

$$\mathrm{Eo_D} = D^2 g(\rho_L - \rho_G) / \sigma = (D/d_b)^2 \mathrm{Eo} \qquad (17.9)$$

Experimentally, it has been confirmed that bubble columns with $\mathrm{Eo_D} > 2{,}700$ are in the "large diameter" regime. For instance, for a water–air bubble column at standard conditions, this corresponds to a critical column diameter of approximately 15 cm.

17.3.4 Important dimensionless numbers for fluidized beds

Finally, let us consider the example of a bubbling fluidized bed. Bubbling fluidized beds are notoriously difficult to scale up because heat and mass transfer characteristics depend on the bubble size and bubble distribution across the bed, which sensitively depend on the scale (diameter and height) of the bed. In order to find relevant conditions for a scaled-up or scaled-down fluidized bed, we can analyze the forces that act on a fluidized bed of hard particles, leading to the particle Reynolds number (ratio of fluid inertial to viscous forces) and bed Froude number (ratio of inertial to gravitational forces):

$$\mathrm{Re_p} = \rho_F d_p v_F / \eta \qquad (17.10)$$

$$Fr_p = v_F^2/(gD) \tag{17.11}$$

where ρ_F is the fluid (gas or liquid) density, d_p is the particle diameter and v_F is the fluid superficial velocity. Besides these two dimensionless numbers, the particle-to-fluid density ratio ρ_P/ρ_F, the particle-to-bed diameter ratio d_p/D and bed height-to-diameter ratio H/D are also important dimensionless parameters that should remain the same. In practice, it is difficult to match all these dimensionless numbers simultaneously in a smaller scale system. For instance, strictly speaking, smaller particles should be used for a smaller scale bed to keep the particle-to-bed ratio the same, but this may lead to a dominance of forces that were not considered in this analysis, such as the cohesive forces. Fortunately, if the particle-to-bed ratio is sufficiently large, the influence of that parameter becomes relatively weak.

17.4 Computational fluid dynamics (CFD)

CFD is the science of predicting fluid flow, heat transfer, mass transfer, chemical reactions and related phenomena by solving the mathematical equations that govern these processes using a numerical algorithm. It is a simulation technique that is increasingly used in process engineering for design and development, mainly due to the increased capacity of computing power. The simulations allow for an in-depth analysis of the fluid mechanics and local effects in process equipment, which can be used to improve performance, better reliability, more confident scale-up, improved product quality and higher plant productivity. Today, many commercial CFD programs are available, each with different capabilities, special physical models, numerical methods, geometric flexibility and user interfaces. Before using such a program, it is important to understand the different types of CFD simulations summarized in Tab. 17.2.

17.4.1 Types of CFD simulations

17.4.1.1 CFD for single-phase flows
For single-phase flows, there are three main types of CFD simulations: direct numerical simulations (DNS), large eddy simulations (LES) and Reynolds-averaged Navier–Stokes (RANS). The difference between these three types is related to the way in which turbulent flow is modeled. Turbulent flows are chaotic flows that appear at higher Reynolds numbers, i.e., they are caused by the inertia of the fluid. In turbulent flows, vortices or eddies of many sizes appear, which interact with each other. The largest eddies appear on a length scale L, which is typically as large as the size of the equipment, while the smallest eddies appear on the Kolmogorov length scale η_K. The ratio of the smallest to the largest eddy is determined by the Reynolds number:

Tab. 17.2: Key features, advantages and disadvantages of the most important types of CFD simulations for single-phase and dispersed multiphase flows.

Single phase	Key feature	Advantage	Disadvantage
Direct numerical simulation (DNS)	Flow fully resolved down to the Kolmogorov scale	High accuracy	Computationally very expensive
Large eddy simulation (LES)	Only larger eddies resolved; smaller eddies modeled through turbulent viscosity	Relatively high accuracy (depending on grid size)	Cheaper than DNS, but still expensive
Reynolds-averaged Navier–Stokes (RANS)	Averaged steady state; turbulence models	Computationally much cheaper than DNS and LES	Steady state; choice of turbulence model is an art
Multiphase	Key feature	Advantage	Disadvantage
Direct numerical simulation (DNS)	Flow fully resolved: grid cell ≪ particles/bubbles; surface tension and contact force models	High accuracy	Computationally very expensive
Discrete particle/ bubble model (DPM/DBM)	Flow unresolved: grid cell > particles/bubbles; contact force models	Relatively high accuracy	Cheaper than DNS, but still expensive; dependency on drag correlations
Two-fluid model (TFM)	Flow unresolved: grid cell > dispersed element; rheological model for dispersed phase	Computationally cheaper than DNS and DPM/DBM	Dependent on drag correlations and rheological models (e.g., KTGF)

$$\frac{\eta_K}{L} \approx \mathrm{Re}_L^{-3/4} \tag{17.12}$$

where $\mathrm{Re}_L = \rho_F v_f L / \eta$ is based on the largest relevant scale L of the vessel and the typical flow velocity v_f of the fluid.

In DNS, eddies of all scales are fully resolved in space and time. DNS is, therefore, the most accurate of all single-phase CFD methods. For larger geometries or larger flow velocities, this quickly becomes computationally extremely expensive because the ratio between the largest and the smallest eddies becomes too large. To overcome this problem, in LES, only the larger eddies that contain the largest amount of turbulent energy are explicitly resolved, while the effects of the smaller eddies on the fluid flow are effectively modeled through dissipative forces, typically through a "turbulent" viscosity.

For very large equipment, even LES simulations can be computationally very expensive, in which case, one often resorts to the RANS solvers. In RANS, averaging operations are applied to the original (Navier–Stokes) equations, allowing them to be solved on very coarse length scales. Several different approaches to effectively model turbulence exist. The turbulence model chosen should be best suited to the particular flow problem. Understanding the limitations and advantages of the selected one is crucial if the best answer is to be obtained, with minimum computation time. The disadvantage of RANS is that it implicitly assumes steady-state operation, so no time-dependent phenomena can be modeled. Unsteady RANS and hybrid forms of LES and RANS exist to overcome this problem.

17.4.1.2 CFD for dispersed multiphase flows

For dispersed multiphase flows, there are also different types of CFD simulations, depending on the level of detail with which the dispersed phase and its coupling to the continuous phase is described: DNS, discrete particle/bubble models (DPM/DBM) and Two-fluid Models (TFM).

In DNS, the flow of the continuous fluid around the dispersed elements (particles, droplets or bubbles) as well as the flow inside the droplets or bubbles are fully re-solved. Pressure and stress boundary conditions are used at the interface between the continuous fluid and dispersed elements. In the case of dispersed flows with bubbles or droplets, surface tension forces must also be modeled. In the case of dispersed flows with solid particles, a contact model must be specified to calculate the interparticle forces, in case the particles come in direct contact with each other. Such DNS simulations are very accurate, but also computationally very expensive, typically allowing for only a few hundred or a few thousand dispersed elements to be simulated.

To overcome these limitations, one may choose to not fully resolve the flow of the continuous fluid around the dispersed elements. Instead, the continuous fluid flow is coarse-grained to a scale that is several (typically, 3–20) times larger than the size of the dispersed elements, and the fluid-dispersed elements are assumed to keep their (typically, spherical) shape. Because the flow-field is no longer resolved on the scale of the dispersed elements, effective correlations are needed to specify the average drag force between the coarse-grained fluid and the dispersed elements. Note that for semi-dilute and dense flows, not only should the drag force on the dispersed elements be modelled, but conversely, the drag force on the fluid due to the presence of the dispersed elements should also be modeled. This is called two-way coupling. Only in very specific applications with very dilute dispersed flows is it allowed to use one-way coupling, in which the results of a single-phase flow simulation is used as a passive field, in which the dispersed elements are moving. Because of the importance of direct interactions between the dispersed elements, especially for dense dispersed flows, their motion is tracked explicitly by solving the Newton's equations of motion, where the force on each element arises from the interactions with the continuous fluid (drag and buoyancy), external forces such as gravity, and direct interaction forces with neighboring dispersed elements. Such methods are called DPM or DBM for dispersed particles and bubbles, respectively. DPM or DBM models allow for the simulation of up to a million dispersed elements, which is often sufficient to accurately predict flow in lab-scale or pilot-scale equipment. Coarse-grained versions of DPM and DBM, in which groups of particles or bubbles are represented by a single larger particle or bubble also exist, but require careful tuning of the effective properties.

For larger-scale equipment, more approximations need to be made. The most expensive part of a DPM/DBM model is the evaluation of the direct interactions between all pairs of particles/bubbles, but in many applications, it is not so important to know the exact location and velocity of each and every particle or bubble. In the two-fluid model, only the local phase fraction of particles or bubbles and the average velocity of that phase (relative to the continuous phase) are tracked. So, the dispersed and continuous phases are treated like they are two interpenetrating continuous fields. Similar to DPM and DBM models, effective correlations for the average drag force between the two phases are needed. Moreover, to capture the effects of direct interactions between the particles or bubbles, an accurate rheological model is required for the dispersed phase that relates the particle or bubble interaction stresses to the rate of deformation of the particle or bubble suspension. For monodisperse spherical particles, a very successful approach is to adopt the kinetic theory of granular flow (KTGF). Unfortunately, it is difficult to extend KTGF to polydisperse particle mixtures or to nonspherical particles. Further, coarse-grained versions of TFM also exist and are known as filtered TFM models.

17.4.2 Basic principles of CFD

Whatever type of CFD method is used, CFD programs usually consist of three main parts: a preprocessor, a solver and a postprocessor (figure 17.15). The preprocessor is used to define the geometry and the properties of the system, the solver performs the actual calculations and the postprocessor analyzes results from the calculations.

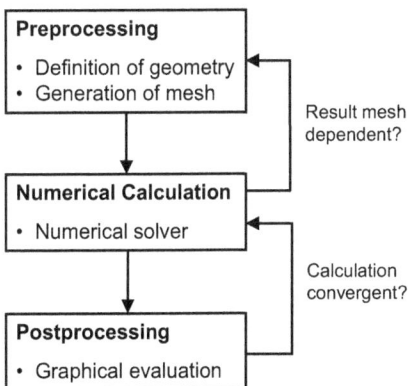

Fig. 17.15: Steps in a CFD simulation.

17.4.2.1 The preprocessor

The first part of the CFD codes is the preprocessor, in which the equipment is defined. If one wants to calculate the extent of mixing in a stirred vessel for example, several properties have to be provided to the computer program. First of all, the geometry of the vessel has to be specified. Often, this can be done by a graphical user interface that allows the user to simply draw the vessel or provide the dimensions for the radius and height of the vessel. Alternatively, the geometry may be imported from the output file of a computer-aided design program. Also, the shape, size and position of the stirrer have to be added, as well as the rotation speed. After the geometry is defined, the physical properties of the liquids, gases, solids or a combination of these have to be provided. Important data are density as a function of temperature and/or composition, viscosity, heat capacity and thermal conductivity (if heat-transfer is involved). These data are often provided as simple first-order equations, possibly by two equations, for all properties, if a phase transition is involved. After these properties are defined, initial conditions such as the initial temperature and pressure have to be given.

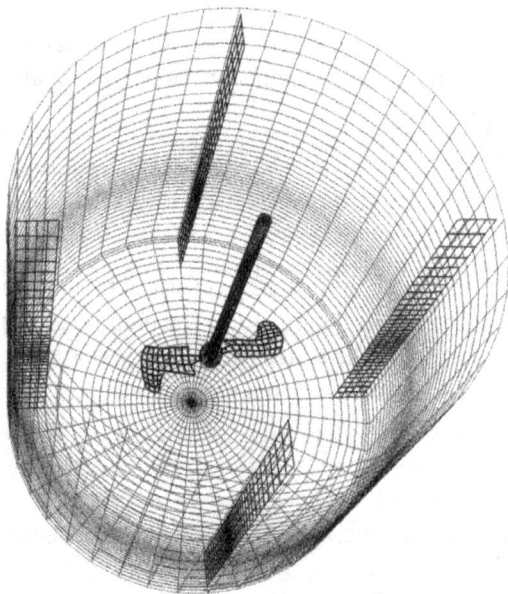

Fig. 17.16: CFD grid structure for an agitated vessel.

Grid generation is the next important step in programming the problem. To apply CFD, the geometry of interest needs to be divided or discretized into a number of computational cells. Discretization is the method of approximating the differential equations by a system of algebraic equations for the variables, at some set of discrete locations in space and time. The discrete locations are referred to as the grid or mesh. The number of grid cells can vary from a few thousand, for a simple two-dimensional problem, to hundreds of millions for very large and complicated ones. The size of the grid cells must be chosen to obtain the desired level of accuracy and to match the relevant length scales of the problem. Laminar flow, for example, requires a less detailed grid than turbulent flow. Often the grid consists of cells of different sizes and orientations, of which an example is given for a stirred tank in figure 17.16. At the height of the stirrer, the flow of the liquid is more likely to be turbulent and therefore a more finely divided grid is used. Further away from the stirrer, the fluid flow will be less complicated and a larger grid can be used.

Once the grid has been created, boundary conditions need to be applied. Pressures, velocities, mass flows and scalars, such as temperature may be specified at the inlets; temperature, wall shear rates, or heat fluxes may be set at the walls, and pressure or normal velocity gradients may be fixed at the outlets. The final step in the preprocessor, before calculations can start, is selecting the physical models for the phenomena that have to be considered by the computer. For example, special attention needs to be paid to the accurate modeling of the turbulence.

17.4.2.2 The solver

With the grid created, the boundary conditions set, physical properties defined and models chosen, the simulation can start. The code will solve the appropriate equations for all the grid cells using an iterative procedure. These equations are basically the same momentum, heat and mass conservation laws as we have applied in Chapter 3, which take the following general form:

$$\text{Accumulation} = \text{Inflow} - \text{Outflow} + \text{Production} \qquad (17.13)$$

In Chapter 3, the conservation laws were applied to macroscopic volumes, leading to ordinary differential equations (e.g., for a batch reactor) or algebraic equations (e.g., for a steady-state continuously ideal stirred tank reactor) that can be solved to predict the average quantities for the entire reactor. Here, the same conservation laws are applied to microscopic volumes, leading to partial differential equations that can be solved to predict time-dependent quantities at all locations inside the vessel. It is these partial differential equations that are solved numerically on a grid in CFD. Different numerical algorithms are available to solve the partial differential equations on a grid. The best known are the finite difference, finite volume and finite element methods. Usually, for each time step, the solution is obtained by an iterative method. Since the geometry can be divided into millions of cells, each with equations containing up to 20 unknowns, it is easy to imagine that a lot of computing power is needed to solve these systems of equations.

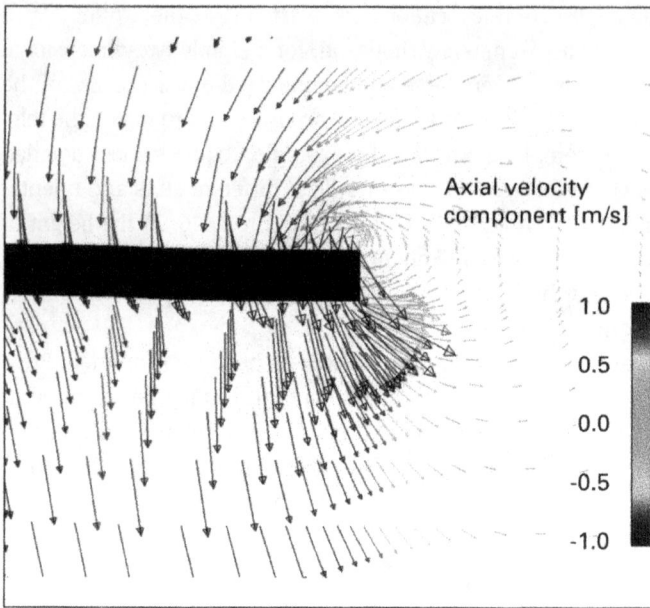

Fig. 17.17: Local formation of vortices at the blade tip.

17.4.2.3 The postprocessor

During the simulation, an enormous amount of data is generated, which will be hard to understand in this form. Therefore, a postprocessor is usually incorporated in CFD-codes to analyze the results. Different methods are used for the visualization of the problem: vector plots, shaded contour plots or streamlines and/or particle tracking. Temperature and pressure profiles are usually depicted in shaded contour plots and velocities can be given either as a vector plot, as illustrated in figure 17.17 for single-phase flow around the tip of a propeller blade, or as color-coded streamlines.

17.4.3 Applications

Apart from process engineering, CFD is extensively used in the automotive indus-try, airplane design and racing car development. In all these areas, CFD is used to minimize the air resistance, increase top speed or minimize fuel consumption. To illustrate the successful application to many types of process equipment, a stirred tank reactor and a fluidized bed are discussed.

Stirred tank reactors are one of the most widely used pieces of processing equip-ment. Traditionally, their design is based on parameter correlations, such as power draw and impeller pumping capacity. Mixing time correlations are available, but these are often difficult to extend outside of the experimentally studied parameter range.

Stirred tank reactors are a prime example of a hydrodynamically controlled process and have been the focus of extensive CFD modeling. Figure 17.18 shows the simulated global flow conditions of a 500 m³ fermenter, equipped with three stirrers. These single-phase flow simulations were used to determine the local flow velocity against the cooling coil tubes, at the level of the radial-flow impeller.

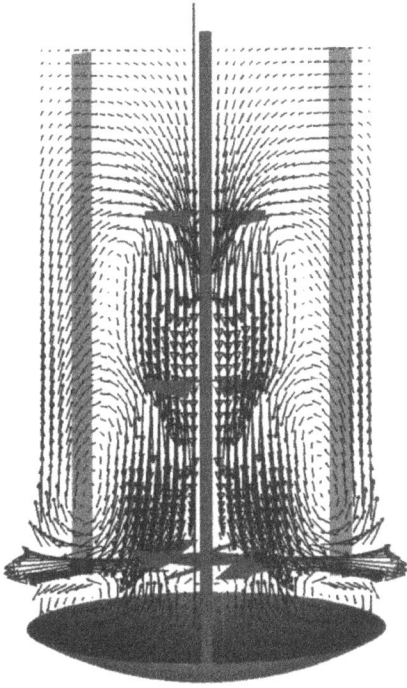

Fig. 17.18: Example of a single-phase CFD simulation of a stirred tank fermenter.

Fluidized beds are essentially tanks filled with solid particles, supported by a fine distributor grid, below which, the fluid (liquid or gas) is introduced. If the amount of fluid is high enough, the bed will become fluidized, i.e., the solid particle content will begin to behave like a fluid. Multiphase CFD simulations are often used in modeling fluidized beds, where complex interactions take place between the solid particles and the upward flowing fluid. It is used to predict the formation of fluid-bubbles, velocities of fluid and solids and the volume fraction of solids, and is able to indicate if dead zones are present. Figure 13.19 shows an example of a CFD-DPM simulation of the start-up of a gas-solid fluidized bed containing relatively large rod-like particles. It shows how the particles burst out of the bed when the first large bubble reaches the surface of the bed.

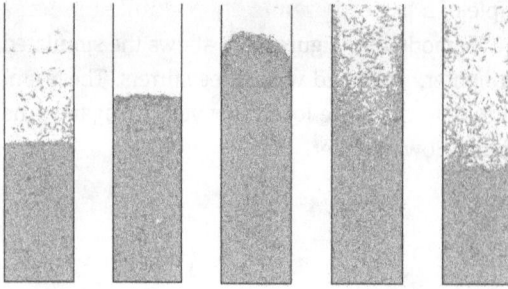

Fig. 17.19: Example of a CFD-DPM simulation of the start-up of a gas-solid fluidized bed containing rod-like particles. Time progresses from left to right.

Nomenclature

A	Heat transfer area	$[m^2]$
d	Pipe diameter	$[m]$
d_b	Bubble diameter	$[m]$
d_p	Particle diameter	$[m]$
D	Impeller diameter or column diameter	$[m]$
E	Power density	$[J/kg]$
Eo	Eötvös number	$[-]$
Fr	Froude number	$[-]$
L	Length scale of largest eddy	$[m]$
Mo	Morton number	$[-]$
N	Rotation speed	$[rpm]$
P	Power consumption	$[J/s]$
ΔP	Pressure drop	$[Pa, bar]$
Q	Flow rate	$[m^3/s]$
r	Radius	$[m]$
Re	Reynolds number	$[-]$
u	Superficial gas or liquid velocity	$[m/s]$
v	Fluid velocity	$[m/s]$
V	Volume	$[m^3]$
We	Weber number	$[-]$
α	Phase volume fraction	$[-]$
η	Viscosity	$[Pa\ s]$
η_k	Kolmogorov length scale	$[m]$
ρ_f	Fluid phase density	$[kg/m^3]$
ρ_G	Gas phase density	$[kg/m^3]$
ρ_L	Liquid phase density	$[kg/m^3]$
ρ_P	Particle density	$[kg/m^3]$
ϕ	Flow number	$[-]$

18 Process safety

18.1 Safety problems in chemical plants

Large-scale chemical production is one of the most important industrial activities. Although its products play a key role in human nutrition, health, quality of life and welfare, many people regard chemical production as dangerous. Despite the good safety records of chemical plants when compared to other industries, grave accidents have occurred, endangering human life and the environment, and causing considerable damage. The names Flixborough (1974), Seveso (1976), Bhopal (1984) and Basel (1986) call dramatic events to mind. Three types of events are traditionally associated with the chemical industry:

– Releases and spills (Seveso, Bhopal): An uncontrolled release from a closed facility requires an undesired opening of the containment or damage to the walls of vessels, process equipment, fittings or piping. The release of liquids and solids generally poses a danger to the environment by penetrating the soil. Gaseous substances are transported by the motion of ambient air and may spread over large areas.
– Fires (Basel): Fires in chemical plants can occur if flammable substances are released and come in contact with an ignition source. The effects of fires relate to the action of heat and the release of pollutants.
– Explosions (Flixborough): The best known is the instantaneous decomposition of sensitive chemical compounds to gaseous species. The second chemical explosion is the rapid oxidation of gases or finely dispersed particles in oxidizing gases. A highly exothermic reaction heats the gases to high temperature so quickly that the pressure in confined systems increases dramatically. The third type of chemical explosion is the "thermal runaway" explosion, initiated by a homogeneous exothermic reaction that goes out of control. The accompanying temperature rise can lead to a dangerous pressure buildup by raising the vapor pressure of reactants or by initiating another reaction that liberates the gases.

Damage caused by these events comes about through direct action of chemicals on humans and the environment or indirect action of liberated energy. In the Bhopal accident, many people died after inhaling a dangerous gas that had escaped from a tank storage facility. Most of the damage from the Basel warehouse fire was due to the firefighting water carrying the starting material, decomposition products and combustion products out of the storage facility, harming the ecosystem of the Rhine. The consequences of the Flixborough explosion occurred largely through the destructive action of the pressure wave, initiated by the ignition of a cloud of escaped flammable gases.

https://doi.org/10.1515/9783110712445-018

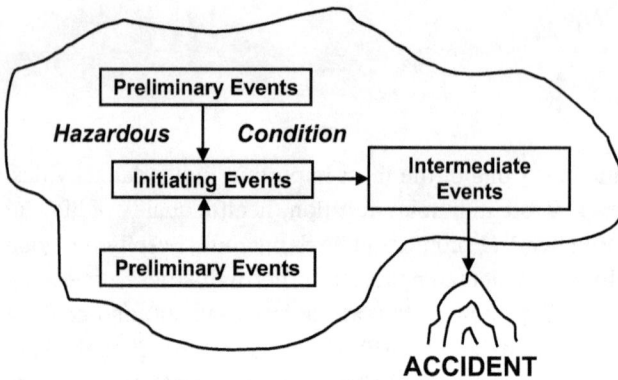

Fig. 18.1: Events that lead to an accident (adapted from [199]).

However, these accidents do not just happen. They are the result of a long process with many steps. Figure 18.1 illustrates the events that lead to an accident. Preliminary events can be anything that influences the initiating event. They set the stage for a hazardous condition. The initiating event is the actual mechanism or condition that causes the accident to occur. Intermediate events can have two effects. They may propagate or ameliorate the accident. Many times, all of these steps have to be completed before an accident can occur. This is also illustrated by the so-called "accident triangle" shown in Fig. 18.2. The principle has been to consider the worst-case scenario and relate it to other major injuries, loss-of-time injuries, minor injuries and non-injury accidents. It should be clear that for each major accident, an enormous amount of minor and non-injury accidents happen, which can eventually add up to create a major incident. If the engineer cannot prevent one or more of these accident steps from occurring, then he can either prevent the mishap or at least mitigate its effects.

18.2 Development, design and construction of safe plants

18.2.1 Introduction

It is obvious that most of the dangers in the operation of chemical plants have to do with the reactivities of the substances present and the way they are treated in the plant. A conflict arises because chemical processes are impossible without substances that show hazardous properties and effects. Although chemistry requires reactive substances, it is this same reactivity that represents the danger they pose. The substances must therefore be reliably contained in the process equipment, and their reactivity must be governed so that uncontrolled chemical reactions cannot take place. The principal hazards of substances are caused by releases and spills

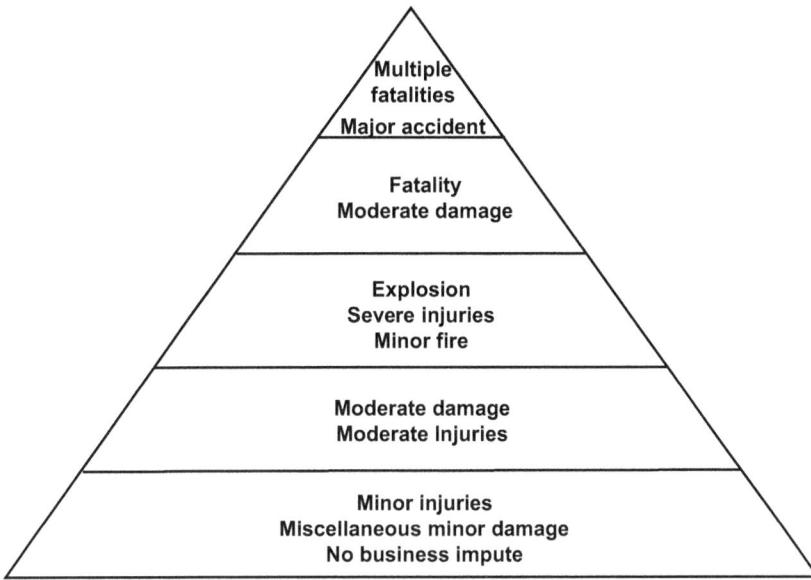

Fig. 18.2: The accident triangle.

outside the plant, and by uncontrolled chemical reactions between them. The larger the quantities of substances and energy released, greater is the danger. Therefore, plant and process safety efforts should have the objective to minimize the quantities of substances contained and finally control the potential risks that remain.

18.2.2 Safety assessment

Chemical processes and their associated technical facilities are developed in steps. After process development in the laboratory, testing on a pilot scale is commonly practiced before the project goes through the various planning stages. Each phase involves questions about the safety of the process and the plant, all of which must be answered before going to the next phase. One of the most important steps in safety assessment is the systematic identification of possible disturbing influences, their initiating events and effects during each stage of the process development. In this process, the main safety tasks for the process designer, engineer and constructor are the correct identification and assessment of all hazards, followed by appropriate steps to reduce and control them. These safety tasks concern the process itself, and the safe design and operation of the technical facility required for the process:

1. Safe process designs by identifying all types of hazards, assessing their hazard potentials, minimizing the hazard potentials and deactivating the hazard potentials.

2. Safe plant design and operation by systematically analyzing the danger sources, evaluating their probabilities of occurrence, minimizing sources of trouble and error, and employing a fault-tolerant design

If the industrial plant is to be constructed for a process that has been safety optimized in this way, two analytical tasks followed by two design tasks remain to be performed:
1. The plant system must be systematically searched for danger sources that can activate the deactivated hazard potential
2. When possible faults are identified, their frequencies or probabilities of occurrence must be evaluated so that appropriate safety measures can be taken
3. All possibilities for minimizing the sources of trouble and error must be exhausted
4. As far as possible, the facility must be designed and equipped so that faults are "forgiven" without resulting in harm.

The most expedient way of creating a safe plant is, thus, to plan for safety studies. At each step in the process and plant development, safety analyses must be done in order to pose the right questions and immediately seek solutions to the problems identified. Four phases of safety assessment can be identified:
1. Create safety principles by compiling and determining safety, toxicological and ecological data, identifying the sources of danger in the process, examining the possible safety solutions and establishing the safety concept for the process
2. Define the safety concept for the plant by performing systematic analysis and identifying technical protective measures
3. Perform a detailed safety analysis by analyzing all plausible forms of trouble as to the cause, effect and corrective measures, and adopting the final detailed safety concept
4. Conduct the safety acceptance of the plant by doing a nominal/actual comparison and carrying out functional tests

Appropriate safety engineering involves assessing the hazards as to both the possible scope and the probability of occurrence. The methodological aids available for use in these tasks are all characterized by clear and easily understood structures and systematic procedures. These methods are similar in that they are characterized as "deviation analysis"; they look for possible hazards that can arise in the process, in the plant or in the plant operation if an error occurs, or the state or sequence of actions that deviate from the prescribed state of sequence. The selection of the method should depend on the aim of the study. If the aim is a purely qualitative evaluation, focusing on a very small balance volume, the checklist method will be sufficient in most cases. A checklist enumerates points that, according to experience, are associated with hazards in the

handling of substances and mixtures of substances, or in the performance of a technical process. It can be as detailed as desired, and must be suitable for the kind of analysis being carried out. Checklists can be used to ensure the completeness of the safety
concept in later phases of a plant project to ensure that all possible events are included
in the safety concept. They have the obvious advantage that they can be adapted to
any problem. Their drawback is that things not included may not always be recognized
and dealt with. Therefore, more sophisticated methods have to be applied if the character of a system is more complex.

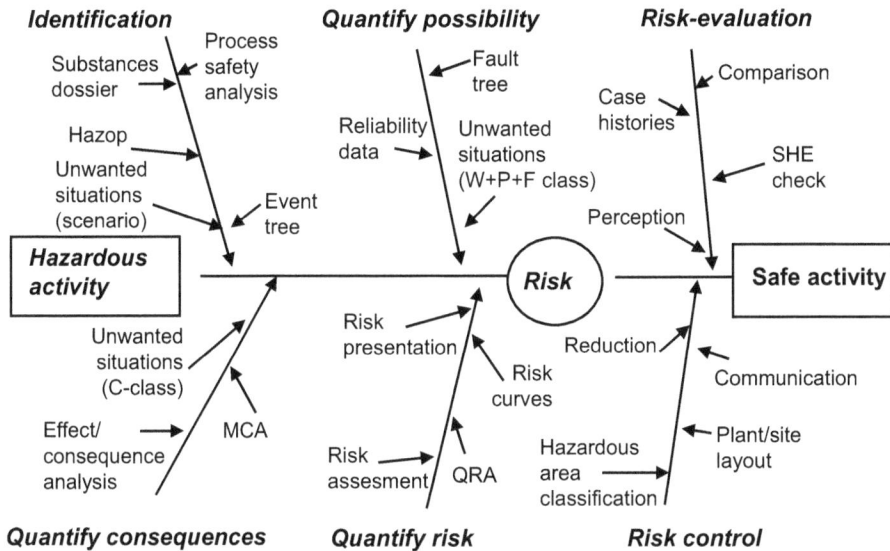

Fig. 18.3: The fishbone model.

18.2.3 Structure of safety studies

Safety studies can be classified on the basis of the fishbone model shown in Fig. 18.3.
They are commonly subdivided into hazard identification studies, effect quantification
studies, hazard probability quantification studies, risk quantification studies, and risk
assessment and risk control studies.

Hazard identification studies include the process safety analysis (BOWTIE)
and case studies of incidents during the development/conceptual engineering, substances dossier, design-phase hazard (HAZID, FMEA, DOW FEI, DOW CEI) and environmental (ENVID) identification studies during basic engineering and the HAZOP
(Hazard and Operability analysis) and process hazard review before commissioning
of the newly built plant. The process safety analysis is a systematic investigation
into the acute inherent hazards (explosion, fire, chemical reactions, toxicity, etc.) of

a process performed during the conceptual engineering phase. The analysis takes place at the level of flow diagrams and identifies locations where enough ingredients are present for a hazard to occur. The substances dossier is a collection of all physical and SHE (safety, health and environment) data on the raw materials, auxiliary materials, by-products and final products to be used in the plant. It is the basis for starting a process safety analysis. The design-phase hazard study is performed to ensure that the plant design is such as to allow safe operation in terms of SHE, under all circumstances. It concerns a systematic investigation during the engineering phase into all foreseeable deviations from the normal process operation. Unwanted situations are found and classified into the effect categories.

The HAZOP study is a final systematic inspection with regard to the SHE aspects of the design of a plant or part of a plant, before commissioning. It is most probably the best known and most widely acknowledged of the qualitative methods. The method is very systematic and reaches a high degree of completeness with respect to the identification of possible process deviations. The prerequisite to the conduction of a HAZOP analysis is the existence of a thorough process description, including detailed information on the design and process data as well as complete piping and installation diagrams. The necessary effort may become extremely high for complex plants and is already quite significant for smaller units. The process hazard review is a HAZOP study for existing plants.

Effect quantification studies include the maximum credible accident (MCA) analysis and the classification of unwanted situations into the effect categories. The MCA analysis is performed during conceptual engineering to determine the unwanted event in a plant that, while still being credible, has the most serious effects on the surrounding area. In principle, an MCA study covers only the effects on humans. For a number of scenarios, the maximum effects on the surroundings are calculated. These effects are pressure waves upon an explosion, fatalities upon exposure to an acutely toxic substance and heat radiation from a fire. The scenario involving the largest effect distance is the MCA. Classification of the unwanted situations into the effect categories uses a flow chart to classify the effects into effect categories, to enable identification of risks to individuals, the environment and economic interests to be controlled in a structured manner during basic engineering.

Hazard probability quantification studies include the classification of hazardous areas and risk (fault tree) analysis. Basic engineering classification of the hazardous areas is used to classify plant areas where an ignitable or explosive atmosphere may be present so that adequate measures can be taken with regard to the potential ignition sources. Fault-tree analysis is one of the most important methods to clarify the logical connection of a disturbance and the events that may have caused it. The construction of a fault-tree always begins with the top event. Having identified this top event, the immediate causes that will lead to it are searched for. In order to use fault trees for the quantitative determination of the probability of occurrence of the top-event, probabilities and frequencies are attributed to the individual states and the malfunctioning components.

Risk quantification is done through quantitative risk analysis that provides information during conceptual engineering on the probability that a person staying at a given place outside the fence dies as a consequence of an incident in the plant. This is called the individual risk. In addition, the group risk is determined for the plant and this is done on the basis of the actual population density. Examples of risk assessment and risk control studies are the design basis for plants and buildings, minimum distances between process units, pressure-relieving and depressurizing venting systems, explosion hazards in spaces and equipment and SHE assessment. For instance, the design basis for plants and buildings is made to ensure that in the event of a disaster in a process plant, buildings continue to perform their vital protective functions and adjacent plants are adequately protected to prevent domino effects. It is investigated, for instance, whether buildings must be pressure-resistant and/or gas-tight.

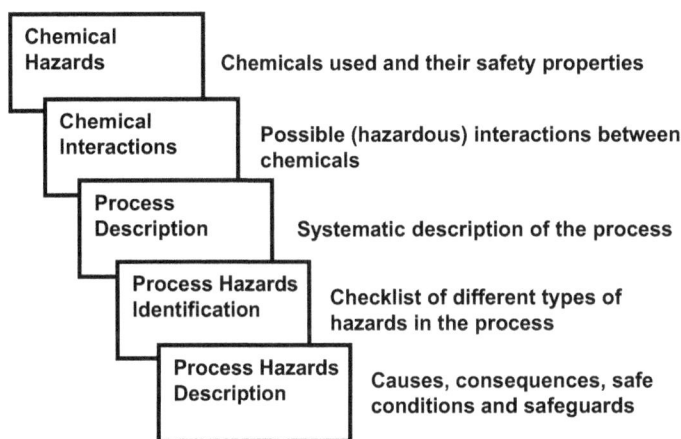

Chemical Hazards	Chemicals used and their safety properties
Chemical Interactions	Possible (hazardous) interactions between chemicals
Process Description	Systematic description of the process
Process Hazards Identification	Checklist of different types of hazards in the process
Process Hazards Description	Causes, consequences, safe conditions and safeguards

Fig. 18.4: Various steps in process hazard determination.

18.3 Identification of hazardous properties of substances

18.3.1 Pure components

As illustrated by Fig. 18.4, the identification of the potential hazardous effects of a substance is the first step in the identification of hazards. The risk assessment of a chemical substance is a complex procedure based on a comparison of its potential adverse effects with the reasonably foreseeable exposure of people and the environment to that substance. Hazard identification is mainly based on a critical evaluation of the intrinsic properties of a substance, such as physicochemical, toxicological and eco-toxicological properties. For each of these groups, it is possible to define a number of specific end points, and for each of them, select internationally recognized guidelines to determine the specific parameters. Interpretation of such parameters leads to the hazard identification of the chemicals. An important aid in the identification of potentially hazardous pure component properties is the so-called Material Safety Data Sheet (MSDS).

Physicochemical properties play an important role in the potential of a substance to produce adverse effects. Each physicochemical property plays a specific role. For instance, the molecular mass of the molecule gives an idea of whether the substance is bioavailable. A comparatively small molecule can enter an organism and produce its effect much more readily than a large molecule. The melting point characterizes the physical state of the substance. If the substance is a solid at room temperature, it is less mobile than a substance in the liquid or gaseous state. Like the melting point, the boiling point characterizes the physical state of the substance under normal pressure conditions. The lower the boiling point, the more easily the substance can be transferred into the air and larger is the risk of inhalatory effects. Relative density represents the mass ratio between a solid or liquid substance and water or a gaseous substance and air. When referred to water, its density indicates whether the substance can be expected to stratify on the surface or sink to the bottom. Similarly, for air, substances that are denser then air tend to stratify around the source of emission, while gases with a density lower than air diffuse easier into the air. Vapor pressure is also an important way to find out whether the substance may readily become volatile. A substance with a high vapor pressure chiefly affects inhalatory exposure, transfer potential and risk of explosion. Water solubility has direct implications for the possibility of dissolution in surface waters, and, therefore, the mobility in water and soil.

Flammability provides direct information on the possibility of the so-called physical effect. The risk of flammability may derive from a low vapor ignition temperature in the presence of an ignition source or from the possibility of the substance igniting spontaneously in air with no ignition point. For liquid products, the flash point is the parameter that allows quantification of the flammability risk. It is defined as the lowest temperature at which a liquid emits vapors in such a quantity as to produce a flammable vapor/air mixture. Figure 18.5 shows a setup for the experimental determination of flash points. The substance is slowly heated and its vapors are exposed to a spark at regular intervals. The lowest temperature at which the vapor/air mixture flashes is the flash point.

A gas explosion is a rapid chemical reaction (oxidation) in the gas phase, which propagates through the explosive mixture in a self-sustaining process. The heat released leads rapidly to high temperatures, and can result in the buildup of high pressures. A gas explosion requires the presence of an explosive mixture consisting of a finely dispersed fuel in an oxidizer (oxygen, ozone, nitrogen dioxide)

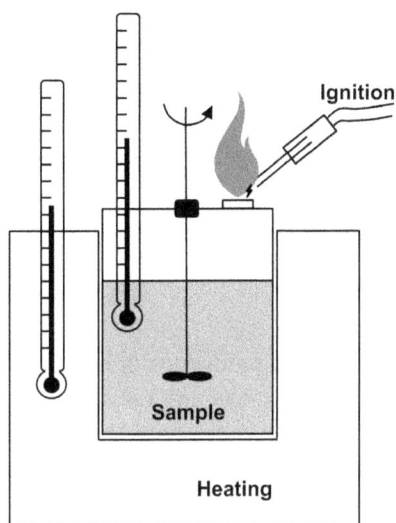

Fig. 18.5: Setup for flash point determination.

and the simultaneous presence of an effective ignition source. Any gaseous system containing mixed fuel and an oxidizer is explosive only within a certain concentration range, the explosion range. In practice, there is often only one fuel and the oxidizer is air. Here, the explosion range can be characterized by simply stating the concentration limits of the fuel in air. These bounds, the lower explosion limit (LEL) and the upper explosion limit (UEL) have been tabulated for a number of combustible gases and vapors. Nonexplosive mixtures in the range below the LEL are said to be too lean and those above the UEL are said to be too rich.

A substance with oxidizing properties may bring about dangerous reactions, causing fire, explosion or the formation of other hazardous substances. Oxidizing properties are related to the potential of the substance to act as an oxidant in an oxidation–reduction reaction, such as organic peroxides.

Toxicological properties are of direct interest in assessing the risk of chemical substances. They include a range of effects understood either as general poisoning of the organism or aimed at one or more particular target organs. The dosage administered and the duration of the toxicological tests are inversely proportional with reference to acute, sub-acute, sub-chronic and chronic studies. Acute toxicity includes those effects that may cause direct harm to health, following a single exposure. It is common to distinguish irreversible effects, such as death, from reversible effects, such as irritation of the eyes and skin, or corrosive and sensitizing properties. With regard to acute toxic effects, substances are classified as very toxic, toxic or harmful. This description is based on the median lethal dose LD_{50}, at which half the experimental animals die. Information on the more hidden effects that the substances may cause by exposure to low dosages over a period of time is provided by medium- or long-term toxicity tests. A basic aim of these tests is to identify the maximum dose level at which the substance does not present toxic effects. A substance is considered mutagenic if its adsorption may produce heritable genetic damage or increase their frequency. Similarly, substances are considered carcinogenic if they may produce cancer or increase its frequency. Teratogenic substances cause non-heritable birth defects. Eco-toxicological properties are indicators of the potential effects that a substance may have on the environment.

18.3.2 Exothermic chemical reactions

The majority of chemical reactions are accompanied by the production of heat (exothermic), which heats up the material itself, the reaction vessel and the surroundings. The degree of heat dissipation depends on the heat capacity of the system itself, the heat removal capacity provided by the reaction vessel design and/or the latent heat of the phase transitions. If the heat production rate exceeds the heat removal capacity of the system, the resulting temperature raise of the system will lead to a runaway reaction due to self-acceleration. As the reaction velocity increases, more heat is produced per unit time, and the system continues to heat up more and more rapidly till the maximum possible reaction velocity is reached. The consequences of a runaway event can be manifold. One possibility is that the accompanying temperature rise may lead to solvent evaporation, with the subsequent possibility of a vapor cloud explosion or a dangerous pressure buildup, by raising the vapor pressure of reactants. Another possibility is the formation of gaseous products from a decomposition reaction, which leads to a pressure increase and the risk of vessel rupture.

Tab. 18.1: Typical heats of reaction for common chemical reactions.

Reaction type	ΔH (kJ/mol)	Reaction type	ΔH (kJ/mol)
Neutralization (HCl)	−55	Hydrogenation	−560
Neutralization (H_2SO_4)	−105	Amination	−120
Diazotation	−65	Combustion	−900
Sulfonation	−105	Diazo-decomposition	−140
Nitration	−130	Nitro-decomposition	−400
Epoxidation	−96		

The overall heat of reaction and the rate of heat production determine the hazardous character of a chemical reaction. Some typical values for heats of reaction are summarized in Tab. 18.1. The chemical heat release of a reaction is controlled by the ability of the system to dissipate this energy. In mathematical terms, this is formulated as a heat balance, including an accumulation term as well as the heat generation and dissipation terms. A more simplified approach to illustrate the phenomena of runaway reaction systems is the Semenov model. Based on a pseudo-zero-order kinetic approach, the heat production rate, Q_R, is governed by an exponential dependence on temperature:

$$Q_R = V(-\Delta H_r)\, k_\infty\, C_{A,\infty}\, \exp\left(-\frac{E_A}{RT}\right) \tag{18.1}$$

The rate of heat removal Q_C depends linearly on the driving temperature difference between the uniform reaction mass temperature and the ambient (jacket) temperature:

$$Q_C = U A\,(T_R - T_A) \tag{18.2}$$

The controllability of the heat production rate can be explained by plotting the two heat flows as a function of temperature. As illustrated in Fig. 18.6, three different cases can be distinguished. In case 1, small deviations from the steady state, represented by the lower of the two intersections, automatically result in a return to the origin. This is rated as a stable operating point. In the upper steady state, the original operation conditions are never reached again once a temperature deviation occurs. In the case of a temperature decrease, the process quickly approaches the lower steady state. For a temperature increase, the heat production rate always exceeds the heat removal capacity of the system. This leads to an unhindered self-acceleration of the reaction rate and, thereby, of the heat production rate. The same is true for all operating conditions of case 3. Case 2 represents the limiting case of the first occurrence of an unstable operating point.

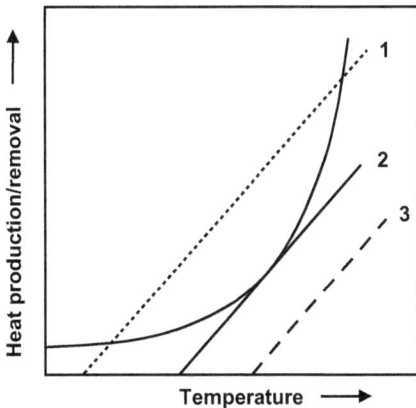

Fig. 18.6: Semenov plot.

In the runaway scenario, it is assumed that a reactor is operated under normal operating conditions and a cooling failure occurs. In the case of a cooling failure or when the heat removal is not sufficient to compensate for the heat production, the temperature increases proportionally to the heat of reaction. Thus, the reaction energy is a direct measure of the severity of a runaway. It is usually expressed as the adiabatic temperature increase ΔT_{ad} that the reaction mixture exhibits when the full heat of reaction is accommodated by the reaction mixture temperature increase:

$$\Delta T_{ad} = \frac{-\Delta H_r \, C_A^0}{\rho \, C_p} \tag{18.3}$$

The adiabatic temperature increase is a valuable tool to estimate the dynamics of a runaway. As a general rule, high energies ($\Delta T_{ad} > 100$ K) result in fast runaway or thermal explosion, while lower energies result in slower temperature increase rates. In practice, three levels are used to classify the runaway hazard of a chemical reaction:

Low: $\quad\quad\quad\quad\;$ $\Delta T_{ad} < \;$ 50 K $\quad$ No pressure buildup
Medium: $\;$ 50 K $< \Delta T_{ad} <$ 200 K
High: $\quad\quad\quad\quad$ $\Delta T_{ad} >$ 200 K

18.4 Inherently safer plant design

18.4.1 The concept and its benefits

The term, inherently safer, implies that the process is safe because of its very nature and not because equipment has been added to make it safer. Traditional plant designs try to reduce the risk by adding protective equipment and following safe methods of working. Nevertheless, it should be our second best choice. Before we install safety equipment to control the consequences of a hazard, we should ask if the hazard can be eliminated. Plants should therefore be designed so that they are user-friendly, can tolerate departures from ideal performance and equipment failure without serious effects on safety, output or efficiency.

The essence of the inherently safer approach to plant design is the avoidance of hazards rather than their control by adding protective equipment. To achieve this goal, the planner should replace hazardous substances by less hazardous ones wherever possible. Large inventories should be avoided as far as possible. Applying the concept of inherently safer design at the very beginning of a project, we may be able to choose a safe product instead of a hazardous one. When a process is being chosen, we may be able to choose a route that avoids the use of hazardous raw materials or intermediates. Once the chemistry has been decided and we are developing a flow sheet, we may be able to choose or develop intensified equipment, such as reactors and heat exchangers that do not require large quantities of

materials in progress. When we come to the detailed design, we may be able to reduce inventories by the application of well-known methods.

Inherently, safer plants are usually cheaper than conventional ones because
they do not need so much additional protective equipment. Although it is difficult
to say how much cheaper, on most plants, we can expect to save at least 5% of the
capital cost for a new plant if we could reduce our inventories of hazardous materials. Equally important would be reductions in the cost of testing and maintaining
the equipment. Finally, the biggest savings from reducing inventories will probably
come from a reduction in the size of the plant items and a corresponding reduction
in the size of the piping, structures and foundations.

18.4.2 The road to friendlier plants

The first constraint on the road to friendlier plants is that company procedures do not
usually ask for safety studies to be carried out early in design. Safety advisors do
not get involved and safety studies are not carried out till comparatively late in design. Then, it is too late to make major changes and avoid hazards. All we can do is
control them by adding protective equipment. However, when recognized in time, intensification, substitution, attenuation and simplification, schematically illustrated
in Fig. 18.7, are valuable techniques that may result in an inherently safer design
because hazards are avoided instead of controlled.

18.4.2.1 Intensification or minimization

The best way to prevent large leaks of hazardous materials is to not have so much
hazardous material about. Friendly plants contain low inventories of hazardous materials; so little that it does not matter if the entire inventory leaks out. What you do
not have cannot leak. This may seem obvious but till recently, little thought was
given to ways of reducing the amount of hazardous material in a plant. Engineers
simply designed a plant and accepted whatever inventory the design required, confident that they could keep it under control. No unit operation offers more scope for
reduction of inventory than the reaction. Many continuous reactors, such as liquid-
phase oxidation reactors, contain large inventories of highly flammable liquids. As a
rule, reactors, of all types, are not large because a large output is desired but because
the conversion is low, reaction is slow or both. When conversion is low, most of the
throughput has to be recovered and recycled, which increases the plant inventory.
For example, nitroglycerin used to be made in batch reactors containing about one
ton. Now, it is made in small, continuous reactors containing a few hundred grams.

18.4.2.2 Substitution

If intensification is not possible, then an alternative is substitution using a safer material in the place of a hazardous one. Thus, it may be possible to replace flammable refrigerants and heat transfer media by nonflammable ones, hazardous products by safer ones, and processes that use hazardous raw materials or intermediates by processes that do not. For instance, water under pressure can be used as a heat transfer medium to replace flammable oils.

18.4.2.3 Attenuation

Another alternative to intensification is attenuation by using a hazardous material under less hazardous conditions. For example, gaseous propylene is used to produce polypropylene, instead of liquid propylene dissolved in a flammable solvent. Attenuation is sometimes the reverse of intensification, for, if we make reaction conditions less extreme, we may need a longer residence time and a larger inventory.

18.4.2.4 Simplify

Modern plants are very complicated; this makes them expensive and provides too many opportunities for error. Simpler plants are friendlier than complex plants

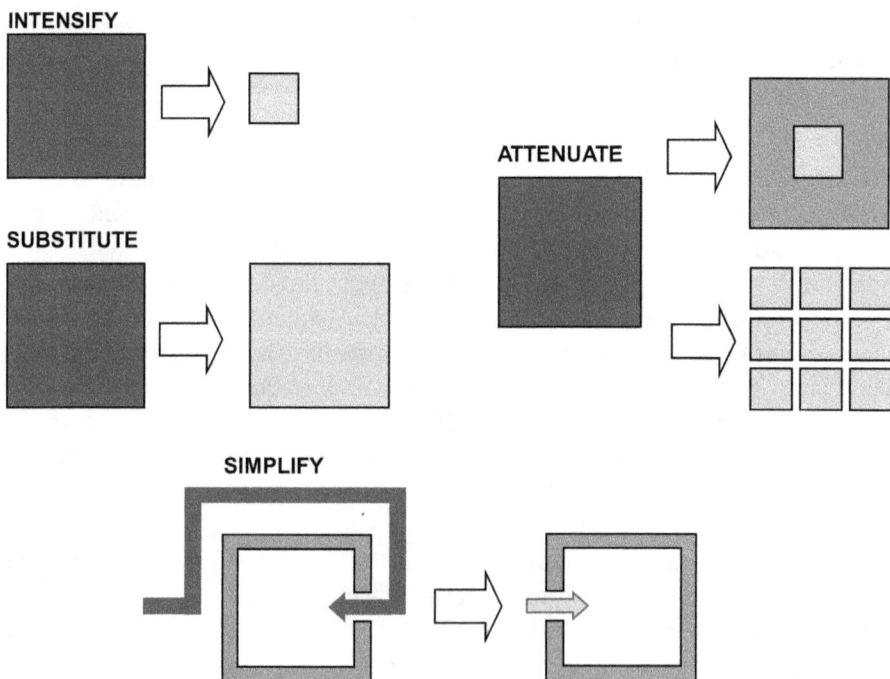

Fig. 18.7: The various roads to inherently safer plants (adapted from [205]).

because they provide fewer opportunities for error and less equipment that can fail and/or leak. They are usually also cheaper. The main reason for complexity in plant design is the need to add equipment to control hazards. Inherently safer plants are, therefore, also simpler plants.

Nomenclature

A	Heat transfer area	$[m^2]$
C	Concentration	$[mol/m^3]$
DOW CEI	DOW Chemical Exposure Index	
DOW FEI	DOW Fire and Explosion Index	
E_A	Activation energy	$[J/mol\ K]$
ENVID	ENVironmental hazard IDentification	
FMEA	Failure mode effect analysis	
HAZID	HAZard IDentification	
ΔH_R	Heat of reaction	$[kJ\ mol]$
k_∞	Preexponential factor	$[s^{-1}]$
Q_C	Heat removal rate	$[kJ/s]$
Q_R	Heat production rate	$[kJ/s]$
R	Gas constant	$[8.314\ J/mol\ K]$
T	Temperature	$[K]$
U	Heat transfer coefficient	$[W/m^2\ K]$
V	Volume	$[m^3]$

A Base chemicals

A.1 Ammonia

A.1.1 General description

In this chapter, the conversion of natural gas to ammonia is discussed. Natural gas mainly consists of methane CH_4, which is the actual reactant. The overall reaction takes place according to

$$CH_4 + H_2O + N_2 \rightarrow 4\,NH_3 + CO_2 + H_2 \tag{A.1}$$

As with most processes, this reaction is not carried out in one step. First the natural gas has to be desulfurized and steam is added. Next the methane is thermally cracked to hydrogen, CO and CO_2 and in a second step the majority of the CO is converted to CO_2. After removal of the CO_x, the hydrogen is dried and a reaction with nitrogen takes place to produce the ammonia, which is cooled and purified. The corresponding flow sheet is given in Fig. A.1.

A.1.2 Desulfurization

The supplied natural gas contains some sulfur (as H_2S and COS), which is a catalyst poison. To remove this contamination hydrogen is added and in the presence of a NiO-catalyst almost all of the sulfur is converted to H_2S and absorbed on NiO, which is converted to NiS. After some time the catalyst bed has to be replaced, because all the NiO is consumed. For large amounts of sulfur removal, an acid gas absorption stripper system is used for bulk removal and after that the described NiO system is used.

A.1.3 Primary and secondary reforming

The purified natural gas, which contains about 80% methane, is converted to hydrogen and CO as given in equations (A.2a) and (A.2b). The latter is called the water-gas shift reaction:

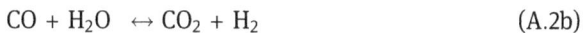

$$CH_4 + H_2O \leftrightarrow CO + 3H_2 \tag{A.2a}$$

$$CO + H_2O \leftrightarrow CO_2 + H_2 \tag{A.2b}$$

https://doi.org/10.1515/9783110712445-019

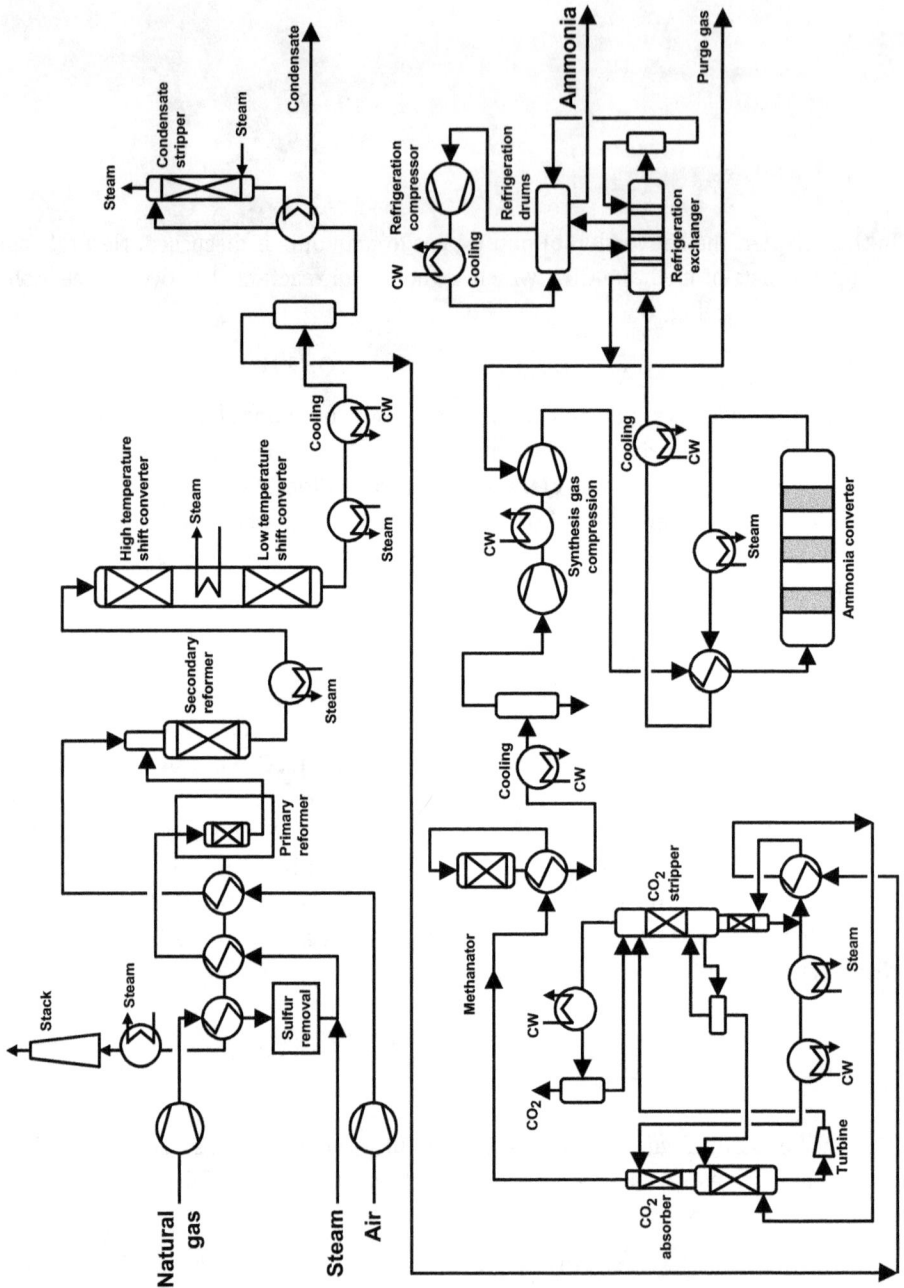

Fig. A.1: Flow sheet of an ammonia plant (adapted from [208]).

Both of these reactions are endothermic, which means a lot of heat is required to let these reactions take place. High temperature, low pressure and high steam to carbon ratios favor the reactions.

In the primary reformer the feed of methane and steam is led through nickel alloy tubes, placed in a furnace to heat the gas to 750–850 °C at a pressure of 30–40 atmosphere. The nickel content of the alloy, which also acts as a catalyst, is in the range of 20–30%. Most of the methane is converted, but there is still a large amount of CO in the outlet gas. Side reactions taking place are carbon formation by decomposition of higher hydrocarbons and by CO reduction and disproportionation. Most of this takes place in the first section of the reformer, when the hydrogen content is still low. Promoters, such as potash, are used to suppress the carbon formation, which decreases the cracking of higher hydrocarbons. The reformer type used is a so-called Kellogg box type. In the box reformer, the catalyst tubes are arranged in parallel, single-width rows heated from both sides, by either gas or liquid fired burners, located in the furnace arch.

In the secondary reformer the reaction is continued at higher temperatures (up to 1,000 °C) to achieve a low methane slip (0.2–0.3 vol%). Again the catalyst used is nickel, this time on an alumina support to withstand the high temperatures.

A.1.4 Shift conversion

Because the ammonia synthesis catalyst is very sensitive for poisoning by CO and CO_2, these gasses have to be removed. The easiest way to do this is to convert the carbon monoxide to carbon dioxide and absorption of the dioxide with a suitable absorbent. Unfortunately, the conversion of CO leads to equilibrium. At 300 °C only 85% conversion is obtained, which even decreases at higher temperatures. Usually the conversion is separated into two parts: a high-temperature conversion (HTS) to remove the bulk of the CO and a low-temperature conversion (LTS) for the remainder. The HTS takes place at 350–400 °C, over a magnetite catalyst stabilized with chromium oxide. The main reason to apply high temperatures is the high reaction speed, so the conversion of the bulk of the CO_2 takes place very rapidly. The LTS converter operates at temperatures of 200–250 °C in the presence of a CuO catalyst supported on alumina and zinc oxide. The converter design is based on an already used catalyst system, so when the catalyst load is fresh (and thereby more active) the operating temperature can be lower than design temperatures. Gradually raising the temperature compensates for the lower activity of the used catalyst and thus a constant outlet composition is achieved.

A.1.5 Carbon dioxide removal

Carbon dioxide is an acidic gas, which reacts reversibly with aqueous solutions of al-kanol amines. In general, solutions of weak bases (15–40 wt %) in water are applied for the removal of CO_2 from gases. Strong bases are inapplicable, because they would give irreversible absorption. The most used solvent is methyl-di-ethanol-amine (MDEA) and the CO_2 content can be lowered to 50 ppm. In most processes, absorption is af-fected between 20 and 40 °C in countercurrent operation in a column, preferably at the highest possible pressure. The amount of non-acid gas components absorbed is so small that the quantity released upon pressure reduction is too small to return them to the absorber; moreover, expansion turbines cannot be applied here. The regeneration is carried out at atmospheric pressure by applying heat. Since much water is removed from the solution along with desorbed CO_2, it is customary to cool the top product of the regenerator with a cooler/condenser and return the water to the column as reflux. A simple flow sheet of this absorption stripper system is given in Fig. A.2.

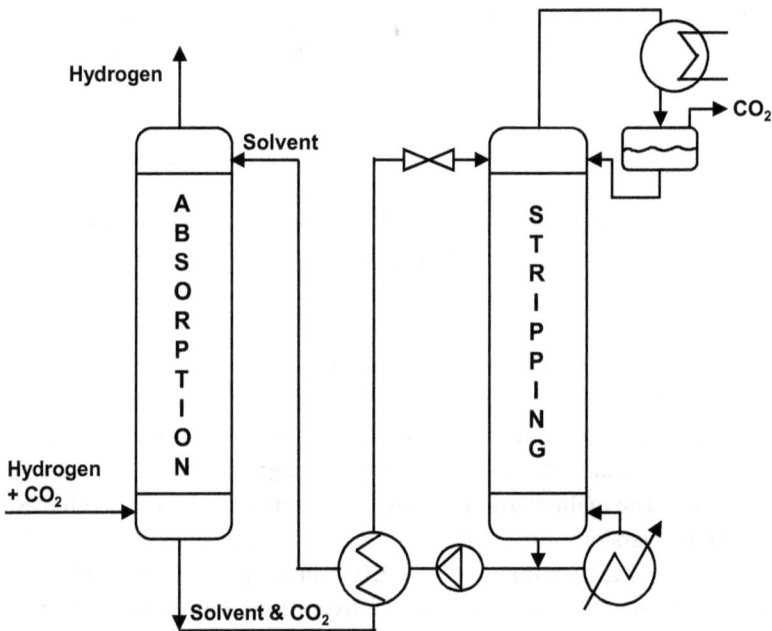

Fig. A.2: CO_2 removal system.

A.1.6 Final purification

Because oxygen containing compounds poison the ammonia syntheses catalyst, they must be efficiently removed. This is done in three steps, methanation, dehydration and excess nitrogen removal. The reactions taking place in the methanation are exactly the reverse of the reforming reactions at a temperature of 300–400 °C. They almost go to completion, reducing the CO and CO_2 content to a few ppm. Dehydration is done by the use of molecular sieves, usually at the interstage of the syntheses gas compressor to reduce volume requirement. Excess nitrogen removal is necessary because there are some processes, which use excess air (and thus excess nitrogen) and the nitrogen has a negative influence on the ammonia production. This is usually carried out by cryogenic purification, which involves cooling and stripping of the partly liquefied stream. Even here the CO content is lowered, about 50%.

A.1.7 Ammonia syntheses and recovery

The purified gas consists mainly of hydrogen and nitrogen (with a molar ratio of 3:1) and some inert components like argon, methane and sometimes helium. As can be seen from equation (A.3), the reaction will lead to an equilibrium, and therefore not all of the raw material can be converted in one pass. A higher conversion can be reached by introducing a recycle loop. The exact recycle stream in proportion to the product stream has to be determined after economic considerations. After mixing with the recycle loop, compression takes place to 220 bar and a temperature of 500 °C. By doing this, the speed of the reaction is increased and the equilibrium is shifted to a more favorable ammonia production. As discussed before, the syngas is dried at this stage using molecular sieves, after which the ammonia from the recycle is recovered. This reduces the required refrigeration, because the converter effluent is not saturated by the syngas addition, and cooling water can be used instead of more complicated refrigeration methods. This method leads to a very pure syngas, which increases catalyst lifetime (even up to 20 years) due to the lack of contaminants. To maintain a high partial pressure of the reactants, the inert components entering with the syngas have to be removed using a purge stream, as will be discussed later on:

$$N_2 + 3H_2 \leftrightarrow 2NH_3 + heat \tag{A.3}$$

The actual reaction takes place in an ammonia converter (see Fig. A.3) in which the gas flows through three adiabatic fixed beds and part of the heat of reaction is utilized to generate steam in the two intermediate heat exchangers. Most of the conversion takes place in the first bed, which has the highest driving force to equilibrium. The gas mixture is cooled and the excess heat is used to produce steam (40 bar) and to heat up the H_2/N_2 mixture before it enters the converter. For structural reasons the heat exchanger is incorporated between the inflow and outflow in the lowest part of the pressure casing. The reaction gas then flows upward in the annular gap between the pressure casing and the fixed beds, whereby it is further heated and at the same time protects the pressure-bearing structural components against excessively high fixed bed temperatures to prevent hydrogen embrittlement.

The mixture now enters the recycle loop, from which the ammonia is recovered. This is done by cooling the mixture to about 6 °C, from which the liquid ammonia is removed. The remaining gas is heated and enters the converter again, together with the fresh feed. The purge stream is lead through a post-converter, in which the remaining hydrogen and nitrogen is converted to ammonia. Again the mixture is cooled to remove the ammonia. The remaining gas is purged.

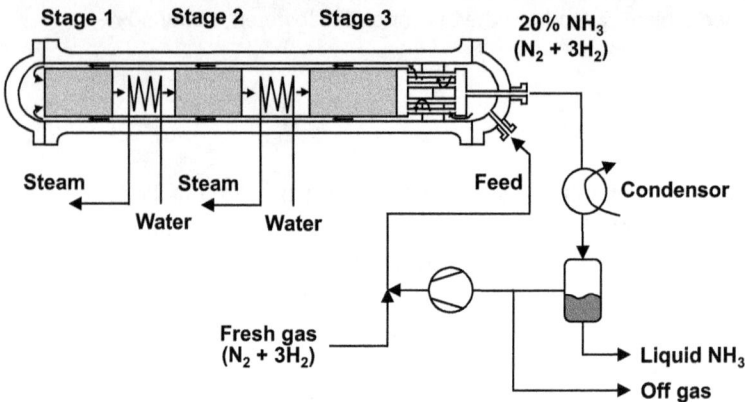

Fig. A.3: Schematic of a multistage ammonia converter (adapted from [208]).

A.2 Inorganic acids

In almost all chemical processes acids are needed, either as raw material or as a catalyst. Two of the most used are nitric acid and sulfuric acid. If a company has a high acid consumption, it might be economically attractive to install production plants of its own.

A.2.1 Nitric acid

Almost all commercial processes are based on the oxidation of ammonia to nitrogen oxide, which in its turn is reacted with water to form nitric acid. The overall reaction taking place is

$$12\,NH_3 + 21\,O_2 \rightarrow 8\,HNO_3 + 4\,NO + 14\,H_2O \tag{A.4}$$

Historically, different design philosophies between the United States and Europe have led to the development of two basic types of weak acid plants: the high monopressure and the dual-pressure processes. The high monopressure process has been favored in the United States because of its lower capital cost and traditionally lower energy and ammonia prices. In Europe the dual-pressure process is the most commonly used (see Fig. A.4).

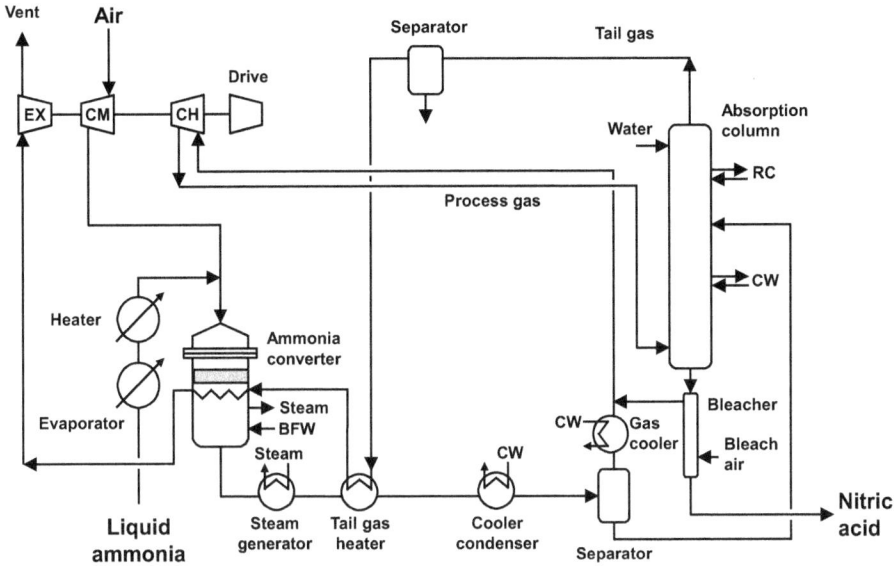

Fig. A.4: Dual-pressure process flow sheet (adapted from [207]).

The ammonia, which is supplied at a 16 bar pressure is vaporized at 6 bar and heated to 45 °C. Impurities like oil and water are removed by a purge and the ammonia is led through a candle filter where impurities larger then 3 micron are removed. Air is passed through a filter and compressed to 5.2 bar, temperatures can rise up to 220 °C. After compression, the air is split into two streams; a primary stream for ammonia incineration and a secondary stream for oxidation of NO. The compressor is powered by the energy recovery from the expansion of the spent process gasses, which are expanded from 10 bar to almost atmospheric pressure. Ammonia and primary air are then mixed in a 9:1 ratio and directed to the burners.

The ammonia/air mixture is passed over a platinum/rhodium (95/5%) gauze to produce nitric oxide, water vapor and a lot of heat. The resulting gasses are cooled, thus producing 79 bar steam and heating the off-gasses. The steam is used to power the compressor. The initial catalyst gauzes contain smooth wires and after some time catalyst migration takes place, resulting in a rough surface. Due to the increased surface area the ammonia conversion increases at first, but a gradual decline with time is observed because of catalyst loss. Because the amount of precious metal lost is financially significant, glass fiber filters or palladium catchment gauzes are used to recover lost catalyst particles.

As the gasses produced are cooled further to 50 °C, the nitric oxide is oxidized to nitrogen dioxide. The water formed condenses at this stage and together with part of the NO_2 it forms a 35 wt% nitric acid solution. Because hot liquid nitric acid is very corrosive, expensive materials are needed which are resistant to corrosion by hot acid. The remaining gas is mixed with the secondary air and compressed further to 10 bar. Temperature is increased by this procedure to 170 °C. The steam needed to drive the compressor is generated completely in the plant itself. After cooling in two separate heat exchangers, the temperature is decreased to about 50 °C and 65 wt% nitric acid is formed, which is fed to the absorption tower.

In the absorption tower, process gasses are brought in contact with water in a counter current flow. The NO_2 is absorbed and the NO is oxidized with the remaining oxygen to form NO_2. Typical dimensions are a diameter of 6 m, a height of 70 m and a number of trays ranging from 30 to 50. The heat generated by the oxidation is removed using internal cooling spirals. The acid produced is led through a Raschig-ring bed and the remaining nitric gasses are stripped from the solution. The de-gassed product, a 60 wt% nitric acid solution, is first stored before it is pumped to customers.

A.2.2 Sulfuric acid

Together with nitric acid, sulfuric acid is one of the most used industrial acids. One of its uses is as absorbent for water in the production of nitric acid. The first step in the process is burning the liquid sulfur, followed by further oxidation of the SO_2 to SO_3 and the actual formation of sulfuric acid:

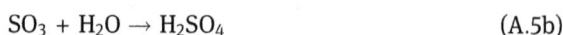

$$S + 1 \tfrac{1}{2} O_2 \rightarrow SO_3 \qquad\qquad\qquad (A.5a)$$

$$SO_3 + H_2O \rightarrow H_2SO_4 \qquad\qquad\qquad (A.5b)$$

The process mostly used is the double absorption process (Fig. A.5), which is described below. It was developed because conventional single absorption plants could not satisfy air pollution regulations, concerning the sulfur oxides content of the vented gasses. An additional advantage is the higher conversion that can be reached with the double absorption process. The basic difference between the two processes is the interstage SO_3 absorption, which is absent in the single absorption process.

Usually a sulfuric acid plant consists of a single train, regardless of the production capacity. A horizontal brick-lined combustion chamber is applied to burn the liquid sulfur. Air is first filtered and dried before it enters the combustion chamber. The molten sulfur, at a temperature ranging from 135 to 155 °C, is atomized by spray nozzles. These operate at a typical pressure of about 28 bar to achieve a very fine distribution of the sulfur. The self-sustaining ignition temperature of pure sulfur is 260 °C and a source of ignition is not needed if the chamber is preheated to operating temperatures

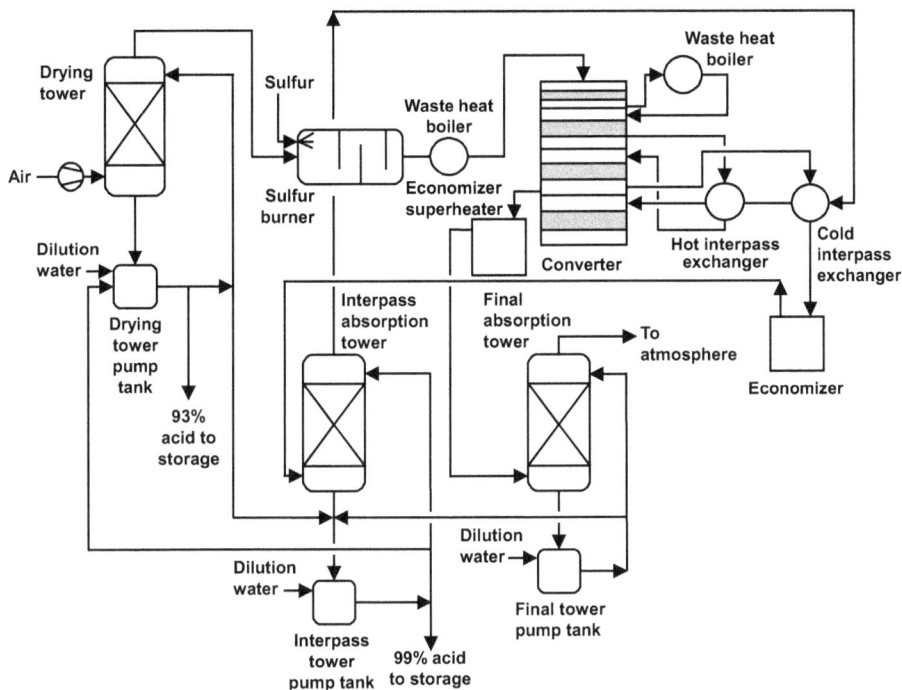

Fig. A.5: Double absorption process flow sheet (adapted from [207]).

of 400–425 °C. The sulfur dioxide concentration in the off-gas of the combustion chamber can range from 3 to 14 vol%, but most of the plants employ a concentration of about 10–11 vol%. The SO_2 produced is led over multiple catalyst beds to achieve an almost complete conversion to SO_3. Most of the plants operating now have four beds. The catalyst used is either platinum or vanadium pentoxide, of which the latter is used almost exclusively. The oxide is supported on kieselguhr or zeolite and promoted with potassium sulfate. It is supposed that the oxygen from the surface of the catalyst is transferred to the SO_2, after which the surface of the catalyst is re-oxidized by the oxygen from the feed.

In the double absorption process, the conversion of SO_2 to SO_3 takes place in four catalyst beds, with an interstage SO_3 absorption after the third bed. This way the SO_2 conversion is raised to 99.7% instead of 98%. In the interstage absorption, 90–95% of the total amount of sulfur trioxide produced in the whole process is absorbed. Absorption takes place in a packed absorption tower where the gas is brought in contact with concentrated sulfuric acid (98.5%). The typical inlet temperature for the acid ranges from 70 to 80 °C, resulting in an outlet temperature of 100–125 °C depending on the acid circulation rate. The rise in temperature is due to the sensible heat of the process gas and the heat of hydration from SO_3. The acid concentration in the recycle to the absorption is maintained at 98.5% to minimize the total vapor pressure, thus preventing the formation of H_2SO_4 mists in the absorption tower. This is done by the addition of process water and product removal from the recycle. After absorption the remaining process gas has to be heated again from approximately 80 to 425 °C. Now the gas is led through a single catalyst bed to convert the last SO_2 to SO_3. This is absorbed in a column equal to the interstage absorber, except that the last column outlet temperature is lower than 105 °C. This is due to the lower amount of gas absorbed.

A.3 Ammonia-based products

A lot of chemicals are made from ammonia, including, caprolactam, acrylonitrile, urea and melamine, which are also base chemicals. The greater part of the production is used for the production of fibers.

A.3.1 Caprolactam

Caprolactam is mostly used as raw material for nylon 6, a commonly used polymer. The actual formation of caprolactam is performed in a hydroxylamine/phosphoric acid buffer solution. The solution has to be regenerated by addition of hydrogen, which in the oximation is used to form cyclohexanon oxime, an intermediate in the production of caprolactam. The oxime rearranges to caprolactam in oleum. Yields of 98% can be reached by fine-tuning the process. Scheme A.1 presents the overall reaction scheme.

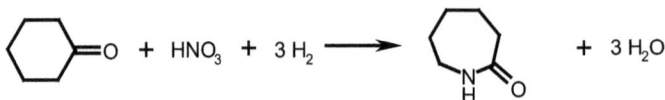

Scheme A.1: Caprolactam formation.

A.3.1.1 Oxime formation
As illustrated in Fig. A.6, the oximation takes place in a hydroxylamine/phosphoric acid buffer solution. After separation of the oxime, the remaining ammonium phosphate buffer solution is recycled to hydroxylamine synthesis and concentrated. The hydroxylamine formation takes place over a palladium catalyst on a carbon or alumina support by addition of hydrogen. Unreacted hydrogen is separated from the catalyst and recycled. After reaction with the phosphoric acid in the solution, the hydroxylamine phosphate is formed at pH = 1.8.

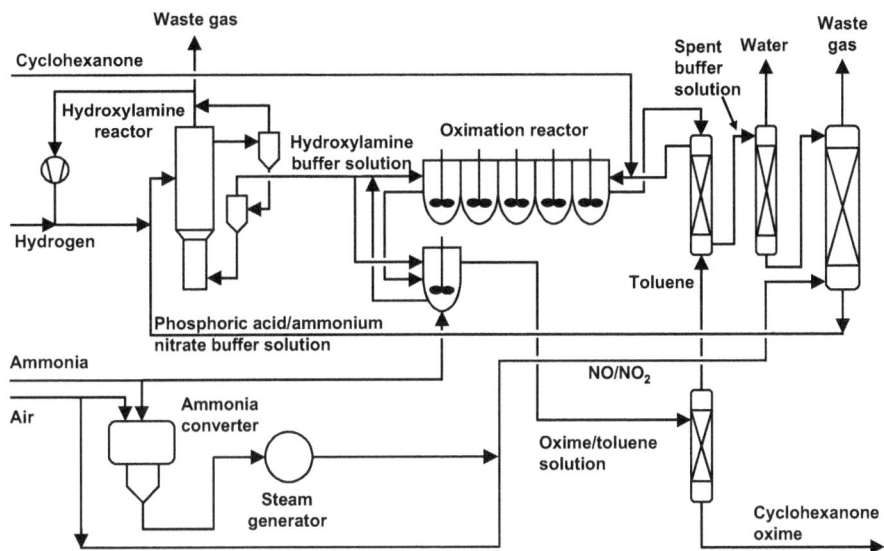

Fig. A.6: Cyclohexanone oxime production flow sheet (adapted from [208]).

The hydroxylamine phosphate reacts with the cyclohexanone to form cyclohexanone oxime in the oximation reactor. Water is used as a solvent and the phosphoric acid is liberated. The oximation reactor consists of a cascade of mixers and separators in a countercurrent system at pH = 2. The extracting agent used is toluene. A conversion of up to 98% can be achieved. The remaining 2% of cyclohexanone is oximated with about 3% of the mainstream hydroxylamine phosphate and some added ammonia. The oxime solution (about 30% in toluene) is separated from the water and the oxime is separated from the toluene. The purified toluene is recycled to the oximation reactor to extract the fresh oxime from the buffer solution. The oxime is fed to the Beckham rearrangement process, which is discussed later on.

To avoid poisoning of the palladium catalyst, cyclohexanone and oxime are extracted from the spent buffer solution. This is done using toluene as an extracting agent in a packed column. After removal of this contamination, the toluene is stripped from the buffer solution with steam. The next step before recycling the buffer is the removal of excess ammonium ions. To achieve this, ammonia is burned to NO_x in an incinerator. The nitrogen oxide formed is brought in contact with the buffer in the decomposition column and the nitrogen produced is vented.

A.3.1.2 Beckham rearrangement process
To obtain caprolactam from the produced oxime, almost all producers use the Beckham rearrangement process. The rearrangement is a very rapid and exothermic reaction and therefore fresh oxime solution in oleum is added to a relatively large

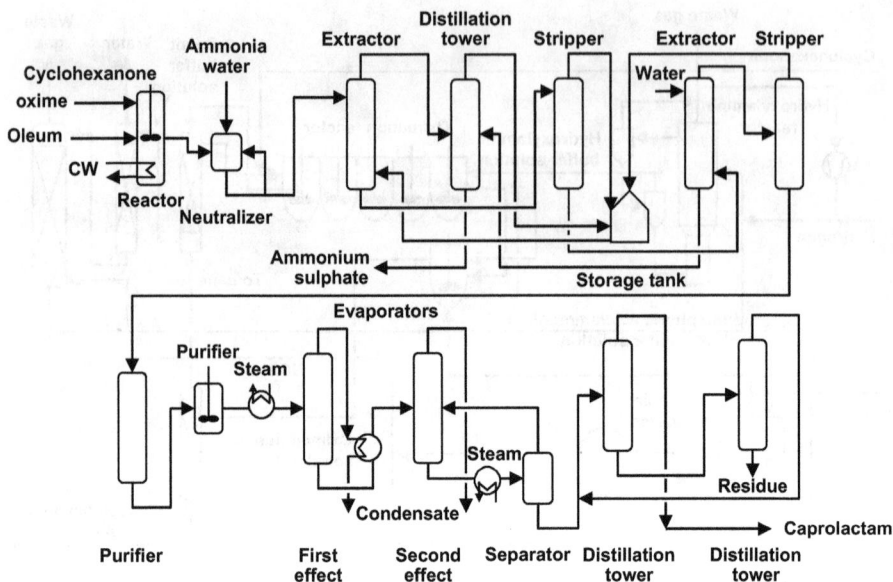

Fig. A.7: Flow sheet of the bisulfate lactam process (adapted from [207]).

amount of already rearranged product. Additional cooling is still needed, which is done by applying a heat exchanger, to keep the temperature at 125 °C. Sometimes sulfur dioxide is added to remove water, which is also formed in the reaction. The sulfuric acid formed catalyzes the reaction and the sulfate of caprolactam is obtained. After neutralization with ammonia a caprolactam solution is formed. In the bisulfate lactam process schematically shown in Fig. A.7, the ammonia consumption is only half that of the conventional processes. The crude caprolactam is removed in a separator drum and extracted with solvents such as benzene or toluene (in conventional processes). In the bisulfate process a commercially available chlorinated hydrocarbon is used, preferably an aliphatic hydrocarbon with one or two carbon atoms. The caprolactam solution is led through an extractor, where the lactam is extracted in a counter current process with the chlorinated hydrocarbon. The organic stream is transferred to an atmospheric pressure stripper and the lactam concentration is increased to about 60% by evaporating some solvent overhead. The solvent is condensed and recycled. The concentrated lactam solution is again led through an extractor, this time demineralized water is used as extracting agent. The 30% aqueous solution contains traces of solvent, which are stripped, condensed and stored for recycle. The solution is delivered to a conventional caprolactam purification system. This includes two distillation steps to remove any solvent present in the caprolactam. The lactam is obtain in molten form and can be stored as such or be solidified in a flaker.

The ammonium bisulfate solution from the first extractor and the impure solvent from the second are fed to a distillation column. Pure solvent leaves overhead and the bottom product is an approximately 50% aqueous solution of ammonium bisulfate. The solvent is stored and recycled and the water from the aqueous solution is evaporated. The remainder of the solution is incinerated to provide a dilute sulfur dioxide solution. This is further oxidized to trioxide over a vanadium pentoxide catalyst and absorbed in sulfuric acid to produce the oleum used as a solvent in the Beckham rearrangement.

A.3.2 Acrylonitrile

Acrylonitrile is almost exclusively produced by a process designed by Sohio (Standard Oil of Ohio), which basically consists of vapor-phase ammoxidation of propylene over a catalyst. The reaction taking place is given by

$$C_3H_6 + NH_3 + 1\frac{1}{2}\,O_2 \rightarrow C_3H_3N + 3\,H_2O \tag{A.6}$$

The process itself is simple and conventional in every aspect; no extreme pressures, temperatures, reactors or materials are needed. Therefore, the flow sheet and process description, given in Fig. A.8, appear rather simple.

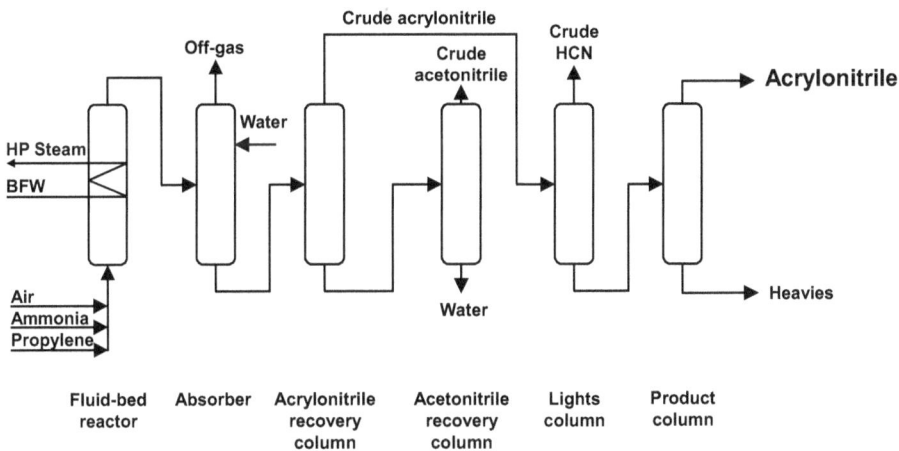

Fig. A.8: Acrylonitrile production (adapted from [207]).

In a single pass fluid bed reactor, propylene, ammonia and air are brought in contact with a catalyst. Modern catalysts are based on a bismuth-molybdenum oxide; formerly used catalysts were iron-antimony oxide, uranium-antimony oxide and tellurium-molybdenum oxide. Under normal operating conditions, no catalyst regeneration has to take place. But when needed, circulating the reactor with air for a short period can do it very easily. The reaction takes place at 400–450 °C and 2 bar with a yield of 75–80%. Byproducts are HCN, acetonitrile, CO_x and nitrogen. The hot reactor effluent is quenched with water in a counter current absorber, where any unreacted ammonia is neutralized with sulfuric acid. The absorber off-gas contains N_2, CO, CO_2 and unreacted hydrocarbons, which can either be vented directly or led through an incinerator before venting. The acrylonitrile solution is led to a recovery column, which produces a crude acrylonitrile (containing HCN) as a top product and a bottom product, which is led to a second recovery column. In this recovery column a small amount of oxalic acid is used to stabilize the carbonylic cyanohydrins. As a result of this, no volatile contaminants are formed from them and they leave the column as heavy impurities. The crude acrylonitrile is distilled in a next column, separating the HCN and acrylonitrile. The top product of the second recovery column is a crude acetonitrile, which can either be incinerated or be further purified to produce solvent quality acetonitrile.

A.3.3 Urea

Urea is used as a raw material for the production of melamine and for fertilizers. Since the introduction of the Stamicarbon CO_2 stripping process more than 100 plants of this type have been built all over the world, indicating the efficiency of this process.

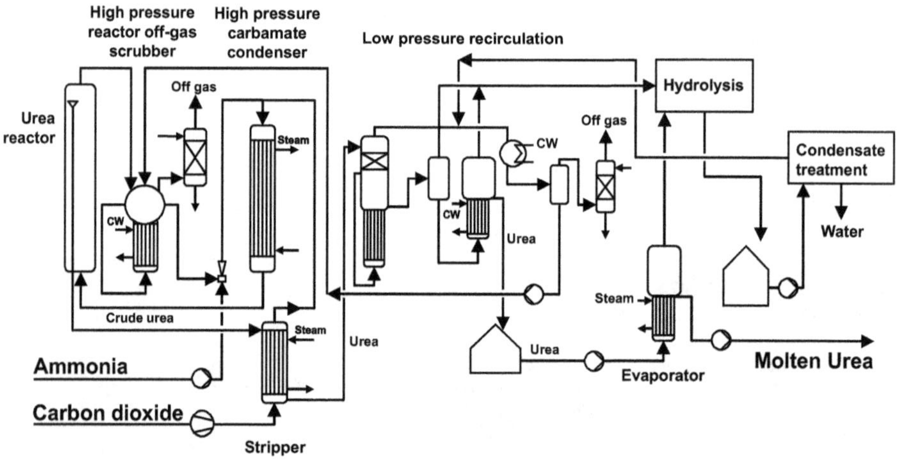

Fig. A.9: Flow sheet of the Stamicarbon CO_2 stripping process (adapted from [208]).

The synthesis stage of the Stamicarbon process (Fig. A.9) consists of a urea reactor, a stripper for unconverted reactants, a high-pressure carbamate condenser and a high-pressure reactor off-gas scrubber. To achieve maximum urea yield per pass through the reactor at the stipulated optimum pressure of 140 bar, an $NH_3 : CO_2$ molar ratio of 3:1 is applied. The greater part of the unconverted carbamate is decomposed in the stripper, where ammonia and carbon dioxide are stripped off. This stripping action is accomplished by countercurrent contact between the urea solution and fresh carbon dioxide at synthesis pressure. Low ammonia and carbon dioxide concentrations in the stripped urea solution are obtained, such that the recycle from the low-pressure recirculation stage is minimized. These low concentrations of both ammonia and carbon dioxide in the stripper effluent can be obtained at relatively low temperatures of the urea solution because carbon dioxide is only sparingly soluble under such conditions. Condensation of ammonia and carbon dioxide gases leaving the stripper occurs in the high-pressure carbamate condenser at synthesis pressure. As a result, the heat liberated from ammonium carbamate formation is at a high temperature. This heat is used for the production of 4.5 bar steam for use in the urea plant itself. The condensation in the high-pressure carbamate condenser is not effected completely. Remaining gases are condensed in the reactor and provide the heat required for the dehydration of carbamate, as well as for heating the mixture to its equilibrium temperature.

In a recent improvement to this process, the condensation of off-gas from the stripper is carried out in a pre-reactor, where sufficient residence time for the liquid phase is provided. As a result of urea and water formation in the condensing zone, the condensation temperature is increased, thus enabling the production of steam at a higher pressure level. The carbon dioxide feed, invariably originating from an associated ammonia plant, always contains hydrogen. To avoid the formation of explosive hydrogen-oxygen mixtures in the tail gas of the plant, hydrogen is catalytically removed from the carbon dioxide feed. Apart from the air required for this purpose, additional air is supplied to the fresh carbon dioxide input stream. This extra portion of oxygen is needed to maintain a corrosion-resistant layer on the stainless steel in the synthesis section. Before the inert gases, mainly oxygen and nitrogen, are purged from the synthesis section, they are washed with carbamate solution from the low-pressure recirculation stage in the high-pressure scrubber to obtain a low ammonia concentration in the subsequently purged gas. Further washing of the off-gas is performed in a low-pressure absorber to obtain a purge gas that is practically ammonia free. Only one low-pressure recirculation stage is required due to the low ammonia and carbon dioxide concentrations in the stripped urea solution. Because of the ideal ratio between ammonia and carbon dioxide in the recovered gases in this section, water dilution of the resulting ammonium carbamate is at a minimum despite the low pressure (about 4 bar). As a result of the efficiency of the stripper, the quantities of ammonium carbamate for recycle to the synthesis section are also minimized, and no separate ammonia recycle is required. The urea solution coming from the recirculation stage contains about 75 wt% urea. This solution is concentrated in the evaporation section. If the process is combined with a prilling tower for final product shaping, the final moisture content of urea from the evaporation section is ca. 0.25 wt%. If the process is combined with a granulation unit, the final moisture content may vary from 1 to 5 wt%, depending on granulation requirements. Higher moisture contents can be realized in a single-stage evaporator, whereas low moisture contents are economically achieved in a two-stage evaporation section. When urea with an extremely low biuret content is required (at a maximum of 0.3 wt%), pure urea crystals are produced in a crystallization section. These crystals are separated from the mother liquor by a combination of sieve bends and centrifuges and are melted prior to final shaping in a prilling tower or granulation unit. The process condensate emanating from water evaporation from the evaporation or crystallization sections contains ammonia and urea. Before this process condensate is purged, urea is hydrolyzed into ammonia and carbon dioxide, which are stripped of with steam and returned to urea synthesis via the recirculation section. This process condensate treatment section can produce water with high purity, thus transforming this "wastewater" treatment into the production unit of a valuable process condensate, suitable for, e.g., cooling tower or boiler feed water makeup.

A.3.4 Melamine

For the low pressure production there are three (commercial) processes available: the BASF, the Chemie Linz and the Stamicarbon process. Because the Stamicarbon process is the most commonly used, it is discussed here. Scheme A.2 gives the reaction of urea into melamine.

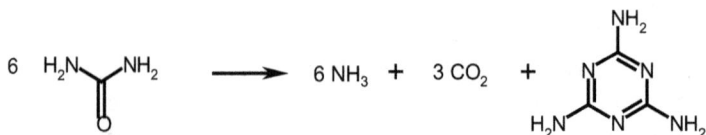

$$6 \; H_2N\text{-}CO\text{-}NH_2 \longrightarrow 6 \; NH_3 + 3 \; CO_2 + \text{melamine}$$

Scheme A.2: Melamine formation from urea.

Like the BASF process, the DSM Stamicarbon process (Fig. A.10) involves only a single catalytic stage. However, it differs from the former in that it is operated at 7 bar, the fluidizing gas is pure ammonia, the catalyst is of the silica alumina type, and melamine is recovered from the reactor outlet gas by water quench and recrystallization. Urea melt is fed into the lower part of the reactor. The silica alumina catalyst is fluidized by preheated (150 °C) ammonia, which enters the reactor at two points: at the bottom of the reactor to fluidize the catalyst bed, and at the urea nozzles to atomize the urea feed. The reaction is maintained at 400 °C by circulating molten salt through heating coils within the catalyst bed. The melamine-containing reaction mixture from the reactor is quenched first in a quench cooler and then in a scrubber with recycled mother liquor from the crystallization section. The resulting melamine suspension is concentrated to ca. 35 wt% melamine in a hydrocyclone, after which it is fed to a desorption column where part of the ammonia and carbon dioxide dissolved in the suspension is stripped off and returned to the scrubber. The preceding steps are all carried out at reaction pressure; for the following stages, the pressure is reduced. The suspension leaving the bottom of the desorber is diluted with recycled and preheated mother liquor and water. Activated carbon and filter aids may also be added. The melamine dissolves completely, although separate dissolving vessels are necessary to allow sufficient time for dissolution. The resulting solution is filtered using precoat-type filters. Crystallization of melamine is carried out in a vacuum crystallizer, and crystals are separated from the mother liquor by hydrocyclone and centrifuge. The crystals are dried in a pneumatic dryer and then conveyed to product bins. Surplus ammonia must he recovered as fluidizing gas from the wet ammonia carbon dioxide mixture leaving the desorption column and the scrubber. The hot gas mixture is partly condensed by heat-exchange with the mother liquor from melamine dissolution. The condensate and uncondensed gas are then passed at 7 bar to an absorption column. Liquid makeup ammonia is led to the top of this column to condense any carbon dioxide remaining in the ammonia gas. The ammonia is then compressed and recycled as fluidizing and urea-atomization gas for the reactor.

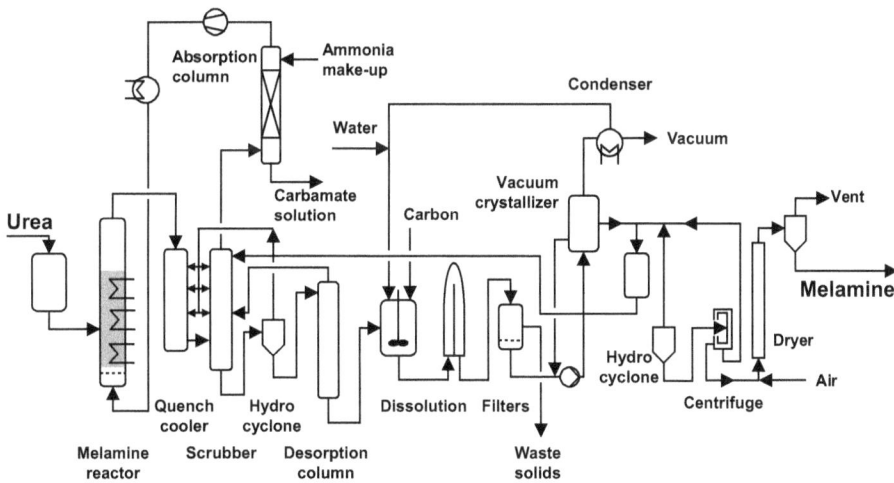

Fig. A.10: Flow sheet of the Stamicarbon melamine process (adapted from [208]).

A.4 Naphtha cracking

In this chapter the thermal cracking of naphtha to a number of unsaturated hydrocarbons, such as ethylene, propylene and styrene is discussed. These are all very important monomers for polymer production, like polyethylene and polypropylene. One advantage of the produced aliphatic compounds is that they are more reactive than the saturated naphtha. Naphtha is chosen because it is the most used raw material for the production of ethylene and propylene. Ethylene and propylene have lowered the demand for acetylene as an intermediate for polymer production, because of the higher production cost of the latter. Steam cracking primarily produces ethylene, but also propylene, and, as secondary products, depending on the feedstock employed, a C_4 fraction rich in butadiene and a C_{5+} fraction with a high content of aromatics, mostly benzene. The variety of products obtained by steam cracking makes this a key process, around which a complex of user installations can be build.

A.4.1 Basic principles

Cracking can be done in several different ways; at high temperatures (750–850 °C, thermal cracking) or at lower temperatures with the use of a catalyst and by reaction with hydrogen (hydrocracking). The following description concerns a process for thermal cracking. The basic reactions governing the cracking of higher fractions comprise the cracking of a saturated hydrocarbon to a paraffin (containing only single bonds) and an olefin (one double bond). This step is called primary cracking. By secondary cracking subsequent reactions at various points of their hydrocarbon chain of the species thus formed yield a number of light products that are rich in olefins, whose composition depends on the feedstock and the operating conditions. An example is shown in Scheme A.3, illustrating the formation of ethylene from heptane through a combination of primary and secondary cracking.

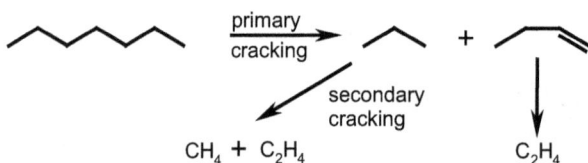

Scheme A.3: Primary and secondary cracking reactions.

The compounds thus formed, if subjected to subsequent intense dehydrogenation, are capable of forming a number of aromatic compounds and particularly benzene. Further dehydrogenation will lead to the formation of coke and tar products, which have a negative impact on the operation of the naphtha cracker. Cracking reactions

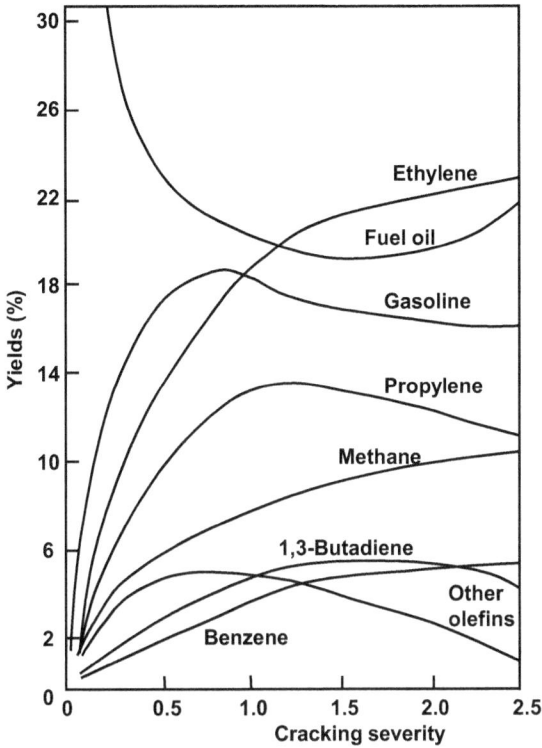

Fig. A.11: Temperature dependence of product composition (vol%).

become significant at temperatures of about 700 °C, whereas dehydrogenation only takes place substantially at temperatures above 800–850 °C. The temperature at which the cracker is operated, higher temperature means higher cracking severity, has a significant influence on the composition of the products, as is illustrated in Fig. A.11.

Because naphtha is a mixture of hydrocarbons, with a boiling point varying from 100 to 220 °C, and the large variety of products obtained, a naphtha cracker is one of the more complicated crackers. The most important parts are the furnace, in which the actual cracking takes place, and a complex system of fractionation columns, used to separate the product mixture. These two sections are referred to as the hot and cold section. A flow sheet is given in Fig. A.12.

A.4.2 Hot section

The necessary heat for cracking the naphtha is supplied by burning natural gas in several furnaces. The feedstock enters the furnace in the convection zone, where it is preheated and mixed with steam, which is also preheated in this section of the

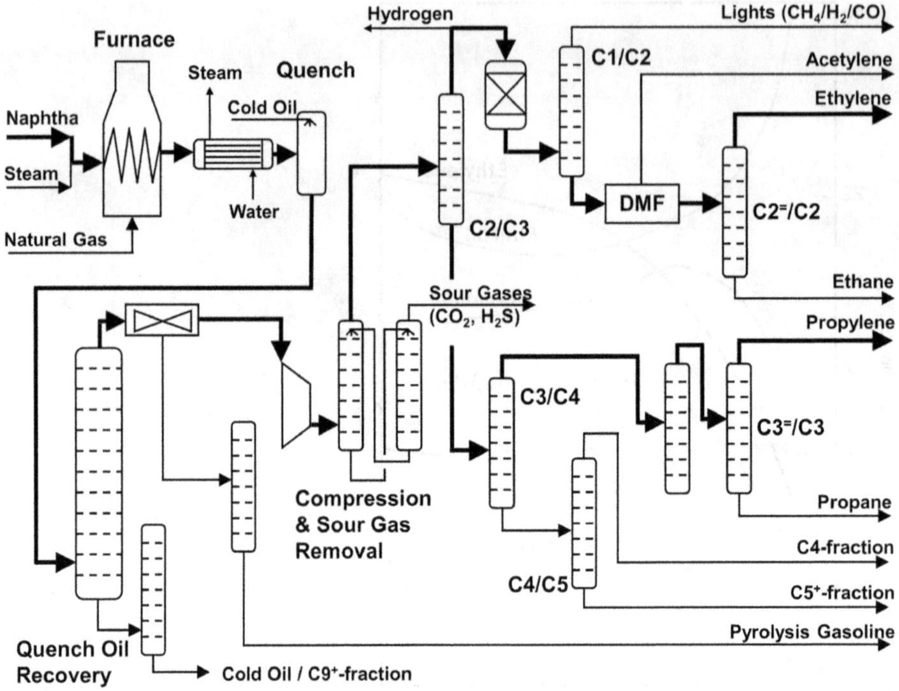

Fig. A.12: Flow sheet of a naphtha cracker.

furnace. Here heat transfer takes place by the rising of gases, being the reason why it is called the convection zone. Next the feedstock is lead through the radiation zone of the furnace, where a rapid rise in temperature occurs. The feedstock temperature in this section reaches about 750–800 °C and the actual cracking takes place. A sketch of a typical furnace is given in Fig. A.13.

The residence time in this zone is shorter than one second, because of the side reactions, such as dehydrogenation, which proceed rapidly at the cracking temperatures. To avoid unwanted side and/or subsequent reactions the effluents are cooled very rapidly, a so-called quench, and the composition of the mixture is preserved. Quenching generally takes place in two steps: first an indirect quench using a water-cooled heat exchanger and a second direct quench by adding the heavy pyrolysis byproducts. The heat transfer in the water-cooled heat exchanger is so high that high-pressure steam is generated, which can be used for cracking.

After this the effluents are transferred to a primary fractionating column, where separation takes place between quench oil (C₉₊ fraction), a gasoline side stream and light pyrolysis products. The primary fractionating column operates at a pressure of 1.5 bar and temperatures of 110/170 °C (top/bottom). The steam still present in the furnace effluent condenses in the fractionating column and is recycled with makeup process water. The pyrolysis gasoline has a very high percentage of aromatics, mostly

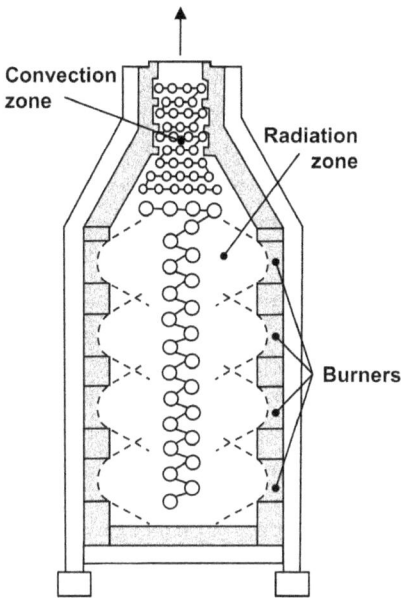

Fig. A.13: Example of a steam-cracking furnace (adapted from [201]).

benzene, which can be used in several other plants. Styrene production for example, which is discussed later on, is one of those processes. The light fraction contains a wide range of gaseous products, which have to be separated and purified in the cold section. Before entering the cold section the top product is compressed, resulting in a liquid phase, which is required for distillation.

A.4.3 Cold section

The liquid leaving the compressor still contains some sour gasses, like CO_2 and H_2S. For downstream applications it is required to remove these. This can be done by an absorber-stripper combination, using an alkanolamine as solvent for the sour gasses. Because the distillation takes place at high pressure and low temperatures, it is necessary to reduce the water content to about 5 ppm (e.g., drying by molecular sieves) to prevent the formation of ice crystals. This is usually done before the last (of 4–5) compression step.

The first distillation column (33 bar, −40/ +95 °C) has a light fraction containing C_2 and lower hydrocarbons and a heavy fraction of C_{3+}. The light fraction enters another fractionating column (14 bar, −115/−40 °C), separating the mixture in a CH_4/ H_2/CO-mixture (light end) and a heavy fraction containing acetylene, ethylene and ethane. Before distilling the heavy fraction, acetylene is absorbed with di-methyl-formamide (DMF). The last separation step for this end is between ethylene and ethane, which takes place at 8 bar and −60/−30 °C. This is a very difficult separation, which is performed in the C_2 splitter. This is a distillation column with approximately 120 trays and a relatively high reflux ratio. A C_2 splitter can be as high as 90 meters due to the large number of trays needed to separate the ethane from the ethylene. As a result of this, a high refrigerating capacity is needed in the condenser. The ethylene liquid is drawn from tray 5 (close to the top of the column). Ethane is first used to supply part of the condenser duty in the C_2-splitter and then recycled to the furnace to achieve a higher ethylene yield. This leads to a lower energy consumption in two ways: first the external cooling in the condenser is reduced and second the heating duty required is lower than without this kind of heat transfer.

The heavy fraction of the first distillation column (C_{3+} fraction) also undergoes several distillations, first separating C_3 and C_4 fractions (at 15 bar, 40–100 °C) and after that further separation to propylene, propane (from the C_3 fraction), a C_4 and a C_{5+} fraction. The most important side-products are butadiene, benzene and styrene, which can all be used in several other plants, e.g., for polymerization (butadiene and styrene).

A.4.4 Coke formation

In the description above one of the most important problems, the coke formation, was only shortly mentioned, and will be discussed in greater detail here. Coke is a term used here for high molecular weight hydrocarbons with a very low hydrogen content (less than 4%). These compounds range from tar-like fluids, to fluffy depositions and even very hard carbon depositions. Despite the precaution taken, coke formation will always occur in this section, mainly at the end of the furnace tube, the inlet of the quench and at the end of the quench, where heavy hydrocarbons from the mixture condense and slowly dehydrogenate. In order to maintain a constant production level, most plants have multiple furnaces, so they can be cleaned one by one while the others keep the production up to the desired level. In practice, cleaning of the furnace has to take place once every two or three months. This can be done by taking the furnace off-line and leading steam with 1 volume% of air through the tubes at an operating temperature ranging from 600 to 800 °C. This is called slow burning of the coke, which can take up to 30 h. Addition of extra air can speed up the process, using 20 vol% of air, for example, the procedure requires only 5 h. The coke left can be mechanically removed by means of high-pressure water.

A.5 Oxidation processes

In this chapter three oxidation reactions are described: oxidation of toluene to benzoic acid and phenol, cyclohexane to cyclohexanol and cyclohexanone and the formation of maleic anhydride from butane. All these processes give more than one product (wanted or unwanted), so again there is a lot of separation technology involved. Oxidation is a type of reaction in which a raw material is reacted with oxygen, which usually is taken from the air. Sometimes, for example, in case of equilibrium, the oxygen content of air is not high enough and pure oxygen has to be used.

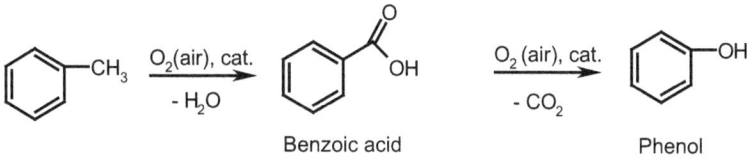

Scheme A.4: Oxidation of toluene to benzoic acid and phenol.

A.5.1 Toluene oxidation

Benzoic acid is an important raw material in the production of caprolactam, which is discussed in Chapter 4. Benzoic acid can also be seen as an intermediate in the production of phenol, as it is the product of the first production step in phenol production shown in Scheme A.4. Between the first and second step benzoic acid can be (partially) extracted from the process to serve as product or as intermediate for caprolactam production.

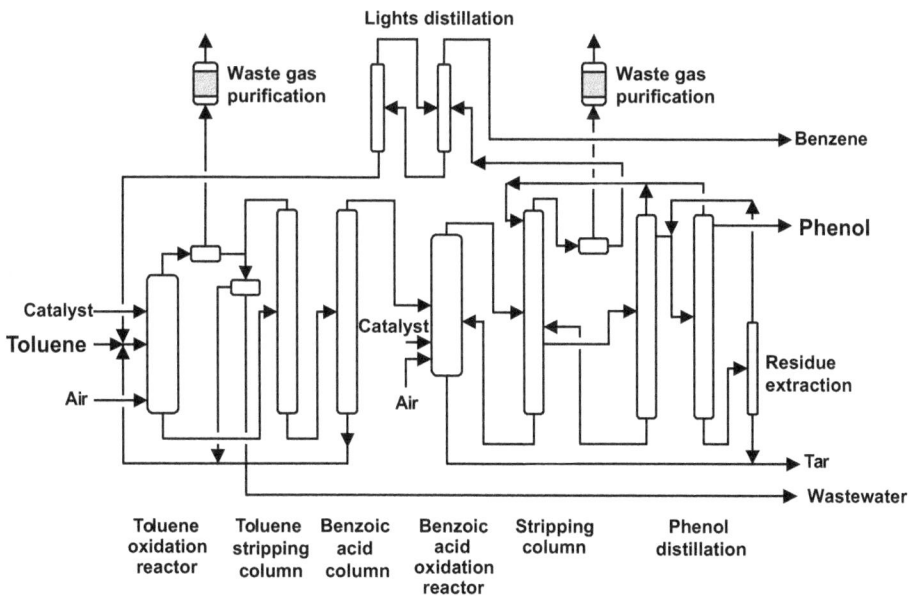

Fig. A.14: Toluene oxidation flow sheet (adapted from [208]).

The first stage of the production process (Fig. A.14) is the liquid-phase oxidation of toluene with atmospheric oxygen over a cobalt catalyst. The second step is the decarboxylation of the isolated benzoic acid, using a copper catalyst. Despite the two-step process, yields of 60–70 mol% can be reached, with a product purity of almost 100%. The impurities are mostly organic traces, but besides these, the color and color stability, which are considered an indication for the stability and quality of the product, are often important for customers. Various techniques are utilized to improve the color and stability, but most of them are regarded trade secrets.

Fresh toluene is fed into the oxidation reactor together with recycled toluene and the corresponding quantity of catalyst (usually 100–300 ppm). The catalyst used is cobalt naphthenate or benzoate, which are soluble in toluene. Air is added using air sprinklers. During the reaction, the oxygen content of the air is lowered from 20 to 4 vol%. The waste gas from the reactor, containing toluene and water is cooled and toluene absorption takes place on activated carbon. The reactor is operated at 136–160 °C and 1–2 atm, yielding a benzoic acid concentration in the reactor of 40–60 wt%. The reactor is utilized with heat exchangers because the reaction is highly exothermic. The toluene from the absorber is recycled. The benzoic acid formed is separated from the toluene using a toluene-stripping column and again the toluene is recycled. The bottom product of the column, which consists of benzoic acid and higher boiling byproducts is distilled again in the benzoic acid column, where the benzoic acid is distilled overhead. The bottom product is disposed or can be lead back to the oxidation reactor. One of the byproducts is benzaldehyde, which can be commercially obtained from the mixture by stripping from the reactor effluent.

The purified benzoic acid is led to the decarboxylation reactor, together with 1–5% of a copper catalyst; this stage also takes place in the liquid phase. The copper is reduced from Cu^{2+} to Cu^+, which is oxidized by the oxygen from the air. To improve the selectivity of the catalyst often metal salts are added, mainly magnesium salts. The temperature in the reactor is 220–250 °C and the pressure is only slightly above atmospheric (2.5 atm). Again sprinklers introduce air and water vapor is introduced. The phenol formed is extracted from the reactor as a vapor and separated from inert gasses (which contain some benzene and toluene) in the water-hydrocarbon stripping column. The waste gasses are freed from aromatics and purged. The phenol from the stripping column is distilled in the crude phenol column and for final purification in the phenol column, leaving tar as a bottom product. The bottom product from the stripping column is also tar, but with a high benzoic acid content (high enough to extract and recycle). The benzoic acid is extracted from the tar with water and recycled to the reactor. All the tar streams are combined and incinerated, where complete removal of the catalyst from the flue gas has to be ensured.

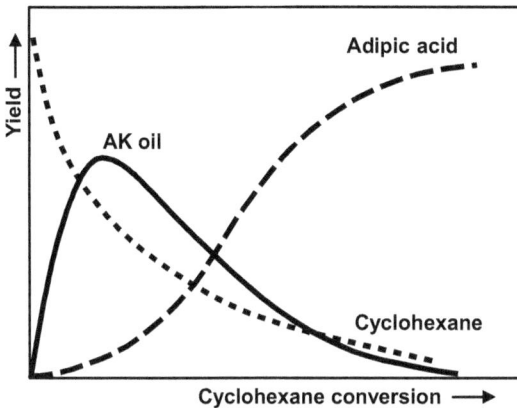

Fig. A.15: AK-oil concentration as a function of cyclohexane conversion.

A.5.2 Cyclohexane oxidation

Cyclohexane oxidation is used to produce cyclohexanol and cyclohexanone. Until 1940 this was done almost exclusively by hydrogenation of phenol. This process has been completely replaced by the liquid-phase oxidation of cyclohexane. The mixture of cyclohexanol and cyclohexanone is also known as AK-oil (from Alkohol-Ketone oil). Unfortunately the formation of adipic acid cannot be prevented, so this becomes a secondary product, which can be extracted from the reactor effluent. Adipic acid is one of the products obtained after oxidation of cyclohexanol. Therefore an optimum in AK-oil yield will be present, as shown in Fig. A.15. Adipic acid is used as a raw material in the production of nylon 4,6. Because the reaction products can be oxidized as easily as cyclohexane, the conversion per passage has to be kept low (about 10%), to prevent the subsequent oxidation of the desired products. Due to the application of a reflux the yield is still 70–90%. If possible the feed should be free of aromatics, because these are resistant to oxidation under the circumstances in the reactor. Therefore they will accumulate as inerts and a purge has to be installed, resulting in product and raw material losses.

As illustrated in Fig. A.16, the oxidation is usually carried out with a soluble cobalt catalyst in a series of stirred autoclaves at 140–180 °C and 8–20 bar. Nowadays, it is even possible to operate the process without a catalyst. Before entering the oxidizers, the feed is preheated and up to 15% of water is added to reduce production of cyclohexanol esters. The heat produced in the autoclaves is usually enough to heat the feedstock, so no external heat production is needed. In the stirred autoclaves, air is bubbled through the liquid phase, and part of the water is recycled. The rest of the mixture is led through several oxidizers, and after that through several distillation columns to remove most of the cyclohexane and part of

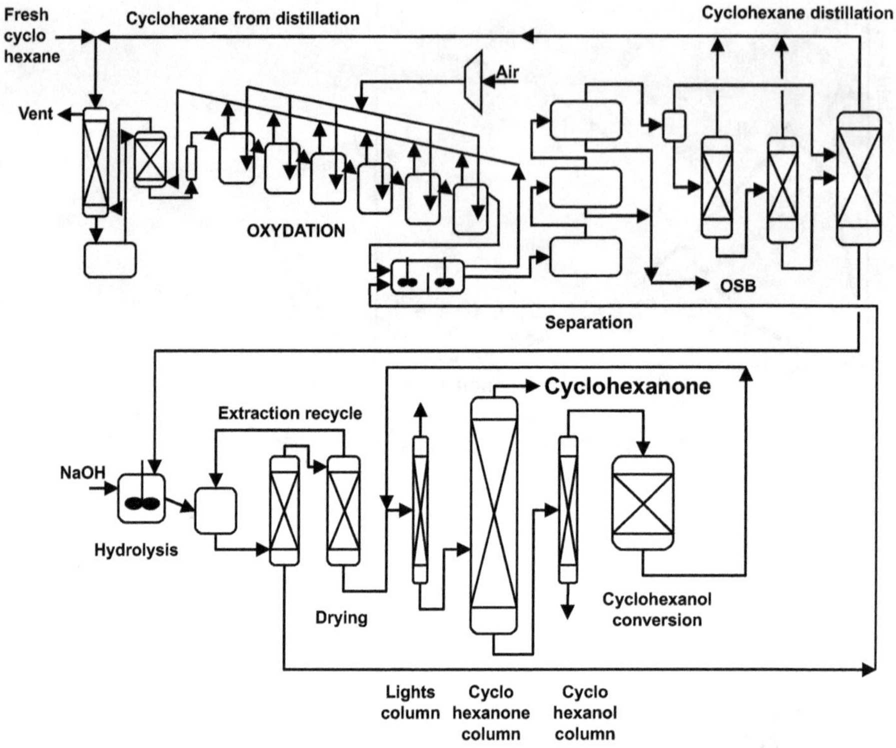

Fig. A.16: Flow sheet of cyclohexane oxidation.

the added water. The bottom product, which contains AK-oil and several byproducts like adipic acid and esters of cyclohexanol, is mixed with sodium hydroxide solution to hydrolyze the formed esters back to cyclohexanol. After drying, the mixture is distilled again and the cyclohexanone is obtain as the top product. The cyclohexanol is converted by dehydrogenation to cyclohexanone and recycled.

A.5.3 n-Butane oxidation

This process was originally carried out in a fixed bed reactor, but since 1989 fluid bed technology has been used in most of the plants. This is called the ALMA process. The basic principle of a fluid bed is a reactor with a finely divided (catalyst) powder, through which a gas is directed. The powder will start moving (in other words, is fluidized) at a certain gas flow and behaves like a fluid. Advantages of this method are the up to 30% lower investments, the production of high-pressure steam and the very easy temperature control. Due to the homogenous mixing in the reactor, the heat produced by the exothermal reaction (see Scheme A.5) will be equally divided in the reactor and no hot spots will be formed.

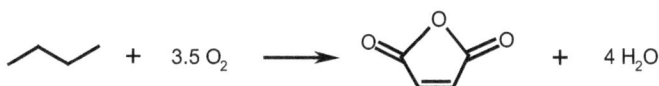

Scheme A.5: Oxidation of n-butane.

Before the introduction of the ALMA process, the crude maleic anhydride (MA) was distilled to remove contaminants and byproducts, which required a lot of energy (supplied by steam). The process described uses an organic solvent to recover the MA instead of distillation. MA is used as raw material for unsaturated polyester resin, fumaric and maleic acid and as a lubricant oil additive.

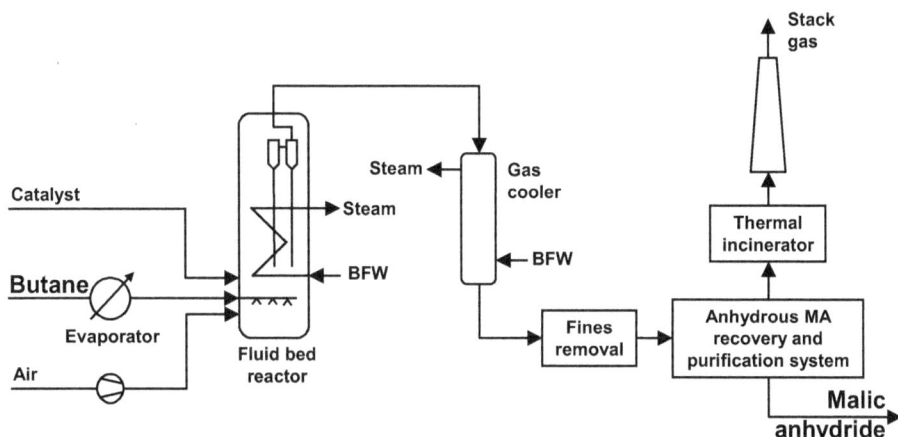

Fig. A.17: N-Butane oxidation flow sheet (adapted from [210]).

As can be seen from the flow sheet in Fig. A.17, the butane is first evaporated and then enters the fluid bed reactor. The air is compressed and enters the reactor below the catalyst support to fluidize the catalyst bed. The bed has a temperature of 420–430 °C and a pressure of 3.3 bar. In the conventional process butane and air where mixed and precautions had to be taken to prevent ignition of the very explosive mixture. Butane content had to be lower than 2 mol% and huge flows where necessary to achieve an acceptable production. The catalyst used is vanadium phosphate (VPO), which is really imported in this process. Normally a catalyst is used to speed up the reaction, but in this case the reaction would hardly take place without the catalyst. The catalyst can be easily replaced, either partially or totally, or the amount can be decreased or increased to meet the desired production capacity (within certain limits of course). Separation of the catalyst from the gas flow takes place partly in the reactor. The last

catalyst particles are removed in two cyclones after the gases leave the fluidize bed. Because the reaction taking place is exothermic, heat has to be removed to prevent a temperature rise. Heat is removed by a cooling system consisting of tubes in the reactor, through which water is led. This way high-pressure steam can be produced. The air and butane are supplied in stoichiometric concentrations, which have the following advantages compared to the fixed bed technology:
- Reduction of the compressor investment and utility due to the smaller volume
- Smaller reactor size and
- Reduced diameter of the MA absorber

MA is not the only product formed, unwanted byproducts are CO, CO_2, acetic acid, fumaric acid and acrylic acid. The total yield of MA is only 40–50%, with a product purity of 98.5%. A second improvement is the use of an MA recovery system using an organic solvent instead of water. This lowers the steam consumption significantly because no water evaporation has to take place and MA separation can be done in one (absorption) step. The absorption takes place in an absorber–stripper system, using a patented solvent. In the absorber a reaction takes place between the solvent and the MA, this is called reactive absorption. Because the bond formed is very weak, the MA can be easily recovered from the solvent. Molten MA (55–56 °C) can be stored for weeks in stainless steel tanks under an inert gas atmosphere without any change in quality.

A.6 Fischer-Tropsch

In the coal and natural gas liquefaction technology, coal or natural gas is converted into a liquid product containing hydrocarbons and oxygenates. The commercially used technologies involve three main process steps:
1. Synthesis gas production and purification
2. Fischer-Tropsch synthesis
3. Product upgrading

These three steps are schematically represented in the block diagram of Fig. A.18 and will be the subject of further discussion in the following paragraphs. For synthesis gas production the focus is on coal gasification, although the conversion of natural gas by steam reforming (see Section A1.3) is also widely used.

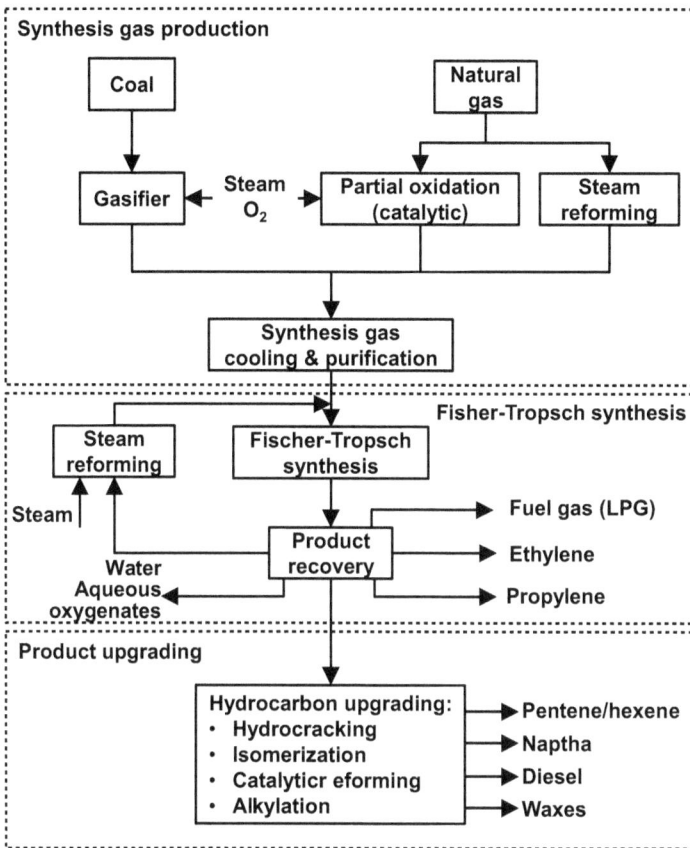

Fig. A.18: Overall process scheme of Fischer-Tropsch (adapted from [212]).

A.6.1 Synthesis gas production

A.6.1.1 Coal gasification

The process starts in the gasification plant where coal is converted into crude synthesis gas under pressure and at high temperature in the presence of steam:

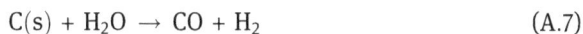

$$C(s) + H_2O \rightarrow CO + H_2 \tag{A.7}$$

As an example, Lurgy dry ash gasifiers can be used, which cope well with the 20–40% ash containing low-grade coal. The normal operating pressure of the Lurgi gasifiers is 27 atm. The gasifier is a countercurrent operation with hot ash exchanging heat with the gasification agent at the bottom and the hot product gases heating, devolatizing and drying the coal fed in at the top (Fig. A.19).

Coal
feed

Coal
lock

Water

Scrubbing
cooler

Gasification
chamber

Syngas
Quench water

Rotating
grate

Steam
Oxygen

Ash
lock

Ash

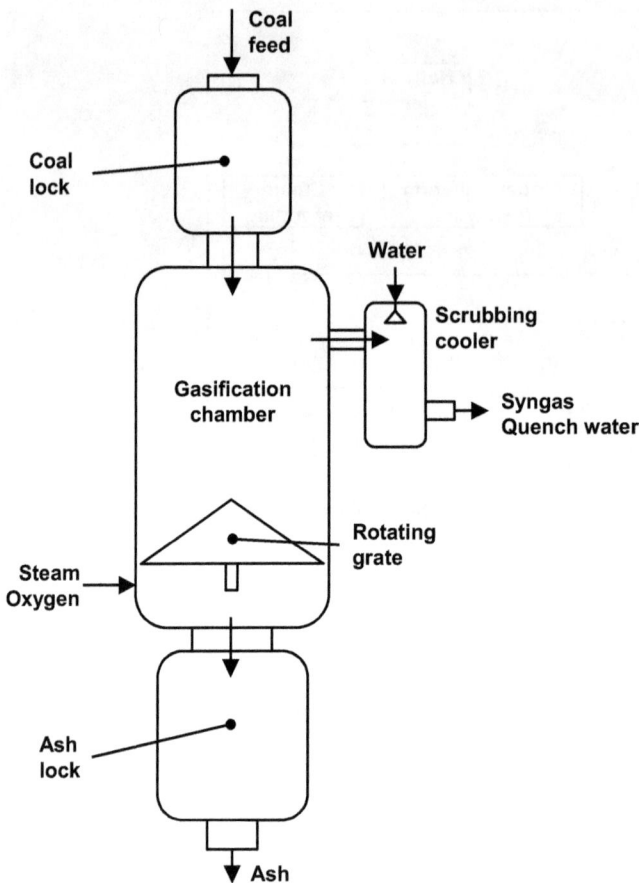

Fig. A.19: Lurgi dry ash gasifier (adapted from [212]).

A.6.1.2 Gas purification

The gasifier effluent gas contains impurities that interfere with synthesis and is first cooled to condense out tars, oils and excess steam. The lighter oil is hydrofined and added to the gasoline pool.

The raw synthesis gas is further purified in the Rectisol process. The gas is washed in stages with methanol down to −55 °C to remove most of the CO_2 and H_2S. The composition of the purified gas is about 13% CH_4, 1%(Ar + N_2), 1% CO_2 and 85% (H_2 + CO). The total sulfur level of the gas is typically about 0.03 mg/m^3. In addition to decreasing the sulfur to the low levels required, the Rectisol process also removes the remaining contaminants such as tar naphtha vapor, ammonia and cyanide.

The phenols and cresols dissolved in the steam condensate are recovered by countercurrent solvent extraction (butyl acetate or diisopropyl ether) at the Phenol-solvan plant, and the NH_3 is then steam stripped from the water. The phenols are refined and sold, and the NH_3 is converted to fertilizer. The remaining water is biologically treated and reused at the complex.

A.6.2 Fischer-Tropsch synthesis

A.6.2.1 Principles

After purification, the clean syngas is catalytically converted to a wide range of products, such as hydrocarbons, alcohols, aldehydes, ketones and acids. Production of hydrocarbons and alcohols by the Fischer-Tropsch synthesis can be represented as follows:

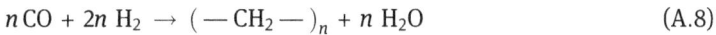

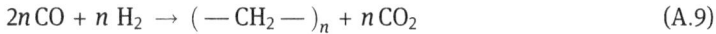

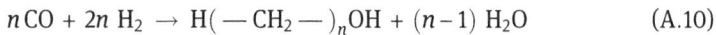

$$n\,CO + 2n\,H_2 \rightarrow (-CH_2-)_n + n\,H_2O \tag{A.8}$$

$$2n\,CO + n\,H_2 \rightarrow (-CH_2-)_n + n\,CO_2 \tag{A.9}$$

$$n\,CO + 2n\,H_2 \rightarrow H(-CH_2-)_nOH + (n-1)\,H_2O \tag{A.10}$$

Reaction (A.8) represents the formation of olefins, reaction (A.8) paraffins and reaction (A.10) alcohols. The proper selection of catalyst and reaction conditions, hydrocarbons and oxygenates ranging from methane and methanol through high (>10,000) molecular weight paraffin waxes can be synthesized. To date, iron catalysts are most frequently used in commercial reactors. Not only are they much cheaper to manufacture than their cobalt and ruthenium equivalents, the products are also more olefinic.

A.6.2.2 Classical reactors

Classically two types of reactors were used. The fixed bed Arge reactors (Fig. A.20) produce mainly heavy liquid hydrocarbons and waxes, and the transported fluidized bed reactors (Fig. A.21) make predominantly gaseous hydrocarbons and gasoline. The circulating fluid bed reactors are known as Sasol Synthol reactors. They have a much higher gas throughput than the fixed bed reactors. The throughputs of the individual reactors were increased about threefold by increasing the reactor diameters, as well as by raising the operating pressure.

A.6.2.2.1 Low-temperature fixed bed reactors (Arge)

A flow sheet representing one fixed bed reactor train is given in Fig. A.20. The synthesis gas enters at the top of the reactor where it is preheated and then flows through the reactor tubes that are surrounded by water. By controlling the steam pressure above the water the desired reactor temperature is maintained. The normal operating pressure is 27 atm, and the temperature can vary from 220 to 250 °C. The Fischer-Tropsch reaction is highly exothermic and to ensure a high rate of heat exchange between the catalyst particles and the tube walls a high linear gas velocity is essential. To obtain a high degree of conversion as well as a high linear gas velocity, part of the tail gas is recycled. A large fraction of the hydrocarbon product is in the liquid phase inside the reactor.

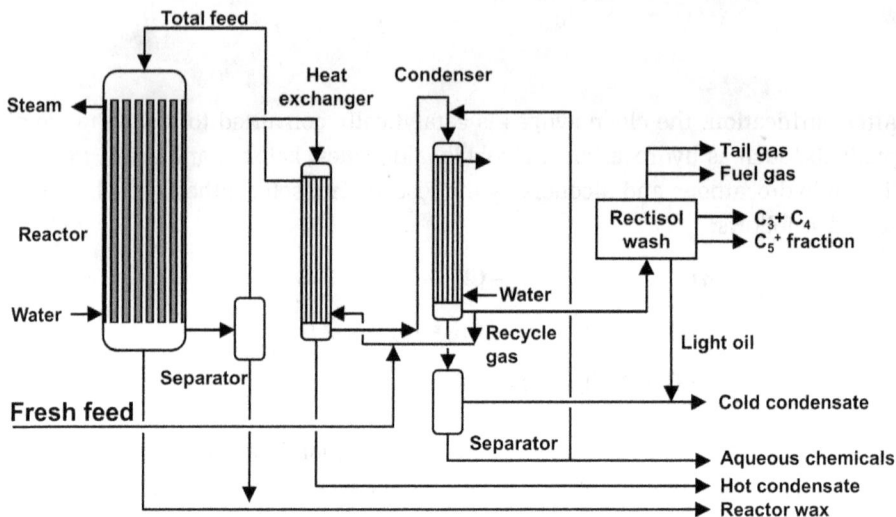

Fig. A.20: Flow sheet of ARGE fixed bed reactor (adapted from [207]).

A.6.2.2.2 High-temperature synthol reactors

The Synthol reactors can be described as transported or circulating fluidized beds. The overall height of the reactors is 50 m. As depicted in Fig. A.21, the gas (fresh feed plus recycle) is introduced into the bottom of the reactor where it meets a stream of hot catalyst flowing down the standpipe. This preheats the gas to its ignition temperature. Gas plus catalyst then flows up through the right-hand-side reaction zones. The two banks of heat exchangers inside the reactor remove a large portion of the reaction heat, the balance being absorbed by the recycle and product gases. The catalyst disengages from the gas in the wider settling hopper and flows down the standpipe to continue the cycle. The unreacted gas, together with the hydrocarbon product vapors, leaves the reactor via cyclones, which remove the entrained finer catalyst particles and return them to the hopper. The reactor exit temperature is typically around 340 °C. An important operational constraint of the process is that heavy, waxy hydrocarbons are detrimental because they condense on the fine catalyst particles and cause bed defluidization.

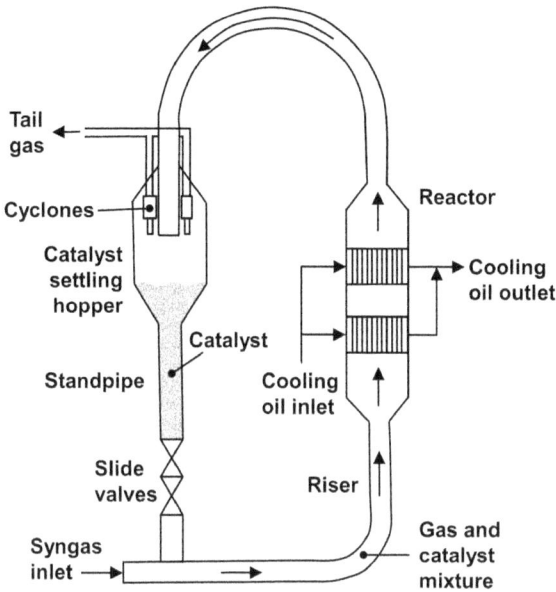

Fig. A.21: Synthol reactor (adapted from [207]).

A.6.2.3 Advanced reactor designs

A.6.2.3.1 Slurry-phase reactors

In the slurry-phase reactor the finely divided catalyst is suspended in a heavy oil (Fig. A.22). The liquid medium is usually a Fischer-Tropsch wax with low viscosity at the reaction temperature. The preheated synthesis gas is fed to the bottom of the reactor where it is distributed into the slurry and bubbles upward through the slurry. As the gas bubbles upward through the slurry it diffuses into the slurry and is converted into more wax by the Fischer-Tropsch reaction. The liquid medium surrounding the catalyst greatly improves heat transfer and avoids overheating. The heat generated from this reaction is removed through the reactor's cooling coils, which generate steam and allow high conversions to be reached. Conversion of synthesis gas in a single pass can be as high as 90% with a high selectivity for liquids in the gasoline boiling range. The wax product is separated from the slurry containing the catalyst particles in a specially developed process. The lighter, more volatile fractions leave in a gas stream from the top of the reactor, which is cooled to recover the lighter cuts and water.

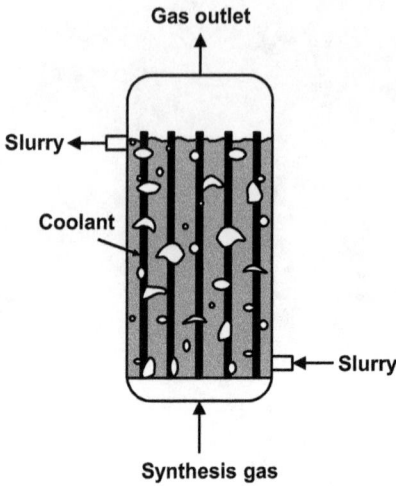

Fig. A.22: Slurry-phase distillate reactor.

A.6.2.2.2 Advanced synthol reactor

In this fixed fluidized bed (FFB) reactor the gas passes upward through the catalyst bed at velocities in the region of 10 to 60 cm/s. The catalyst bed, though expanded, is not transported and remains in a fixed position (Fig. A.23). Such a bed is simpler to operate than the circulating bed and, being much smaller, less costly. The percentage of conversion was found to be independent of the pressure. The fresh feed and recycle flows always increase in proportion to the increase in pressure. Thus FFB reactors have a potential for increased synthesis gas throughputs.

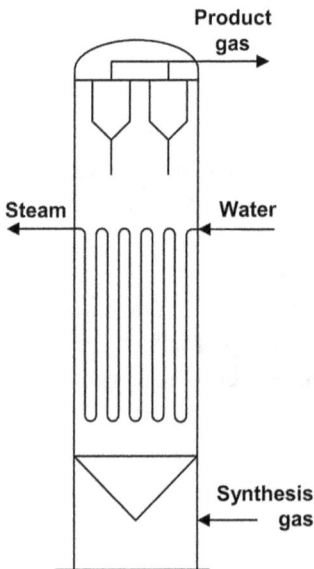

Fig. A.23: Fixed fluidized bed reactor.

A.6.2.4 Product selectivities

The ARGE fixed bed reactor operates at relatively low temperature and is intended to produce a large quantity of wax and a minimum of methane. The objective of the Synthol reactor, on the other hand, is to maximize production of materials in the boiling range of gasoline. Hydrocarbons from the Synthol reactors are substantially more olefinic than those produced in the ARGE system and the fluid-bed reactors produced considerably more oxygenated species, with ethanol being the primary product.

A.7 Industrial electrochemical processes

Electrochemical synthesis of base chemicals on an industrial scale is currently limited to only a few chemicals. Important examples treated here are the production of chlorine gas, hydrogen peroxide and adiponitrile. However, the number of base chemicals that will be produced electrochemically in the near future is expected to rise dramatically, driven by the global transition to renewable energy sources, most of which produce energy in the form of electricity. Although details will differ for each base chemical, the basic principles of these electrochemical conversion processes will be similar to those of the three mentioned examples.

A.7.1 Chlorine gas production

The chloralkali process is one of the most commonly applied electrochemical conversion processes, and currently the largest in tonnage. It is the main technology used to produce chlorine gas (Cl_2), which is widely used in the chemical industry. Useful co-products are sodium hydroxide (NaOH) and hydrogen gas (H_2). A typical chloralkali electrochemical cell is shown in Fig. A.24. At the cathode (the negative potential electrode), positive hydrogen ions are pulled from water and reduced to hydrogen gas, releasing hydroxide ions:

$$2H_2O + 2e^- \rightarrow H_2 + 2OH^- \tag{A.11}$$

Saturated brine, containing predominantly sodium chloride NaCl at a concentration of almost 25%, is passed along the anode (positive potential electrode), where the chloride ions are oxidized to chlorine gas:

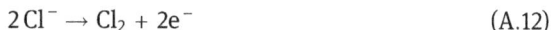

$$2Cl^- \rightarrow Cl_2 + 2e^- \tag{A.12}$$

An ion-exchange membrane allows for sodium ions Na^+ to pass from anolyte to catholyte, thus closing the electric circuit and simultaneously forming the sodium hydroxide solution in the catholyte compartment.

In commercial Chloralkali processes the membrane is often Nafion, a synthetic polymer material that can tolerate high temperatures and corrosive environments for

Chlorine   Hydrogen

Brine

Chlorine
gas

Ion-exchange
membrane

Hydrogen
gas

Na⁺

Nickel
cathode

Titanium
anode

NaOH
solution

+   −

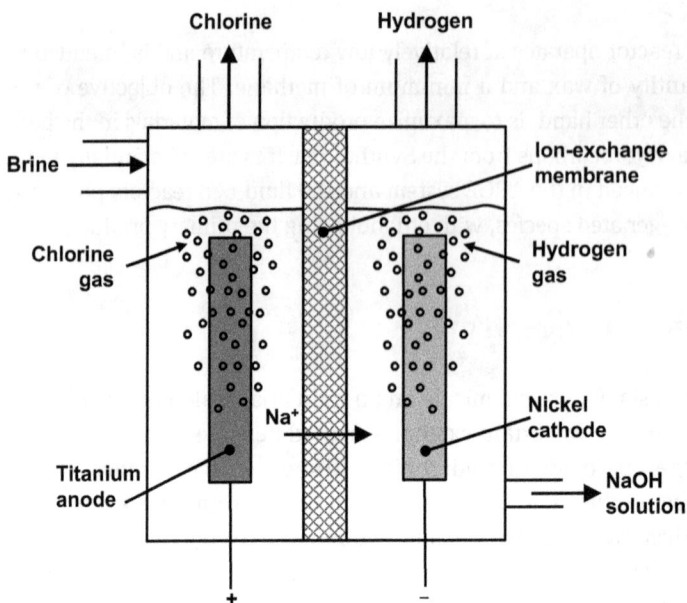

Fig. A.24: Production of chlorine gas and sodium hydroxide by an ion-exchange membrane Chloralkali cell; (adapted from [215]).

several months of operation. The anodes are typically made of titanium, coated with noble metal oxides such as RuO_2, while the cathodes are typically made of nickel. These choices represent an optimal combination of corrosion resistant electrodes with relatively low activation overpotentials. To maximize mass transfer, and thereby minimize concentration overpotentials, the gaps between electrodes and membrane are made as narrow as possible. However, the gaps cannot be too narrow: chlorine and hydrogen gas bubbles are formed during the process, which may get stuck if the gaps are too narrow, leading to an increase in electrical resistance of the electrolyte-gas mixture and incapacitation of parts of the electrodes covered by gas instead of liquid electrolyte. Chloralkali cells are operated at high temperature slightly below the boiling temperature of the electrolyte, typically 80 to 95 °C, to maximize reaction rates at the electrodes and diffusivities of ions through the electrolyte and membrane.

Despite all optimizations, Chloralkali cells still require a significant overpotential to produce chlorine gas in significant amounts. To reach commercially desired current densities of 200–1,000 mA/cm², the necessary applied potential is typically 3.1–3.4 V, which is much higher than the equilibrium potential of the total reaction, equal to 2.2 V at 90 °C. The most important factor determining this overpotential, which is dissipated in the form of heat, is the electric resistance of the membrane. Significant efforts are therefore made to further reduce the membrane resistance while keeping it mechanically and chemically stable.

A.7.2 Hydrogen peroxide production

Hydrogen peroxide (H_2O_2) solutions are used as oxidizing or bleaching agents in many industries. Currently, hydrogen peroxide is produced on an industrial scale through non-electrochemical methods, such as the indirect anthraquinone (AO) process. A disadvantage is that these production processes involve large and complex installations for (catalytic) hydrogenation, oxidation and extraction, with excessive use of expensive solvents and negative environmental impact.

Direct electrochemical production of hydrogen peroxide from oxygen and water has not been developed as much, but offers advantages in terms of environmental impact. It is currently employed by industries that need on-site generation of relatively small quantities of hydrogen peroxide, such as the pulp and paper industry. Although commercialization of an electrochemical process for direct electrochemical hydrogen peroxide production has been slow, considerable progress has been made.

The direct formation of hydrogen peroxide starts with reduction of dissolved oxygen at the cathode:

$$O_2 + H_2O + 2e^- \rightarrow HO_2^- + OH^- \tag{A.13}$$

The formed HO_2^- ions react with water to hydrogen peroxide and hydroxide ions:

$$HO_2^- + H_2O \rightarrow H_2O_2 + OH^- \tag{A.14}$$

At the anode, the hydroxide ions are reacting to oxygen and water:

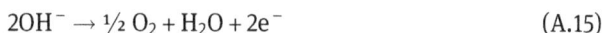

$$2OH^- \rightarrow \tfrac{1}{2}O_2 + H_2O + 2e^- \tag{A.15}$$

The major issues hampering large-scale deployment of electrochemical hydrogen peroxide reactors are the slow reaction kinetics, occurrence of competing reactions and $NaHO_2$ precipitation. A number of these problems have been overcome by use of an alkaline electrolyte, operation at relatively low (ambient) temperature, and use of porous carbon-based electrodes. An example of an electrochemical reactor for direct hydrogen peroxide production is shown in Fig. A.25. In this design, oxygen gas is pumped into a porous cathode consisting of a bed of particles through which the electrolyte is trickling down. Each particle consists of a graphite core with a carbon black and PTFE shell, offering both high surface-area for reactions and good electrical conductivity. The trickling bed operation reduces mass transfer limitation, and thus increases reaction rates, by offering three-phase conditions where solid electrode, gaseous oxygen and liquid electrolyte come together at close distances. Under atmospheric conditions, this above process leads to formation of 1% hydrogen peroxide. Higher concentrations of around 3% can be reached by pressurizing the electrolyte to approximately 100 bar, which increases the oxygen dissolution in the catholyte.

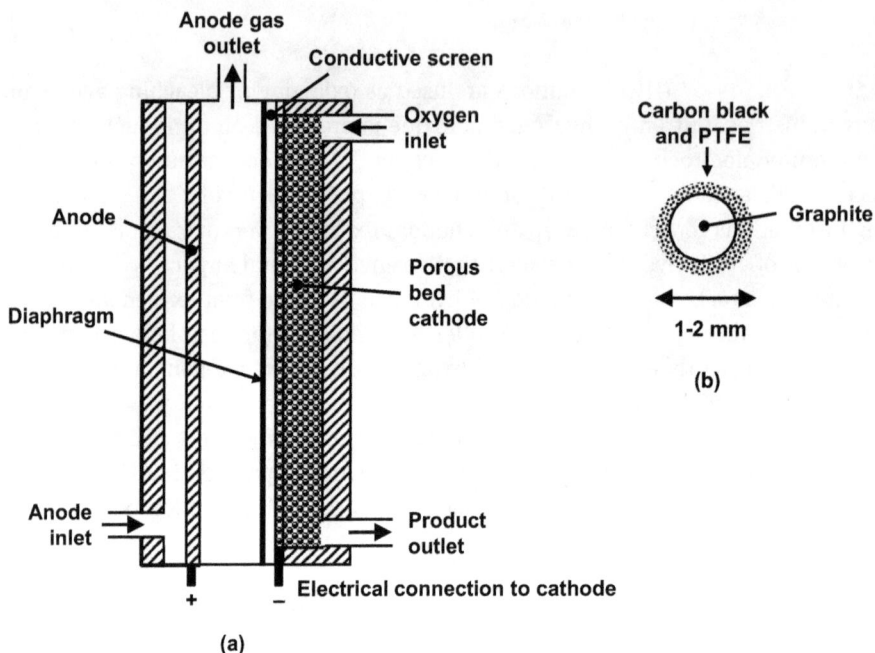

Fig. A.25: (a) Electrochemical reactor for hydrogen peroxide production, based on a trickle flow porous bed cathode, filled with (b) composite carbon-black-PTFE-graphite particles (adapted from [216]).

A.7.3 Adiponitrile production

The most successful electrochemical conversion process of an *organic* compound is the production of adiponitrile from acrylonitrile by electrohydrodimerization. Adiponitrile is an important intermediate for the production of nylon 6,6 polymers. In 2000, the global production of adiponitrile was over 1.5 million metric tons per year, of which almost a third was produced electrochemically. The cathodic reaction takes place at a very negative potential on a cadmium cathode:

$$2CH_2 = CHCN + 2H^+ + 2e^- \rightarrow NC(CH_2)_4 CN \tag{A.16}$$

while the anodic reaction is a simple oxygen evolution reaction on a steel anode:

$$H_2O \rightarrow 2H^+ + \frac{1}{2} O_2 + 2e^- \tag{A.17}$$

Curiously, the cathodic reaction takes place at a more negative potential compared to that where the hydrogen evolution reaction is expected to occur. However, because in practice the overpotential for hydrogen production is high on a cadmium cathode, the hydrogen evolution reaction is inhibited and *effectively* shifted to much more negative potentials. Moreover, quaternary ammonium salts are strongly and preferentially (compared to water and protons) adsorbed on the cathode, leading to a high selectivity for adiponitrile instead of hydrogen.

Figure A.26 gives the general flow sheet of electrochemical conversion of acrylonitrile to adiponitrile. To avoid side reactions, it is necessary to have a high catholyte flow rate, which means that the conversion per pass is very low, of the order of only 0.2%. Therefore, the acrylonitrile containing catholyte is continuously fed from a reservoir to the electrochemical cells, and the reacted mixture is fed back to the catholyte reservoir. A fraction of the mixture in this reservoir is bled continuously to a series of separation steps involving extraction columns and vacuum distillation.

A disadvantage of the above process is that it requires high cell voltages and a quite complex cell design, including membranes. Modern electrochemical adiponitrile production therefore takes place in so-called undivided cells (Fig. A.27), in which there is a single electrolyte, flowing in between bipolar electrodes in a bipolar stack without any membranes. The bipolar electrodes have a dual role, one side acting as anode and the other side acting as cathode. They are created by cadmium-plating one side (the side facing the positive endplate) of each steel plate electrode.

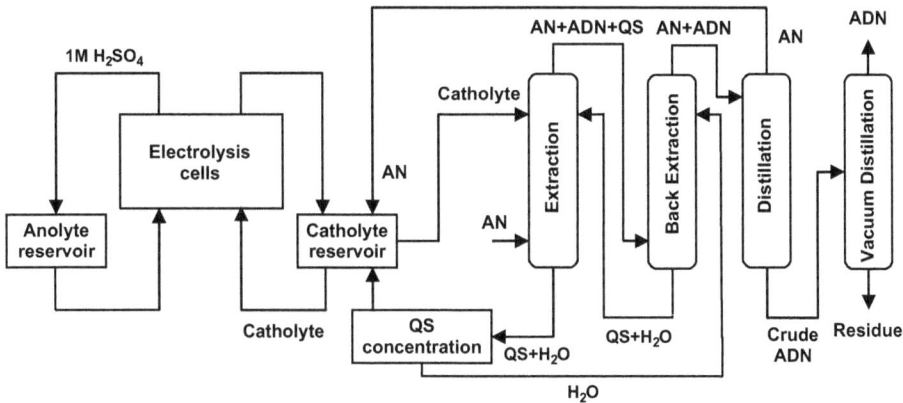

Fig. A.26: Flow sheet of electrochemical conversion of acrylonitrile to adiponitrile (AN, acrylonitrile; ADN, adiponitrile; QS, tetraethylammonium ethylsulfate (adapted from [216]).

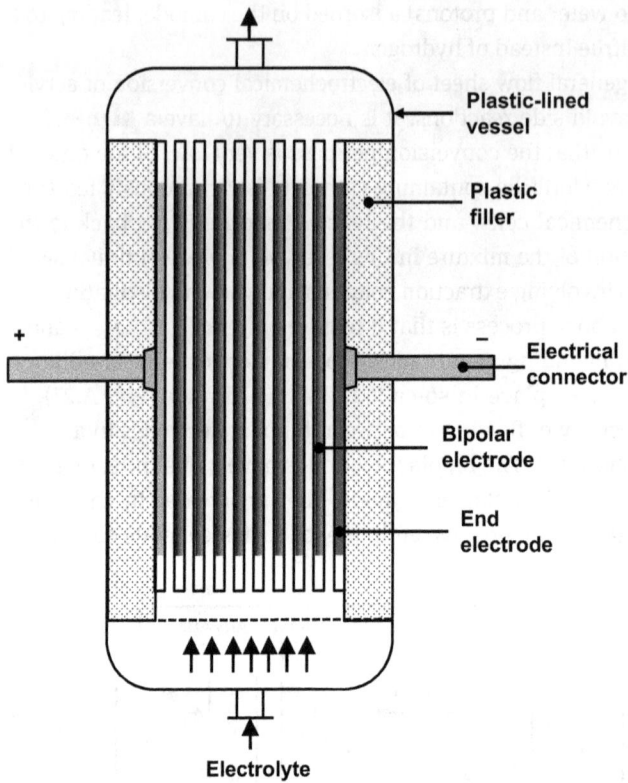

Fig. A.27: Schematic of an undivided electrochemical cell for adiponitirile production (adapted from [217]). The bipolar electrodes are made of carbon steel plates, where one side of each plate is cadmium-plated.

B Polymer manufacturing

B.1 Polyethylene

B.1.1 High-pressure process

The high-pressure process is used to manufacture almost 20% of the worldwide polyethylene (PE) production. It was the first commercially available process. The first plant (owned by ICI) began operation in 1939 and produced LDPE (low-density polyethylene) at 2,000 bar, which was used in cables and other devices for radar because of its excellent insulation properties at high frequencies. High-pressure LDPE has an important advantage concerning today's emphasis on the environment and recycling of plastics: it is catalyst-free. Incineration for energy recovery is one approach to plastics recycling, and incinerating high-pressure LDPE does not release metal catalyst particles into the atmosphere. Because of this there has been an upswing in interest for LDPE with the market still growing. The largest single-train high-pressure plant ever built has a production capacity of 185 kt/year and offers considerable economy of scale. For a flow sheet of this plant see Fig. B.1.

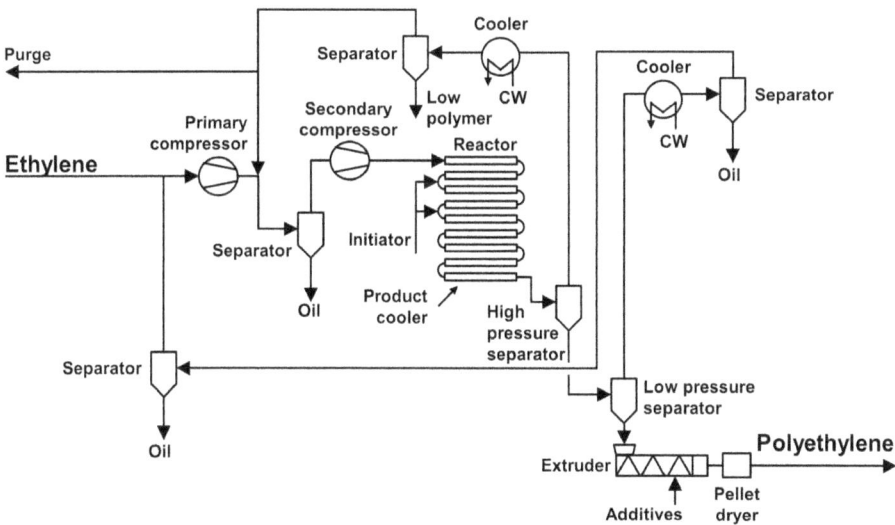

Fig. B.1: Flow sheet of a high-pressure polyethylene plant (adapted from [207]).

Ethylene feedstock is compressed to 250 bar by a primary compressor and then to between 2,000 and 2,500 bar by a secondary compressor. It is preheated to an initiation temperature of 140 °C and enters the reactor in a plug flow regime where

https://doi.org/10.1515/9783110712445-020

polymerization is started by adding peroxide initiator. The reaction being exothermic, the temperature rises until all initiator is consumed. Temperature peaks of up to 350 °C have been recorded. This way conversion rates up to 37% can be obtained, depending on product grade and required quality. Reactor tubes are cooled, using a jacket filled with high-pressure water at 200 °C. When the temperature has dropped sufficiently, fresh initiator is added. The tubular reactor can be as long as 0.5–1.5 km, usually divided into a number of straight sections connected by 180° curves. The inside diameter of the tubes can range from 25 to 75 mm and due to the high pressure applied a ratio of outside to inside diameter of about 2.5 is needed to provide the necessary strength. With polymerization is complete the LDPE and unconverted ethylene are expanded to around 250 bar through a product cooler into a high-pressure separator, from which the ethylene can be recovered, cooled and recycled to the suction side of the secondary compressor. The remaining ethylene dissolved in the LDPE is recovered in a low-pressure separator, which also serves as hopper for the homogenizing extruder. The extruder has one or more degassing sections for further devolatilization of the LDPE melt.

LDPE pellets from this process can be produced with densities from 919 to 928 kg/m^3. This reactor design offers benefits in product quality, running costs and safety. Because of its high and stable heat-transfer coefficient the reactor can be smaller than in other tubular processes.

B.1.2 Solution polymerization

The process described was designed by Stamicarbon and is called the "compact" polyethylene solution process, because there are fewer steps involved than in conventional systems. It was originally designed to produce HDPE (high-density polyethylene), but after a few small modifications it can be used for LLDPE (linear-low-density polyethylene) and every other density PE. Figure B.2 gives a flow sheet of the process described below.

First of all, the (polymer grade) ethylene is absorbed by an (inert) solvent under the removal of absorption heat. Hexane, for example, can be used as a solvent, which is recycled from the first flash vessel. The catalyst, diluted in the solvent, is added to an agitated vessel-type reactor, together with the ethylene solution. Reactor feed is cooled to maintain a constant temperature of 150–250 °C in the reactor, depending on the desired product. The pressure in the reactor is between 50 and 100 bar. The conversion per pass is relatively high, compared to the other processes. Mixing requirements for the reactor are severe, because of the short residence time (<10 min.), small amount of catalyst and high conversion. The catalyst is composed of several commercially available components, but proprietary.

The heat of the reaction is used to flash almost all of the unreacted ethylene and most of the solvent. The overhead is condensed and recycled to the reactor. By

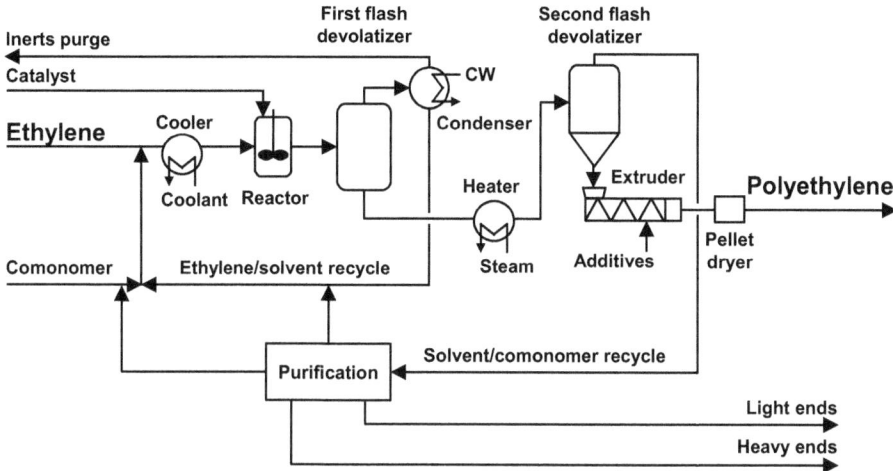

Fig. B.2: Flow sheet of the Stamicarbon polyethylene solution process (adapted from [213]).

controlling the flash, the solution concentration in the reactor vessel can be adjusted. Only a small purge is needed to prevent inerts buildup. The polymer obtained is very pure, but the remaining catalyst is still deactivated by a catalyst deactivator. In order to reduce the content of volatiles, the solution is heated and subsequently flashed. The required additives are added to the molten polymer before pelletizing and solidifying in water. The pellets are dried in a so-called pneumatic impact drier, which has the advantage that, apart from the air blower, no rotating parts are present and energy and maintenance costs are low.

The solvent from the second flash vessel has to be purified before recycling. This is done using a combination of distillation, extraction and drying steps. These are used to remove light ends, heavy ends and water respectively. As an extra precaution, the solvent is led over molecular sieve to remove even more water.

B.1.3 Slurry process

This process is capable of producing the whole range of HDPE resins, from low molecular weight waxes to resins with high molecular weights. Even the production of UHMWPE (ultra-high-molecular-weight PE) is possible. This type of PE production can be subdivided into four different types. Two of them employ loop reactors with a diluent circulating at high speed: one uses a low boiling hydrocarbon as solvent, the other a high boiling one. The third type uses a continuously stirred tank reactor with a high boiling solvent; the last is a liquid pool process, which uses propane or butane as a solvent.

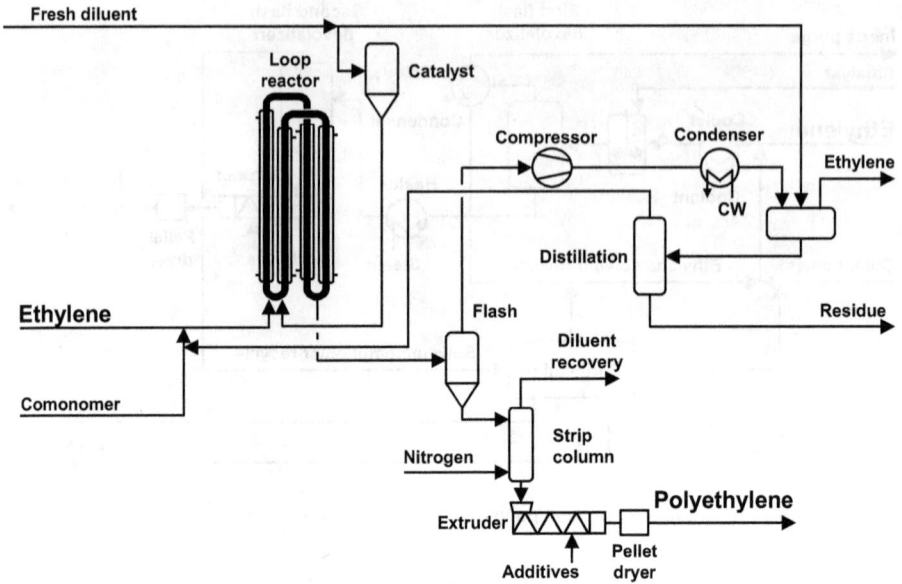

Fig. B.3: The Phillips Petroleum slurry polyethylene process (adapted from [213]).

The Phillips Petroleum Company designed the process described in 1961 (see Fig. B.3). Because it is one of the most efficient, a major part of all the HDPE is produced by this process. It produces PE ranging in density from 920 to 970 kg/m^3 and ranging from very high to very low molecular weight. The process uses two types of catalyst. A chromium catalyst which has to be activated by fluidization in hot air. Activating conditions are important because, to some extent, they determine the properties of the PE. The melt index can be varied as well as the molecular weight distribution, from medium to broad. The second catalyst is organometallic, producing a narrow molecular weight distribution. The feedstock is ethylene, to which a number of comonomers can be added, such as 1-hexene. The hydrocarbon carrier is isobutane. Hydrogen is used for molecular weight control. The reactor is operated at temperatures of 95–110 °C and pressures of 30–45 bar. The reactor takes the form of either a pipe loop or a number of loops, with an axial pump to circulate the components. The pipe has vertical sections up to 50 meters, connected by horizontal pipes of about 5 meters. The pipe diameter is in the range of 0.5–1.0 meter. Typical gas velocities are in the range of 5–12 m/s, to ensure a turbulent flow regime. This is done to achieve good heat exchange with the cooling water and to prevent polymer deposition on the walls.

As the polymerization reaction is exothermic, the reactor is jacketed with a cooling water system. Temperature is controlled accurately throughout the reactor, precise temperature control is important for consistent finished product. A plant will have a number of reactors, each reactor being designed for a specific product. The number of reactors will depend on the operator's requirements. In the reactor particles of PE are produced, suspended in the carrier, with a total PE content of up to 40 wt%. The polyethylene is formed in the catalyst pores and because of the internal pressure the particle pulverizes. This way the catalyst is evenly divided throughout the polymer product. The slurry is allowed to settle in settling legs and a polymer slurry of about 55 wt% is removed from here. A small amount of diluent is added through the settling legs in countercurrent flow with the falling polymer particles. This is done to remove unreacted ethylene from the product and thereby achieving a higher monomer conversion.

After leaving the reactor the slurry travels to a flash chamber where the components are separated: gases and light hydrocarbons volatilize and exit through the top and the PE powder through the bottom. The exiting vapor stream is filtered to remove entrained PE and then passes, via a compressor, to a fractionation column to recover and recycle isobutane. Vapor from the column passes via an accumulator either to an ethylene plant for ethylene recovery or to a boiler to be incinerated. Meanwhile the PE powder passes to a purge column, where nitrogen is used to further reduce residual hydrocarbons. The powder then passes to an extruder for incorporation of additives, melt mixing and pelletizing. Phillips indicated that the process is cheap to build and operate. There is no polymer buildup on the reactor walls, so it is not necessary to schedule shutdowns for reactor cleaning. Waste from the process is low in quantity and limited to a few chemicals; liquid hydrocarbon waste comes from the bottom of the fractionator and can be discharged to a flare or a recovery system. Small emissions of gaseous hydrocarbons are flared.

B.1.4 Gas-phase process

Several patented processes have been designed to produce PE in the gas phase. One of the most efficient is the UNIPOL process, originally developed by Union Carbide in 1968. Its original design was only capable of producing HDPE, but after a few modifications it can now produce LLDPE as well. In addition Union Carbide joined with Shell Chemical Company to use the UNIPOL-process with Shell's highly active SHAC catalyst for polypropylene production. The process is widely used because of its simplicity, cost efficiency and flexibility.

The UNIPOL process for HDPE or LLDPE is shown in Fig. B.4 and uses a fluidized bed reactor. Use of the process is growing rapidly because of its simplicity and favorable economics. Its success depends heavily on the development of a highly active and selective catalyst. The first catalyst consisted of chromium compounds and optional modifiers on dehydrated silica. This catalyst is fed as a dry powder to the lower region of the fluidized bed reactor. The operating pressure (set by economic factors) is ca 20 bar. During reactor operation, this pressure is maintained by controlling the flow of makeup ethylene into the system. The reactor can be rather high (up to 30 m) with a length to diameter ratio of approximately 7.

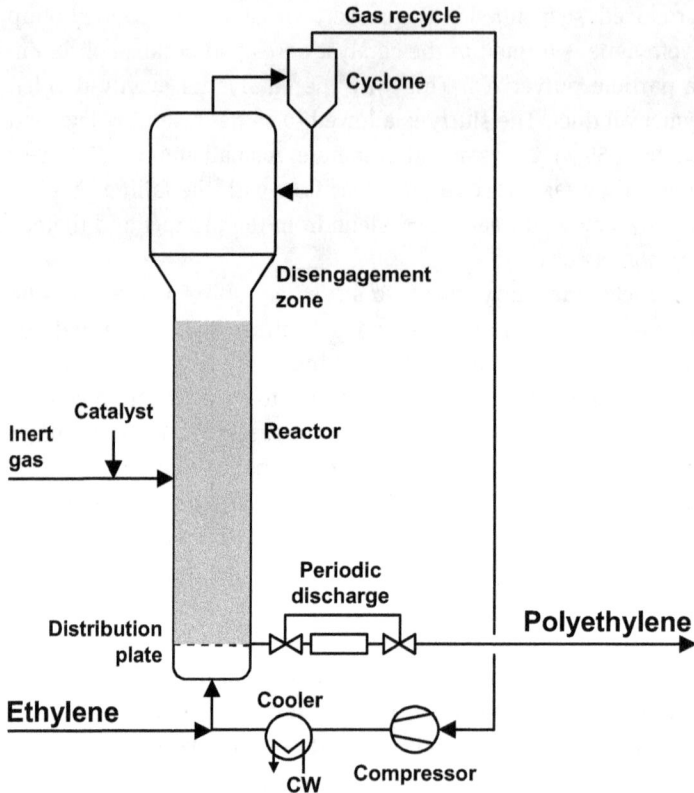

Fig. B.4: Flow sheet of the UNIPOL process (adapted from [207]).

There is however, a larger disengagement zone at the top, where the gas velocity drops, allowing entrained particles to fall back. Solvents have been eliminated, and the recirculating ethylene gas fluidizes and transports powder, providing heat and mass transfer and raw material. Because of the rapid recycling of ethylene, which is needed for fluidization and for heat removal, the conversion per pass is only 2%. Ethylene, heated by the powder, is taken overhead through a solids-disengaging zone in the upper part of the fluid bed reactor. After pressurizing with a blower, the overhead gas is forced through a cooler designed to minimize deposition of the small amount of entrained reactive fines. Makeup monomer (and comonomer, if LLDPE is desired) is added, and the cooled gas is directed back into the reactor through a gas-distribution nozzle system designed to support the bed, promote turbulence, and allow entrained fines to return to the bed. Reaction temperatures are 70–105 °C; a compromise between achieving high catalyst activity with rising temperature and keeping a safe margin below the softening point of the polymer, which is 125 °C for HDPE or 100 °C for LLDPE. If the gas cooler uses typical cooling-tower water, the available temperature difference for cooling the gas is 40–50 °C.

Under these conditions, the heat balance requires that the gas-recirculation rate is 40–50 times larger than the gas-makeup or production rate of the reactor. Sensors determine the heat gained by the gas traversing the bed and adjusts the catalyst feed rate to control the production rate.

After an average residence time of 3–5 h in the fluid bed reactor, the powder particles have an average diameter of 500 micrometer. The powder is released inter- mittently near the bottom of the fluidized bed through specially designed abrasion- resistant valves and is sent with the included gas to a two-stage separation section. This separation section is shown in Fig. B.5. The system uses gravity and the pres- sure drop across the bed to transfer resin into the upper tank while venting gas back into the top of the reactor. After the upper tank is filled, it is isolated from the reactor and the resin is discharged into the lower tank. This forces gas in the lower tank from the previous discharge into the upper tank, where it can be recycled to the reactor on the next discharge to the upper tank. With a release to nearly ambi- ent pressure, up to 90% of the included (and some dissolved) gas is readily sepa- rated from the powder, compressed, and returned to the reactor loop without any need for purification. After the powder passes through a gas-lock valve system, nearly all of the remaining monomer is purged from it with nitrogen, and the pow- der can be air-conveyed to blending, finishing, and storage. The UNIPOL process eliminates all solvent separation and recovery. In addition, high catalyst activity generates so much polymer per mass of catalyst that hardly any residue remains, i.e., perhaps 1 ppm chromium and neither de-ashing nor deactivation is required. Catalyst selectivity has been improved, and low molecular weight fractions, which can make the powder sticky, are nearly eliminated. Polymer deposits fouling the walls are almost totally absent in the fluid bed reactor, and the product contains no gels. No extraction step is needed. In the absence of moisture, mild steel is the ma- terial of construction, except for powder conveying in finishing and storage, where aluminum is used to avoid iron contamination.

B.2 Polypropylene

B.2.1 Gas-phase process

The process most used in gas-phase polymerization of propylene is the Amoco/Chisso process, utilizing a horizontally stirred bed (Fig. B.6). It can be used for a variety of polymerizations, but is particularly suitable for the polymerization of ethylene and propylene. In this process, the monomer is added to several separate sections of the polymerization reactor, optionally together with a certain amount of hydrogen. The temperature in the sections may vary from each other and is usually between 40 °C and the softening point of the content of the reactor. Each section contains a certain amount of polymer particles with a certain degree of polymerization and particle size.

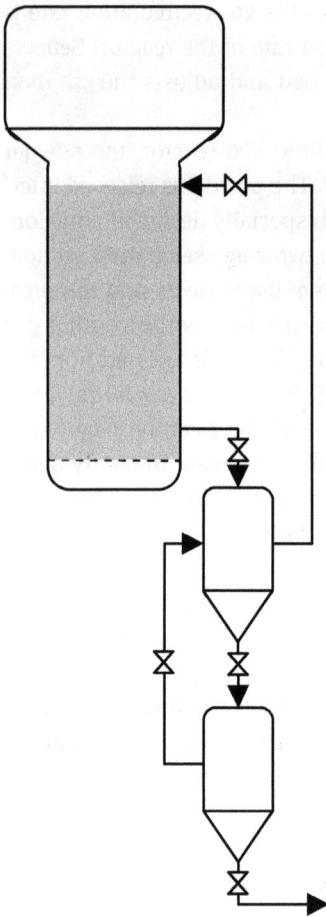

Fig. B.5: UNIPOL product discharge/separation system (adapted from [213]).

Polymer product passes through all the beds to obtain a product with the desired properties. The particles in every section are severely agitated using impellers with two paddles each, the paddles of two adjoining impellers placed in 90° position relative to the other. The paddles are designed to stir the bed while not moving the particles to the adjacent bed. Stirring takes place to achieve mass- and heat-transfer at a sufficient level. The polymer product is transferred to the next bed through a take-off barrier opening as shown in Fig. B.7. The reactor is operated isobaric and continuous cooling and product removal takes place. All the beds are continuously quenched for cooling purposes; the liquid used for quenching is usually isobutane or isopentane. Often the catalyst is supplied together with the quench, either to all beds or to a smaller number of beds. The catalyst consists of two compounds: one being a tri-alkylaluminum chloride, the other an activated titanium trichloride. The amount of catalyst added can be used to control the particle size of the polymer.

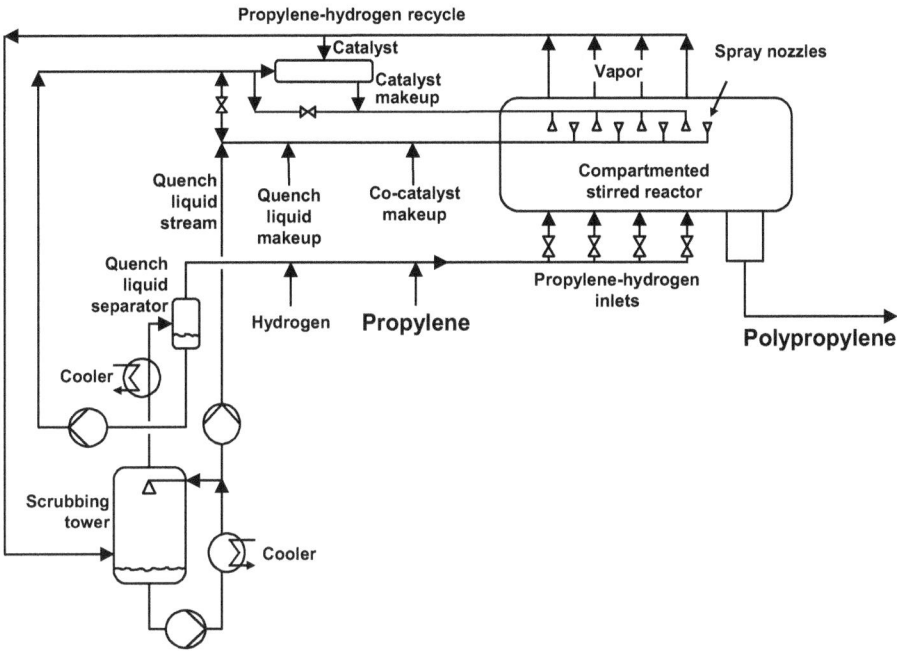

Fig. B.6: Flowsheet of the Amoco/Chisso polymerization of propylene (adapted from [207]).

The off-gasses are purified before recycling to the reactor again. This step includes the removal of polymer fines and catalyst components. After their removal, the off-gasses are led to a scrubber to remove the quench-liquid, which is partially recycled after cooling. The major part of the liquid is used to scrub the incoming off-gasses. The overhead of the scrubber is combined with the fresh monomer feed and enters the reactor below the polymer particle bed.

The produced polymer is transferred to the take-off vessel, almost without a pressure drop. In this vessel, the polymer is reacted with an extra amount of monomer. This operation takes place adiabatically and the heat produced is used for melting the polymer. Together with external heat this results in a molten polymer, which is easier to handle than in the solid state. The polymer is devolatilized and converted to the appropriate commercial size product. Water is added to the molten polymer, thereby killing the catalyst to prevent unwanted reactions after the polymer has left the plant. This is also the point were several additives and coloring agents are added to the polymer.

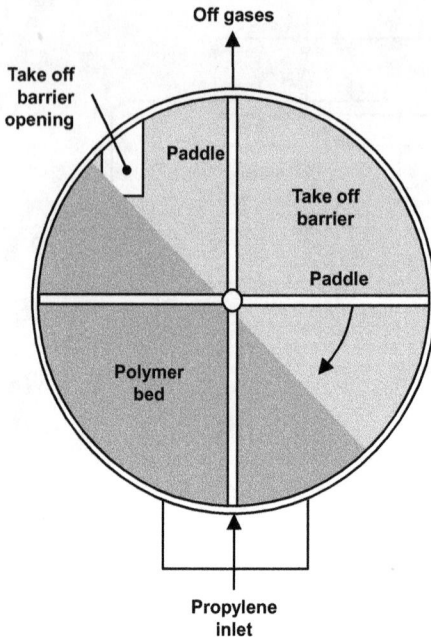

Fig. B.7: Cross section of the horizontal stirred bed polymerization reactor.

B.2.2 Slurry process

The process described, and shown in Fig. B.8, was originally designed by Montedison S.p.A. and Mitsui petrochemical industries, Ltd. Due to the improved activity of the catalyst, it was possible to simplify the original slurry process, which involved polymer washing to remove catalyst particles and alcohol removal. The process consists of four steps: polymerization, preceded by catalyst preparation and monomer and catalyst metering, phase separation and drying, compounding and pelletizing including the addition of stabilizers and solvent recovery.

B.2.2.1 Catalyst preparation

In the polymerization reaction, which is carried out in a solution in heptane, a catalyst is needed. The catalyst used in this process is a two-component catalyst: one consisting of a tri-alkyl-aluminum compound and the other mainly containing activated magnesium or manganese chloride. This system has proven to be a highly active and stereospecific catalyst, and yields can be as high as 1 million grams of polypropylene (PP) per gram of catalyst. Due to this high activity, there is no need for a catalyst removal step, adding to the simplicity of the process.

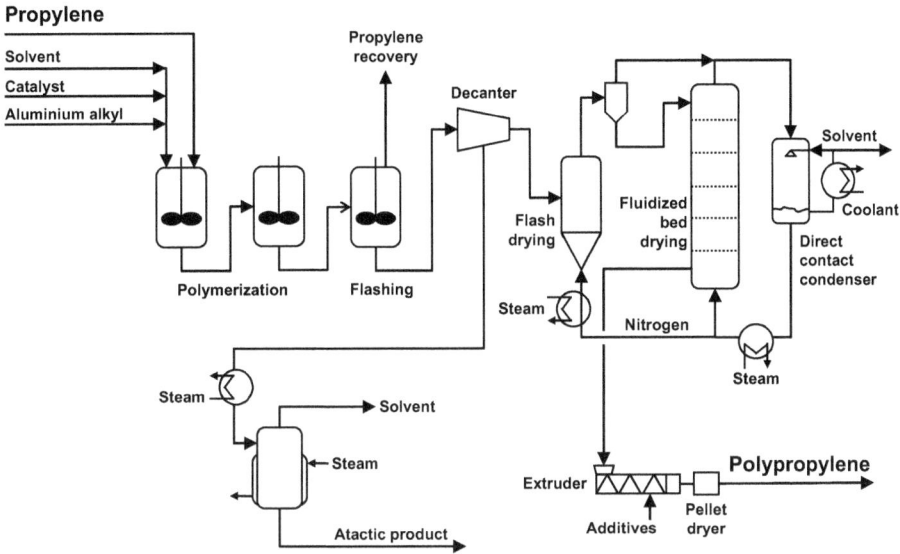

Fig. B.8: Polypropylene slurry process (adapted from [213]).

The catalyst is prepared just before it is used to ensure high activity and fresh catalyst. The aluminum-containing compound of the catalyst can be made by various methods, all starting with a tri-alkyl-aluminum compound. Addition of a Lewis base, such as pyridine, diethylether and tetrahydrofuran, yields the desired product. The magnesium or manganese containing compound can also be made by several methods, one of which is a combination with an alkyl-aluminum compound. The magnesium chloride, which is used most often as a support, is dry-milled to achieve a surface area greater than 3 m²/g and preferably greater than 20 m³. During the dry-milling a titanium containing substance, such as $TiCl_4C_6H_5COOC_2H_5$, is added. The Al/Ti ratio is between 10 and 1,000, depending on the desired product specification. To ensure a good stereoregulation, a weak stereoregulator is added to the tri-alkylaluminum compound. An example of this stereoregulator is dimethylether, which is used most often. In addition to this weak regulator, a strong stereoregulator is added directly to the reactor. The most used strong regulator is p-ethoxybenzoate.

B.2.2.2 Polymerization

The polymerization takes place in a cascade of two identical reactors, which are operated at 75 °C and 16 bar. The reactors have a length-to-diameter ratio of 2:1 and are equipped with high-intensity agitators and internal cooling coils. The walls of the reactor are also cooled to ensure a high enough heat transfer, this way avoiding the melting of the polymer particles formed. During operation, the reactors are completely filled with liquid. The solid's concentration in the outlet of the first reactor is about 32.5 wt% and is used as the inlet for the second reactor. Because of the lower monomer concentration in the second reaction vessel, the conversion is lower. The overall conversion is about 97%, of which 83% is achieved in the first reactor.

B.2.2.3 Phase separation and drying

First of all the unreacted propylene is recovered by flashing and storage for further use. This is done in a flash-vessel, which operates at 1 bar, followed by a small distillation unit. The propylene stream also contains some propane, which can be present in the propylene feed. The remaining slurry is centrifuged to remove the bulk of the diluent. After this step the volatiles content of the polymer is about 30 wt%. It is transferred to the primary flash dryer, which operates under a closed nitrogen circulation at 1 bar and 116 °C. The volatiles content of the polymer is lowered to about 5 wt%. The temperature of the inlet gasses is limited to 130 °C to prevent polymer melting. The partially dry polymer is transferred to a final drying stage. This is done in a fluidized bed, where virtually all hydrocarbon diluent is removed from the polymer. The fluidized bed also operates under a continuous nitrogen flow, from which the evaporated diluent is removed in a scrubber. To achieve the low volatiles content in the polymer, the scrubber operates at low temperatures (−18 °C) to remove all diluent from the nitrogen recycle.

B.2.2.4 Compounding and pelletizing

This step is conducted in a conventional way, as discussed earlier. The crude polymer is added to an extruder; molten and additives are introduced if desired. In this section, homogenization takes place as well.

B.2.2.5 Solvent recovery

The solvent from the several drying stages has to be purified before recycling. This is done in a simple two-column distillation unit to remove heavy ends and monomer and diluent recovery. After drying the diluent can be recycled to the diluent vessel.

B.3 EPDM

Because of the limited production of natural rubber, or the difficulties in obtaining them, the search for synthetic rubbers began. Several types of synthetic rubber have been developed, among which is the EPDM (ethylene–propylene–diene monomer). In EPDM production a third monomer is usually added to introduce unsaturated bonds, which can be used for vulcanization (cross-linking) of the rubber. The third monomer often used is ENB (5-ethylidene-bicylo(2,2,1)heptene-2), which is the un-conjugated diene shown in Scheme B.1. Important uses of EPDM include tires, foot-wear and plastic modification.

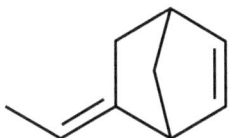

Scheme B.1: 5-Ethylidene-bicyclo[2.2.1]heptene-2 (ENB).

EPDM is produced in a continuous process, of which the solution process in the most commonly used. A process also used by a number of producers is the slurry pro-cess, which has lower catalyst and steam cost, but provides a lower flexibility in product specification and a lower homogeneity of the EPDM due to diffusion limita-tions. The details of the process vary a lot for each manufacturer and are highly pro-prietary, as can be seen from the number of patents related to this production process. The catalyst used can be divided into two parts. The first is a transition metal halide, such as $TiCl_4$, VCl_4 and $VOCl_3$, which is used mostly. The second part is a metal alkyl component such as $(C_2H_5)AlCl_2$ or $(C_2H_5)_2AlCl$. Both components are very sensitive to water and therefore only a few ppm of water is allowed in the feed streams. This catalyst system is completely soluble in hexane and has therefore a higher activity than heterogeneous catalyst systems. This can be explained by the lack of polymer film formation on the catalyst surface; there is no real catalyst surface at all. The product of the reactivity ratios for ethylene and propylene is almost unity, indicating that the polymer formed will be almost perfectly random.

A flow sheet of a typical EPDM production process is given in Fig. B.9. Most manu-facturers use a solution in hexane. In this solution process ethylene and propylene are mixed, a common composition uses 40 wt% of the latter. Before mixing the monomers are purified, dried and cooled to −30 °C. The gas mixture is added to an organic sol-vent, to a concentration varying from 3 to 8 wt%. Optionally the third monomer is added and the solution is led to the reactor. If ENB is used, it is first purified to remove the polymerization inhibitors. This is done by washing with a caustic solution, wash-ing with water and drying. The reaction takes place in two well-stirred vessels at about 60 °C and 25 bar in the presence of a catalyst system. The vessels operate in series and optionally fresh monomers and/or catalyst is added to the second vessel. The polymer-ization itself is exothermic and very fast, with the lifetime of a single growing chain being a few seconds at most. The heat produced has to be removed to ensure the right reaction conditions for obtaining a product with the desired molecular weight and dis-tribution. This is done by applying a cooling jacket filled with chilled water.

The exit stream of the second reactor is mixed with polypropylene glycol to ter-minate the reaction. In a flash vessel almost all of the unreacted ethylene is removed and recycled to the reactor together with recycled hexane. The rubber solution contains about 10% of EPDM and some unreacted olefins, all in homogenous solution. The solution is mixed with superheated water to remove the catalyst from the EPDM. The aqueous layer that settles out is partly discarded (10%) and the remainder is recycled. The dispersion is then steam stripped to remove all of the solvent and the rubber crumbs are formed. To obtain a dry product, the slurry of rubber crumbs is pumped over a shaker screen to remove excess water. The crumb, which now has a water content of 50 wt%, is then fed to the first stage of a mechanical-screw dewatering and drying press. Here the water content is lowered to 3–6% as the rubber is pushed through a perforated plate. After another drying stage, where the temperature of the rubber reaches 150 °C, the remaining water is flashed off. A final step in the drying process takes place in a fluidized bed at a temperature of 110 °C. After this step the remaining volatile matter is lower than 0.3 wt%. The EPDM crumb is then weighed, pressed into bales and packed for storage and shipment.

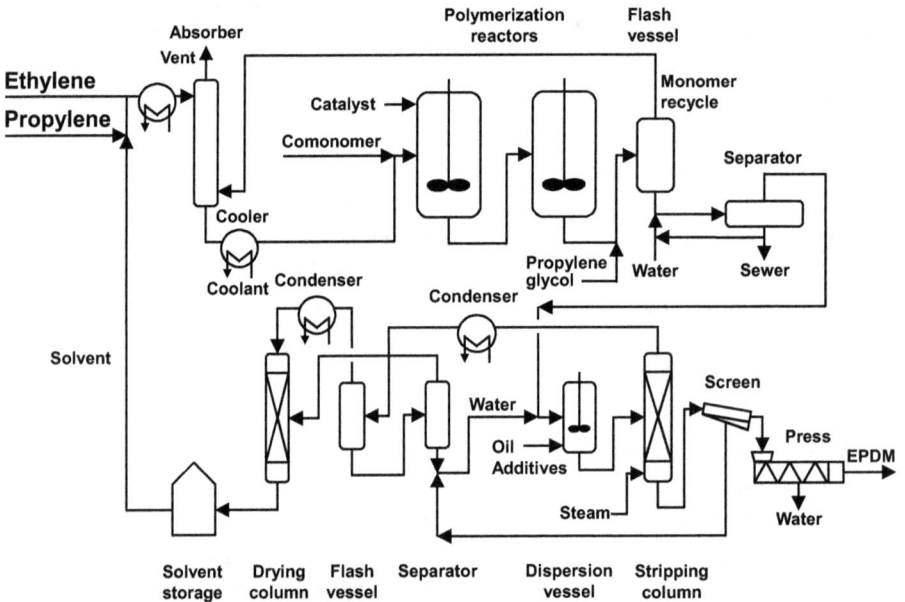

Fig. B.9: Typical EPDM process flow sheet.

The vapors from the flash vessel are partly condensed and separated into two layers. Both layers are sent to a distillation section for recovery of monomers and solvent. The organic layer is distilled and the resulting solvent-rich and ENB-rich streams are sent to various other distillation columns. The solvent is first freed of water and after that 1-hexene is removed. After drying over molecular sieves, the solvent is recycled. The heptenes and higher olefins are removed from the ENB by distillation and before recycling. The ENB dimer is removed by vacuum distillation.

B.4 Polyamides

B.4.1 Introduction

Polyamides (nylons) can be produced in a wide range of compositions, properties and prices. Because of this they are a group of polymers that are commonly used. Nylons can be divided into to types: homopolymers (like nylon 6) made from one monomer, and heteropolymers (like nylon 4,6) made of two different monomers. These two types are illustrated in Scheme B.2. The numbers added to "nylon" indicate the number of carbon atoms in the monomers used and thus to the type of nylon. Nylon 4,6 has a very high melting point (290 °C) and a high strength compared to other polymers. Therefore, it is used as a heat-resistant material and for the production of cogs applied in a high-temperature environment.

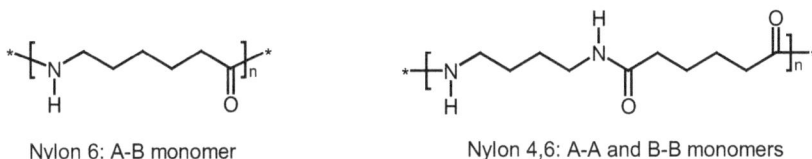

| Nylon 6: A-B monomer | Nylon 4,6: A-A and B-B monomers |

Scheme B.2: Different types of nylons.

B.4.2 Nylon 6

The production of nylon 6 is one of the few large-scale industrial exploitations of a ring-opening polymerization. The monomer used is caprolactam, the production of which has been discussed in Chapter 4. For the initiation of the polymerization, the presence of water is essential, but for the completion of the reaction water has to be removed in order to obtain a maximum yield of nylon 6. Almost all major nylon 6 producers use a continuous process know as the VK process, which is shown in Fig. B.10. Most of the producers have modified the process to their liking, thus producing the desired quality, composition and properties. The process was originally developed in Germany by BASF and its full name is the Vereinfacht Kontinuierlich (= simplified continuous) tube process.

  In this process caprolactam, titanium dioxide (TiO_2) and steam enter the VK tube at the top. The water content can be raised up to 15 wt% in modern processes to ensure an acceptable polymerization speed. Today, the prepolymerization is carried out in two VK tubes in series, the first operating at a pressure of more than 1.5 bar, the last operating at atmospheric pressure. This is done to shorten the reaction time relative to the one VK tube process. The upper part of the reactor has a temperature of 240–280 °C at atmospheric pressure. Usually the upper part of the

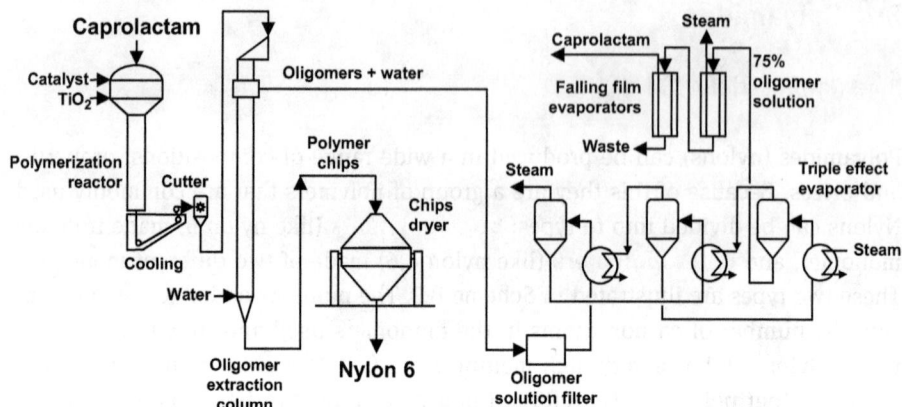

Figure B.10: Nylon-6 VK-type process flow sheet (adapted from [208]).

reactor has a 40–60% larger diameter and the reactants are introduced near the outer walls of the tube. After a quarter of the reactor (the part with the larger diameter), almost all of the water has been removed from the mixture and carried off overhead. The water is purified and returned to the reactor. The resulting prepolymer, usually containing 10–14 monomer molecules, is fed to the lower part of the reactor. This section also operates at temperatures of 240–280 °C, resulting in a product leaving the reactor in a molten state. Residence time in this section of the reactor can be several hours. After passing a filter to remove unwanted particles, the product is cast and granulated after cooling in water. Before solid state after-polymerization, the pellets have to be carefully dried in order to obtain high molecular weights in the after-polymerization. This is done using oxygen-free nitrogen to remove any water vapor. Solid-phase condensation takes place at 130–180 °C, either in vacuum or in an inert gas-stream.

B.4.3 Nylon 4,6

In the 1980s, DSM introduced a new commercial aliphatic polyamide, nylon 4,6. It was first studied in the 1930s, but no further research was done because of the difficulties in obtaining high molecular weights and degradation of the product. The two-step process developed by DSM (Fig. B.11), however, settles these problems. Raw materials used are adipic acid and diaminobutane or the salt of those materials. First of all the salt of diaminobutane and adipic acid has to be produced. This is done by mixing molten diaminobutane with the salt-solution. To this solution, a 47% solution of adipic acid in water is added. The major part of the solution formed is passed through a heat exchanger and added to a reaction vessel at a temperature of 98 °C. The minor part is diluted with water for pH monitoring, which is the major

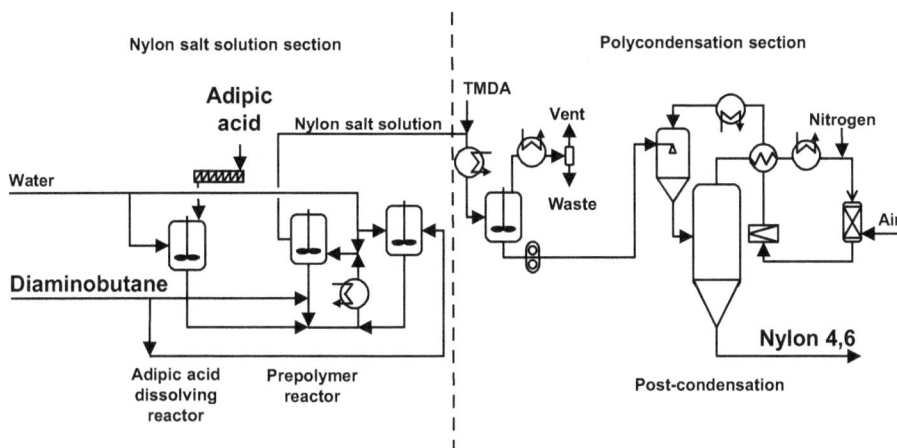

Fig. B.11: Nylon 4,6 production (adapted from [213]).

control parameter for diaminobutane addition. The salt formed in the reactor is removed at the rate it is formed by an overflow in the reactor.

In the first step of the polymerization, a prepolymer is prepared as a slurry in water. The amount of water added is preferably limited to 20 wt%, whereas the temperature is kept between 190 and 220 °C, usually at 210 °C. The use of a small amount of excess diamine has a positive influence on the properties of the polymer obtained after solid-phase polymerization. The excess ranges from 0.2 to 6 wt%, depending on the desired properties. The reaction is carried out in a closed vessel and the reactor is operating under the elevated autogenous pressure which is generally between 8 and 15 bar. Reaction time is as short as 30 min, depending on the conditions applied. Although the degree of polymerization in this stage is very low, between 6 and 14, this is the most advantageous pre-polymer for the polycondensation. The latter process can take place at relatively moderate conditions and still produce a high molecular weight polyamide, which is very stable. Discharging the reaction mixture from the reactor under a nitrogen atmosphere terminates the pre-polymerization. The water present in the mixture is flashed off by slowly reducing the pressure to atmospheric. The discharge can also take place by spray dying techniques or other, similar, processes. These have the significant advantage that the prepolymer can be readily obtained in powdered form. If spray drying is applied, the drying agent is hot nitrogen, which is largely recycled after drying and purification.

The post-condensation reaction takes place in the solid phase, usually in a fluidized bed at temperatures of 225–275 °C. The bed is kept in an inert (nitrogen) atmosphere with a water content of 20–50 vol% at an almost atmospheric pressure. This step will take four hours at maximum, which is relatively short. In order to obtain a uniform molecular weight distribution, it is important that the feed to the fluidized bed consists of particles with a uniform diameter. This is important for almost all polymers, but especially for nylon 4,6 because it is not kept in the melt long enough for transamidation to establish a uniform molecular weight distribution. The excess diamine is removed from the nitrogen stream and recycled. The high molecular weight pellets leave the solid-phase reactor and are cooled and stored under nitrogen to prevent degradation. The diaminobutane that results from the post-polymerization is removed by hot nitrogen, condensed after dust removal and disposed. After the post condensation, the nylon is led to an extruder, pelletized and quenched with water. To remove the water, the slurry passes a pre-separator and a dewatering screen. The pellets are then led to a fluidized bed dryer, where water is removed by hot air. The product is obtained at an overflow at the top of the fluidized bed.

B.5 Saturated and unsaturated polyesters

The well-known soda bottle is made of PET (poly-ethylene-terephtalate). This is only one of the many applications in which polyester is used. Others include clothing, audio- and videotapes and some packaging applications. Polyesters are a class of polymers characterized by the so-called ester bond. A very important difference in the properties of saturated and unsaturated polyesters is their behavior at elevated temperatures. Saturated polyesters are thermoplastics, they melt at certain temperatures, whereas unsaturated polyesters are thermoharders. Unsaturated polyesters form cross-links at higher temperatures, forming a network and thus hardening the polyester.

B.5.1 Saturated polyesters

The most produced saturated polyesters PET (poly-ethylene-terephtalate) and PBT (poly-butylene-terephtalate); both are illustrated in Scheme B.3.

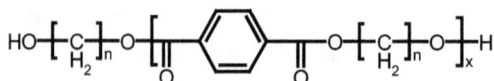

Scheme B.3: Main structure of PET ($n = 2$) and PBT ($n = 4$).

Both polyesters are produced in a similar batch process, which will be discussed in this chapter, with some minor differences. Still the properties of PET and PBT are quite different, so some attention is paid to the relation between structure and properties. Production of polyesters can be done in two ways, either by transesterification or by direct esterification.

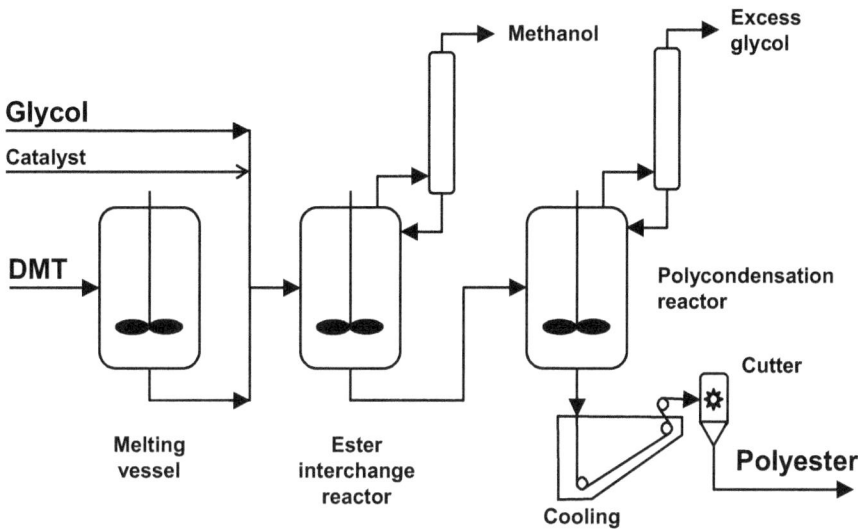

Fig. B.12: Batch process for polyester production (adapted from [207]).

B.5.1.1 Transesterification

First, the process for PET production is described, after that, the differences with PBT production are discussed. The typical layout of a batch process for the production of polyesters is given in Fig. B.12. The production process can be divided into three steps: melting, transesterification and polycondensation. Dimethyl terephtalate (DMT) is melted in a stirred tank at 150–160 °C under a nitrogen atmosphere, heating takes place by steam, carrier oil or electricity. The molten DMT and ethylene glycol are re-acted in heated, stirred trans-esterification reactors at 150–200 °C, operated at normal pressure in nitrogen atmosphere. At the start of the reaction, a lower temperature is favored to minimize the sublimation of DMT. The methanol, which is formed during the reaction, is continuously distilled off from the reaction mixture. Normally a glycol excess of 0.5–1 mol/mol of DMT is used to control the end groups and molecular weight. To achieve reasonable reaction rates, catalysts are a necessity. Although cata-lyst listed in patent literature cover practically the whole periodic table, weak basic compounds, such as amines, oxides, alkoxides and acetates, are used as transesterifi-cation catalysts. Ca, Mg, Zn, Cd, Pb and Co are reported to be the most effective. The transesterification product is added to the polycondensation reactor, which has to be equipped with a very efficient stirrer. Temperature is slowly increased to 250 °C, in the meantime decreasing the pressure. This way, the excess glycol is distilled off and polycondensation can take place. Sufficiently high molecular weights can be obtained at temperatures of 270–280 °C and a pressure lower than 1 mbar. In principle the transesterification catalyst can be used here also, but the properties of the product obtained are not satisfactory. Therefore the transesterification catalyst is masked with a phosphorous compound and replaced by a polycondensation catalyst, such as antimony, germanium, titanium or lead compounds. Polycondensation is stopped when a predefined melt viscosity (and thus a defined molecular weight) is reached. The power consumption of the stirrer motor is often used as an indication for this, as it will increase with increasing melt viscosity. Removing the vacuum with oxygen-free nitrogen stops the polycondensation. This is done to prevent degradation of the polymer by oxidation. The melt is directly quenched with water and comminuted into chips or pellets. Before further processing the PET must be carefully dried at 80–130 °C to reduce the water content to below 0.01 wt%. During the drying process crystallization of the PET occurs, which is necessary in order to prevent agglomeration of the particles during prolonged storage. The comminuted PET is often subjected to solid-phase post-condensation for a fairly long time (up to 20 h) at temperatures of up to 250 °C in an inert gas stream, resulting in an increase of molecular weight.

B.5.1.2 Direct esterification

The direct esterification of terephtalic acid with ethylene glycol became important when economic processes were developed for producing fiber-grade acid. In new plants the direct esterification is the preferred method because of the higher reaction rate, higher molecular masses can be obtained and no transesterification catalyst is needed. The process is almost identical to the process for powder coating production, which is discussed later. The main difference in the process is the addition of organic solvent after part of the water is removed. This is a so-called solvatation step, in which bubbles are formed and the partial pressure of water is lowered, thus promoting the evaporation of the remainder of the water from the reaction mixture. The vapor is cooled and separated, after which the solvent is returned and the water disposed. Saturated polyesters can be produced at lower pressures than powder coatings, due to the composition of the reaction mixture.

B.5.1.3 Differences between PET and PBT production

As stated earlier, the production of PBT and PET are almost similar. They differ only in raw material, catalyst and some operating conditions. For PBT production the transesterification is still preferred, whereas direct esterification is gaining popularity in the production of PET. For both products DMT is used as a raw material, but the comonomer for PBT is 1,4-butanediol instead of glycol. A significant byproduct in the production of PBT is tetrahydrofuran, which is formed from the diol. In PET production loss due to side-product formation is less than in its PBT counterpart. In PBT production, only one catalyst is added, which is used in transesterification and in polycondensation. Usually a tetra-alkyl titanate is used as a catalyst, because it is both very active and soluble.

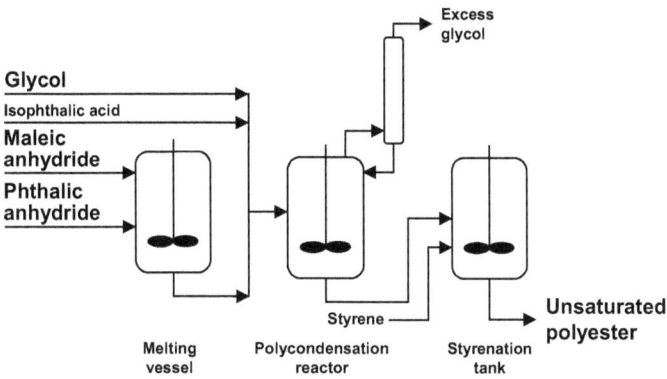

Fig. B.13: Batch-wise production of UP resins (adapted from [207]).

B.5.2 Unsaturated polyesters

Usually, unsaturated polyesters (UP) are produced by batch-wise condensation of di-carboxylic acids or anhydrides with diols. An often used polyester is made of maleic anhydride, styrene and diols. Even today the greater part of the UP resin is used in the production of glass fiber reinforced plastic. Di-carboxylic acids are esterified with diols at 180–230 °C in a nitrogen atmosphere in heatable and coolable reactors while stirring (see Fig. B.13). The capacity of the stainless steel reactors used ranges from 8 to 20 m². Esterification is generally performed with a slight excess (up to 10%) of diols. Usually the maleic anhydride is measured and added in the liquid state. The water formed in the reaction is distilled off. This is a necessary step because of the use of volatile glycols, which are separated from the water in the distillation column and recycled to the reactor. This is done in almost all polyester production processes.

Passage of nitrogen through the reaction mixture promotes removal of water and increases the esterification rate considerably. The end point of the condensation is determined by measuring the acid number and, possibly, also the hydroxyl number and the viscosity. Commercial UP resins have a mean molecular mass of 2,000–4,000. The liquid melt is cooled and then dissolved, while stirring, in styrene in coolable mixers. The styrene contains inhibitors such as hydroquinone or tert-butylcatechol. Also a catalyst is added, often a tin salt such as hydrated mono-butyl tin oxide. This catalyst is preferred because of the resulting product stability during storage. The esterification is reversible, and addition of steam will lower the degree of polymerization. This has the advantage that fine-tuning of the molecular weight can be easily done. Even if the same raw materials are used, the resin products may differ as regards their processing properties and end properties depending on the amount of excess glycol, the temperature, and reaction conditions. In the case of raw materials with widely differing esterification rates, two-stage condensation must be used to obtain uniform buildup of the polyester chain and thus favorable end properties. Azeotropic esterification is important only for certain paint resins. An entraining agent (e.g., toluene or xylene) is added and during condensation the water is distilled off together with the toluene. The toluene is then separated from the water with a dephlegmator and returned to the reactor. At the end of the reaction the toluene must be removed from the resin (e.g., with a thin-film evaporator).

B.5.3 Powder coatings

A special application of polyester resins is in powder coatings. These are used for protection of underlying materials, decoration and modification of the surface properties. A powder coating mainly consists of a (fine) polyester powder, to which several substances are added to obtain the necessary properties. Examples of these substances are coloring agents, hardeners, stabilizers and sometimes paraffin. Powder coatings can be applied in several ways; by a fluid bed process, by flaking or by spraying.

Production of powder coatings is usually done batch-wise, although very large batches are needed to produce a constant quality powder. After the initial production, it is hard to modify the properties of the coating, so all process parameters have to be kept constant. Large batches are also needed because cleaning of the equipment is time consuming and expensive. It has to be done very thoroughly, because it might be possible that two coatings are incompatible. The applied batch reactors are designed for a wide pressure range (vacuum to several bars), equipped with internal heating/cooling coils and a multitask agitator. First of all the glycols, acids and catalyst are added to the reactor, this is then closed and heated at high speed. Due to the poor solubility of the acids in glycol, the reactor content is slurry. This results in a rising pressure and a slowly starting reaction. When a predefined pressure level is reached, a valve to a distillation column is opened. In this column, separation of water and glycol takes place, water is disposed and the glycol is recycled. Due to the evaporation of water, the temperature rise is slower. The pressure is kept at a constant level to ensure good operation of the distillation column, e.g., to avoid flooding. Another reason to keep the pressure at an elevated level is the formation of oligomers in the reactor, thus reducing the amount of volatiles. When the temperature reaches a value dictated by the desired product specification, pressure is slowly reduced. This is possible because the amount of volatiles has decreased enough for safe operation of the distillation column. When atmospheric pressure is reached, it is important to remove the water as fast as possible to preserve the conversion obtained. If the water would not be removed, the polyester would slowly degrade because of the equilibrium reaction. Most of the water is removed at atmospheric pressure; the remainder is removed at reduced pressure. After cooling the reactor contents to a suitable temperature, the desired additives can be added. After mixing, the polymer is sent to a cooling belt, where further cooling to room temperature takes place and flakes are formed.

C Life science products

C.1 Benzaldehyde-based products

For the production of cinnamon aldehyde and cinnamyl alcohol, the main commercial process uses benzaldehyde as a raw material. D-Phenylglycine and p-hydroxyphenyl-glycine are important raw materials in antibiotic production. Phenylalanine is used in the production of aspartame, a synthetic sweetener.

C.1.1 Amino acids

All amino acids exist in two optically active forms, but in most cases only one enantiomer is commercially interesting. Therefore, a method has to be found to separate the optical isomers. This is usually done by forming a salt with another optically active compound. The trick is to find a solvent in which one of the salts formed is much less soluble than the salt of the other isomer. After filtration, either the precipitate or the remaining solution can be processed to obtain the desired product.

Optically active amino acids are normally produced by the enzymatic resolution of D,L-amino-acid amides. Phenylglycine amide and phenylglycine production takes place this way, starting with benzaldehyde. The amides are made by an efficient procedure, which starts with the commercially available aldehyde. The aldehyde is converted under alkaline conditions in the presence of catalytic amounts of acetone (see Scheme C.1). The reaction is usually carried out at normal pressure and temperature and the acetone is also used as a solvent in the last synthesis steps. For the conversion of one mole of nitrile, one mole of water is needed. The acetone can be recycled by distillation and the remaining D,L-amino acid amide can be used in further processing, e.g., optical resolution.

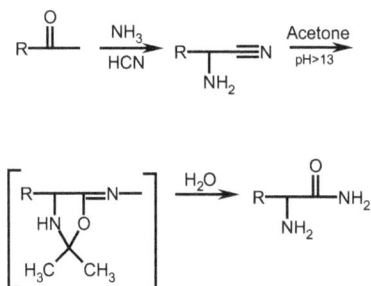

Scheme C.1: Production of amino acid amide.

https://doi.org/10.1515/9783110712445-021

The resolution of the amino acid amide is carried out enzymatically with the use of the complete microorganism (*Pseudomonas putida* ATCC 12,633) around room temperature. The reason to use this micro organism is the high relative activity and the broad spectrum of substrates which can be used. The stereoselectivity of the enzyme is almost 100% in hydrolyzing the L-amino acid. The optimum pH range of the enzyme is between 8 and 10, of which the lower value is preferred, due to the unwanted hydrolysis at higher pH values. After the reaction, the mixture contains L-amino acid and D-amino acid amide, which have to be separated. This is done by addition of benzaldehyde, resulting in the Schiff base of the amino acid amide (see Scheme C.2), which is completely insoluble in water. After filtering the mixture, the amino acid amide is acid hydrolyzed at about 100 °C and the resulting D-amino acid is racemized and recycled.

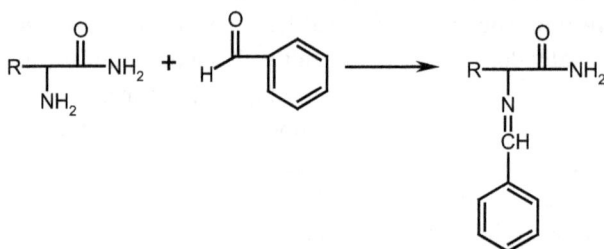

Scheme C.2: Schiff base of benzaldehyde and D-amino acid amide.

Phenylalanine is also produced as described above (starting with phenylacetaldehyde), but can also be produced by a chemical method, which is described below. The reactions taking place in the production of phenylalanine are given in Scheme C.3.

As can be seen from the reactions taking place, the process can be divided into four main steps. Most of them are operated batch-wise, with exception of the last two (neutralization and washing). First of all, a condensation reaction takes place. This is done by mixing benzaldehyde and hydantoin in a mixture of MEA (monoethanol amine) and water at 88 °C. BZH (benzylideenhydantoin) crystals and heat are formed in this reaction, so cooling has to take place. In the next step, the BZH is separated from the reaction mixture in a centrifuge after washing with water at 15 °C. After centrifugation, an extra amount of water is added to the BZH cake to produce easy manageable slurry. The slurry is heated to about 90 °C with steam and enters the hydrogenation section. This hydrogenation takes place at 70 °C and high pressure in the presence of a Pd/C catalyst. After completion of the hydrogenation, the catalyst is removed by filtering and an additional amount of NaOH is added, which is used in the hydrolysis step. This takes place at elevated temperature (140 °C) and pressure, and lasts for several hours. More NaOH is added as a 50 wt% solution. The NH_3 formed during this period is removed, as well as part of the water, by flashing. The

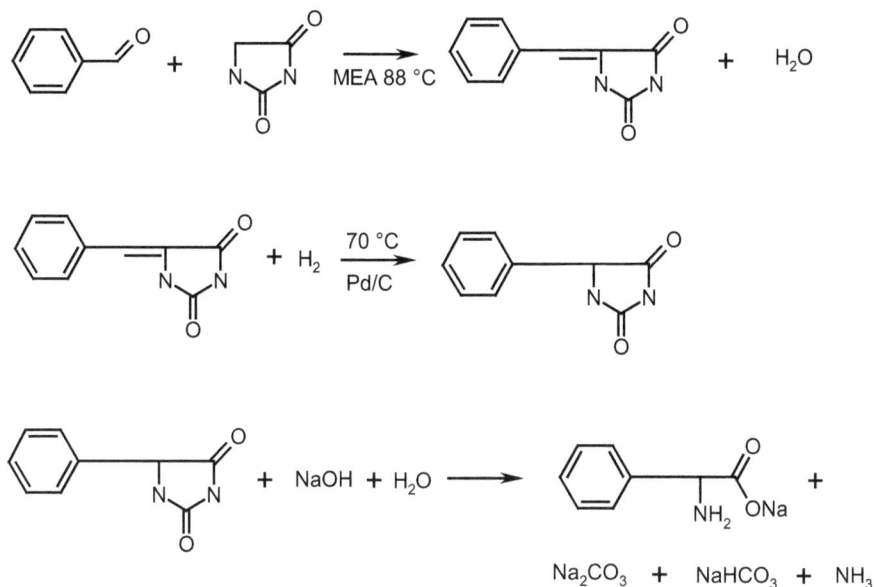

Scheme C.3: Industrial production of phenylalanine.

sodium salt, which is formed, is stored in a buffer storage to make it possible to switch from batch to continuous mode of operation.

After hydrolysis the dissolved sodium salt is neutralized with diluted sulfuric acid and D,L-phenylalanine crystals are formed. The solids are separated from the liquid by centrifugation and washed with water. After removal of the last traces of the mother liquor, the crystals are washed with methanol to remove water. Depending on the desires of the customer, the phenylalanine is dried or supplied as slurry in methanol.

L-Valine is produced from isobutyraldehyde in a similar process as described before, but D-valine is highly soluble in water, which makes separation complicated. Therefore the D-valine amide is separated from the mixture by ion-exchange and converted to D-valine using a non-stereospecific microorganism. D-Valine is obtained after spray drying the reaction mixture.

C.1.2 Cinnamon aldehyde and cinnamyl alcohol

Another series of products made from benzaldehyde are cinnamon aldehyde and cinnamyl alcohol through the industrial reactions are given in Scheme C.4. Cinnamon aldehyde is produced by the crossed aldol condensation reaction of benzaldehyde and acetaldehyde. A diluted solution of sodium hydroxide is used as catalyst. The benzaldehyde and hydroxide are added to the reactor and small

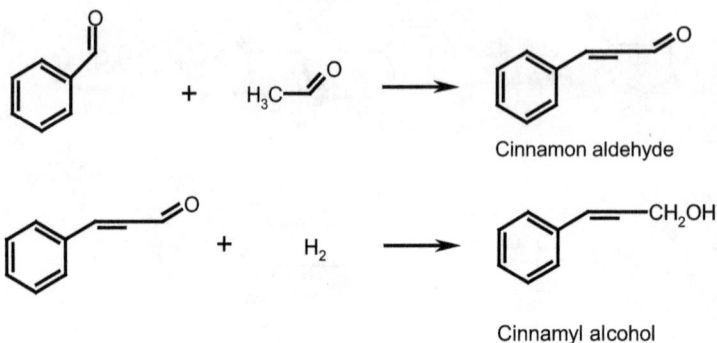

Cinnamon aldehyde

Cinnamyl alcohol

Scheme C.4: Benzaldehyde reaction products.

portions of acetaldehyde are introduced continuously in order to keep the acetaldehyde concentration low. This way the selectivity toward cinnamon aldehyde is increased. When all the acetaldehyde is added, the reaction is completed. The cinnamon aldehyde is separated from most of the hydroxide by phase separation and neutralization. A coalescer removes the water formed in the neutralization. The crude cinnamon aldehyde is purified by distillation. Cinnamyl alcohol is usually made by the hydrogenation of cinnamon aldehyde. It is one of the major components of cassia oil, from which it is also extracted commercially. This is done to produce "natural" benzaldehyde for several applications in the production of food, because only natural benzaldehyde is allowed as an additive. Production of cinnamyl alcohol is operated batch-wise.

The hydrogenation takes place in the liquid phase, usually in a liquid system of water and toluene. The amount of water is preferably between 0.2 and 1 g/g of toluene. The amount of toluene has to be 0.5–5 g/g of aldehyde to be converted. The reaction takes place in the presence of a platinum catalyst, used in most of the commercial hydrogenation processes. The amount used is preferably 0.05–2.5 mg/g of aldehyde. The water used contains 0.2–0.5 mol alkali hydroxide per 100 g. The reaction has to take place at a temperature below 60 °C and preferably between −5 and 45 °C. Hydrogen pressure has no significant effect on the hydrogenation speed, but it is usually maintained between 5 and 300 bar in commercial processes. The conversion of the aldehyde is usually almost complete and upon completion, the mixture is separated in an organic layer and a water fraction. The water fraction can be reused in the hydrogenation of an extra amount of aldehyde. The organic layer contains the desired product, which can be separated from the solvent by distillation, for example.

C.1.3 Benzyl alcohol

The production of benzyl alcohol from benzaldehyde is also operated batch-wise. The process can be divided into a production and a purification section. Production takes place in a hydrogenation reactor at 10 bar. To speed up the reaction a Ra/Ni-catalyst is added and the reactor is operated at 85 °C. The hydrogen gas supplied to the reactor has an inert content of about 10%, so a purge is installed to prevent buildup of inert gasses. After cooling the mixture to a temperature below 40 °C, the catalyst is removed by filtering. The catalyst has been severely deactivated during the reaction, because of the hydrogen absorbed. Therefore the catalyst will have to be regenerated before further use.

The crude benzyl alcohol still contains sodium, which is present as a contamination in the feed, and has to be removed before further processing the product. Removal is done by ion exchange and the purified benzyl alcohol is stored. Final purification of the benzylalcohol is done by batch distillation and product grades differ depending on the distillation. If the sodium is not removed before distillation of the mixture, a slurry of sodium benzoate will form in the distillation section. This way severe fouling takes place and the distillation will not take place properly.

C.2 α-Picoline

Picoline is a pyridine derivate that is often used in the production of medicines for cancer curing. These medicines are often a complex of a precious metal with an aromatic organic compound. There are several types of picolines, as can be seen from Fig. C.1, but α-picoline is the most used.

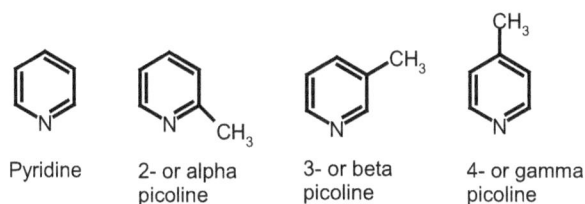

Fig. C.1: Structures of pyridine and picolines.

Most processes for the production of picoline result in a mixture of pyridine and several picolines. Another thing they have in common is that they are either trade secrets or patented. Most of them use ammonia, acetaldehyde and formaldehyde as raw materials. An alternative process, designed by Stamicarbon, uses acrylonitrile, acetone and hydrogen. The first step in this process is the production of 5-OHN (5-oxohexane nitrile), which is subsequently hydrogenated. Raw materials for the production of

5-OHN as shown in Fig. C.2 are acetone, acrylonitrile and small amounts of acid and a primary amine. The primary amine and acid are used as catalysts with benzoic acid being the preferred acid. All raw materials have to be (almost) free of water and above all oxygen. Oxygen initiates tar production and fouling of the reactor by polymerization of acrylonitrile. An inert gas is added to lower the oxygen content below 20 ppm, whereas molecular sieves are used to remove the water. The reaction is carried out at temperatures ranging from 160–200 °C at elevated pressure. Pressures of about 20 bar are common practice. The primary amine and acid act as a catalyst and can be chosen from a wide range of commercially available products. The amine preferably has a boiling point lower than that of the 5-OHN to make it possible to reuse the amine after separation by distillation. The amount of amine can vary between 0.01 and 0.2 mol/mol of acetonitrile, depending on the amine. Because the amine is in a solution with acetone a Schiff base is formed, which is the actual catalyst. The acid can be either a carboxylic acid, such as formic or acetic acid, or a mineral acid such as hydrochloric or phosphoric acid. The acid is added in an amount of 0.002–0.01 mol/mol of acrylonitrile. A small amount of water is added to prevent some unwanted side reactions. Water content of the reaction mixture can be varied from 1 to 3 wt%.

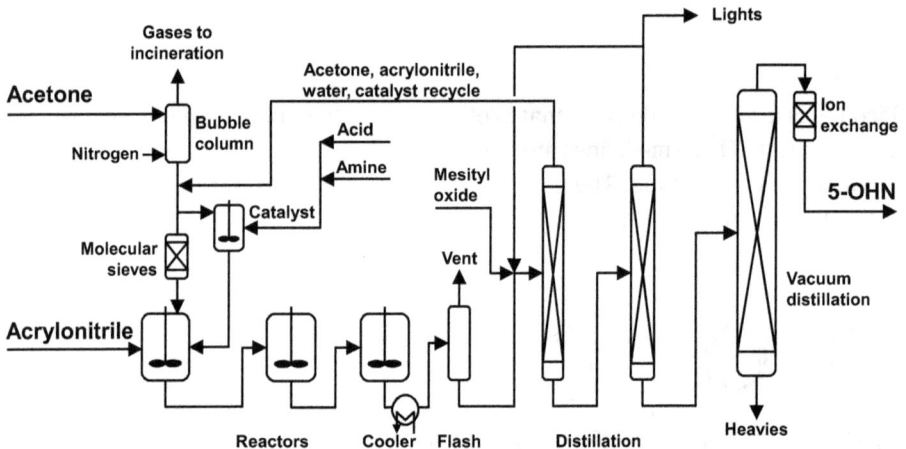

Fig. C.2: 5-Oxohexane nitrile (5-OHN) production (adapted from [214]).

Above acrylonitrile conversions of 90% unwanted side reactions may occur, due to prolonged reaction time and higher reaction temperatures needed. Therefore, the conversion is kept below 90%, usually around 85%. In addition, an excess acetone is added to keep to conversion low. The amount of acetone used is generally 2–12 times that of acrylonitrile on molar bases. The total residence time in the reactors is about 3 h, spread out evenly over three reactors. A relatively large portion of the used raw materials is converted to side products a selectivity toward 5-OHN (based on acrylonitrile, being the highest) of only 86%. After reaction the mixture is allowed to cool and expanded to atmospheric pressure. Before entering the first distillation column, mesityl oxide is added to keep the bottom temperature of the reactor below 195 °C. At the top of this column acetone, acrylonitrile, water and catalyst are recovered and recycled. The residue is led to a second column that operates at reduced pressure (0.2 bar). The overhead stream, consisting of mesityl oxide and 3-isopropyl aminopropionitrile (20 wt%) is recycled to the first column. The bottom stream is led to the last distillation column with an overhead pressure of about 30 mbar. The 5-OHN is taken overhead at a 95% purity. Further purification by passing the product stream through an ion exchange bed results in a 99.9% purity.

The 5-OHN produced in the preceding step is heated to 240 °C at atmospheric pressure and mixed with hydrogen. (Fig. C.3) Hydrogen is added in a 5 mol/mol of nitrile ratio, partially supplied from a recycle of treated reactor off-gas. Hydrogenation of the nitrile takes place over a catalyst bed in a tubular reactor, which contains a palladium catalyst on an alumina support. The palladium content of the catalyst particles is about 0.5 wt%. The off-gasses are cooled and part of the reaction mixture is condensed. The liquid formed mainly consists of 2-methylpyridine (α-picoline) and 2-methyl-piperidine. The remaining gas mainly consists of hydrogen, with 5–20% other gasses (reaction product, raw material and impurities). It has been shown that recycling the remaining hydrogen containing gas without further treatment would cause a significant loss in 5-OHN conversion and is thus economically unfavorable. For this reason a treatment procedure for the gas-recycle was developed.

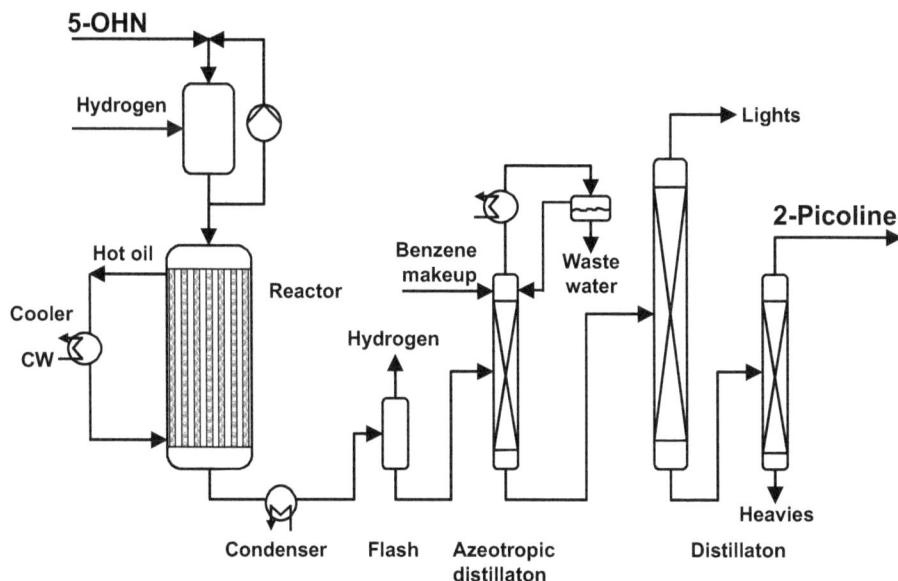

Fig. C.3: Dehydrocyclization of 5-OHN to α-picoline (adapted from [214]).

To remove water the cooled mixture is subjected to an azeotropic distillation with benzene. The water/benzene mixture that is taken overhead is separated, the benzene is recycled and the water is disposed. Two additional distillation columns are needed to remove light and heavy ends respectively. The nitrile conversion is almost a hundred percent, the yield of α-picoline is about 84% and almost 5% of the nitrile is converted to 2-methylpiperidine.

C.3 Aspartame

Aspartame is a synthetic sweetener, with a low energy content (in other words: has a low caloric value). Therefore, it can be used in light sodas as a replacement for natural sugar. For the sweetness, it is important to synthesize only the α-variant of aspartame. The β-variant is not sweet at all and has no commercial use. DSM has produced aspartame by the enzymatic coupling of D,L-phenylalanine methyl ester with Z-L-Asp (L-N-benzyloxycarbonyl-aspartic acid). This is a very specific reaction, in which only the L-isomer of phenylalanine methyl ester is involved, the D-isomer remains unchanged. The enzyme used, thermolysin, was discovered in a hot water pool in Japan, and has excellent thermostability and resistance to organic chemicals. The reactions involved in the enzymatic production are illustrated in Fig. C.4.

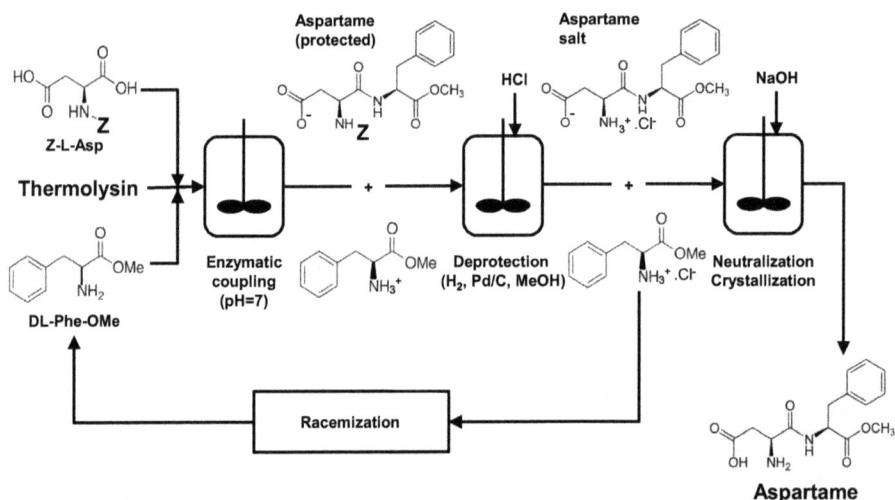

Fig. C.4: Enzymatic production of aspartame.

The addition product of the condensation product with D-phenylalanine methyl ester almost completely precipitates, shifting the equilibrium to the condensation product instead of the hydrolysis product. The salt formed is easily separated into its components by HCl treatment enabling recycling of the D-phenylalanine methyl ester after racemization.

The *N*-formyl-L-aspartyl-L-phenylalanine obtained is hydrolyzed to remove the protecting formyl group. This can be accomplished by addition of methanol and hydrochloric acid. The concentration of methanol should be between 3 and 5 wt% and the acid concentration between 9 and 18 wt%. An extra advantage of the addition of methanol is the conversion of acetic acid and formic acid to components with a lower boiling point. Thereby, the separation is simplified. This reaction step takes place at 35 °C and lasts for about 6 days. The resulting hydrochloride salt of α-aspartame is easily separated from its isomer because the solubility of the former is considerably lower than that of the latter. Separation takes place by filtration, centrifugation, decantation or another conventional method. The solid aspartame hydrochloride is neutralized and aspartame is obtained by crystallization.

C.4 Penicillin

Penicillin was first discovered in 1929 by Fleming, and it was found to kill a wide range of bacteria in experiments during the 1930s. Penicillin is not just one substance; it is a family of related chemicals. All the penicillins have the same main structure, but different side-chains. An example is given in Fig. C.5.

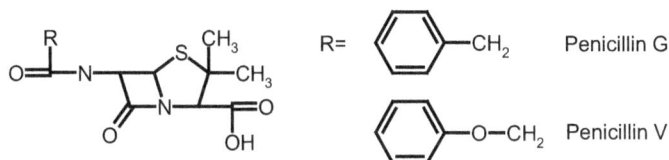

Fig. C.5: Penicillin main structure and some side chains.

The various types have particular antibiotic properties and are formed in various concentrations from the activity of a particular nutrient under particular circumstances. Several types of penicillin have been synthesized on laboratory scale, but the costs for duplicating the synthesis to plant scale are too high. Microbes often accomplish in a single step an economically feasible molecular change that otherwise can only be achieved by long chemical synthesis. Fermentation may also permit the use of cheaper raw materials and results in a greater specificity and in a purer form. Therefore, all of the penicillin is produced by fermentation. Because fermentation involves working with living organisms, it requires some different techniques and has several specific characteristics. Therefore, it will be discussed here shortly. It has several advantages, but it is often only possible to obtain a product in a very low concentration. Therefore, several concentration and separation steps are required to obtain a pure substance or concentrated solution.

The chronological development of the fermentation industry can be divided into five stages. Prior to 1900, the fermentation industry consisted of breweries, vinegar and wine production. In addition, yogurt, soy sauce and other fermented foods were produced. Between 1900 and 1940 chemicals such as lactic acid, glycerol and some organic solvents, such as acetone, were made by fermentation. Some advances in fermentation technology were made, including the fed batch cultures in yeast production, to overcome the initial problems with oxygen shortage. During this stage, the concept of sterile media, pure culture technique and the need to understand the immense role played by microbial physiology and biochemistry in fermentations, arose. During the next 20 years, many developments took place and a large impulse came from the wartime need for penicillin. The penicillin process required many developments, mainly in the field of aerating large quantities of viscous mycelial cultures and the need to isolate a valuable product, which was present only in low concentrations. Mutation-selection programs arose and sparging with sterilized air became a common practice. The number of different products increased very rapidly from 1960 to 1970. About 25 antibiotics, several important enzymes, polysaccharides, microbial insecticides and flavor-enhancing products were introduced. This era also saw the emergence of products such as flavors and fragrances, which are made in small quantities, but continued to command relatively high prices. A number of large companies began investigating the opportunities of microbial biomass to serve as protein source for poultry, cows and pigs. The fifth stage was announced by several genetic engineering discoveries, i.e., the in vitro manipulation of microorganisms. From then on, it was possible to produce new products by fermentation or produce them in higher concentrations.

The microorganism used is the most valuable part of the whole operation and it has to be preserved in some way to ensure its availability at all times. This is usually done by keeping a supply of the strain, which is used in production, in liquid nitrogen. This way mutation is kept to a minimum and integrity of the strain will remain intact. Most companies do not attempt to patent microorganisms, because it might alert competitors to areas of interest. The chance that another company discovers the same strain for the desired product is relatively small. It has been reported that companies found one useful organism in 10,000 investigated cultures (or even worse: 3 out of 400,000). Most organisms only show their potential at exactly the right circumstances and in the right medium. But after a strain development program, the production for penicillin-G is now at a maximum of 75 g/L of medium/microorganism mix. This is a 15,000-fold increase compared to the natural level of penicillin-G production.

Most existing fermentation processes can be broken down into several distinct operations. The first stage is the formulation of a medium to grow the culture during inoculum propagation and fermentation. Second is the production of a sterile medium, fermenters and related equipment. Next is the propagation of a pure and active culture to feed the production fermenter, followed by culturing the microorganisms at the optimum conditions for product formation. A next step is the separation of the product from the original culture and subsequent concentrating and purification of the product. The last step is the treatment and disposal of the cellular and effluent by-products of the process.

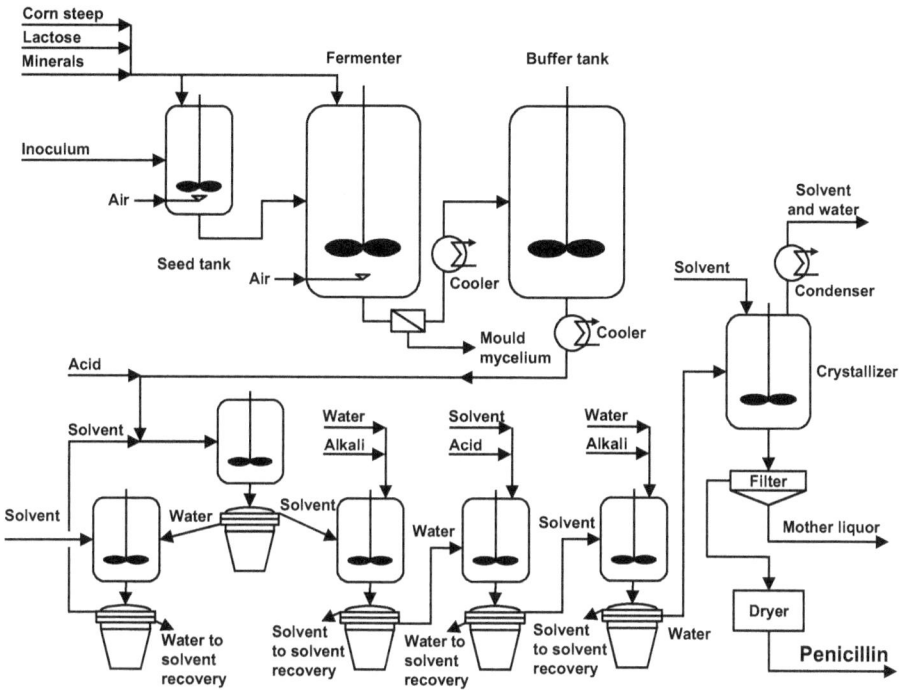

Fig. C.6: Flow sheet of penicillin production (adapted from [210]).

The penicillin production process is schematically depicted in Fig. C.6. Before the actual fermentation four laboratory scale development steps are needed to produce the seed mold culture in a series of successive batch procedures. In the first three of these, glassware vessels are loaded with the culture and nutrients and are placed in a shaker at constant temperature for the desired time. Every separate stage will take about a day; all apply a temperature of 25 °C to encourage mold growth. Usually the mold is kept in a nutrient consisting of corn steep (as a source of nitrogen), lactose and mineral salts such as ammonium sulfate (as a sulfur source). Later on it was discovered that corn steep not only acted as a source of nitrogen, but also contained phenylalanine and phenethylamine, which are precursors for phenyl penicillin (penicillin-G). Today, an extra amount of phenyl acetate is added to increase the production of penicillin-G. The last stage is a tank-sized "seed" fermenter where the mycelia are grown, but in which little penicillin is formed. In the "seed" fermenter, air is blown through a violently agitated mixture containing the culture formed earlier, corn steep, lactose, and mineral salts. To prevent foam formation in the fermenter, oil is added. Examples of oil used are soy bean oil, cotton seed oil and lard. The products from this tank are fed to a larger production fermenter after about a day of operation. This fermenter can be 200 m³ in volume and is agitated with a motor driven stirrer. Corn steep, lactose, and mineral salts are added. The first day of cultivation is used to build up the concentration of penicillin-forming mold. Then the penicillin precursor, the compound that is altered to form the penicillin side chain, is added. For penicillin-G this would be phenyl acetic acid; to get penicillin-V, phenoxy acetic acid would be used. The penicillin is formed in this fermenter during the next three to seven days. As the process goes on, the sugar content of the mix drops and the pH increases modestly. All of these stages are carried out at a pH of about 6.5 and a temperature ranging from 23 to 28 °C. Care has to be taken that no contamination of the reactor, feed or culture can take place. Three to five production fermenters might be fed from the output of a single seed fermenter. In the stages described before, it is very important to maintain a sterile environment. This means that all the equipment, feedstock, air and cultures have to be sterilized. Sterilizing the feed is done by applying heat exchangers with sufficient residence time. Air is led through various filters for sterilization. These include glass wool packed columns, stainless steel mesh and bacteriological filters.

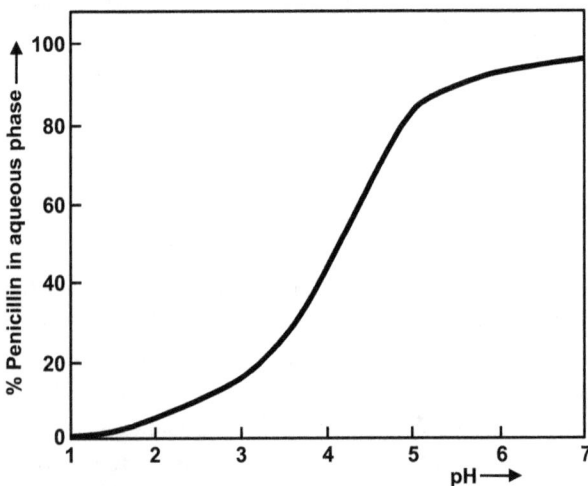

Fig. C.7: Effect of the pH on the penicillin distribution.

When the fermentation is complete, the fermentation mixture is fed to a filter where the mold mycelia are filtered from the broth in a continuous rotary vacuum filter.

The liquid (called broth) is cooled to about 4 °C by a refrigerated cooler and sent to a storage tank. Up to this point, the process has operated completely in a batch mode. From here on, a semicontinuous mode is used. Mineral acid and solvent (butyl acetate) are added to the cold broth, and the solvent collects the penicillin transferred from the broth in the mixing tank. Then the solvent and broth are separated in a disc centrifuge. The broth is contacted with a second batch of solvent to complete the removal of the penicillin. This second batch of solvent, containing small quantities of penicillin, is used in the first contact stage with the fresh broth. Extraction efficiency is almost 100%. The depleted broth is sent to a distillation unit, which is not shown in the figure, for the removal of traces of solvent. Figure C.7 shows the effect of pH on the distribution of penicillin between butyl acetate and water. After phase separation the pH is shifted several times in order to make the penicillin distribute alternately to the aqueous and solvent phase leaving the impurities behind. Additionally activated carbon might be used to remove any traces of coloring agents. After the last addition of alkali the solvent and water are evaporated and a precipitate of the penicillin potassium salt is formed and separated from the remaining mother liquor on a vacuum band filter, where washing takes place as well as drying in hot air. Crystal purity at the end of the filter is 99.5%.

C.5 Synthetic antibiotics

C.5.1 Introduction

The semisynthetic cephalosporins (SSCs) are a subgroup of the β-lactam antibiotics, they all have the basic structure given in Fig. C.8. Differences between the several cephalosporins are the substitutes at the C3 and C7 positions, which are shown in Fig. C.8, together with an example of the substitutes. The SSCs most used have a CH_3 substitute at the C3 position. An overview of the production of SSCs and SSAs (semisynthetic amoxillins) is given in Fig. C.9.

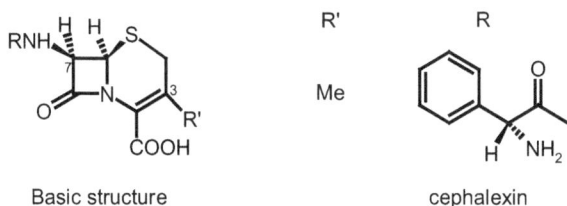

Basic structure   cephalexin

Fig. C.8: Cephalosporin basic structure and a specific example.

Production of SSCs usually starts from one of the following raw materials; penicillin G or V, 7-ACA (7- amino-cephalosporanic acid) or cephalosporin C (the structure

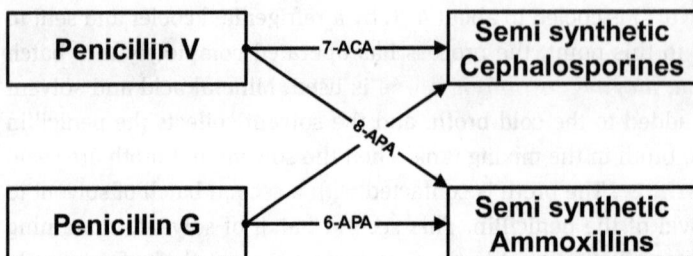

Fig. C.9: Production of SSCs and SSAs.

of the latter two is given in Fig. C.10). SSCs, like all other β-lactam antibiotics, exert their antibacterial effect by interfering with the synthesis of the bacterial cell wall. These antibiotics tend to be irreversible inhibitors of the cell wall, thereby disturbing the membrane function of the cell wall. Ultimately this leads to the death of the bacteria.

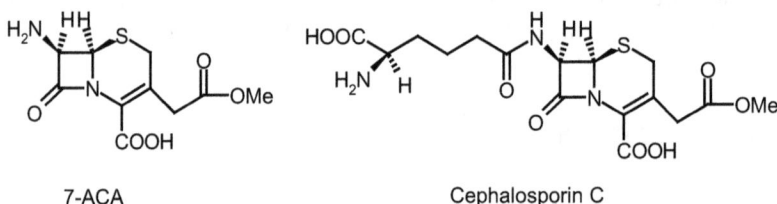

Fig. C.10: Raw materials for SSC synthesis.

C.5.2 Cephalosporins

The production of SSCs can be divided into several steps, which are usually present. These steps include:

1. Selection of the starting material, penicillin has to be rearranged to the cephalexine nucleus
2. Cleavage of the acyl side chain of the precursor
3. Synthesis of the C7 and C3 side chain precursors
4. Acylation of the C7 amino function to introduce the desired acyl-amino side chain
5. Introduction of the C3 side chain and
6. Protection and/or activation of functional groups may be required

The order in which these steps are performed depends on the manufacturer. The most important factors in the decision are usually of an economic nature. It is a common practice to introduce the side chains that use the most expensive raw material in (one of) the last step(s). Usually this comes down to the complete production of the side chain, before attachment to the cephalosporin basis (usually 7-ACA). Both 7-ACA and the penicillins are made by fermentation and after purification, they can be used as a raw material for the production of the SSCs.

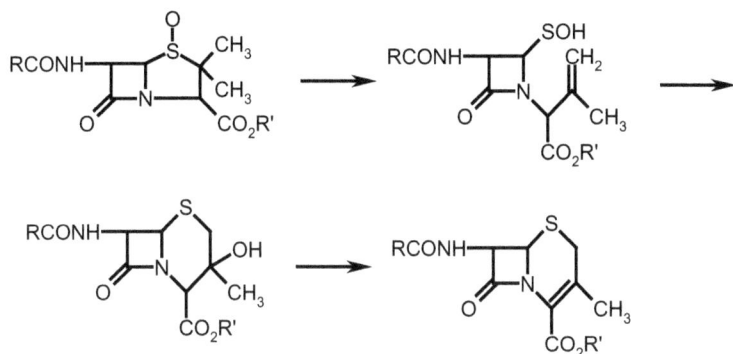

Scheme C.5: 7-ADCA formation.

As an example of the production of SSCs the production of cephalexin will be discussed. One of the aforementioned penicillins can be used as a starting material together with phenyl glycyl chloride hydrochloride. First of all the penicillin has to be converted to a cephalexin basis called 7-amino-desacetoxi-cephalosporanic acid (7-ADCA) by ring expansion of the corresponding penicillin sulfoxide ester, see Scheme C.5. In the first step, the carboxylic acid group of the penicillin is protected by addition of a group such as p-nitro-benzyl, 2,2,2-trichloroethyl or trimethylsilyl. Usually penicillin G is used, because this component can be enzymatically converted to 7-ADCA, thus avoiding several (expensive) chemical conversion steps. If the conversion takes place at ambient temperatures in the presence of phenyl acetic acid, phenylacetyl-7-ADCA is formed. The advantage of this compound is the efficient recovery from the reaction mixture by adding an immiscible organic solvent and a pH shift, like described in the penicillin production. If the amount of organic solvent is kept low, volumetric flows can be significantly reduced, as well as the cost for the downstream processing equipment. After removal of the phenylacetyl side chain, the 7-ADCA can be obtained at high purity. This is an intermediate, which is used in the production of several SSCs, because of its favorable economics. It has been shown that the stereochemistry of the R-group in 7-ADCA is not effected by the ring expansion, which is important to maintain the antibiotic activity of the product.

The 7-ADCA may be converted to cephalexin according to the following procedure, which is one of the first commercially available processes. In order to protect the carboxyl group, the 7-ADCA is silylated first. This is done by a reaction with a silyl halide, trimethyl-chlorosilane, for example, in a solution with dimethylformamide $(HCON(CH_3)_2)$. Best results are obtained if an excess of silylating agent of two molar equivalents is used. After this reaction between the silylated 7-ADCA and phenyl-glycyl-chloride hydrochloride can take place in the same reaction medium. A tertiary nitrogen base is added, preferably with a pK_a between 4.5 and 5.5, pyridine can be used very well. The amount used is about 0.5 equivalents. The reaction takes place at low temperatures, usually between −20 and −40 °C, while continuously stirring. If pyridine is used, best results are obtained at −30 °C. A slight excess of phenylglycyl-chloride hydrochloride is used, usually about 1.1 equivalents. The reaction is usually completed within 1 h. After completion of the reaction, a hydrogen containing component, such as water, is added and the reaction product is desilylated by this. The cephalexin can now be separated from the reaction mixture as a bis-dimethylformamide solvate by diluting the mixture with water. The amount used is about 1–2 parts per 10 parts by volume of dimethylformamide and adjusting the pH to about 6.9 by the addition of aqueous ammonia. The ammonia is added portion-wise over a period of 1 h. The resulting precipitate of cephalexin bis-dimethylformamide can then be separated by filtration or centrifugation. The resulting precipitate can be converted to cephalexin by addition of a dilute aqueous solution of a mineral acid, hydrogen chloride for example, and heating the mixture to 60 °C. After addition of a nitrogen base, such as triethylamine, the cephalexin precipitates can be separated by filtration or another suitable technique.

C.5.3 Amoxillins

Again, production will be discussed based on an example. This time, the production of ampicillin is discussed, which is one of the first semisynthetic penicillin derivatives. Another process is the production of amoxillin, which only differs from the ampicillin-process in the raw material used. For ampicillin, phenyl glycine is used as one of the starting materials, whereas in the production of amoxillin p-hydroxy-phenylglycine is used. The production is preferably carried out as a batch process, but can also be operated continuously. The raw materials used in the production of ampicillin are 6-APA (6-aminopenicillanic acid) and phenyl glycine derivatives. 6-APA can be obtained from penicillin G or V by removal of the side chains at the amino-group. It is advantageous to maintain a low concentration of 6-APA in solution in the reactor, because of a higher conversion obtained and a considerably better stir ability of the mixture. The total amount of 6-APA needed is added to an empty reactor, partially dissolved in water, at a pH of 6.0–6.8 to ensure a concentration of dissolved 6-APA lower than 250 mM. The total concentration of 6-APA *and* ampicillin present in the reactor is kept above 350 mM. An additional method to keep the concentration below the stated maximum is the slow addition of the phenyl glycine derivative. It has been found that this way the amount of 6-APA dissolved can be kept low also. If desired, up to 30 vol% of an organic solvent can be added to the solution. Suitable solvents are lower aliphatic alcohols, such as methanol and ethanol, which are soluble in water.

The phenyl glycine derivative is added in an activated form, usually as an amide (PGA) or an ester, of which the first is preferred. To obtain a high solubility of the PGA, it is useful to add it to the reactor as a salt. Usually PGA ½H$_2$SO$_4$ is used, because of its very high solubility and ease in handling. To obtain the mentioned salt, PGA is dissolved in water and concentrated H$_2$SO$_4$ is added slowly, while keeping the temperature below 25 °C. For economical reasons the molar ratio of PGA relative to 6-APA is kept below 2.5 and usually between 1.2 and 1.8. Higher molar ratios have the disadvantage of the hydrolysis of large amounts of PGA. The PGA salt is slowly added to the 6-APA solution, while keeping the pH between 6.0 and 6.8 and the temperature between −5 and 35 °C. The reaction is catalyzed by an enzyme, which is usually immobilized to simplify the removal from the reaction mixture. All of the enzymes that are known as penicillin amidase or penicillin acylase are suitable catalysts for this process. An enzyme often used is penicillin G acylase. A suitable technology for this process involves immobilizing the enzyme on a carrier, which contains a gelling agent, such as gelatin, combined with a polymer with free amino groups. When the reaction is almost complete, it is favorable to stop the reaction. This can be done by lowering the pH to 5.0–5.7, lowering the temperature or a combination of both methods. After stopping the reaction, the contents of the reactor are a combination of solids, such as ampicillin, D-phenylglycine and immobilized enzyme, and mother liquor. The solids can be separated from the mother liquor by filtration and the valuable components therein can be recovered by a pH shift and subsequent (controlled) crystallization. The mother liquor contains only a few byproducts after filtration and can be recycled without further purification.

C.6 Glyoxylic acid

Glyoxylic acid (see Fig. C.11) is one of the maleic anhydride-based products and is also known as glyoxalic or oxoacetic acid. It is often made by the oxidation of glycoxal with nitric acid, but it can also be produced by the ozonolysis of maleic anhydride. It is a stable organic compound with two functional groups in a small unit structure. Uses of glyoxylic acid are as a raw material in the production of amoxillin, benzaldehydes, vanillin and cross-linking agents. It is commercially available only as a 50% aqueous solution, due to the very hygroscopic character of the monohydrate.

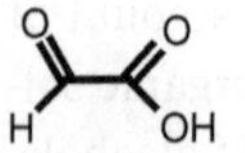

Fig. C.11: Glyoxylic acid.

Advantages of the ozonolysis of maleic anhydride for the production of glyox-ylic acid are the moderate conditions, which can be applied. First, the maleic anhy-dride used as a raw material has to be esterified. This is usually done by dissolving the anhydride in a lower aliphatic alcohol, methanol or ethanol for example, and addition of an acid catalyst. Examples of catalyst used are strong acid ion exchang-ers in the H-form and small amounts of mineral acid. For the further course of the process, it is important to ensure that the maleic anhydride is (almost) completely esterified. The ozonolysis is carried out at −10 to 5 °C in, for example, a bubble col-umn. The solution obtained at the esterification step can be used without the need for purification and the ozone is added in equivalent amount, often diluted with air to reduce heat production to a controllable amount. A typical dilution value is 10 g of ozone/m³. Because the ozonolysis is an exothermic process, cooling is applied to maintain a favorable reaction temperature.

The next step in the process is the hydrogenation of the ozonolysis products, among which are several peroxides. Care has to be taken that the total amount of peroxides in the hydrogenation reactor does not exceed 0.1 mol/L and is preferably kept below 0.02 mol/L. Because of the low concentration of peroxides, poisoning of the catalyst and loss in activity is prevented. To ensure this a suspension of catalyst in methanol is introduced to the (empty) reactor and only after this, the ozonolysis solution is (slowly) added to the reaction mixture. Several traditional hydrogenation catalysts can be used, with or without a support. Catalysts most used are palladium and platinum, of which the latter is preferred. The amount of catalyst used should be about 0.5–2 wt% based on the total amount of ozonized maleic anhydride fed per hour. In the reactor, a pressure of 1 to 3 atmosphere is applied at a temperature of 35–40 °C and a pH value of 2–4. Because a small amount of acids is produced as by-products in the reaction, continuous addition of a base, such as sodium hydrox-ide, is necessary. After hydrogenation, the catalyst can be removed from the reaction mixture and recycled without any regeneration or purification steps. For the proce-dure described, the yield of methyl glyoxylate after hydrogenation is 94% of the theoretical value.

The process uses a recycle, as shown in Fig. C.12, to obtain a maximum conver-sion. The products from the ozonolysis are first stored in a storage tank, before they are led over the catalyst bed. The peroxide concentration in the solution should be kept below 1.5 mol/L, because peroxides tend to decompose explosively in relatively high concentrations. Before passing over the bed, hydrogen is added to a hydrogen pressure of about 1.2 bar. The reacted solution is then collected in the collection ves-sel, where temperature control takes place. If necessary cooling or heating is applied to maintain the solution temperature between room temperature and 50 °C, this is

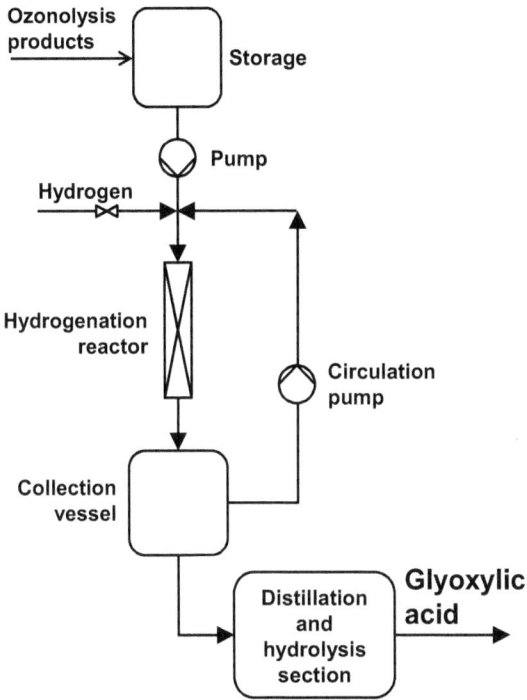

Fig. C.12: Hydrogenation of the ozonolysis products.

the optimum temperature range. Because the conversion in one pass is not sufficient, the mixture is recycled by the circulation pump. In the recycle loop, samples are taken to measure the amount of peroxides and products in the solution. If the product content is sufficient, the mixture is discharged from the collection vessel and sent to a distillation section for purification. After this step, a solution of hemi-acetals of glyoxylates is obtained, which can be converted to glyoxylic acid by hydrolysis. This can be done by addition of water and heating the mixture, while continuously distilling off the alcohol produced in the reaction. Acceleration of the hydrolysis is effectuated by the addition of catalytic amounts of either an acid or a base. There is no need to purify the mixture, obtained after hydrogenation, before hydrolyzing the products present in the mixture. After distillation, a high purity 50% glyoxylic acid solution in water is obtained.

C.7 Food additives

Another important category in life science products are the food additives. In general, these substances are used for improving several properties of food, stabilizers, coloring agent and aromatic substances are all part of this category.

C.7.1 Quinine

Quinine is a so-called alkaloid, a family to which caffeine, morphine and cocaine also belong. It is used as a medicine for malaria treatment led by the native medicine, made from bark, used for treating malarial fever. It is a compound which is commercially extracted from bark instead of chemical synthesis. This is due to its complex structure, which is shown in Fig. C.13.

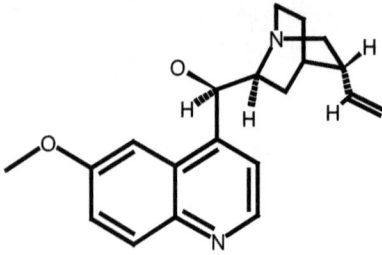

Fig. C.13: Quinine structure.

The production of quinine can be divided into an extraction and a purification step. During the extraction the finely powdered chinchona bark is treated with dissolved lime and 5% sodium hydroxide solution. The quinine is extracted with aromatic hydrocarbons at 60 °C. The alkaloids are recovered from the organic solvent with dilute sulfuric acid; the resulting solution is heated to its boiling point and neutralized with sodium hydroxide. Depending on the pH the mono- or bisulfate crystallizes upon cooling. Purification can be done by recrystallization of either the mono- or bisulfate obtained in the extraction step.

C.7.2 Enzymes

Until about 1950, the predominant method of producing industrial enzymes was by extraction from animal or plant resources, but nowadays this production method accounts for only 10% of the total production. Most of the enzymes are produced in microorganisms by fermentation, either in a continuous or in a batch process. Batch processes are usually carried out with the use of a support for the growing culture, whereas continuous processes are usually carried out in suspension.

The batch-wise operated processes can be divided into two different classes, one using relatively small trays for cultivation, the other using a deep bed. In the tray cultivation process (given in Fig. C.14) a layer of 2–4 cm of substrate is placed on a tray (hence the name), which is placed in an air-conditioned room. Incubation takes also place in these rooms. The seed for the fermentation is stored, either freeze-dried or frozen at temperatures below −80 °C. Inoculation takes place by the use of spores in dry or suspended form. The media used are usually based on wheat bran, because of its high content of nutrients and large surface area. Rice, corn and soybeans are also used, but these require treatment for surface increasing or hull removal in order to allow the fungus to penetrate. The moisture content of the medium should be between 30 and 70% in order to reach optimal production. The bran has to be sterilized to prevent contamination of the product. This is usually done by subjecting the bran to an acidic solution at 95 °C for 15–30 min. The heat produced by the microorganisms is usually removed by moistened cool air.

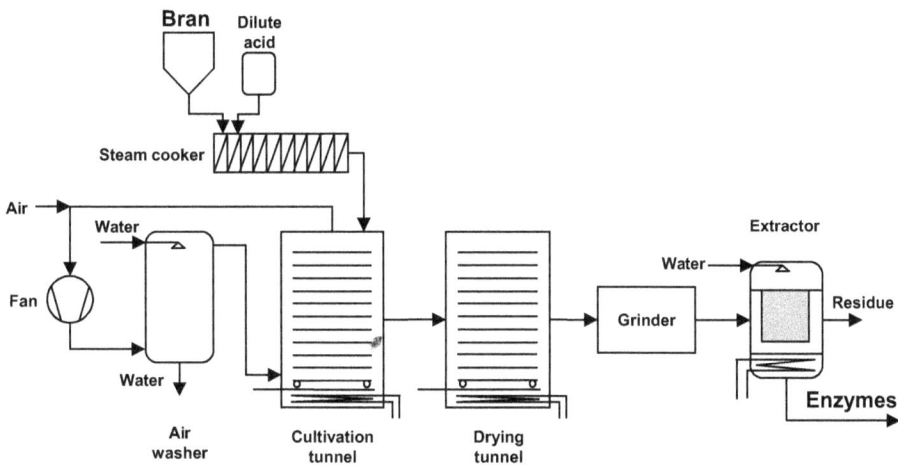

Fig. C.14: Tray cultivation (adapted from [210]).

The deep bed process was developed in order to meet the enormous demands for enzymes for soybean fermentation. In this process a substrate layer with a depth varying from 0.6 to 1.8 m (see Fig. C.15). The rectangular beds can be quite large, dimensions of 5×60 m are not uncommon. The most important advantage of the deep bed process is that it can be fully automated and operated continuously. This way the purification section can also be operated continuously. The equipment used is almost similar to that used in the malting industry. Temperatures applied in all process steps are between 0 and 10 °C except drying, where higher temperatures are applied. For enzymes with a seasonal utilization it is convenient to dry the bran before extraction. This way the enzyme can be produced all year round in relatively small equipment and the enzyme can be extracted at the time of demand. Enzymes are extracted with water, which, however, may contain acids, salts, buffer or stabilizers.

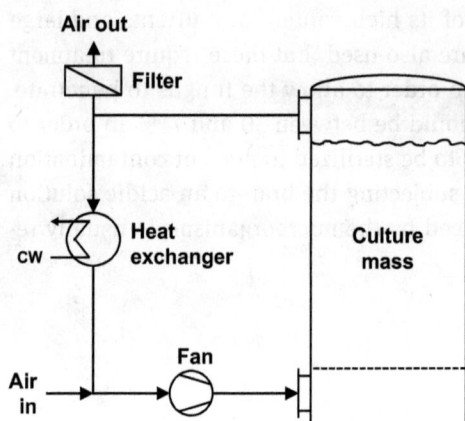

Fig. C.15: Deep bed process (adapted from [210]).

References and further reading

General

[1]   *Perry's Chemical Engineers' Handbook*, 9th edn., (D.W. Green and M.Z. Southard, Editors), McGraw Hill, New York, 2018.
[2]   *Coulson and Richardson's Chemical Engineering*, Volume 1–6, 3rd–7th edn., Butterworth-Heinemann, Oxford, 1994–2019.
[3]   *Handbook of Chemical Engineering Calculations*, 4th edn., (T.G. Hicks and N.P. Chopey, Editors), McGraw Hill, New York, 2012.
[4]   J.R. Couper, et al., *Chemical Process Equipment, Selection and Design*, 2nd edn., Gulf Professional Publishing, Burlington, 2005.
[5]   A.B. de Haan, H.B. Eral, B. Schuur, *Industrial Separation Processes*, 2nd edn., De Gruyter, Berlin, 2020.
[6]   P.C. Wankat, *Separation Process Engineering: Includes Mass Transfer Analysis*, 4th edn., Prentice Hall, Upper Saddle River, NJ, 2016.
[7]   *Kirk-Othmer Separation Technology*, 2nd edn., (R. Clavier, Editor), Wiley, 2008.
[8]   *Handbook of Separation Techniques for Chemical Engineers*, 3rd edn., (P.A. Schweitzer, Editor), McGraw Hill, New York, 1997.
[9]   D.W. Rousseau, *Handbook of Separation Process Technology*, John Wiley & Sons, New York, 1987.
[10] *Kirk-Othmer Encyclopedia of Chemical Technology*, 5th edn., (A. Seidel, Editor), John Wiley & Sons, Hoboken, 2004.
[11] *Ullmann's Encyclopedia of Industrial Chemistry*, 7th edn., Wiley-Interscience, New York, 2002–2021.

Chapter 1

[12] M.A. Benvenuto, *Industrial Chemistry*, De Gruyter, Berlin, 2013.
[13] J.A. Moulijn, M. Makkee, A.E. van Diepen, *Chemical Process Technology*, 2nd edn., Wiley VCH, New York, 2013.
[14] J.A. Kent, *Handbook of Industrial Chemistry and Biotechnology*, 13th edn., Springer, New York, 2017.
[15] P.J. Chenier, *Survey of Industrial Chemistry*, 3rd edn., Springer, New York, 2002.
[16] P.N. Sharrat, *Handbook of Batch Process Design*, Blackie Academic & Professional, London, 1997.
[17] J.A. Tyrrel, *Fundamentals of Industrial Chemistry: Pharmaceuticals, Polymers and Business*, 1st edn., Wiley, 2017.
[18] D.V. Chaudhary, *Industrial* Chemistry, The New Era International Publishing House, 2015.
[19] U. Onken, A. Behr, *Chemische Prozeßkunde*, Georg Thieme Verlag, Stuttgart, 1996.

Chapter 2

[20] W.L. McCabe, J.C. Smith, P. Harriot, *Unit Operations of Chemical Engineering*, 7th edn., McGraw Hill, New York, 2005.
[21] M. Martin, *Industrial Chemical Process Analysis and Design*, Elsevier, Amsterdam, 2016.

https://doi.org/10.1515/9783110712445-022

[22] W.L. Luyben, L.A. Wenzel, *Chemical Process Analysis: Mass and Energy Balances*, Prentice Hall International, London, 1988.

[23] U. Diwerkar, *Batch Processing: Modeling and Design*, CRC Press, Boca Raton, 2014.

[24] T. Majozi, E.R. Seid, J.-Y. Lee, *Understanding Batch Chemical Processes*, CRC Press, Boca Raton, 2019.

[25] P.N. Sharrat, *Handbook of Batch Process Design*, Blackie Academic & Professional, London, 1997.

Chapter 3

[26] M. Schmall, *Chemical Reaction Engineering: Essentials, Exercises and Examples*, CRC Press, Boca Raton, 2014.

[27] L.K. Doraiswamy, D. Uner, *Chemical Reaction Engineering: Beyond the Fundamentals*, CRC Press, Boca Raton, 2013.

[28] T.O. Salmi, J.-P. Mikkola, J.P. Warna, *Chemical Reaction Engineering and Reactor Technology*, CRC Press, Boca Raton, 2019.

[29] T.O. Salmi, J.P. Warna, J.R. Hernández Carucci, C.A. de Araújo Filho, *Chemical Reaction Engineering*, De Gruyter, Berlin, 2020.

[30] D.Y. Murzin, *Chemical Reaction Technology*, De Gruyter, Berlin 2015.

[31] H.S. Fogler, *Elements of Chemical Reaction Engineering*, Pearson Education, Boston, 2020.

[32] O. Levenspiel, *Chemical Reaction Engineering*, 3rd edn., John Wiley & Sons, New York, 1999.

[33] G. Marin, G. Yablonsky, D. Constales *Kinetics of Chemical Reactions, Decoding Complexity*, Wiley-VCH, Weinheim, 2019.

[34] J. Ancheyta, *Chemical Reaction Kinetics: Concepts, Methods and Case Studies*, Wiley, New York, 2017.

[35] D. Murzin, *Engineering Catalysis*, 2nd edn., De Gruyter, Berlin, 2020.

[36] J. Hagen, *Industrial Catalysis: A Practical Approach*, 3rd edn., Wiley VCH, Weinheim, 2015.

[37] U. Hanefeld, L. Lefferts, *Catalysis*, Wiley-VCH, Weinheim, 2018.

[38] A. Behr, P. Neubert, *Applied Homogeneous Catalysis*, Wiley-VCH, Weinheim, 2012.

[39] P. van Leeuwen, *Homogeneous Catalysis: Understanding the Art*, Kluwer Academic, Dordrecht, 2004.

[40] G. Ertl, et al., *Handbook of Heterogeneous Catalysis*, Volume 1–8, 2nd edn., Wiley VCH, Weinheim, 2008.

[41] J. Ross, *Heterogeneous Catalysis: Fundamentals and Applications*, Elsevier, Amsterdam, 2011.

[42] J.M. Thomas, W.J. Thomas, *Principles and Practice of Heterogeneous Catalysis*, Wiley-VCH, Weinheim, 2014.

Chapter 4

[43] J.A. Conesa, *Chemical Reactor Design*, Wiley-VHC, Weinheim, 2019.

[44] R.E. Hayes, J.P. Mmbaga, *Introduction to Chemical Reactor Analysis*, 2nd edn., CRC Press, Boca Raton, 2012.

[45] G.F. Froment, K.B. Bischoff, *Chemical Reactor Analysis and Design*, 3rd edn., John Wiley & Sons, New York, 2010.

[46] E.B. Naumann, *Chemical Reactor Design, Optimization and Scale Up*, 2nd edn., John Wiley Sons, New York, 2008.

[47] A. Cybulski, J.A. Moulijn, *Structured Catalysts and Reactors*, 2nd edn., CRC Press, Boca Raton, 2005.

[48] P. Trambouze, J.-P. Euzen, *Chemical Reactors, from Design to Operation*, Institut Francais du Petrole Publications, 2004.

[49] D. Nardin, *Trends and Opportunities with Modern Buss Loop Reactor Technology*, Proceedings Chemspec Europa 95 BACS Symposium, 1995, 87–92.

[50] J.M. Valverde Millan, *Fluidization of Fine Powders*, Springer, Dordrecht, 2013.

[51] D. Kunii, O. Levenspiel, *Fluidization Engineering*, 2nd edn., Butterworth-Heinemann, Boston, 1991.

[52] J.G. Yates, P. Lettieri, *Fluidized Bed Reactors: Processes and Operating Conditions*, Springer, 2016.

[53] D. Geldart, *Gas Fluidization Technology*, John Wiley & Sons, Chichester, 1986.

[54] J.R. Grace, X. Bi, N. Ellis, *Essentials of Fluidization Technology*, Wiley-VCH, Weinheim, 2020.

[55] D. Pletcher, F.C. Walsh, *Industrial electrochemistry*, 2nd edn., Chapman and Hall, London, 1990.

[56] S.C. Perry, C. Ponce de León, F.C. Walsh, *Review – The Design, Performance and Continuing Development of Electrochemical Reactors for Clean Electrosynthesis*, Journal of the Electrochemical Society, 2020, 167, 155525.

[57] E.M. Stuve, *Electrochemical Reactor Design and Configurations*, in: G. Kreysa, K. Ota, R.F. Savinell (eds.), *Encyclopaedia of Applied Electrochemistry*, Springer, New York, 2014.

[58] K. Jackowska, P. Krysiński, *Applied Electrochemistry*, De Gruyter, Berlin, 2020.

Chapter 5

[59] C.F. Mandenius, *Bioreactors: Design, Properties and Applications*, Wiley-VCH, Weinheim, 2016.

[60] S. Liu, *Bioprocess Engineering: Kinetics, Sustainability and Reactor Design*, 3rd edn., Elsevier, Amsterdam, 2020.

[61] J. Villadsen, et al., *Bioreaction Engineering Principles*, Springer, New York, 2011.

[62] T. Panda, *Bioreactors: Analysis and Design*, Tata McGraw Hill, New Delhi, 2011.

[63] A. Moser, P. Manor, *Bioprocess Technology: Kinetics and Reactors*, Springer Verlag, New York, 2011.

[64] A. Wiseman, *Handbook of Enzyme Biotechnology*, 2nd edn., Pharma Book Syndicate, 2010.

[65] I.J. Dunn, et al., *Biological Reaction Engineering*, 3rd edn., Wiley VCH, Weinheim, 2021.

[66] C. Wittmann, J.C. Liao, *Industrial Biotechnology, Products and Processes*, Wiley-VCH, Weinheim, 2016.

[67] B.R. Maiti, *Principles of Bioreactor Design*, Viva Books, 2018.

[68] K. Schügerl, *Bioreaction Engineering, Volume 1, Reactions Involving Microorganisms and Cells*, John Wiley & Sons, Chichester, 1990.

[69] K. Schügerl, *Bioreaction Engineering, Volume 2, Characteristic Features of Bioreactors*, John Wiley & Sons, Chichester, 1990.

Chapter 6

[70] *Distillation: Fundamentals and Principles* (A. Gorak and E. Sorensen, Editors), Elsevier, Amsterdam, 2014.
[71] *Distillation: Equipment and Processes* (A. Gorak and Z. Olujic, Editors), Elsevier, Amsterdam, 2014.
[72] *Distillation: Operation and Applications* (A. Gorak and H. Schoenmakers, Editors), Elsevier, Amsterdam, 2014.
[73] A. Vogelpohl, *Distillation the Theory*, 2nd edn., De Gruyter, Berlin, 2021.
[74] A.A. Kiss, *Advanced Distillation Technologies: Design, Control and Applications*, John Wiley & Sons, Chichester, 2013.
[75] Z. Lei, et al., *Special Distillation Processes*, 2nd edn., Elsevier, Amsterdam, 2021.
[76] J.G. Stichlmair, J.R. Fair, *Distillation: Principles and Practice*, Wiley-VCH, New York, 1998.
[77] Z. Tadmor, C.G. Gogos, *Principles of Polymer Processing*, 2nd edn., Wiley, New York, 2005.
[78] R.J. Albalak, *Polymer Devolatilization*, Marcel Dekker, New York, 1996.
[79] J.A. Biesenberger, *Devolatilization of Polymers*, Hanser, Munich, MacMillan, New York, 1983.

Chapter 7

[80] J. Rydberg, C. Musikas, G.R. Choppin, *Principles and Practices of Solvent Extraction*, 2nd edn., Marcel Dekker, New York, 2004.
[81] J.C. Godfrey, M.J. Slater, *Liquid-Liquid Extraction Equipment*, John Wiley & Sons, New York, 1994.
[82] J.D. Thornton, *Science and Practice of Liquid-Liquid Extraction, Vol. 1&2*, Oxford Science Publications, Oxford, 1992.
[83] C. Poole, *Liquid Phase Extraction*, Elsevier, Amsterdam, 2019.
[84] J.P. Duroudier, *Liquid-Liquid and Solid-Liquid Extractors*, ISTE Press Ltd, London, 2016.
[85] F. Chemat, M.A. Vian, *Alternative Solvents for Natural Products Extraction*, Springer, New York, 2014.
[86] J. Lindey, *Supercritical Fluid Extraction: Technology, Applications and Limitations*, Nova Science Pub, 2014.
[87] G. Brunner, *Supercritical Fluids as Solvents and Reaction Media*, Elsevier, Amsterdam, 2004.

Chapter 8

[88] D. Eimer, *Gas Treating: Absorption Theory and Practice*, John Wiley & Sons, New York, 2019.
[89] P. Chattopadhyay, *Absorption and Stripping*, Asian Books, New Delhi, 2007.
[90] R. Zarzycki, A. Chacuk, *Absorption Fundamentals & Applications*, Pergamon Press, Oxford, 1993.

Chapter 9

[91] F. Rouquerol, J. Rouquerol, K. Sing, *Adsorption by Powders & Porous Solids: Principles, Methods and Applications*, 2nd edn., Academic Press, San Diego, 2013.
[92] E. Worch, *Adsorption Technology in Water Treatment: Fundamentals, Processes and Modeling*, De Gruyter, Berlin, 2012.

[93] C. Tien, *Introduction to Adsorption*, Elsevier, Amsterdam, 2018.

[94] R.T. Yang, *Adsorbents: Fundamentals and Applications*, Wiley Interscience, 2007.

[95] D.M. Ruthven, S. Farooq, K.S. Knaebel, *Pressure Swing Adsorption*, Wiley-VCH, New York, 1993.

[96] M. Mahadeva Swami, *Adsorption and its Applications*, Lambert Academic Publishing, Chisinau, 2020.

[97] D.M. Ruthven, *Principles of Adsorption and Adsorption Processes*, John Wiley, New York, 1984.

[98] A.K. Sengupta, *Ion Exchange in Environmental Processes*, John Wiley & Sons, Hoboken, 2017.

[99] *Ion Exchange Technology* (Inamuddin and M. Luqman, Editors), Springer, New York, 2012.

[100] A.A. Zagorodni, *Ion Exchange Materials: Properties and Applications*, Elsevier, Amsterdam, 2006.

[101] C.E. Harland, *Ion Exchange: Theory & Practice*, 2nd edn., RSC Publishing, 1994.

Chapter 10

[102] S. Tarleton, R.J. Wakeman, *Solid Liquid Separation: Equipment Selection and Process Design*, Elsevier Science, Oxford, 2006.

[103] R.J. Wakeman, S. Tarleton, *Solid Liquid Separation: Principles of Industrial Filtration*, Elsevier Science, Oxford, 2005.

[104] R.J. Wakeman, S. Tarleton, *Solid Liquid Separation: Scale Up of Industrial Equipment*, Elsevier Science, Oxford, 2005.

[105] L. Svarovsky, *Solid-Liquid Separation*, 4th edn., Butterworths Heinemann, Boston, 2001.

[106] Rushton, A.S. Ward, R.G. Holdich, *Solid-Liquid Filtration and Separation Technology*, 2nd edn., Wiley-VCH, 2000.

[107] P.N. Chermisinoff, *Solids/Liquids Separation*, 2nd edn., Butterworths Heinemann, Boston, 1998.

[108] D.B. Purchas, *Solid/Liquid Separation Technology*, Uplands Press, Croydon, 1981.

[109] K. Sutherland, *Filters and Filtration Handbook*, 5th edn., Elsevier Science, Oxford, 2008.

[110] T. Spark, G. Chase, *Filters and Filtration Handbook*, 6th edn., Elsevier Science, Oxford, 2015.

[111] R. Di Felice, R. Kehlenbeck, *Sedimentation Velocity of Solids in Finite Size Vessels*, Chemical Engineering and Technology, 2000, 23, 1123–1126.

Chapter 11

[112] R.J. Robichaux, *Gas-Liquid Separation: Liquid Droplet Development Dynamics and Separation*, Outskirts Press, Parker, 2009.

[113] J.-P. Duroudier, *Liquid-Gas and Solid-Gas Separators*, Elsevier, Amsterdam, 2016.

[114] M. Steward, K. Arnold, *Gas-Liquid and Liquid-Liquid Separators*, Elsevier, Amsterdam, 2008.

[115] A. Bürkholz, *Droplet Separation*, VCH-Verlag, Weinheim, 1989.

[116] L. Svarovsky, *Solid-Gas Separation*, Elsevier Scientific, Amsterdam, 1981.

Chapter 12

[117] E. Drioli, L. Giorno, F. Macedonio, *Membrane Engineering*, De Gruyter, Berlin, 2018.
[118] *Encyclopedia of Membrane Science and Technology* (E.M.V. Hoek and V.V. Tarabara, Editors), Wiley-VCH, Weinheim, 2013.
[119] R.W. Baker, *Membrane Technology and Applications*, 3rd edn., Wiley, 2012.
[120] K. Mohanty, M.K. Purkait, *Membrane Technologies and Applications*, CRC Press, Boca Raton, 2011.
[121] A.K. Pabby, S.S.H. Rizvi, A.M. Sastre, *Handbook of Membrane Separations: Chemical, Pharmaceutical, Food and Biotechnological Applications*, 2nd edn., CRC Press, Boca Raton, 2015.
[122] *Membrane Technology in the Chemical Industry* (S. Pereira Nunes and K.-V. Peinemann, Editors), Wiley-VCH, Weinheim, 2006.

Chapter 13

[123] A.G. Jones, *Crystallization Process Systems*, 2nd edn., Butterworth Heinemann, Boston, 2015.
[124] *Crystallization: Basic Concepts and Industrial Applications* (W. Beckmann, Editor), Wiley-VCH, Weinheim, 2013.
[125] M.R.B. Andreeta, *Crystallization: Science and Technology*, Intech, Rijeke, 2012.
[126] *Handbook of Industrial Crystallization* (A.S. Myerson, Editor), 3rd edn., Cambridge University Press, Cambridge, 2019.
[127] J.W. Mullin, *Crystallization*, 4th edn., Butterworth Heinemann, New York, 2001.
[128] *Crystallization Technology Handbook*, 2nd edn., (A. Mersmann, Editor), Marcel Dekker, New York, 2001.
[129] N.S. Tavare, *Industrial Crystallization, Process Simulation Analysis and Design*, Plenum Press, New York, 1995.
[130] H.M. Omar, S. Rohani, *Crystal Population Balance Formulation and Solution Methods: A Review*, Crystal Growth & Design 2017, 17, 4028–4041.

Chapter 14

[131] M.J. Rhodes, *Particle Technology*, 2nd edn., Wiley, New York, 2008.
[132] *Powder Technology Handbook*, 4th edn., (K. Higashitani, et al., Editors), CRC Press, Boca Raton, 2020.
[133] M.E. Fayed, L. Otten, *Handbook of Powder Science & Technology*, 2nd edn., Chapman & Hall, New York, 1997.
[134] D. Chulia, M. Deleuil, Y. Pourcelot, *Powder Technology and Pharmaceutical Processes*, Elsevier, Amsterdam, 1994.
[135] *Modern Drying Technology* (E. Tsotsas, A.S. Mujumdar, Editors), Wiley, New York, 2014.
[136] *Handbook of Industrial Drying*, 4th edn., (A.S. Mujumdar, Editor), CRC Press, Boca Raton, FL, 2014.
[137] C.M. van't Land, *Industrial Drying Equipment, Selection and Application*, Marcel Dekker, New York, 1991.

[138] C. Strumillo and T. Kudra, *Drying: Principles, Applications and Design*, Topics in Chemical Engineering, Volume 3, Gordon & Breach Science Publishers, New York, 1986.

[139] K. Masters, *Spray Drying in Practice*, 3rd edn., SprayDryConsult, 2002.

[140] T.C. Hua, B.L. Liu, H. Zhang, *Freeze Drying of Pharmaceutical and Food Products*, Elsevier, Oxford, 2010.

[141] *Granulation* (A. Salman et al., Editors), Elsevier, Oxford, 2006.

[142] J. Litster, B. Ennis, *The Science and Engineering of Granulation Processes*, Springer, New York, 2004.

[143] *Production, Handing and Characterization of Particulate Materials* (H. Merkus and G.M.H. Meesters, Editors), Springer, Heidelberg, 2016.

[144] B. Abulnaga, *Slurry Systems Handbook*, 2nd edn., McGraw Hill, New York, 2021.

[145] D. Mills, *Pneumatic Conveying Design Guide*, 3rd edn., Butterworth Heinemann, Burlington, 2015.

[146] G.E. Klinzing, et al., *Pneumatic Conveying of Solids*, Springer, New York, 1997.

Chapter 15

[147] K.T. Ulrich, S.D. Eppinger, *Product Design & Development*, 5th edn., McGraw Hill, Boston, 2011.

[148] E.L. Cussler, G.D. Moggridge, *Chemical Product Design*, 2nd edn., Cambridge University Press, Cambridge, 2011.

[149] *Product Design and Engineering: Best Practices* (U. Brocker, W. Meier and G. Wagner, Editors), Wiley VCH, New York, 2007.

[150] A.A. Tracton, *Coating Technology Handbook*, 3rd edn., CRC Press, Boca Raton, 2005.

[151] *Product Design and Engineering: Formulations of Gels and Pastes* (U. Brockel, W. Meier and G. Wagner, Editors), Wiley VCH, New York, 2013.

[152] H. Mollet, A. Grubenmann, *Formulation Technology*, Wiley VCH, New York, 2008.

[153] Th. F. Tadros, *Formulation of Disperse Systems*, Wiley VCH, New York, 2014.

[154] J. Krijgsman, *Formulation and Application*, in Product Recovery in Bioprocess Technology, Butterworth-Heinemann, Oxford, 1992.

[155] E. Smulders, *Laundry Detergents*, Wiley VCH, New York, 2001.

[156] M.S. Showell, *Powdered Detergents*, Marcel Dekker, New York, 1998.

[157] C. Defonseka, *Processing of Polymers*, De Gruyter, Berlin, 2020.

[158] C. Rauwendaal, *Polymer Extrusion*, 2nd edn., Hanser Gardner, Cincinatti, 2014.

[159] J.L. White, E.K. Kim, *Twin Screw Extrusion*, 2nd edn., Hanser Gardner, Cincinatti, 2010.

[160] M.J. Stevens, J. Covas, *Extruder Principles and Operation*, 2nd edn., Springer, New York, 1995.

[161] *Encyclopedia of Polymer Blends* (I. Avraam, Editor), Wiley-VDH, New York, 2011.

[162] I. Manaz-Zloczower, *Mixing and Compounding of Polymers*, 2nd edn., Hanser Gardner, Cincinnati, 2009.

[163] D.B. Todd, *Plastics Compounding: Equipment and Processing*, Hanser Gardner, Cincinnati, 1998.

[164] Z. Tadmor, C.G. Gogos, *Principles of Polymer Processing*, 2nd edn., Wiley, New York, 2005.

[165] A.N. Wilkinson and A.J. Ryan, *Polymer processing and structure development*, Kluwer Academic Publishers, Dordrecht, 1999.

[166] N.P. Cheremisinoff, P.N. Cheremisinoff, *Handbook of Applied Polymer Processing Technology*, Marcel Dekker, New York, 1996.

[167] D.H. Morton-Jones, *Polymer Processing*, Chapman and Hall, London, 1989.

[168] C.A. Harper, *Handbook of Plastic Processes*, Wiley, Hoboken, 2006.

Chapter 16

[169] J. Harmsen, A.B. de Haan, P.L.J. Swinkels, *Product and Process Design*, De Gruyter, Berlin, 2018.
[170] D. Erwin, *Industrial Chemical Process Design*, 2nd edn., McGraw Hill, New York, 2013.
[171] W.D. Seider, et al., Product and Process Design Principles: Synthesis, Analysis and Evaluation, 4th edn., John Wiley & Sons, New York, 2017.
[172] R. Smith, *Chemical Process: Design and Integration*, 2nd edn., Wiley-Blackwell, Chichester, 2016.
[173] A.B. Koltuniewicz, *Sustainable Process Engineering – Prospects and Opportunities*, De Gruyter, Berlin, 2014.
[174] E. Zondervan, *Product Driven Process Design*, De Gruyter, Berlin, 2020.
[175] J. Harmsen, J.B. Powel, *Sustainable Development in the Process Industries*, John Wiley, New York, 2010.
[176] J. Harmsen, *Industrial Process Scale-Up*, Elsevier, Oxford, 2013.
[177] G.H. Vogel, *Process Development*, Wiley-VCH, Weinheim, 2005.
[178] L.A. Hulshof, *Right First Time in Fine-Chemical Process Scale-Up*, Scientific Update, Mayfield, 2013.
[179] N.G. Anderson, *Practical Process Research & Development*, 2nd edn., Academic Press, San Diego, 2012.
[180] U. Diwekar, *Batch Processing, Modelling and Design*, CRC Press, Boca Raton, 2014.
[181] T. Meyer, G. Reniers, *Engineering Risk Management*, De Gruyter, Berlin, 2013.

Chapter 17

[182] *Handbook of Industrial Mixing* (E.L. Paul, et al., Editors), John Wiley & Sons, New York, 2003.
[183] G.B. Tatterson, *Scaleup and Design of Industrial Mixing Processes*, 2nd edn., Tatterson, 2003.
[184] M. Zlokarnik, Stirring, *Theory and Practice*, Wiley VCH, Weinheim, 2001.
[185] E.K. Todtenhaupt, G. Zeiler, *Handbook of Mixing Technology*, 3rd edn., Ekato Holding GmbH, Freiburg, 2012.
[186] N. Harnby, M.F. Edwards, A.W. Nienow, *Mixing in the Process Industries*, 2nd edn., Butterworth Heinemann, Oxford, 1992.
[187] M. Zlokarnik, *Scale-Up in Chemical Engineering*, 2nd edn., Wiley-VCH, Weinheim, 2006.
[188] T. Szirtes, P. Rosza, *Applied Dimensional Analysis and Modelling*, 2nd edn., Butterworth Heinemann, Burlington, 2007.
[189] M. Zlokarnik, *Dimensional Analysis and Scale-Up in Chemical Engineering*, Springer Verlag, Berlin, 1991.
[190] A. Bisio, R.L, Kabel, *Scale-Up of Chemical Processes, Conversion from Laboratory Scale Tests to Successful Commercial Scale Design*, John Wiley & Sons, New York, 1985.
[191] H. Versteeg, W. Malalasekra, *An Introduction to Computational Fluid Dynamics*, Pearson, India, 2010.
[192] P. Wesseling, *Principles of Computational Fluid Dynamics*, Springer, Berlin, 2009.
[193] T.J. Chung, *Computations Fluid Dynamics*, 2nd edn., Cambridge University Press, New York, 2010.
[194] J. Jung, *Design and Understanding of Fluidized Bed Reactors: Application of CFD Techniques to Multi-phase Flows*, VDM Verlag, Saarbrucken, 2009.

Chapter 18

[195] N.J. Bahr, *System Safety Engineering and Risk Assessment*, 2nd edn., CRC Press, Boca Raton, 2015.

[196] F.P. Lees, *Loss Prevention in the Process* Industries, *Hazard* Identification, *Assessment and Control*, 4th edn., Butterworth & Heinemann, Boston 2012.

[197] D.A. Crowl, J.F. Louvar, *Chemical Process Safety: Fundamentals with Applications*, 4th edn., Pearson, 2019.

[198] T. Kletz, *What Went Wrong?: Case Histories of Process Plant Disasters*, 5th edn., Gulf Publishing Company, Burlington, 2009.

[199] J. Atherton, F. Gil, *Incidents that Define Process Safety*, John Wiley & Sons, Hoboken, 2008.

[200] R.E. Sanders, *Chemical Process Safety: Learning from Case Histories*, 4th edn., Butterworth-Heinemann, Boston, 2015.

[201] F. Stoessel, *Thermal Safety of Chemical Processes*, 2nd edn., Wiley-VCH, Weinheim, 2020.

[202] M.K. Purkait, et al., *Hazards and Safety in Process Industries: Case Studies*, CRC Press, Boca Raton 2021.

[203] F. Crawly, B.J. Tyler, *Hazop, Guide to Best Practices*, 3rd edn., IChemE, London, 2015.

[204] T. Kletz, *Hazop and Hazan, Identifying and Assessing Process Industry Hazards*, 4th edn., Institution of Chemical Engineers, 1999.

[205] T. Kletz, P. Amoytte, *Process Plants: A Handbook for Inherently Safer Design*, 2nd edn., CRC Press, Boca Raton, 2010.

Appendices

[206] P.J. van den Berg and W.A. de Jong, *Introduction to Chemical Process Technology*, Delft University Press, 1980.

[207] *Kirk-Othmer Encyclopedia of Chemical Technology*, 5th edn., (A. Seidel, Editor), John Wiley & Sons, Hoboken, 2004.

[208] *Ullmann's Encyclopedia of Industrial Chemistry*, 7th edn., Wiley-Interscience, New York, 2002-.

[209] U. Onken, A. Behr, *Chemische Prozeßkunde*, Georg Thieme Verlag, Stuttgart, 1996.

[210] J.A. Kent, *Handbook of Industrial Chemistry and Biotechnology*, 13th edn., Springer, New York, 2017.

[211] P.J. Chenier, *Survey of Industrial Chemistry*, 3rd edn., Springer, New York, 2002.

[212] M.E. Dry, *The Sasol Fischer-Tropsch Processes*, in Applied Industrial Catalysis, Volume 2, 1983.

[213] *Encyclopedia of Polymer Science and Technology*, 4th edn., (H.F Mark, Editor), Wiley, New York, 2014.

[214] *Ullmann's Fine Chemicals*, Wiley VCH, Weinheim, 2014.

[215] C.A.C. Sequeira, D.M.F. Santos, *Electrochemical Routes for Industrial Synthesis*, Journal of the Brazilian Chemical Society 2009, 20, 387–406.

[216] D. Pletcher, F.C. Walsh, *Industrial Electrochemistry*, 2nd edn., Chapman and Hall, London, 1990.

[217] *Organic Electrochemistry*, 4th edn., (H. Lund and O. Hammerich, Editors), Marcel Dekker, Inc.: New York, 2001, pp. 1259–1308.

Index

https://doi.org/10.1515/9783110712445-023

www.ingramcontent.com/pod-product-compliance
Lightning Source LLC
Chambersburg PA
CBHW060944210326
41598CB00031B/4712